Cosmology

The Origin and Evolution of Cosmic Structure

Cosmology

The Origin and Evolution of Cosmic Structure

Second Edition

Peter Coles
School of Physics & Astronomy,
University of Nottingham, UK

Francesco Lucchin
Dipartimento di Astronomia,
Università di Padova, Italy

JOHN WILEY & SONS, LTD

Telephone (+44) 1243 779777

Email (for orders and customer service enquiries): cs-books@wiley.co.uk

Visit our Home Page on www.wileyeurope.com or www.wiley.com

Reprinted April 2003, January 2008

Other Wiley Editorial Offices

John Wiley & Sons Inc., 111 River Street, Hoboken, NJ 07030, USA

Jossey-Bass, 989 Market Street, San Francisco, CA 94103-1741, USA

Wiley-VCH Verlag GmbH, Boschstr. 12, D-69469 Weinheim, Germany

John Wiley & Sons Australia Ltd, 33 Park Road, Milton, Queensland 4064, Australia

John Wiley & Sons (Asia) Pte Ltd, 2 Clementi Loop #02-01, Jin Xing Distripark, Singapore 129809

John Wiley & Sons Canada Ltd, 22 Worcester Road, Etobicoke, Ontario, Canada M9W 1L1

Library of Congress Cataloging-in-Publication Data

(applied for)

British Library Cataloguing in Publication Data

A catalogue record for this book is available from the British Library

ISBN 13: 978 0 471 48909 2 (H/B)

Contents

Preface to First Edition

This is a book about modern cosmology. Because this is a big subject – as big as the Universe – we have had to choose one particular theme upon which to focus our treatment. Current research in cosmology ranges over fields as diverse as quantum gravity, general relativity, particle physics, statistical mechanics, nonlinear hydrodynamics and observational astronomy in all wavelength regions, from radio to gamma rays. We could not possibly do justice to all these areas in one volume, especially in a book such as this which is intended for advanced undergraduates or beginning postgraduates. Because we both have a strong research interest in theories for the origin and evolution of cosmic structure – galaxies, clusters and the like – and, in many respects, this is indeed the central problem in this field, we decided to concentrate on those elements of modern cosmology that pertain to this topic. We shall touch on many of the areas mentioned above, but only insofar as an understanding of them is necessary background for our analysis of structure formation.

Cosmology in general, and the field of structure formation in particular, has been a 'hot' research topic for many years. Recent spectacular observational breakthroughs, like the discovery by the COBE satellite in 1992 of fluctuations in the temperature of the cosmic microwave background, have made newspaper headlines all around the world. Both observational and theoretical sides of the subject continue to engross not only the best undergraduate and postgraduate students and more senior professional scientists, but also the general public. Part of the fascination is that cosmology lies at the crossroads of many disciplines. An introduction to this subject therefore involves an initiation into many seemingly disparate branches of physics and astrophysics; this alone makes it an ideal area in which to encourage young scientists to work.

Nevertheless, cosmology is a peculiar science. The Universe is, by definition, unique. We cannot prepare an ensemble of universes with slightly different parameter values and look for differences or correlations in their behaviour. In many branches of physical science such experimentation often leads to the formulation of empirical laws which give rise to models and subsequently theories. Cosmology is different. We have only one Universe, and this must provide the empirical laws we try to explain by theory, as well as the experimental evidence we use to test the theories we have formulated. Though the distinction between them is, of course, not completely sharp, it is fair to say that physics is predominantly characterised by experiment and theory, and cosmology by observation and paradigm.

(We take the word 'paradigm' to mean a theoretical framework, not all of whose elements have been formalised in the sense of being directly related to observational phenomena.) Subtle influences of personal philosophy, cultural and, in some cases, religious background lead to very different choices of paradigm in many branches of science, but this tendency is particularly noticeable in cosmology. For example, one's choice to include or exclude the cosmological constant term in Einstein's field equations of general relativity can have very little empirical motivation but must be made on the basis of philosophical, and perhaps aesthetic, considerations. Perhaps a better example is the fact that the expansion of the Universe could have been anticipated using Newtonian physics as early as the 17th century. The Cosmological Principle, according to which the Universe is homogeneous and isotropic on large scales, is sufficient to ensure that a Newtonian universe cannot be static, but must be either expanding or contracting. A philosophical predisposition in western societies towards an unchanging, regular cosmos apparently prevented scientists from drawing this conclusion until it was forced upon them by 20th century observations. Incidentally, a notable exception to this prevailing paradigm was the writer Edgar Allan Poe, who expounded a picture of a dynamic, cyclical cosmos in his celebrated prose poem *Eureka*. We make these points to persuade the reader that cosmology requires not only a good knowledge of interdisciplinary physics, but also an open mind and a certain amount of self-knowledge.

One can learn much about what cosmology actually means from its history. Since prehistoric times, man has sought to make sense of his existence and that of the world around him in some kind of theoretical framework. The first such theories, not recognisable as 'science' in the modern sense of the word, were mythological. In western cultures, the Ptolemaic cosmology was a step towards the modern approach, but was clearly informed by Greek cultural values. The Copernican Principle, the notion that we do not inhabit a special place in the Universe and a kind of forerunner of the Cosmological Principle, was to some extent a product of the philosophical and religious changes taking place in Renaissance times. The mechanistic view of the Universe initiated by Newton and championed by Descartes, in which one views the natural world as a kind of clockwork device, was influenced not only by the beginnings of mathematical physics but also by the first stirrings of technological development. In the era of the Industrial Revolution, man's perception of the natural world was framed in terms of heat engines and thermodynamics, and involved such concepts as the 'Heat Death of the Universe'.

With hindsight we can say that cosmology did not really come of age as a science until the 20th century. In 1915 Einstein advanced his theory of general relativity. His field equations told him the Universe should be evolving; Einstein thought he must have made a mistake and promptly modified the equations to give a static cosmological solution, thus perpetuating the fallacy we discussed. It was not until 1929 that Hubble convinced the astronomical community that the Universe was actually expanding after all. (To put this affair into historical perspective, remember that it was only in the mid-1920s that it was demonstrated – by Hubble and others – that faint nebulae, now known to be galaxies like our own Milky Way, were actually outside our Galaxy.) The next few decades saw considerable theoretical and observational developments. The Big Bang and steady-state cosmologies were proposed and their respective advocates began a long and acrimonious debate about which was correct, the legacy of which lingers still. For many workers this debate was resolved by the discovery in 1965 of the cosmic microwave background radiation, which was immediately seen to be good evidence in favour of an evolving Universe which was hotter and denser in the past. It is reasonable to regard this discovery as marking the beginning of 'Physical Cosmology'. Counts of distant galaxies were also showing evidence of evolution in the properties of these objects at this time, and the first calculations had already been made, notably by Alpher and Herman in the late 1940s, of the elemental abundances expected to be produced by nuclear reactions in the early stages of the Big Bang. These, and other, considerations left the Big Bang model as the clear victor over the steady-state picture.

By the 1970s, attention was being turned to the question that forms the main focus of this book: where did the structure we observe in the Universe around us actually come from? The fact that the microwave background appeared remarkably uniform in temperature across the sky was taken as evidence that the early Universe (when it was less than a few hundred thousand years old) was very smooth. But the Universe now is clearly very clumpy, with large fluctuations in its density from place to place. How could these two observations be reconciled? A 'standard' picture soon emerged, based on the known physics of gravitational instability. Gravity is an attractive force, so that a region of the Universe which is slightly denser than average will gradually accrete material from its surroundings. In so doing the original, slightly denser region gets denser still and therefore accretes even more material. Eventually this region becomes a strongly bound 'lump' of matter surrounded by a region of comparatively low density. After two decades, gravitational instability continues to form the basis of the standard theory for structure formation. The details of how it operates to produce structures of the form we actually observe today are, however, still far from completely understood.

To resume our historical thread, the 1970s saw the emergence of two competing scenarios (a terrible word, but sadly commonplace in the cosmological literature) for structure formation. Roughly speaking, one of these was a 'bottom-up', or hierarchical, model, in which structure formation was thought to begin with the collapse of small objects which then progressively clustered together and merged under the action of their mutual gravitational attraction to form larger objects. This model, called the isothermal model, was advocated mainly by American researchers. On the other hand, many Soviet astrophysicists of the time, led by Yakov B. Zel'dovich, favoured a model, the adiabatic model, in which the first structures to condense out of the expanding plasma were huge agglomerations of mass on the scale of giant superclusters of galaxies; smaller structures like individual galaxies were assumed to be formed by fragmentation processes within the larger structures, which are usually called 'pancakes'. The debate between the isothermal and adiabatic schools never reached the level of animosity of the Big Bang versus steady-state controversy but was nevertheless healthily animated.

By the 1980s it was realised that neither of these models could be correct. The reasons for this conclusion are not important at this stage; we shall discuss them in detail during Part 3 of the book. Soon, however, alternative models were proposed which avoided many of the problems which led to the rejection of the 1970s models. The new ingredient added in the 1980s was non-baryonic matter; in other words, matter in the form of some exotic type of particle other than protons and neutrons. This matter is not directly observable because it is not luminous, but it does feel the action of gravity and can thus assist the gravitational instability process. Non-baryonic matter was thought to be one of two possible types: hot or cold. As had happened in the 1970s, the cosmological world again split into two camps, one favouring cold dark matter (CDM) and the other hot dark matter (HDM). Indeed, there are considerable similarities between the two schisms of the 1970s and 1980s, for the CDM model is a 'bottom-up' model like the old baryon isothermal picture, while the HDM model is a 'top-down' scenario like the adiabatic model. Even the geographical division was the same; Zel'dovich's great Soviet school were the most powerful advocates of the HDM picture.

The 1980s also saw another important theoretical development: the idea that the Universe may have undergone a period of inflation, during which its expansion rate accelerated and any initial inhomogeneities were smoothed out. Inflation provides a model which can, at least in principle, explain how such homogeneity might have arisen and which does not require the introduction of the Cosmological Principle *ab initio*. While creating an observable patch of the Universe which is predominantly smooth and isotropic, inflation also guarantees the existence of small fluctuations in the cosmological density which may be the initial perturbations needed to feed the gravitational instability thought to be the origin of galaxies and other structures.

The history of cosmology in the 20th century is marked by an interesting interplay of opposites. For example, in the development of structure-formation theories one can see a strong tendency towards *change* (such as from baryonic to non-baryonic models), but also a strong element of *continuity* (the persistence of the hierarchical and pancake scenarios). The standard cosmological models have an expansion rate which is decelerating because of the *attractive* nature of gravity. In models involving inflation (or those with a cosmological constant) the expansion is accelerated by virtue of the fact that gravity effectively becomes *repulsive* for some period. The Cosmological Principle asserts a kind of large-scale *order*, while inflation allows this to be achieved locally within a Universe characterised by large-scale *disorder*. The confrontation between steady-state and Big Bang models highlights the distinction between *stationarity* and *evolution*. Some variants of the Big Bang model involving inflation do, however, involve a large 'metauniverse' within which 'miniuniverses' of the size of our observable patch are continually being formed. The appearance of miniuniverses also emphasises the contrast between *whole* and *part*: is our observable Universe all there is, or even representative of all there is? Or is it just an atypical 'bubble' which just happens to have the properties required for life to evolve within it? This brings into play the idea of an Anthropic Cosmological Principle which emphasises the *special* nature of the conditions necessary to create observers, compared with the *general* homogeneity implied by the Cosmological Principle in its traditional form.

Another interesting characteristic of cosmology is the distinction, which is often blurred, between what one might call cosmology and metacosmology. We take cosmology to mean the scientific study of the cosmos as a whole, an essential part of which is the testing of theoretical constructions against observations, as described above. On the other hand, metacosmology is a term which describes elements of a theoretical construction, or paradigm, which are not amenable to observational test. As the subject has developed, various aspects of cosmology have moved from the realm of metacosmology into that of cosmology proper. The cosmic microwave background, whose existence was postulated as early as the 1940s, but which was not observable by means of technology available at that time, became part of cosmology proper in 1965. It has been argued by some that the inflationary metacosmology has now become part of scientific cosmology because of the COBE discovery of fluctuations in the temperature of the microwave background across the sky. We think this claim is premature, although things are clearly moving in the right direction for this to take place some time in the future. Some metacosmological ideas may, however, remain so forever, either because of the technical difficulty of observing their consequences or because they are not testable even in principle. An example of the latter difficulty may be furnished by Linde's chaotic inflationary picture of eternally creating miniuniverses which lie beyond the radius of our observable Universe.

Despite these complexities and idiosyncrasies, modern cosmology presents us with clear challenges. On the purely theoretical side, we require a full integration of particle physics into the Big Bang model, and a theory which treats gravitational physics at the quantum level. We also need a theoretical understanding of various phenomena which are probably based on well-established physical processes: nonlinearity in gravitational clustering, hydrodynamical processes, stellar formation and evolution, chemical evolution of galaxies. Many observational targets have also been set: the detection of candidate dark-matter particles in the galactic halo; gravitational waves; more detailed observations of the temperature fluctuations in the cosmic microwave background; larger samples of galaxy redshifts and peculiar motions; elucidation of the evolutionary properties of galaxies with cosmic time. Above all, we want to stress that cosmology is a field in which many fundamental questions remain unanswered and where there is plenty of scope for new ideas. The next decade promises to be at least as exciting as the last, with ongoing experiments already probing the microwave background in finer detail and powerful optical telescopes mapping the distribution of galaxies out to greater and greater distances. Who can say what theoretical ideas will be advanced in light of these new observations? Will the theoretical ideas described in this book turn out to be correct, or will we have to throw them all away and go back to the drawing board?

This book is intended to be an up-to-date introduction to this fascinating yet complex subject. It is intended to be accessible to advanced undergraduate and beginning postgraduate students, but contains much material which will be of interest to more established researchers in the field, and even non-specialists should find it a useful introduction to many of the important ideas in modern cosmology. Our book does not require a high level of specialisation on behalf of the reader. Only a modest use is made of general relativity. We use some concepts from statistical mechanics and particle physics, but our treatment of them is as self-contained as possible. We cover the basic material, such as the Friedmann models, one finds in all elementary cosmology texts, but we also take the reader through more advanced material normally available only in technical review articles or in the research literature. Although many cosmology books are on the market at the moment thanks, no doubt, to the high level of public and media interest in this subject, very few tackle the material we cover at this kind of 'bridging' level between elementary textbook and research monograph. We have also covered some material which one might regard as slightly old-fashioned. Our treatment of the adiabatic baryon picture of structure formation in Chapter 12 is an example. We have included such material primarily for pedagogical reasons, but also for the valuable historical lessons it provides. The fact that models come and go so rapidly in this field is explained partly by the vigorous interplay between observation and theory and partly by virtue of the fact that cosmology, in common with other aspects of life, is sometimes a victim of changes in fashion. We have also included more recent theory and observation alongside this pedagogical material in order to provide the reader with a firm basis for an understanding of future developments in this field. Obviously, because ours is such an exciting field, with advances being made at a rapid rate, we cannot claim to be definitive in all areas of contemporary interest. At the end of each chapter we give lists of references - which are not intended to be exhaustive but which should provide further reading on the fundamental issues - as well as more detailed technical articles for the advanced student. We have not cited articles in the body of each chapter, mainly to avoid interrupting the flow of the presentation. By doing this, it is certainly not our intention to claim that we have not leaned upon other works for much of this material; we implicitly acknowledge this for any work we list in the references. We believe that our presentation of this material is the most comprehensive and accessible available at this level amongst the published works belonging to the literature of this subject; a list of relevant general books on cosmology is given after this preface.

The book is organised into four parts. The first, Chapters 1–4, covers the basics of general relativity, the simplest cosmological models, alternative theories and introductory observational cosmology. This part can be skipped by students who have already taken introductory courses in cosmology. Part 2, Chapters 5–9, deals with physical cosmology and the thermal history of the universe in Big Bang models, including a discussion of phase transitions and inflation. Part 3, Chapters 10–15, contains a detailed treatment of the theory of gravitational instability in both the linear and nonlinear regimes with comments on dark-matter theories and hydrodynamical effects in the context of galaxy formation. The final part, Chapters 16–19, deals with methods for testing theories of structure formation using statistical properties of galaxy clustering, the fluctuations of the cosmic microwave background, galaxy-peculiar motions and observations of galaxy evolution and the extragalactic radiation backgrounds. The last part of the book is at a rather higher level than the preceding ones and is intended to be closer to the ongoing research in this field.

Some of the text is based upon an English adaptation of *Introduzione alla Cosmologia* (Zanichelli, Bologna, 1990), a cosmology textbook written in Italian by Francesco Lucchin, which contains material given in his lectures on cosmology to final-year undergraduates at the University of Padova over the past 15 years or so. We are very grateful to the publishers for permission to draw upon this source. We have, however, added a large amount of new material for the present book in order to cover as many of the latest developments in this field as possible. Much of this new material relates to the lecture notes given by Peter Coles for the Master of Science course on cosmology at Queen Mary and Westfield College beginning in 1992. These sources reinforce our intention that the book should be suitable for advanced undergraduates and/or beginning postgraduates.

Francesco Lucchin thanks the Astronomy Unit at Queen Mary & Westfield College for hospitality during visits when this book was in preparation. Likewise, Peter Coles thanks the Dipartimento di Astronomia of the University of Padova for hospitality during his visits there. Many colleagues and friends have helped us enormously during the preparation of this book. In particular, we thank Sabino Matarrese, Lauro Moscardini and Bepi Tormen for their careful reading of the manuscript and for many discussions on other matters related to the book. We also thank Varun Sahni and George Ellis for allowing us to draw on material co-written by them and Peter Coles. Many sources are also to be thanked for their willingness to allow us to use various figures; appropriate acknowledgments are given in the corresponding figure captions.

Peter Coles and Francesco Lucchin
London, October 1994

Preface to Second Edition

While we were putting together the material for the first edition of this book we had vague worries that some parts of it might date reasonably rapidly. We were not, however, prepared for the amazing developments over the last few years that have rocked the foundations of the subject.

For example, when we wrote the first edition we decided not to include detailed discussions of the cosmological constant, an addition to Einstein's theory of general relativity that, in common with many cosmologists, we considered to be a curiosity rather than a necessity. Over the last few years, however, studies of the behaviour of high-redshift supernovae have provided compelling evidence that the expansion of the Universe is accelerating. As far as we know this requires the existence of a cosmological constant, or something very like it.

As well as this unforeseen breakthrough, there has been a steadily accelerating accumulation of extragalactic observational data that have told us much more about the state of the Universe than was known in 1995 when the first edition was published. Redshift maps containing hundreds of thousands of galaxy positions are now available, with more in the pipeline. The first stunning high-resolution maps of the cosmic microwave background from the balloon-borne experiments Boomerang and Maxima have trumped the much-vaunted satellite experiments Planck and MAP. These maps provide near-conclusive evidence that we live in a Universe in which space is flat. These are just a couple of examples we have included in this edition, and more can be anticipated in the future.

Theory too has moved forward in impressive fashion. Massive parallel supercomputers have been deployed to simulate the formation of structure *ab initio* and refine the details of a basic theoretical framework that has largely withstood the challenges placed on it by recent observational breakthroughs. Progress has also been made in modelling the complexities of galaxy formation, with all the messy business of star formation, feedback and merging that this entails. These ideas are used nowadays to help plan observational programmes, so that over the last few years theory and observation have moved closer together. Cosmology is at last becoming an empirically based, testable science.

To reflect these changes and also to improve the presentation of the basic material we have made significant alterations to the book compared with the first edition. Obviously we have added new material on the supernovae, cosmic microwave background and galaxy surveys we mentioned above. We have also introduced a new chapter on gravitational lensing, another 'hot' topic for today's generation of cosmologists. We also changed the structure of the first part of the book to make a gentler introduction to the subject instead of diving straight into general relativity. We also added problems sections at the end of each chapter and reorganised the references.

We decided to keep our account of the basic physics of perturbation growth (Chapters 10–12) while other books concentrate more on model-building. Our reason for this is that we intended the book to be an introduction for physics students. Models come and models go, but physics remains the same. To make the book a bit more accessible we added a sort of 'digest' of the main ideas and summary of model-building in Chapter 15 for readers wishing to bypass the details.

Other bits, such as those covering theories with variable constants and inhomogeneous cosmologies, were added for no better reason than that they are fun. On the other hand, we missed the boat in a significant way by minimising the role of the cosmological constant in the first edition. Who knows, maybe we will strike it lucky with one of these additions!

Because of the dominance that observation has assumed over the last few years, we decided to add a chapter at the end of the book exploring some of the planned developments in observation technology (gravitational wave detectors, new satellites, ground-based facilities, and so on). Experience has shown us that it is hard to predict the future, but this final chapter will at least point out some of the possibilities.

We are grateful to everyone who helped us with this second edition and to those who provided constructive criticism on the first. In particular, we thank (in alphabetical order) George Ellis, Richard Ellis, Carlos Frenk, Andrew Liddle, Sabino Matarrese, Lauro Moscardini and Bepi Tormen for their comments and advice. We also acknowledge the help of many students who helped us correct some of the (regrettably numerous) errors in the original book.

Peter Coles and Francesco Lucchin
Padua, January 2002

PART 1

Cosmological Models

1

First Principles

In this chapter, our aim is to provide an introduction to the basic mathematical structure of modern cosmological models based on Einstein's theory of gravity, the General Theory of Relativity or general relativity for short. This theory is mathematically challenging, but fortunately we do not really need to use its fully general form. Throughout this chapter we will therefore illustrate the key results with Newtonian analogies. We begin our study with a discussion of the Cosmological Principle, the ingredient that makes relativistic cosmology rather more palatable than it might otherwise be.

1.1 The Cosmological Principle

Whenever science enters a new field and is faced with a dearth of observational or experimental data some guiding principle is usually needed to assist during the first tentative steps towards a theoretical understanding. Such principles are often based on ideas of symmetry which reduce the number of degrees of freedom one has to consider. This general rule proved to be the case in the early years of the 20th century when the first steps were taken, by Einstein and others, towards a scientific theory of the Universe. Little was then known empirically about the distribution of matter in the Universe and Einstein's theory of gravity was found to be too difficult to solve for an arbitrary distribution of matter. In order to make progress the early cosmologists therefore had to content themselves with the construction of simplified models which they hoped might describe some aspects of the Universe in a broad-brush sense. These models were based on an idea called the *Cosmological Principle.* Although the name 'principle' sounds grand, principles are generally introduced into physics when one has no data to go on, and cosmology was no exception to this rule.

The Cosmological Principle is the assertion that, on sufficiently large scales (beyond those traced by the large-scale structure of the distribution of galaxies), the Universe is both homogeneous and isotropic. Homogeneity is the property of being identical everywhere in space, while isotropy is the property of looking the same in every direction. The Universe is clearly not exactly homogeneous, so cosmologists define homogeneity in an average sense: the Universe is taken to be identical in different places when one looks at sufficiently large pieces. A good analogy is that of a patterned carpet which is made of repeating units of some basic design. On the scale of the individual design the structure is clearly inhomogeneous but on scales larger than each unit it is homogeneous.

There is quite good observational evidence that the Universe does have these properties, although this evidence is not completely watertight. One piece of evidence is the observed near-isotropy of the cosmic microwave background radiation. Isotropy, however, does not necessarily imply homogeneity without the additional assumption that the observer is not in a special place: the so-called *Copernican Principle.* One would observe isotropy in any spherically symmetric distribution of matter, but only if one were in the middle of the pattern. A circular carpet bearing a design consisting of a series of concentric rings would look isotropic only to an observer standing in the centre of the pattern. Observed isotropy, together with the Copernican Principle, therefore implies the Cosmological Principle.

The Cosmological Principle was introduced by Einstein and subsequent relativistic cosmologists without any observational justification whatsoever. Indeed, it was not known until the 1920s that the spiral nebulae (now known to be galaxies like our own) were outside our own galaxy, the Milky Way. A term frequently used to describe the entire Universe in those days was metagalaxy, indicating that it was thought that the Milky Way was essentially the entire cosmos. The Galaxy certainly does not look the same in all directions: it presents itself as a prominent band across the night sky.

In advocating the Cosmological Principle, Einstein was particularly motivated by ideas associated with Ernst Mach. Mach's Principle, roughly speaking, is that the laws of physics are determined by the distribution of matter on large scales. For example, the value of the gravitational constant G was thought perhaps to be related to the amount of mass in the Universe. Einstein thought that the only way to put theoretical cosmology on a firm footing was to assume that there was a basic simplicity to the global structure of the Universe enabling a similar simplicity in the local behaviour of matter. The Cosmological Principle achieves this and leads to relatively simple cosmological models, as we shall see shortly.

There are various approaches one can take to this principle. One is philosophical, and is characterised by the work of Milne in the 1930s and later by Bondi, Gold and Hoyle in the 1940s. This line of reasoning is based, to a large extent, on the aesthetic appeal of the Cosmological Principle. Ultimately this appeal stems from the fact that it would indeed be very difficult for us to understand the Universe if physical conditions, or even the laws of physics themselves, were to vary dramatically from place to place. These thoughts have been taken further, leading to the *Perfect Cosmological Principle*, in which the Universe is the same not only in all places and in all directions, but also at all times. This stronger version of the Cosmological Principle was formulated by Bondi and Gold (1948) and it subsequently led Hoyle (1948) and Hoyle and Narlikar (1963, 1964) to develop the steady-state cosmology. This theory implies, amongst other things, the continuous creation of matter to keep the density of the expanding Universe constant. The steady-state universe was abandoned in the 1960s because of the properties of the cosmic microwave background, radio sources and the cosmological helium abundance which are more readily explained in a Big Bang model than in a steady state. Nowadays the latter is only of historical interest (see Chapter 3 later).

Attempts have also been made to justify the Cosmological Principle on more direct physical grounds. As we shall see, homogeneous and isotropic universes described by the theory of general relativity possess what is known as a 'cosmological horizon': regions sufficiently distant from each other cannot have been in causal contact ('have never been inside each other's horizon') at any stage since the Big Bang. The size of the regions whose parts are in causal contact with each other at a given time grows with cosmological epoch; the calculation of the horizon scale is performed in Section 2.7. The problem then arises as to how one explains the observation that the Universe appears homogeneous on scales much larger than the scale one expects to have been in causal contact up to the present time. The mystery is this: if two regions of the Universe have never been able to communicate with each other by means of light signals, how can they even know the physical conditions (density, temperature, etc.) pertaining to each other? If they cannot know this, how is it that they evolve in such a way that these conditions are the same in each of the regions? One either has to suppose that causal physics is not responsible for this homogeneity, or that the calculation of the horizon is not correct. This conundrum is usually called the *Cosmological Horizon Problem* and we shall discuss it in some detail in Chapter 7.

Various attempts have been made to avoid this problem. For example, particular models of the Universe, such as some that are homogeneous but not isotropic, do not possess the required particle horizon. These models can become isotropic in the course of their evolution. A famous example is the 'mix-master' universe of Misner (1968) in which isotropisation is effected by viscous dissipation involving neutrinos in the early universe. Another way to isotropise an initially anisotropic universe is by creating particles at the earliest stage of all, the Planck era (Chapter 6). More recently still, Guth (1981) proposed an idea which could resolve the horizon problem: *the inflationary universe*, which is of great contemporary interest in cosmology, and which we discuss in Chapter 7.

In any case, the most appropriate approach to this problem is an empirical one. We accept the Cosmological Principle because it agrees with observations. We shall describe the observational evidence for this in Chapter 4; data concerning radiogalaxies, clusters of galaxies, quasars and the microwave background all demonstrate that the level of anisotropy of the Universe on large scales is about one part in 10^5.

1.2 Fundamentals of General Relativity

The strongest force of nature on large scales is gravity, so the most important part of a physical description of the Universe is a theory of gravity. The best candidate we have for this is Einstein's General Theory of Relativity. We therefore begin this chapter with a brief introduction to the basics of this theory. Readers familiar with this material can skip Section 1.2 and resume reading at Section 1.3. In fact, about 90% of this book does not require the use of general relativity at all so readers only interested in a Newtonian treatment may turn directly to Section 1.11.

In *Special Relativity*, the invariant *interval* between two events at coordinates (t, x, y, z) and $(t + \mathrm{d}t, x + \mathrm{d}x, y + \mathrm{d}y, z + \mathrm{d}z)$ is defined by

$$\mathrm{d}s^2 = c^2\,\mathrm{d}t^2 - (\mathrm{d}x^2 + \mathrm{d}y^2 + \mathrm{d}z^2), \tag{1.2.1}$$

where $\mathrm{d}s$ is invariant under a change of coordinate system and the path of a light ray is given by $\mathrm{d}s = 0$. The paths of material particles between any two events are such as to give stationary values of $\int_{\text{path}} \mathrm{d}s$; this corresponds to the shortest distance between any two points being a straight line. This all applies to the motion of particles under no external forces; actual forces such as gravitation and electromagnetism cause particle tracks to deviate from the straight line.

Gravitation exerts the same force per unit mass on all bodies and the essence of Einstein's theory is to transform it from being a force to being a property of space-time. In his theory, the space-time is not necessarily flat as it is in Minkowski space-time (1.2.1) but may be curved. The interval between two events can be written as

$$\mathrm{d}s^2 = g_{ij}\,\mathrm{d}x^i\,\mathrm{d}x^j, \tag{1.2.2}$$

where repeated suffixes imply summation and i, j both run from 0 to 3; $x^0 = ct$ is the time coordinate and x^1, x^2, x^3 are space coordinates. The tensor g_{ij} is the *metric tensor* describing the space-time geometry; we discuss this in much more detail in Section 1.3. As we mentioned above, particle moves in such a way that the integral along its path is stationary:

$$\delta \int_{\text{path}} \mathrm{d}s = 0, \tag{1.2.3}$$

but such tracks are no longer straight because of the effects of gravitation contained in g_{ij}. From Equation (1.2.3), the path of a free particle, which is called a *geodesic*, can be shown to be described by

$$\frac{\mathrm{d}^2 x^i}{\mathrm{d}s^2} + \Gamma^i_{kl}\frac{\mathrm{d}x^k}{\mathrm{d}s}\frac{\mathrm{d}x^l}{\mathrm{d}s} = 0, \tag{1.2.4}$$

where the Γs are called *Christoffel symbols*,

$$\Gamma^i_{kl} = \tfrac{1}{2} g^{im}\left[\frac{\partial g_{mk}}{\partial x^l} + \frac{\partial g_{ml}}{\partial x^k} - \frac{\partial g_{kl}}{\partial x^m}\right], \tag{1.2.5}$$

and

$$g^{im} g_{mk} = \delta^i_k \tag{1.2.6}$$

is the Kronecker delta, which is unity when $i = k$ and zero otherwise. Free particles move on geodesics but the metric g_{ij} is itself determined by the matter. The key factor in Einstein's equations is the relationship between the distribution of matter and the metric describing the space–time geometry.

In general relativity all equations are tensor equations. A general tensor is a quantity which transforms as follows when coordinates are changed from x^i to x'^i:

$$A'^{kl\ldots}_{pq\ldots} = \frac{\partial x'^k}{\partial x^m} \frac{\partial x'^l}{\partial x^n} \cdots \frac{\partial x^r}{\partial x'^p} \frac{\partial x^s}{\partial x'^q} \cdots A^{mn\ldots}_{rs\ldots}, \tag{1.2.7}$$

where the upper indices are *contravariant* and the lower are *covariant*. The difference between these types of index can be illustrated by considering a tensor of rank 1 which is simply a vector (the rank of a tensor is the number of indices it carries). A vector will undergo a transformation according to some rules when the coordinate system in which it is expressed is changed. Suppose we have an original coordinate system x^i and we transform it to a new system x'^k. If the vector $\boldsymbol{A}$ transforms in such a way that $\boldsymbol{A}' = \partial x'^k / \partial x^i \boldsymbol{A}$, then the vector $\boldsymbol{A}$ is a contravariant vector and it is written with an upper index, i.e. $\boldsymbol{A} = A^i$. On the other hand, if the vector transforms according to $\boldsymbol{A}' = \partial x^i / \partial x'^k \boldsymbol{A}$, then it is covariant and is written $\boldsymbol{A} = A_i$. The tangent vector to a curve is an example of a contravariant vector; the normal to a surface is a covariant vector. The rule (1.2.7) is a generalisation of these concepts to tensors of arbitrary rank and to tensors of mixed character.

In Newtonian and special-relativistic physics a key role is played by conservation laws of mass, energy and momentum. Our task is now to obtain similar laws for general relativity. With the equivalence of mass and energy brought about by Special Relativity, these laws can be written

$$\frac{\partial T_{ik}}{\partial x^k} = 0. \tag{1.2.8}$$

The energy–momentum tensor T_{ik} describes the matter distribution: for a perfect fluid, with pressure p and energy density ρ, it is

$$T_{ik} = (p + \rho c^2) U_i U_k - p g_{ik}; \tag{1.2.9}$$

the vector U_i is the fluid four-velocity

$$U_i = g_{ik} U^k = g_{ik} \frac{\mathrm{d}x^k}{\mathrm{d}s}, \tag{1.2.10}$$

where $x^k(s)$ is the world line of a fluid element, i.e. the trajectory in space–time followed by the particle. Equation (1.2.10) is a special case of the general rule for raising or lowering suffixes using the metric tensor.

It is easy to see that the Equation (1.2.8) cannot be correct in general relativity since $\partial T^{ik}/\partial x^k$ and $\partial T_{ik}/\partial x^k$ are not tensors. Since

$$T'_{mn} = \frac{\partial x^i}{\partial x'^m}\frac{\partial x^k}{\partial x'^n}T_{ik},$$

it is evident that $\partial T_{mn}/\partial x'^n$ involves terms such as $\partial^2 x^i/\partial x'^m \partial x'^n$, so it will not be a tensor. However, although the ordinary derivative of a tensor is not a tensor, a quantity called the *covariant derivative* can be shown to be one. The covariant derivative of a tensor $\boldsymbol{A}$ is defined by

$$A^{kl\ldots}_{pq\ldots;j} = \frac{\partial A^{kl\ldots}_{pq\ldots}}{\partial x^j} + \Gamma^k_{mj}A^{ml\ldots}_{pq\ldots} + \Gamma^l_{nj}A^{kn\ldots}_{pq\ldots} + \cdots - \Gamma^r_{pj}A^{kl\ldots}_{rq\ldots} - \Gamma^s_{qj}A^{kl\ldots}_{ps\ldots} - \cdots \tag{1.2.11}$$

in an obvious notation. The conservation law can therefore be written in a fully covariant form:

$$T_i{}^k{}_{;k} = 0. \tag{1.2.12}$$

A covariant derivative is usually written as a ';' in the subscript; ordinary derivatives are usually written as a ',' so that Equation (1.2.8) can be written $T_{ik,k} = 0$.

Einstein wished to find a relation between matter and metric and to equate T_{ik} to a tensor obtained from g_{ik}, which contains only the first two derivatives of g_{ik} and has zero covariant derivative. Because, in the appropriate limit, Equation (1.2.12) must reduce to Poisson's equation describing Newtonian gravity

$$\nabla^2\varphi = 4\pi G\rho, \tag{1.2.13}$$

it should be linear in the second derivative of the metric. The properties of curved spaces were well-known when Einstein was working on this theory. For example, it was known that the *Riemann-Christoffel tensor*,

$$R^i_{klm} = \frac{\partial \Gamma^i_{km}}{\partial x^l} - \frac{\partial \Gamma^i_{kl}}{\partial x^m} + \Gamma^i_{\text{nl}}\Gamma^n_{km} - \Gamma^i_{nm}\Gamma^n_{kl}, \tag{1.2.14}$$

could be used to determine whether a given space is curved or flat. (Incidentally, Γ^i_{km} is not a tensor so it is by no means obvious, though it is actually true, that R^i_{klm} is a tensor.) From the Riemann-Christoffel tensor one can form the *Ricci tensor*:

$$R_{ik} = R^l{}_{ilk}. \tag{1.2.15}$$

Finally, one can form a scalar curvature, the *Ricci scalar*:

$$R = g^{ik}R_{ik}. \tag{1.2.16}$$

Now we are in a position to define the *Einstein tensor*

$$G_{ik} \equiv R_{ik} - \tfrac{1}{2}g_{ik}R. \tag{1.2.17}$$

Einstein showed that

$$G_{i\;;k}^{\;k} = 0. \tag{1.2.18}$$

The tensor G_{ik} contains second derivatives of g_{ik}, so Einstein proposed as his fundamental equation

$$G_{ik} \equiv R_{ik} - \tfrac{1}{2}g_{ik}R = \frac{8\pi G}{c^4}T_{ik}, \tag{1.2.19}$$

where the quantity $8\pi G/c^4$ (G is Newton's gravitational constant) ensures that Poisson's equation in its standard form (1.2.13) results in the limit of a weak gravitational field. He subsequently proposed the alternative form

$$G_{ik} \equiv R_{ik} - \tfrac{1}{2}g_{ik}R - \Lambda g_{ik} = \frac{8\pi G}{c^4}T_{ik}, \tag{1.2.20}$$

where Λ is called the *cosmological constant*; as $g_{i\;;k}^{\;k} = 0$, we still have $T_{i\;;k}^{\;k} = 0$. He actually did this in order to ensure that static cosmological solutions could be obtained. We shall return to be the issue of Λ later, in Section 1.12.

1.3 The Robertson-Walker Metric

Having established the idea of the Cosmological Principle, our task is to see if we can construct models of the Universe in which this principle holds. Because general relativity is a geometrical theory, we must begin by investigating the geometrical properties of homogeneous and isotropic spaces. Let us suppose we can regard the Universe as a continuous fluid and assign to each fluid element the three spatial coordinates x^α ($\alpha = 1, 2, 3$). Thus, any point in space-time can be labelled by the coordinates x^α, corresponding to the fluid element which is passing through the point, and a time parameter which we take to be the proper time t measured by a clock moving with the fluid element. The coordinates x^α are called *comoving coordinates*. The geometrical properties of space-time are described by a metric; the meaning of the metric will be divulged just a little later. One can show from simple geometrical considerations only (i.e. without making use of any field equations) that the most general space-time metric describing a universe in which the Cosmological Principle is obeyed is of the form

$$\mathrm{d}s^2 = (c\,\mathrm{d}t)^2 - a(t)^2\left[\frac{\mathrm{d}r^2}{1 - Kr^2} + r^2(\mathrm{d}\vartheta^2 + \sin^2\vartheta\,\mathrm{d}\varphi^2)\right], \tag{1.3.1}$$

where we have used spherical polar coordinates: r, ϑ and φ are the comoving coordinates (r is by convention dimensionless); t is the proper time; $a(t)$ is a function to be determined which has the dimensions of a length and is called the *cosmic scale factor* or the *expansion parameter*; the *curvature parameter* K is a constant which can be scaled in such a way that it takes only the values 1, 0 or -1. The metric (1.3.1) is called the *Robertson-Walker metric*.

The significance of the metric of a space–time, or more specifically the metric tensor g_{ik}, which we introduced briefly in Equation (1.2.2),

$$ds^2 = g_{ik}(x)\,dx^i\,dx^k \quad (i, k = 0, 1, 2, 3) \tag{1.3.2}$$

(as usual, repeated indices imply a summation), is such that, in Equation (1.3.2), ds^2 represents the space–time interval between two points labelled by x^j and $x^j + dx^j$. Equation (1.3.1) merely represents a special case of this type of relation. The metric tensor determines all the geometrical properties of the space–time described by the system of coordinates x^j. It may help to think of Equation (1.3.2) as a generalisation of Pythagoras's theorem. If $ds^2 > 0$, then the interval is *timelike* and ds/c would be the time interval measured by a clock which moves freely between x^j and $x^j + dx^j$. If $ds^2 < 0$, then the interval is *spacelike* and $|ds^2|^{1/2}$ represents the length of a ruler with ends at x^j and $x^j + dx^j$ measured by an observer at rest with respect to the ruler. If $ds^2 = 0$, then the interval is *lightlike* or *null*; this type of interval is important because it means that the two points x^j and $x^j + dx^j$ can be connected by a light ray.

If the distribution of matter is uniform, then the space is uniform and isotropic. This, in turn, means that one can define a *universal time* (or *proper time*) such that at any instant the three-dimensional spatial metric

$$dl^2 = \gamma_{\alpha\beta}\,dx^\alpha\,dx^\beta \quad (\alpha, \beta = 1, 2, 3), \tag{1.3.3}$$

where the interval is now just the spatial distance, is identical in all places and in all directions. Thus, the space–time metric must be of the form

$$ds^2 = (c\,dt)^2 - dl^2 = (c\,dt)^2 - \gamma_{\alpha\beta}\,dx^\alpha\,dx^\beta. \tag{1.3.4}$$

This coordinate system is called the *synchronous gauge* and is the most commonly used way of slicing the four-dimensional space–time into three space dimensions and one time dimension.

To find the three-dimensional (spatial) metric tensor $\gamma_{\alpha\beta}$ let us consider first the simpler case of an isotropic and homogeneous space of only two dimensions. Such a space can be either (i) the usual Cartesian plane (flat Euclidean space with infinite curvature radius), (ii) a spherical surface of radius R (a curved space with positive Gaussian curvature $1/R^2$), or (iii) the surface of a hyperboloid (a curved space with negative Gaussian curvature).

In the first case the metric, in polar coordinates ρ ($0 \leqslant \rho < \infty$) and φ ($0 \leqslant \varphi < 2\pi$), is of the form

$$dl^2 = a^2(dr^2 + r^2\,d\varphi^2); \tag{1.3.5 a}$$

we have introduced the dimensionless coordinate $r = \rho/a$, which lies in the range $0 \leqslant r < \infty$, and the arbitrary constant a, which has the dimensions of a length. On the surface of a sphere of radius R the metric in coordinates ϑ ($0 \leqslant \vartheta \leqslant \pi$) and φ ($0 \leqslant \varphi < 2\pi$) is just

$$dl^2 = a^2(d\vartheta^2 + \sin^2\vartheta\,d\varphi^2) = a^2\left(\frac{dr^2}{1 - r^2} + r^2\,d\varphi^2\right), \tag{1.3.5 b}$$

where $a = R$ and the dimensionless variable $r = \sin\vartheta$ lies in the interval $0 \leqslant r \leqslant 1$ ($r = 0$ at the poles and $r = 1$ at the equator). In the hyperboloidal case the metric is given by

$$\mathrm{d}l^2 = a^2(\mathrm{d}\vartheta^2 + \sinh^2\vartheta\,\mathrm{d}\varphi^2) = a^2\left(\frac{\mathrm{d}r^2}{1+r^2} + r^2\,\mathrm{d}\varphi^2\right), \tag{1.3.5 c}$$

where the dimensionless variable $r = \sinh\vartheta$ lies in the range $0 \leqslant r < \infty$.

The Robertson–Walker metric is obtained from (1.3.4), where the spatial part is simply the three-dimensional generalisation of (1.3.5). One finds that for the three-dimensional flat, positively curved and negatively curved spaces one has, respectively,

$$\mathrm{d}l^2 = a^2(\mathrm{d}r^2 + r^2\,\mathrm{d}\Omega^2), \tag{1.3.6 a}$$

$$\mathrm{d}l^2 = a^2(\mathrm{d}\chi^2 + \sin^2\chi\,\mathrm{d}\Omega^2) = a^2\left(\frac{\mathrm{d}r^2}{1-r^2} + r^2\,\mathrm{d}\Omega^2\right), \tag{1.3.6 b}$$

$$\mathrm{d}l^2 = a^2(\mathrm{d}\chi^2 + \sinh^2\chi\,\mathrm{d}\Omega^2) = a^2\left(\frac{\mathrm{d}r^2}{1+r^2} + r^2\,\mathrm{d}\Omega^2\right), \tag{1.3.6 c}$$

where $\mathrm{d}\Omega^2 = \mathrm{d}\vartheta^2 + \sin^2\vartheta\,\mathrm{d}\varphi^2$; $0 \leqslant \chi \leqslant \pi$ in (1.3.6 *b*) and $0 \leqslant \chi < \infty$ in (1.3.6 *c*). The values of $K = 1, 0, -1$ in (1.3.1) correspond, respectively, to the hypersphere, Euclidean space and space of constant negative curvature.

The geometrical properties of Euclidean space ($K = 0$) are well known. On the other hand, the properties of the hypersphere ($K = 1$) are complex. This space is closed, i.e. it has finite volume, but has no boundaries. This property is clear by analogy with the two-dimensional case of a sphere: beginning from a coordinate origin at the pole, the surface inside a radius $r_c(\vartheta) = a\vartheta$ has an area $S(\vartheta) = 2\pi a^2(1 - \cos\vartheta)$, which increases with r_c and has a maximum value $S_{max} = 4\pi a^2$ at $\vartheta = \pi$. The perimeter of this region is $L(\vartheta) = 2\pi a \sin\vartheta = 2\pi a r$, which is maximum at the 'equator' ($\vartheta = \frac{1}{2}\pi$), where it takes the value $2\pi a$, and is zero at the 'antipole' ($\vartheta = \pi$): the sphere is therefore a closed surface, with finite area and no boundary. In the three-dimensional case the volume of the region contained inside a radius

$$r_c(\chi) = a\chi = a\sin^{-1} r \tag{1.3.7}$$

has volume

$$V(\chi) = 2\pi a^3(\chi - \tfrac{1}{2}\sin 2\chi), \tag{1.3.8}$$

which increases and has a maximum value for $\chi = \pi$,

$$V_{max} = 2\pi^2 a^3, \tag{1.3.9}$$

and area

$$S(\chi) = 4\pi a^2 \sin^2\chi, \tag{1.3.10}$$

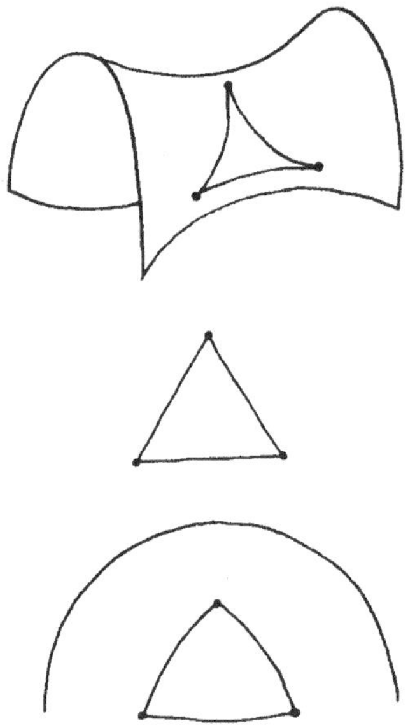

Figure 1.1 Examples of curved spaces in two dimensions: in a space with negative curvature (open), for example, the sum of the internal angles of a triangle is less than 180°, while for a positively curved space (closed) it is greater.

maximum at the 'equator' ($\chi = \frac{1}{2}\pi$), where it takes the value $4\pi a^2$, and is zero at the 'antipole' ($\chi = \pi$). In such a space the value of $S(\chi)$ is more than in Euclidean space, and the sum of the internal angles of a triangle is more than π. The properties of a space of constant negative curvature ($K = -1$) are more similar to those of Euclidean space: the hyperbolic space is open, i.e. infinite. All the relevant formulae for this space can be obtained from those describing the hypersphere by replacing trigonometric functions by hyperbolic functions. One can show, for example, that $S(\chi)$ is less than the Euclidean case, and the sum of the internal angles of a triangle is less than π.

In cases with $K \neq 0$, the parameter a, which appears in (1.3.1), is related to the curvature of space. In fact, the Gaussian curvature is given by $C_{\rm G} = K/a^2$; as expected it is positive for the closed space and negative for the open space. The Gaussian curvature radius $R_{\rm G} = C_{\rm G}^{-1/2} = a/\sqrt{K}$ is, respectively, positive or imaginary in these two cases. In cosmology one uses the term radius of curvature to describe the modulus of $R_{\rm G}$; with this convention a always represents the radius of spatial curvature. Of course, in a flat universe the parameter a does not have any geometrical significance.

As we shall see later in this chapter, the Einstein equations of general relativity relate the geometrical properties of space-time with the energy-momentum tensor describing the contents of the Universe. In particular, for a homogeneous and isotropic perfect fluid with rest-mass energy density ρc^2 and pressure p, the solutions of the Einstein equations are the *Friedmann cosmological equations*:

$$\ddot{a} = -\tfrac{4}{3}\pi G\left(\rho + 3\frac{p}{c^2}\right)a, \tag{1.3.11 a}$$

$$\dot{a}^2 + Kc^2 = \tfrac{8}{3}\pi G\rho a^2 \tag{1.3.11 b}$$

(the dot represents a derivative with respect to cosmological proper time t); the time evolution of the expansion parameter a which appears in the Robertson-Walker metric (1.3.1) can be derived from (1.3.11) if one has an equation of state relating p to ρ. From Equation (1.3.11 b) one can derive the curvature

$$\frac{K}{a^2} = \frac{1}{c^2}\left(\frac{\dot{a}}{a}\right)^2\left(\frac{\rho}{\rho_c} - 1\right), \tag{1.3.12}$$

where

$$\rho_c = \frac{3}{8\pi G}\left(\frac{\dot{a}}{a}\right)^2 \tag{1.3.13}$$

is called the *critical density*. The space is closed ($K = 1$), flat ($K = 0$) or open ($K = -1$) according to whether the *density parameter*

$$\Omega(t) = \frac{\rho}{\rho_c} \tag{1.3.14}$$

is greater than, equal to, or less than unity.

It will sometimes be useful to change the time variable we use from proper time to *conformal time*:

$$\tau = \int \frac{dt}{a(t)}; \tag{1.3.15}$$

with such a time variable the Robertson-Walker metric becomes

$$ds^2 = a(\tau)^2\left[(c\,d\tau)^2 - \left(\frac{dr^2}{1 - Kr^2} + r^2\,d\Omega^2\right)\right]. \tag{1.3.16}$$

1.4 The Hubble Law

The *proper distance*, d_P, of a point P from another point P_0, which we take to define the origin of a set of polar coordinates r, ϑ and φ, is the distance measured by a chain of rulers held by observers which connect P to P_0 at time t. From the Robertson-Walker metric (1.3.1) with $dt = 0$ this can be seen to be

$$d_P = \int_0^r \frac{a\,dr'}{(1 - Kr'^2)^{1/2}} = af(r), \tag{1.4.1}$$

where the function $f(r)$ is, respectively,

$$f(r) = \sin^{-1} r \qquad (K = 1), \tag{1.4.2 a}$$

$$f(r) = r \qquad (K = 0), \tag{1.4.2 b}$$

$$f(r) = \sinh^{-1} r \qquad (K = -1). \tag{1.4.2 c}$$

Of course this proper distance is of little operational significance because one can never measure simultaneously all the distance elements separating P from P_0. The proper distance at time t is related to that at the present time t_0 by

$$d_{\rm P}(t_0) = a_0 f(r) = \frac{a_0}{a} d_{\rm P}(t), \tag{1.4.3}$$

where a_0 is the value of $a(t)$ at $t = t_0$. Instead of the comoving coordinate r one could also define a radial comoving coordinate of P by the quantity

$$d_{\rm c} = a_0 f(r). \tag{1.4.4}$$

In this case the relation between comoving coordinates and proper coordinates is just

$$d_{\rm c} = \frac{a_0}{a} d_{\rm P}. \tag{1.4.5}$$

The proper distance $d_{\rm P}$ of a source may change with time because of the time-dependence of the expansion parameter a. In this case a source at P has a radial velocity with respect to the origin P_0 given by

$$v_{\rm r} = \dot{a} f(r) = \frac{\dot{a}}{a} d_{\rm P}. \tag{1.4.6}$$

Equation (1.4.6) is called the *Hubble law* and the quantity

$$H(t) = \dot{a}/a \tag{1.4.7}$$

is called the *Hubble constant* or, more accurately, the *Hubble parameter* (because it is not constant in time). As we shall see, the value of this parameter evaluated at the present time for our Universe, $H(t_0) = H_0$, is not known to any great accuracy. It is believed, however, to have a value around

$$H_0 \simeq 65 \text{ km s}^{-1} \text{ Mpc}^{-1}. \tag{1.4.8}$$

The unit 'Mpc' is defined later on in Section 4.1. It is conventional to take account of the uncertainty in H_0 by defining the dimensionless parameter h to be $H_0/100 \text{ km s}^{-1} \text{ Mpc}^{-1}$ (see Section 4.2). The law (1.4.6) can, in fact, be derived directly from the Cosmological Principle if $v \ll c$. Consider a triangle defined by the three spatial points O, O′ and P. Let the velocity of P and O′ with respect to O be, respectively, $\boldsymbol{v}(\boldsymbol{r})$ and $\boldsymbol{v}(\boldsymbol{d})$. The velocity of P with respect to O′ is

$$\boldsymbol{v}'(\boldsymbol{r}') = \boldsymbol{v}(\boldsymbol{r}) - \boldsymbol{v}(\boldsymbol{d}). \tag{1.4.9}$$

From the Cosmological Principle the functions $\boldsymbol{v}$ and $\boldsymbol{v}'$ must be the same. Therefore

$$\boldsymbol{v}(\boldsymbol{r} - \boldsymbol{d}) = \boldsymbol{v}'(\boldsymbol{r} - \boldsymbol{d}) = \boldsymbol{v}(\boldsymbol{r}) - \boldsymbol{v}(\boldsymbol{d}). \tag{1.4.10}$$

Equation (1.4.10) implies a linear relationship between $\boldsymbol{v}$ and $\boldsymbol{r}$:

$$v_\alpha = H_\alpha{}^\beta x_\beta \quad (\alpha, \beta = 1, 2, 3). \tag{1.4.11}$$

If we impose the condition that the velocity field is *irrotational*,

$$\nabla \times \boldsymbol{v} = \boldsymbol{0}, \tag{1.4.12}$$

which comes from the condition of isotropy, one can deduce that the matrix $H_\alpha{}^\beta$ is symmetric and can therefore be diagonalised by an appropriate coordinate transformation. From isotropy, the velocity field must therefore be of the form

$$v_i = H x_i, \tag{1.4.13}$$

where H is only a function of time. Equation (1.4.13) is simply the Hubble law (1.4.6).

Another, simpler, way to derive Equation (1.4.6) is the following. The points O, O′ and P are assumed to be sufficiently close to each other that relativistic space-time curvature effects are negligible. If the universe evolves in a homogeneous and isotropic manner, the triangle OO′P must always be similar to the original triangle. This means that the length of all the sides must be multiplied by the same factor a/a_0. Consequently, the distance between any two points must also be multiplied by the same factor. We therefore have

$$l = \frac{a}{a_0} l_0, \tag{1.4.14}$$

where l_0 and l are the lengths of a line segment joining two points at times t_0 and t, respectively. From (1.4.14) we recover immediately the Hubble law (1.4.6).

One property of the Hubble law, which is implicit in the previous reasoning, is that we can treat any spatial position as the origin of a coordinate system. In fact, referring again to the triangle OO′P, we have

$$\boldsymbol{v}_{\mathrm{P}} = \boldsymbol{v}_{\mathrm{O}'} + \boldsymbol{v}'_{\mathrm{P}} = H\boldsymbol{d} + \boldsymbol{v}'_{\mathrm{P}} = H\boldsymbol{r} \tag{1.4.15}$$

and, therefore,

$$\boldsymbol{v}'_{\mathrm{P}} = H(\boldsymbol{r} - \boldsymbol{d}) = H\boldsymbol{r}', \tag{1.4.16}$$

which again is just the Hubble law, this time expressed about the point O′.

1.5 Redshift

It is useful to introduce a new variable related to the expansion parameter a which is more directly observable. We call this variable the *redshift* z and we shall use it extensively from now on in describing the evolution of the Universe because many of the relevant formulae are very simple when expressed in terms of this variable.

We define the redshift of a luminous source, such as a distant galaxy, by the quantity

$$z = \frac{\lambda_0 - \lambda_e}{\lambda_e}, \tag{1.5.1}$$

where λ_0 is the wavelength of radiation from the source observed at O (which we take to be the origin of our coordinate system) at time t_0 and emitted by the source at some (earlier) time t_e; the source is moving with the expansion of the universe and is at a comoving coordinate r. The wavelength of radiation emitted by the source is λ_e. The radiation travels along a light ray (null geodesic) from the source to the observer so that $ds^2 = 0$ and, therefore,

$$\int_{t_e}^{t_0} \frac{c\,dt}{a(t)} = \int_0^r \frac{dr}{(1 - Kr^2)^{1/2}} = f(r). \tag{1.5.2}$$

Light emitted from the source at $t'_e = t_e + \delta t_e$ reaches the observer at $t'_0 = t_0 + \delta t_0$. Given that $f(r)$ does not change, because r is a comoving coordinate and both the source and the observer are moving with the cosmological expansion, we can write

$$\int_{t'}^{t'_0} \frac{c\,dt}{a(t)} = f(r). \tag{1.5.3}$$

If δt and, therefore, δt_0 are small, Equations (1.5.2) and (1.5.3) imply that

$$\frac{\delta t_0}{a_0} = \frac{\delta t}{a}. \tag{1.5.4}$$

If, in particular, $\delta t = 1/\nu_e$ and $\delta t_0 = 1/\nu_0$ (ν_e and ν_0 are the frequencies of the emitted and observed light, respectively), we will have

$$\nu_e a = \nu_0 a_0 \tag{1.5.5}$$

or, equivalently,

$$\frac{a}{\lambda_e} = \frac{a_0}{\lambda_0}, \tag{1.5.6}$$

from which

$$1 + z = \frac{a_0}{a}. \tag{1.5.7}$$

A line of reasoning similar to the previous one can be made to recover the evolution of the velocity $v_p(t)$ of a test particle with respect to a comoving observer. At time $t + dt$ the particle has travelled a distance $dl = v_p(t)\,dt$ and thus finds itself moving with respect to a new reference frame which, because of the expansion of the universe, has an expansion velocity $dv = (\dot{a}/a)\,dl$. The velocity of the particle with respect to the new comoving observer is therefore

$$v_p(t + dt) = v_p(t) - \frac{\dot{a}}{a}\,dl = v_p(t) - \frac{\dot{a}}{a} v_p(t)\,dt, \tag{1.5.8}$$

which, integrated, gives

$$v_{\mathrm{p}} \propto a^{-1}. \tag{1.5.9}$$

The results expressed by Equations (1.5.5) and (1.5.11) are a particular example of the fact that, in a universe described by the Robertson–Walker metric, the momentum q of a free particle (whether relativistic or not) evolves according to $q \propto a^{-1}$.

There is also a simply way to recover Equation (1.5.7), which does not require any knowledge of the metric. Consider two nearby points P and P′, participating in the expansion of the Universe. From the Hubble law we have

$$\mathrm{d}v_{\mathrm{P}'} = H\,\mathrm{d}l = \frac{\dot{a}}{a}\,\mathrm{d}l, \tag{1.5.10}$$

where $\mathrm{d}v_{\mathrm{P}'}$ is the relative velocity of P′ with respect to P and $\mathrm{d}l$ is the (infinitesimal) distance between P and P′. The point P′ sends a light signal at time t and frequency ν which arrives at P with frequency ν' at time $t + \mathrm{d}t = t + (\mathrm{d}l/c)$. Since $\mathrm{d}l$ is infinitesimal, as is $\mathrm{d}v_{\mathrm{P}'}$, we can apply the approximate formula describing the *Doppler effect*:

$$\frac{\nu' - \nu}{\nu} = \frac{\mathrm{d}\nu}{\nu} \simeq -\frac{\mathrm{d}v_{\mathrm{P}'}}{c} = -\frac{\dot{a}}{a}\,\mathrm{d}t = -\frac{\mathrm{d}a}{a}. \tag{1.5.11}$$

The Equation (1.5.11) integrates immediately to give (1.5.5) and therefore (1.5.7).

1.6 The Deceleration Parameter

The Hubble parameter $H(t)$ measures the expansion rate at any particular time t for any model obeying the Cosmological Principle. It does, however, vary with time in a way that depends upon the contents of the Universe. One can express this by expanding the cosmic scale factor for times t close to t_0 in a power series:

$$a(t) = a_0[1 + H_0(t - t_0) - \tfrac{1}{2}q_0H_0^2(t - t_0)^2 + \cdots], \tag{1.6.1}$$

where

$$q_0 = -\frac{\ddot{a}(t_0)a_0}{\dot{a}(t_0)^2} \tag{1.6.2}$$

is called the *deceleration parameter*; the suffix '0', as always, refers to the fact that $q_0 = q(t_0)$. Note that while the Hubble parameter has the dimensions of inverse time, q is actually dimensionless.

Putting the redshift, defined by Equation (1.5.7), into Equation (1.6.1) we find that

$$z = H_0(t_0 - t) + (1 + \tfrac{1}{2}q_0)H_0^2(t_0 - t)^2 + \cdots, \tag{1.6.3}$$

which can be inverted to yield

$$t_0 - t = \frac{1}{H_0}[z - (1 + \tfrac{1}{2}q_0)z^2 + \cdots]. \tag{1.6.4}$$

To find r as a function of z one needs to recall that, for a light ray,

$$\int_t^{t_0} \frac{c\,\mathrm{d}t}{a} = \int_0^r \frac{\mathrm{d}r}{(1 - Kr^2)^{1/2}}, \tag{1.6.5}$$

which becomes, using Equations (1.5.7) and (1.6.3),

$$\frac{c}{a_0}\int_t^{t_0} [1 + H_0(t_0 - t) + (1 + \tfrac{1}{2}q_0)H_0^2(t_0 - t)^2 + \cdots]\,\mathrm{d}t = r + \mathcal{O}(r^3), \tag{1.6.6}$$

and therefore

$$r = \frac{c}{a_0}[(t_0 - t) + \tfrac{1}{2}H_0(t_0 - t)^2 + \cdots]. \tag{1.6.7}$$

Substituting Equation (1.6.4) into (1.6.7) we have, finally,

$$r = \frac{c}{a_0 H_0}[z - \tfrac{1}{2}(1 + q_0)z^2 + \cdots]. \tag{1.6.8}$$

Expressions of this type are useful because they do not require full solutions of the Einstein equations for $a(t)$; the quantity q_0 is used to parametrise a family of approximate solutions for t close to t_0.

1.7 Cosmological Distances

We have shown how the comoving coordinate system we have adopted relates to *proper distance* (i.e. distances measured in a hypersurface of constant proper time) in spaces described by the Robertson–Walker metric. Obviously, however, we cannot measure proper distances to astronomical objects in any direct way. Distant objects are observed only through the light they emit which takes a finite time to travel to us; we cannot therefore make measurements along a surface of constant proper time, but only along the set of light paths travelling to us from the past - our past *light cone*. One can, however, define operationally other kinds of distance which are, at least in principle, directly measurable.

One such distance is the *luminosity distance* d_{L}. This is defined in such a way as to preserve the Euclidean inverse-square law for the diminution of light with distance from a point source. Let L denote the power emitted by a source at a point P, which is at a coordinate distance r at time t. Let l be the power received per unit area (i.e. the flux) at time t_0 by an observer placed at P_0. We then define

$$d_{\mathrm{L}} = \left(\frac{L}{4\pi l}\right)^{1/2}. \tag{1.7.1}$$

The area of a spherical surface centred on P and passing through P_0 at time t_0 is just $4\pi a_0^2 r^2$. The photons emitted by the source arrive at this surface having been redshifted by the expansion of the universe by a factor a/a_0. Also, as we have seen, photons emitted by the source in a small interval δt arrive at P_0 in an interval $\delta t_0 = (a_0/a)\delta t$ due to a time-dilation effect. We therefore find

$$l = \frac{L}{4\pi a_0^2 r^2}\left(\frac{a}{a_0}\right)^2, \tag{1.7.2}$$

from which

$$d_L = a_0^2 \frac{r}{a}. \tag{1.7.3}$$

Following the same procedure as in Section 1.6, one can show that

$$d_L = \frac{c}{H_0}[z + \tfrac{1}{2}(1 - q_0)z^2 + \cdots], \tag{1.7.4}$$

in contrast with the proper distance, d_P, defined by Equation (1.4.1), which has the form $d_P = a_0 r$, with $f(r)$ given by Equations (1.4.2).

Next we define the *angular-diameter distance* d_A. Again, this is constructed in such a way as to preserve a geometrical property of Euclidean space, namely the variation of the angular size of an object with its distance from an observer. Let $D_P(t)$ be the (proper) diameter of a source placed at coordinate r at time t. If the angle subtended by D_P is denoted $\Delta\vartheta$, then Equation (1.2.1) implies

$$D_P = ar\Delta\vartheta. \tag{1.7.5}$$

We define d_A to be the distance

$$d_A = \frac{D_P}{\Delta\vartheta} = ar; \tag{1.7.6}$$

it should be noted that a decreases as r increases for the same D_P and, in some models, the angular size of a source can actually increase with its luminosity distance.

Other measures of distance, less often used, are the *parallax distance*

$$d_\mu = a_0 \frac{r}{(1 - Kr^2)^{1/2}}, \tag{1.7.7}$$

and the *proper motion distance*

$$d_M = a_0 r. \tag{1.7.8}$$

Evidently, for $r \to 0$, and therefore for $t \to t_0$, we have

$$d_p \simeq d_L \simeq d_A \simeq d_\mu \simeq d_M \simeq d_c, \tag{1.7.9}$$

so that at small distances we recover the Euclidean behaviour.

1.8 The m–z and N–z Relations

The general relationship we have established between redshift and distance allows us to establish some interesting properties of the Universe which could, in principle, be used to probe its spatial geometry and, in particular, to test the Cosmological Principle. In fact, there are severe complications with the implementation of this idea, as we discuss in Section 4.7. If celestial objects (such as galaxy clusters, galaxies, radio sources, quasars, etc.) are distributed homogeneously and isotropically on large scales, it is interesting to consider two relationships: the m–z relationship between the apparent magnitude of a source and its redshift and the $N(> l)$–z relationship between the number of sources of a given type with apparent luminosity greater than some limit l and redshift less than z. These relations are also important because, in principle, they provide a way of determining the deceleration parameter q_0.

As we have seen previously,

$$d_{\mathrm{L}} = \frac{c}{H_0}[z + \tfrac{1}{2}(1 - q_0)z^2 + \cdots], \tag{1.8.1}$$

from which

$$l = \frac{L}{4\pi d_{\mathrm{L}}^2} = \frac{LH_0^2}{4\pi c^2 z^2}[1 + (q_0 - 1)z + \cdots]. \tag{1.8.2}$$

Astronomers do not usually work with the absolute luminosity L and apparent flux l. Instead they work with quantities related to these: the *absolute magnitude* M and the *apparent magnitude* m (for more details see Section 4.1). The magnitude scale is defined logarithmically by taking a factor of 100 in received flux to be a difference of 5 magnitudes. The zero-point can be fixed in various ways; for historical reasons it is conventional to take Polaris to have an apparent magnitude of 2.12 in visible light but different choices can and have been made. The absolute magnitude is defined to be the apparent magnitude the source would have if it were placed at a distance of 10 parsec. The relationship between the luminosity distance of a source, its apparent magnitude m and its absolute magnitude M is, therefore, just

$$d_{\mathrm{L}} = 10^{1+(m-M)/5}\ \mathrm{pc}. \tag{1.8.3}$$

The quantity

$$m - M = -5 + 5\log d_{\mathrm{L}}(\mathrm{pc}) \tag{1.8.4}$$

is called the *distance modulus*. Using Equation (1.8.2) we find

$$m - M \simeq 25 - 5\log_{10} H_0 + 5\log cz + 1.086(1 - q_0)z + \cdots, \tag{1.8.5}$$

with H_0 in km s^{-1} Mpc^{-1} and c in km s^{-1}. Here one should remember that 1 Mpc = 10^6 pc and the logarithms are always defined to the base 10. The behaviour of $m(z)$ is sensitive to the value of q_0 only for $z > 0.1$. In reality, as we shall see, there are many other factors which intervene in this type of analysis with the result that we can say very little about q_0, or even its sign. In the regime where it is accurate, that is for $z < z_{max} \simeq 0.2$, Equation (1.8.5) can provide an estimate of H_0, together with a strong confirmation of the validity of the Hubble law and, therefore, of the Cosmological Principle.

Another test of this principle is the so-called *Hubble test*, which relates the number $N(> l)$ of sources of a particular type with apparent luminosity greater than l as a function of l. If the Universe were Euclidean and galaxies all had the same absolute luminosity L, and were distributed uniformly with mean number-density n_0, we would have

$$N(l) = \tfrac{4}{3}\pi n_0 d_l^3, \tag{1.8.6}$$

with d_l given by

$$d_l = \left(\frac{L}{4\pi l}\right)^{1/2}, \tag{1.8.7}$$

from which

$$N(l) \propto l^{-3/2} \tag{1.8.8}$$

and, therefore, introducing the apparent magnitude in the form $m = 2.5 \log_{10} l +$ const.,

$$\log N(l) = 0.6m + \text{const.} \tag{1.8.9}$$

Equation (1.8.9) is also true if the sources have an arbitrary distribution of luminosities around L; in this case all that changes is the value of the constant.

In the non-Euclidean case we have

$$N(l) = 4\pi \int_0^r \frac{n[t(r')]a[t(r')]^3 r'^2}{(1 - Kr'^2)^{1/2}}\, dr', \tag{1.8.10}$$

where $t(r')$ is the time at which a source at r' emitted a light signal which arrives now at the observer. If the galaxies are neither created nor destroyed in the interval $t(r) < t < t_0$, so that $na^3 = n_0 a_0^3$, we see that, upon expanding as a power series, Equation (1.8.10) leads to

$$N(l) = 4\pi n_0 a_0^3 (\tfrac{1}{3} r^3 + \tfrac{1}{10} K r^5 + \cdots). \tag{1.8.11}$$

Recalling that

$$r = \frac{c}{a_0 H_0}[z - \tfrac{1}{2}(1 + q_0)z^2 + \cdots], \tag{1.8.12}$$

Equation (1.8.11) becomes

$$\log N(l) = 3 \log z - 0.651(1 + q_0)z + \text{const.}, \tag{1.8.13}$$

from which one can, in principle, recover q_0. In practice, however, there are many effects (the most important being various evolutionary phenomena) which effectively mean that the constant terms in the above equations all actually depend on z. Nevertheless, Equation (1.8.13) works well for $z < 0.2$, where the term in q_0 is negligible and the constant is, effectively, constant.

1.9 Olbers' Paradox

Having established the behaviour of light in the expanding relativistic cosmology, it is worth revisiting an idea from the pre-relativistic era. Before the development of relativity, astronomers generally believed the Universe to be infinite, homogeneous, Euclidean and static. This picture was of course shattered by the discovery of the Hubble expansion in 1929, which we discuss in Chapter 4. It is nevertheless interesting to point out that this model, which we might call the Eighteenth Century Universe, gave rise to an interesting puzzle now known as *Olbers' Paradox* (Olbers 1826). As a matter of fact, Olbers' Paradox had previously been analysed by a number of others, including (incorrectly) Halley (1720) and (correctly) Loys de Chéseaux (1744). The argument proceeds from the simple observation that the night sky is quite dark. In an Eighteenth Century Universe, the apparent luminosity l of a star of absolute luminosity L placed at a distance r from an observer is just

$$l = \frac{L}{4\pi r^2} \tag{1.9.1}$$

if one neglects absorption. This is the same as Equation (1.7.1). Let us assume, for simplicity, that all stars have the same absolute luminosity and the (constant) number density of stars per unit volume is n. The radiant energy arriving at the observer from the whole Universe is then

$$l_{\text{tot}} = \int_0^\infty \frac{L}{4\pi r^2} 4\pi r^2 \, \mathrm{d}r = nL \int_0^\infty \mathrm{d}r, \tag{1.9.2}$$

which is infinite. This is the Olbers Paradox. It was thought in the past that this paradox could be resolved by postulating the presence of interstellar absorption, perhaps by dust; such an explanation was actually advanced by Lord Kelvin in the 19th century. What would happen if this were the case would be that, after a sufficient time, the absorbing material would be brought into thermodynamic equilibrium with the radiation and would then emit as much radiation as it absorbed, though perhaps in a different region of the electromagnetic spectrum. To be fair to Kelvin, however, one should mention that at that time it was not known that light and heat were actually different aspects of the same phenomenon, so the argument was reasonable given what was then known about the nature of radiation.

In the modern version of the expanding Universe the conditions necessary for an Olbers Paradox to arise are violated in a number of ways we shall discuss later: the light from a distant star would be redshifted; the spatial geometry is not necessarily flat; the Universe may not be infinite in spatial or temporal extent. In fact, the basic reason why an Olbers Paradox does not arise in modern cosmological theories is much simpler than any of these possibilities. The key fact is that no star can burn for an infinite time: a star of mass M can at most radiate only so long as it takes to radiate away its rest energy Mc^2. As one looks further and further out into space, one must see stars which are older and older. In order for them all, out to infinite distance, to be shining light that we observe now, they must have switched on at different times depending on their distance from us. Such a coordination is not only unnatural, it also requires us to be in a special place. So an Olbers Paradox would only really be expected to happen if the Universe were actually inhomogeneous on large scales and the Copernican Principle were violated. The other effects mentioned above are important in determining the exact amount of radiation received by an observer from the cosmological background, but any cosmology that respects the relativistic notion that $E = mc^2$ (and the Cosmological Principle) is not expected to have an infinitely bright night sky. Exactly how much background light there is in the Universe is an observation that can in principle be used to test cosmological models in much the same way as the number-counts discussed in Section 1.8.

1.10 The Friedmann Equations

So far we have developed much of the language of modern relativistic cosmology without actually using the field Equations (1.2.20). We have managed to discuss many important properties of the universe in terms of geometry or using simple kinematics. To go further we must use general relativity to relate the geometry of space-time, expressed by the metric tensor $g_{ij}(x_k)$, to the matter content of the universe, expressed by the energy-momentum tensor $T_{ij}(x_k)$. The *Einstein equations* (without the cosmological constant; see Section 1.12) are

$$R_{ij} - \tfrac{1}{2} g_{ij} R = \frac{8\pi G}{c^4} T_{ij}, \tag{1.10.1}$$

where R_{ij} and R are the Ricci tensor and Ricci scalar, respectively. A test particle moves along a space-time geodesic, that is a trajectory in a four-dimensional space whose 'length' is stationary with respect to small variations in the trajectory.

In cosmology, the energy-momentum tensor which is of greatest relevance is that of a perfect fluid:

$$T_{ij} = (p + \rho c^2) U_i U_j - p g_{ij}, \tag{1.10.2}$$

where p is the pressure, ρc^2 is the energy density (which includes the rest-mass energy), and U_k is the fluid four-velocity, defined by Equation (1.2.10). If the metric is of Robertson–Walker type, the Einstein equations then yield

$$\ddot{a} = -\frac{4\pi}{3} G \left(\rho + 3\frac{p}{c^2} \right) a, \tag{1.10.3}$$

for the time–time component, and

$$a\ddot{a} + 2\dot{a}^2 + 2Kc^2 = 4\pi G \left(\rho - \frac{p}{c^2} \right) a^2, \tag{1.10.4}$$

for the space–space components. The space–time components merely give $0 = 0$. Eliminating $\ddot{a}$ from (1.10.3) and (1.10.4) we obtain

$$\dot{a}^2 + Kc^2 = \tfrac{8}{3} \pi G \rho a^2. \tag{1.10.5}$$

In reality, as we shall see, Equations (1.10.3) and (1.10.5) – the *Friedmann equations* – are not independent: the second can be recovered from the first if one takes the adiabatic expansion of the universe into account, i.e.

$$d(\rho c^2 a^3) = -p\, da^3. \tag{1.10.6 a}$$

The last equation can also be expressed as

$$\dot{p} a^3 = \frac{d}{dt}[a^3(\rho c^2 + p)] \tag{1.10.6 b}$$

or

$$\dot{\rho} + 3\left(\rho + \frac{p}{c^2}\right)\frac{\dot{a}}{a} = 0. \tag{1.10.6 c}$$

1.11 A Newtonian Approach

Before proceeding further, it is worth demonstrating how one can actually get most of the way towards the Friedmann equations using only Newtonian arguments.

Birkhoff's theorem (1923) proves that a spherically symmetric gravitational field in an empty space is static and is always described by the Schwarzschild exterior metric (i.e. the metric generated in empty space by a point mass). This property is very similar to a result proved by Newton and usually known as *Newton's spherical theorem* which is based on the application of Gauss's theorem to the gravitational field. In the Newtonian version the gravitational field outside a spherically symmetric body is the same as if the body had all its mass concentrated at its centre. Birkhoff's theorem can also be applied to the field inside an empty spherical cavity at the centre of a homogeneous spherical distribution of mass–energy, even if the distribution is not static. In this case the metric inside the cavity is the Minkowski (flat-space) metric: $g_{ij} = \eta_{ij}$ ($\eta_{ij} = -1$ for $i = j = 1, 2, 3$; $\eta_{ij} = 1$ for $i = j = 0$; $\eta_{ij} = 0$ for $i \neq j$). This corollary of Birkhoff's theorem also has a Newtonian analogue: the gravitational field inside a homogeneous spherical shell of matter is always zero. This corollary can also be applied if the space outside the cavity is infinite: the only condition that must be obeyed is that the distribution of mass–energy must be spherically symmetric.

A proof of Birkhoff's theorem is beyond the scope of this book, but we will use its existence to justify a Newtonian approach to the time-evolution of a homogeneous and isotropic distribution of material. Let us consider the evolution of the mass m contained inside a sphere of radius l centred at the point O in such a universe. By Birkhoff's theorem the space inside the sphere is flat. If the radius l is such that

$$\frac{Gm}{lc^2} \ll 1, \tag{1.11.1}$$

one can use Newtonian mechanics to describe the behaviour of the particle. Equation (1.11.1) means in effect that the free-fall time for the sphere, $\tau_{\rm ff} \simeq (G\rho)^{-1/2}$, is much greater than the light-crossing time $\tau \simeq l/c$. Alternatively, Equation (1.11.1) means that the radius of the sphere is much larger than the Schwarzschild radius corresponding to the mass m, $r_S = 2mG/c^2$.

As we have seen in Section 1.4, the Cosmological Principle requires that

$$l = d_c \frac{a}{a_0}, \tag{1.11.2}$$

where a is the expansion parameter of the universe which, according to our conventions, has the dimensions of a length, while the comoving coordinate d_c is dimensionless. One can always pick d_c small enough so that at any instant the inequality (1.11.1) is satisfied. We shall see, however, that this quantity actually disappears from the formulae.

Applying a Newtonian approximation to describe the motion of a unit mass at a point P on the surface of the sphere yields

$$\frac{\mathrm{d}^2 l}{\mathrm{d}t^2} = -\frac{Gm}{l^2} = -\frac{4\pi}{3} G\rho l, \tag{1.11.3}$$

or, multiplying by $\dot{l}$,

$$\frac{\mathrm{d}}{\mathrm{d}t}\frac{\dot{l}^2}{2} = -\frac{Gm}{l^2}\dot{l} = \frac{\mathrm{d}}{\mathrm{d}t}\frac{Gm}{l}, \tag{1.11.4}$$

and, integrating,

$$\dot{l}^2 = \frac{2Gm}{l} + C, \tag{1.11.5}$$

which is nothing more than the law of conservation of energy per unit mass: the constant of integration C is proportional to the total energy. From Equations (1.11.2) and (1.11.5) it is easy to obtain the Equation (1.10.4) in the form

$$\dot{a}^2 + Kc^2 = \tfrac{8}{3}\pi G\rho a^2 \tag{1.11.6}$$

by putting

$$C = -K\left(\frac{d_c c}{a_0}\right)^2. \tag{1.11.7}$$

It is clear that, with an appropriate redefinition of d_c, one can scale K so as to take the values 1, 0 or -1. The case $K = 1$ corresponds to $C < 0$ (negative total energy). In this case the expansion eventually ceases and collapse ensues. In the case $K = -1$ the total energy is positive, so the expansion never ends. The case $K = 0$ corresponds to total energy of exactly zero: this represents the 'escape velocity' situation where the expansion ceases at $t \to \infty$.

Equation (1.11.3) implies that there are no forces due to pressure gradients, which is in accord with our assumption of homogeneity and isotropy. Equation (1.11.6) was obtained under the assumption that the sphere contains only non-relativistic matter ($p \ll \rho c^2$). A result from general relativity shows that, in the presence of relativistic particles, one should replace the density of matter in Equation (1.11.3) by

$$\rho_{\text{eff}} = \rho + 3\frac{p}{c^2}, \tag{1.11.8}$$

where ρ now means the energy density (including the rest-mass energy) divided by c^2. In this way, Equation (1.11.3) becomes

$$\ddot{a} = -\frac{4\pi}{3}G\left(\rho + 3\frac{p}{c^2}\right)a. \tag{1.11.9}$$

It is important to note that, from Equation (1.10.6 *a*),

$$\mathrm{d}(\rho c^2 a^3 r_0^3) = -p\,\mathrm{d}(a^3 r_0^3); \tag{1.11.10}$$

from (1.11.9) one obtains (1.11.6) in both the non-relativistic ($p \simeq 0$, $\rho = \rho_{\text{m}}$) and ultra-relativistic ($p \simeq \rho c^2$) cases. In fact Equation (1.11.9), after multiplying by $\dot{a}$, gives

$$\frac{1}{2}\frac{\mathrm{d}}{\mathrm{d}t}\dot{a}^2 = -\frac{4\pi}{3}G\left(\rho a\dot{a} + 3\frac{p}{c^2}a\dot{a}\right). \tag{1.11.11}$$

From (1.11.10) we have

$$3\frac{p}{c^2}a\dot{a} = -3\rho a\dot{a} - \dot{\rho}a^2, \tag{1.11.12}$$

which, substituted in Equation (1.11.11), yields

$$\frac{1}{2}\frac{\mathrm{d}}{\mathrm{d}t}\dot{a}^2 = \frac{\mathrm{d}}{\mathrm{d}t}\left(\frac{4\pi}{3}G\rho a^2\right). \tag{1.11.13}$$

From Equation (1.11.13), by integration, one obtains Equation (1.11.6).

What this shows is that, with Birkhoff's theorem and a reinterpretation of the quantity ρ to take account of intrinsically relativistic effects, we can derive the Friedmann equations using an essentially Newtonian approach.

1.12 The Cosmological Constant

Einstein formulated his theory of general relativity without a cosmological constant in 1916; at this time it was generally accepted that the Universe was static. We outlined the development of this theory in Section 1.2, and the field equations themselves appear as Equation (1.10.1). A glance at the equation

$$\ddot{a} = -\frac{4\pi}{3}G\left(\rho + 3\frac{p}{c^2}\right)a \tag{1.12.1}$$

shows one that universes evolving according to this theory cannot be static, unless

$$\rho = -3\frac{p}{c^2}; \tag{1.12.2}$$

in other words, either the energy density or the pressure must be negative. Given that this type of fluid does not seem to be physically reasonable, Einstein (1917) modified the Equation (1.10.1) by introducing the cosmological constant term Λ:

$$R_{ij} - \tfrac{1}{2} g_{ij} R - \Lambda g_{ij} = \frac{8\pi G}{c^4} T_{ij}; \tag{1.12.3}$$

as we shall see, with an appropriate choice of Λ, one can obtain a static cosmological model. Equation (1.12.3) represents the most general possible modification of the Einstein equations that still satisfies the condition that T_{ij} is equal to a tensor constructed from the metric g_{ij} and its first and second derivatives, and is linear in the second derivative. This modification does not change the covariant character of the equations, and does not alter the continuity condition (1.2.12). The strongest constraint one can place on Λ from observations is that it should be sufficiently small so as not to change the laws of planetary motion, which are known to be well described by (1.10.1).

The Equation (1.12.3) can be written in a form similar to (1.10.1) by modifying the energy-momentum tensor:

$$R_{ij} - \tfrac{1}{2} g_{ij} R = \frac{8\pi G}{c^4} \tilde{T}_{ij}, \tag{1.12.4}$$

with $\tilde{T}_{ij}$ formally given by

$$\tilde{T}_{ij} = T_{ij} + \frac{\Lambda c^4}{8\pi G} g_{ij} = -\tilde{p} g_{ij} + (\tilde{p} + \tilde{\rho} c^2) U_i U_j, \tag{1.12.5}$$

where the effective pressure $\tilde{p}$ and the effective density $\tilde{\rho}$ are related to the corresponding quantities for a perfect fluid by

$$\tilde{p} = p - \frac{\Lambda c^4}{8\pi G}, \qquad \tilde{\rho} = \rho + \frac{\Lambda c^2}{8\pi G}; \tag{1.12.6}$$

these relations show that $|\Lambda|^{-1/2}$ has the dimensions of a length. One can then show that, for a universe described by the Robertson–Walker metric, we can get equations which are analogous to (1.10.3) and (1.10.5), respectively:

$$\ddot{a} = -\frac{4\pi}{3} G \left(\tilde{\rho} + 3\frac{\tilde{p}}{c^2} \right) a \tag{1.12.7}$$

and

$$\dot{a}^2 + Kc^2 = \frac{8\pi G}{3} \tilde{\rho} a^2. \tag{1.12.8}$$

These equations admit a static solution for

$$\tilde{\rho} = -3\frac{\tilde{p}}{c^2} = \frac{3Kc^2}{8\pi G a^2}. \tag{1.12.9}$$

For a 'dust' universe ($p = 0$), which is a good approximation to our Universe at the present time, Equations (1.12.9) and (1.12.6) give

$$\Lambda = \frac{K}{a^2}, \qquad \rho = \frac{Kc^2}{4\pi G a^2}. \tag{1.12.10}$$

Since $\rho > 0$, we must have $K = 1$ and therefore $\Lambda > 0$. The value of Λ which makes the universe static is just

$$\Lambda_E = \frac{4\pi G\rho}{c^2}. \tag{1.12.11}$$

The model we have just described is called the *Einstein universe*. This universe is static (but unfortunately unstable, as one can show), has positive curvature and a curvature radius

$$a_E = \Lambda_E^{-1/2} = \frac{c}{(4\pi G\rho)^{1/2}}. \tag{1.12.12}$$

After the discovery of the expansion of the Universe in the late 1920s there was no longer any reason to seek static solutions to the field equations. The motivation which had led Einstein to introduce his cosmological constant term therefore subsided. Einstein subsequently regarded the Λ-term as the biggest mistake he had made in his life. Since then, however, Λ has not died but has been the subject of much interest and serious study on both conceptual and observational grounds. The situation here is reminiscent of Aladdin and the genie: after he released the genie from the lamp, it took on a life of its own. For more than 60 years the genie lingered, providing neither compelling observational evidence of its existence nor strong theoretical reasons for it to be taken seriously. However, observations do now suggest that it may have been there all along. We shall return to this resurgence of Λ in the next chapter and also in Chapter 7, but in the meantime we shall restrict ourselves to brief comments on two particularly important models involving the cosmological constant, because we shall encounter them again when we discuss inflation.

The *de Sitter universe* (de Sitter 1917) is a cosmological model in which the universe is empty ($p = 0$; $\rho = 0$) and flat ($K = 0$). From Equations (1.12.6) we get

$$\tilde{p} = -\tilde{\rho}c^2 = -\frac{\Lambda c^4}{8\pi G}, \tag{1.12.13}$$

which, on substitution in (1.12.8), gives

$$\dot{a}^2 = \tfrac{1}{3}\Lambda c^2 a^2; \tag{1.12.14}$$

this equation implies that Λ is positive. Equation (1.12.14) has a solution of the form

$$a = A\exp[(\tfrac{1}{3}\Lambda)^{1/2}ct], \tag{1.12.15}$$

corresponding to a Hubble parameter $H = \dot{a}/a = c(\Lambda/3)^{1/2}$, which is actually constant in time. In the de Sitter vacuum universe, test particles move away from each other because of the repulsive gravitational effect of the positive cosmological constant.

The de Sitter model was only of marginal historical interest until the last 20 years or so. In recent years, however, it has been a major component of inflationary universe models in which, for a certain interval of time, the expansion assumes an exponential character of the type (1.12.15). In such a universe the equation of state of the fluid is of the form $p \simeq -\rho c^2$ due to quantum effects which we discuss in Chapter 7.

In the *Lemaître model* (1927) the universe has positive spatial curvature ($K = 1$). One can demonstrate that the expansion parameter in this case is always increasing, but there is a period in which it remains practically constant. This model was invoked around 1970 to explain the apparent concentration of quasars at a redshift of $z \simeq 2$. Subsequent data have, however, shown that this is not the explanation for the redshift evolution of quasars, so this model is again of only marginal historical interest.

1.13 Friedmann Models

Having dealt with a few special cases, we now introduce the standard cosmological models described by the solutions (1.10.3) and (1.10.5). Their name derives from A. Friedmann, who derived their properties in 1922. His work was not well known at that time partly because his models were not static, and the discovery of the Hubble expansion was still some way in the future. His work was in any case not widely circulated in the western scientific literature. Independently, and somewhat later, the Belgian priest George Lemaître obtained essentially the same results and his work achieved more immediate attention, especially in England where he was championed by Eddington. When the work of Lemaître (1927) was published, Hubble's observations were just becoming known, so in the West Lemaître is often credited with being the father of the Big Bang cosmology, although that honour should probably be conferred on Friedmann.

The Friedmann models are so important that we shall devote the next chapter to their behaviour. Here we shall just whet the readers appetite with some basic properties. First, we assume a perfect fluid with some density ρ and pressure p. The form of equation of state giving p as a function of ρ does not concern us for now; we discuss it in Section 2.1. For the moment we also ignore the cosmological constant.

The equations we need to solve are (1.10.3) and (1.10.5), which we rewrite here for completeness:

$$\ddot{a} = -\frac{4\pi}{3}G\left(\rho + 3\frac{p}{c^2}\right)a \tag{1.13.1}$$

and

$$\dot{a}^2 + Kc^2 = \frac{8\pi G}{3}\rho a^2, \tag{1.13.2}$$

as well as the Equation (1.10.6)

$$d(\rho a^3) = -3\frac{p}{c^2}a^2\,da. \tag{1.13.3}$$

The Equations (1.13.1)–(1.13.3) allow one, at least in principle, to calculate the time evolution of $a(t)$ as well as $\rho(t)$ and $p(t)$ if we know the equation of state.

Let us focus for now on Equation (1.13.3), which can be rewritten in a convenient form for $a = a_0$:

$$\left(\frac{\dot{a}}{a_0}\right)^2 - \frac{8\pi}{3}G\rho\left(\frac{a}{a_0}\right)^2 = H_0^2\left(1 - \frac{\rho_0}{\rho_{0c}}\right) = H_0^2(1 - \Omega_0) = -\frac{Kc^2}{a_0^2}, \tag{1.13.4}$$

where $H_0 = \dot{a}_0/a_0$, Ω_0 is the (present) *density parameter* and

$$\rho_{0c} = \frac{3H_0^2}{8\pi G}. \tag{1.13.5}$$

The suffix '0' refers here to a generic reference time t_0 which is also used in the particular case where t is the present time. Equation (1.13.5) is a reminder of the importance of ρ_{0c}: if $\rho_0 < \rho_{0c}$, then $K = -1$, while if $\rho_0 > \rho_{0c}$, $K = 1$; $K = 0$ corresponds to the 'critical' case when $\rho_0 = \rho_{0c}$.

Let us now include the cosmological constant term Λ. In Section 1.12 we showed how one can treat the cosmological constant as a form of fluid with a strange equation of state, as well as a modification of the law of gravity. In that sense, Λ can be thought of as belonging either on the left-hand or right-hand side of the Einstein equations. Either way, the upshot is that Equations (1.13.1) and (1.13.2) become

$$\ddot{a} = -\frac{4\pi}{3}G\left(\rho + 3\frac{p}{c^2}\right)a + \frac{\Lambda c^2 a}{3} \tag{1.13.6}$$

and

$$\dot{a}^2 + Kc^2 = \frac{8\pi G}{3}\rho a^2 + \frac{\Lambda c^2 a^2}{3}, \tag{1.13.7}$$

respectively. If we ignore the original terms in p and ρ we can see that Equation (1.13.7) can be written in a form similar to Equation (1.13.4):

$$\left(\frac{\dot{a}}{a_0}\right)^2 - \frac{\Lambda c^2}{3} = H_0^2\left(1 - \frac{\Lambda}{\Lambda_c}\right) = H_0^2(1 - \Omega_{0\Lambda}) = -\frac{Kc^2}{a_0^2}. \tag{1.13.8}$$

In this case the 'critical' value is

$$\Lambda_c = \frac{3H_0^2}{c^2}, \tag{1.13.9}$$

so that $\Omega_{0\Lambda} = \Lambda c^2/3H_0^2$.

If we now reinstate the 'ordinary' matter we began with, we can see that the curvature is zero as long as $\Omega_{0\Lambda} + \Omega_0 = 1$. The cosmological constant therefore breaks the relationship between the matter density and curvature. Even if $\Omega_0 < 1$, a suitably chosen value of $\Omega_{0\Lambda} = 1 - \Omega_0$ can be invoked to ensure flat space sections.

Bibliographic Notes on Chapter 1

The classic papers of Einstein (1917), de Sitter (1917), Friedmann (1922) and Lemaître (1927) are all well worth reading for historical insights. A particularly erudite overview of the role of observation in expanding world models is given by Sandage (1988). More detailed discussions of the basic background, including the role of general relativity in cosmology, can be found in Berry (1989), Harrison (1981), Kenyon (1990), Landau and Lifshitz (1975), Milne (1935), Misner *et al.* (1972), Narlikar (1993), Peebles (1993), Peacock (1999), Raychaudhuri (1979), Roos (1994), Wald (1984), Weinberg (1972) and Zel'dovich and Novikov (1983).

Problems

1. Suppose that it is discovered that Newton's law of gravitation is incorrect, and that the force F on a test particle of mass m due to a body of mass M has an additional term that does not depend on M and is proportional to the separation r:

$$F = -\frac{GMm}{r^2} + \frac{Amr}{3}.$$

Assuming that Newton's sphere theorem continues to hold, derive the appropriate form of the Friedmann equation in this case and comment on your result.

2. The most general form of a space-time four-metric in the synchronous gauge is

$$ds^2 = c^2\,dt^2 - g_{\alpha\beta}\,dx^\alpha\,dx^\beta = c^2\,dt^2 - dl^2,$$

where $g_{\alpha\beta}$ is the three-metric of the spatial hypersurfaces. By writing the equation of the three-space as that of a constrained surface in four dimensions, show that the most general form of the three-metric compatible with homogeneity and isotropy is given by the Robertson–Walker form.

3. Show that the special-relativistic formula for the Doppler shift,

$$1 + z = \sqrt{\frac{1 + v/c}{1 - v/c}},$$

reduces to $z \simeq v/c$ in the limit of small velocities. Invert the formula to give v/c in terms of z. Calculate the recession velocity of a galaxy at $z = 5$ using the special-relativistic formula.

4. A model is constructed with $\Omega_0 < 1$, $\Lambda \neq 0$ and $k = 0$. Show in this case that

$$q_0 = \tfrac{3}{2}\Omega_0 - 1.$$

5. An object has luminosity distance $d_{\rm L}$ and angular-diameter distance $d_{\rm A}$. Show that

$$\frac{d_{\rm A}}{d_{\rm L}} = \frac{1}{(1+z)^2},$$

independent of cosmology.

2

The Friedmann Models

2.1 Perfect Fluid Models

In this chapter we shall consider a set of homogeneous and isotropic model universes that contain a relatively simple form of matter. In Section 1.13 we explained how a perfect fluid, described by an energy–momentum tensor of the type (1.10.2), forms the basis of the so-called Friedmann models. The ideal perfect fluid is, in fact, quite a realistic approximation in many situations. For example, if the mean free path between particle collisions is much less than the scales of physical interest, then the fluid may be treated as perfect. It should also be noted that the form (1.10.2) is also required for compatibility with the Cosmological Principle: anisotropic pressure is not permitted. To say more about the cosmological solutions, however, we need to say more about the relationship between p and ρ. In other words we need to specify an equation of state.

As we mentioned in the last section of the previous chapter, we need to specify an equation of state for our fluid in the form $p = p(\rho)$. In many cases of physical interest, the appropriate equation of state can be cast, either exactly or approximately, in the form

$$p = w\rho c^2 = (\Gamma - 1)\rho c^2, \tag{2.1.1}$$

where the parameter w is a constant which lies in the range

$$0 \leqslant w \leqslant 1. \tag{2.1.2}$$

We do not use the parameter $\Gamma = 1 + w$ further in this book, but we have defined it here as it is used by other authors. The allowed range of w given in (2.1.2) is often called the Zel'dovich interval. We shall restrict ourselves for the rest of this chapter to cosmological models containing a perfect fluid with equation of state satisfying this condition.

The case with $w = 0$ represents *dust* (pressureless material). This is also a good approximation to the behaviour of any form of non-relativistic fluid or gas. Of course, gas of particles at some temperature T does exert pressure but the typical thermal energy of a particle is approximately $k_B T$ (k_B is the Boltzmann constant), whereas its rest mass is $m_p c^2$, usually very much larger. The relativistic effect of pressure is usually therefore negligible. In more detail, an *ideal gas* of non-relativistic particles of mass m_p, temperature T, density ρ_m and adiabatic index γ exerts a pressure

$$p = nk_B T = \frac{k_B T}{m_p c^2}\rho_m c^2 = \frac{k_B T}{m_p c^2}\frac{\rho c^2}{1 + (k_B T/((\gamma - 1)m_p c^2))} = w(T)\rho c^2, \quad (2.1.3)$$

where ρc^2 is the energy density; a non-relativistic gas has $w(T) \ll 1$ according to Equation (2.1.3) and will therefore be well approximated by a fluid of dust.

At the other extreme, a fluid of non-degenerate, ultrarelativistic particles in thermal equilibrium has an equation of state of the type

$$p = \tfrac{1}{3}\rho c^2. \quad (2.1.4)$$

For instance, this is the case for a gas of photons. A fluid with an equation of state of the type (2.1.4) is usually called a *radiative fluid*, though it may comprise relativistic particles of any form.

It is interesting to note that the parameter w is also related to the adiabatic sound speed of the fluid

$$v_s = \left(\frac{\partial p}{\partial \rho}\right)_S^{1/2}, \quad (2.1.5)$$

where S denotes the entropy. In a dust fluid $v_s = 0$ and a radiative fluid has $v_s = c/\sqrt{3}$. Note that the case $w > 1$ is impossible, because it would imply that $v_s > c$. If $w < 0$, then it is no longer related to the sound speed, which would have to be imaginary. These two cases form the limits in (2.1.2). There are, however, physically important situations in which matter behaves like a fluid with $w < 0$, as we shall see later.

We shall restrict ourselves to the case where w is constant in time. We shall also assume that normal matter, described by an equation of state of the form (2.1.3), can be taken to have $w(T) \simeq 0$. From Equations (2.1.1) and (1.13.3) we can easily obtain the relation

$$\rho a^{3(1+w)} = \text{const.} = \rho_{0w} a_0^{3(1+w)}. \quad (2.1.6)$$

In this equation and hereafter we use the suffix '0' to denote a reference time, usually the present. In particular we have, for a *dust universe* ($w = 0$) or a *matter universe* described by (2.1.3),

$$\rho a^3 \equiv \rho_m a^3 = \text{const.} = \rho_{0m} a_0^3 \quad (2.1.7)$$

(which simply represents the conservation of mass), and for a *radiative universe* ($w = \frac{1}{3}$)

$$\rho a^4 \equiv \rho_{\rm r} a^4 = {\rm const.} = \rho_{0{\rm r}} a_0^4. \tag{2.1.8}$$

If one replaces the expansion parameter a with the redshift z, one finds, for dust and non-relativistic matter,

$$\rho_{\rm m} = \rho_{0{\rm m}}(1+z)^3, \tag{2.1.9}$$

and, for radiation and relativistic matter,

$$\rho_{\rm r} = \rho_{0{\rm r}}(1+z)^4. \tag{2.1.10}$$

The difference between (2.1.9) and (2.1.10) can be understood quite straightforwardly if one considers a comoving box containing, say, N particles. Let us assume that, as the box expands, particles are neither created nor destroyed. If the particles are non-relativistic (i.e. if the box contains 'dust'), then the density simply decreases as the cube of the scale factor, equivalent to (2.1.9). On the other hand, if the particles are relativistic, then they behave like photons: not only is their number-density diluted by a factor a^3, but also the wavelength of each particle is increased by a factor a resulting in a redshift z. Since the energy of the particles is inversely proportional to their wavelength the total energy must decrease as the fourth power of the scale factor.

Notice the peculiar case in which $w = -1$ in (2.1.6), which we demonstrated to be the perfect fluid equivalent of a cosmological constant. The energy density does not vary as the universe expands for this kind of fluid.

Models of the Universe made from fluids with $-\frac{1}{3} < w < 1$ have the property that they possess a point in time where a vanishes and the density diverges. This instant is called the *Big Bang singularity*. To see how this singularity arises, let us rewrite Equation (1.13.4) of the previous chapter using (2.1.6). Introducing the density parameter

$$\Omega_{0{\rm w}} = \frac{\rho_{0w}}{\rho_{0{\rm c}}} \tag{2.1.11}$$

allows us to obtain the equation

$$\left(\frac{\dot{a}}{a_0}\right)^2 = H_0^2\left[\Omega_{0{\rm w}}\left(\frac{a_0}{a}\right)^{1+3w} + (1-\Omega_{0{\rm w}})\right] \tag{2.1.12}$$

or, alternatively,

$$H^2(t) = H_0^2\left(\frac{a_0}{a}\right)^2\left[\Omega_{0{\rm w}}\left(\frac{a_0}{a}\right)^{1+3w} + (1-\Omega_{0{\rm w}})\right], \tag{2.1.13}$$

where $H(t) = \dot{a}/a$ is the Hubble parameter at a generic time t. Suppose at some generic time, t (for example the present time, t_0), the universe is expanding, so that $\dot{a}(t) > 0$. From Equation (1.13.1), we can see that $\ddot{a} < 0$ for all t, provided

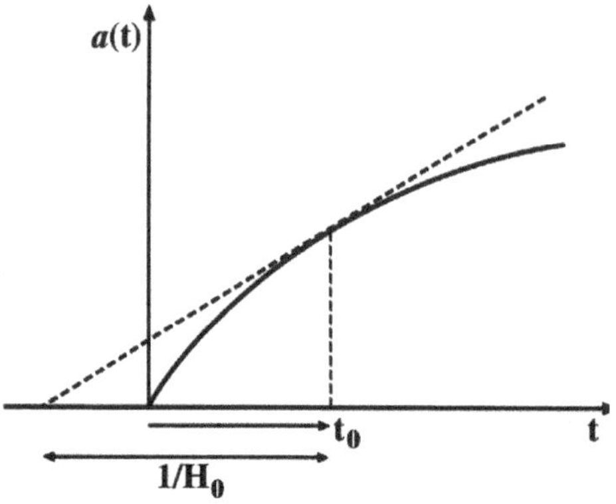

Figure 2.1 The concavity of $a(t)$ ensures that, if $\dot{a}(t) > 0$ for some time t, then there must be a singularity a finite time in the past, i.e. a point when $a = 0$. It also ensures that the age of the Universe, t_0, is less than the Hubble time, $1/H_0$.

$(\rho + 3p/c^2) > 0$ or, in other words, $(1 + 3w) > 0$ since $\rho > 0$. This establishes that the graph of $a(t)$ is necessarily concave. One can see therefore that $a(t)$ must be equal to zero at some finite time in the past, and we can label this time $t = 0$ (see Figure 2.1). Since $a(0) = 0$ at this point, the density ρ diverges, as does the Hubble expansion parameter. One can see also that, because $a(t)$ is a concave function, the time between the singularity and the epoch t must always be less than the characteristic expansion time of the Universe, $\tau_{\rm H} = 1/H = a/\dot{a}$.

The Big Bang singularity is unavoidable in all homogeneous and isotropic models containing fluids with equation-of-state parameter $w > -\frac{1}{3}$, which includes the Zel'dovich interval (2.1.2). It can be avoided, for example, in models with a non-zero cosmological constant, or if the universe is dominated by 'matter' with an effective equation-of-state parameter $w < -\frac{1}{3}$. One might suspect that the singularity may simply be a consequence of the special symmetry of the Friedmann models, and that inhomogeneous and/or anisotropic models would not display such a feature. However, this is not the case, as was shown by the classic work of Hawking an Penrose. We shall return to the unavoidability of the Big Bang singularity later, in Chapter 6.

Note that the expansion of the universe described in the Big Bang model is not due in any way to the effect of pressure, which always acts to *decelerate* the expansion, but is a result of the initial conditions describing a homogeneous and isotropic universe. Another type of initial condition compatible with the Cosmological Principle are those which lead to an isotropic collapse of the universe towards a singularity like a time-reversed Big Bang, often called a *Big Crunch*.

2.2 Flat Models

In this section we shall find the solution to Equation (2.1.12) appropriate to a *flat universe*, i.e. with $\Omega_w = 1$. When $w = 0$ this solution is known as the *Einstein-de Sitter* universe; we shall also give this name to solutions with other values of

$w \neq 0$. For $\Omega_w = 1$, Equation (2.1.12) becomes

$$\left(\frac{\dot{a}}{a_0}\right)^2 = H_0^2\left(\frac{a_0}{a}\right)^{1+3w} = H_0^2(1+z)^{1+3w}, \tag{2.2.1}$$

which one can immediately integrate to obtain

$$a(t) = a_0\left(\frac{t}{t_0}\right)^{2/3(1+w)}. \tag{2.2.2}$$

This equation shows that the expansion of an Einstein–de Sitter universe lasts an indefinite time into the future; Equation (2.2.2) is equivalent to the relation

$$t = t_0(1+z)^{-3(1+w)/2}, \tag{2.2.3}$$

which relates cosmic time t to redshift z. From Equations (2.2.2), (2.2.3) and (2.1.6), we can derive

$$H \equiv \frac{\dot{a}}{a} = \frac{2}{3(1+w)t} = H_0\frac{t_0}{t} = H_0(1+z)^{3(1+w)/2}, \tag{2.2.4 a}$$

$$q \equiv -\frac{a\ddot{a}}{\dot{a}^2} = \frac{1+3w}{2} = \text{const.} = q_0, \tag{2.2.4 b}$$

$$t_{0\mathrm{w}} \equiv t_0 = \frac{2}{3(1+w)H_0}, \tag{2.2.4 c}$$

$$\rho = \rho_{0w}\left(\frac{t}{t_0}\right)^{-2} = \frac{1}{6(1+w)^2\pi G t^2}; \tag{2.2.4 d}$$

in the last expression we have made use of the relation

$$\rho_{0w}t_0^2 \equiv \rho_{0c}t_{0\mathrm{w}}^2 = \frac{3H_0^2}{8\pi G}\left[\frac{2}{3(1+w)H_0}\right]^2 = \frac{1}{6(1+w)^2\pi G}. \tag{2.2.5}$$

Useful special cases of the above relationship are *dust*, or *matter-dominated universes* ($w = 0$),

$$a(t) = a_0\left(\frac{t}{t_0}\right)^{2/3}, \tag{2.2.6 a}$$

$$t = t_0(1+z)^{-3/2}, \tag{2.2.6 b}$$

$$H = \frac{2}{3t} = H_0(1+z)^{3/2}, \tag{2.2.6 c}$$

$$q_0 = \tfrac{1}{2}, \tag{2.2.6 d}$$

$$t_{0\mathrm{m}} \equiv t_0 = \frac{2}{3H_0}, \tag{2.2.6 e}$$

$$\rho_{\mathrm{m}} = \frac{1}{6\pi G t^2}; \tag{2.2.6 f}$$

and *radiation-dominated universes* ($w = \frac{1}{3}$),

$$a(t) = a_0\left(\frac{t}{t_0}\right)^{1/2}, \tag{2.2.7 a}$$

$$t = t_0(1+z)^{-2}, \tag{2.2.7 b}$$

$$H = \frac{1}{2t} = H_0(1+z)^2, \tag{2.2.7 c}$$

$$q_0 = 1, \tag{2.2.7 d}$$

$$t_{0\mathrm{r}} \equiv t_0 = \frac{1}{2H_0}, \tag{2.2.7 e}$$

$$\rho_\mathrm{r} = \frac{3}{32\pi G t^2}. \tag{2.2.7 f}$$

A general property of flat-universe models is that the expansion parameter a grows indefinitely with time, with constant deceleration parameter q_0. The comments we made above about the role of pressure can be illustrated again by the fact that increasing w and, therefore, increasing the pressure causes the deceleration parameter also to increase. Conversely, and paradoxically, a negative value of w indicating behaviour similar to a cosmological constant corresponds to negative pressure (tension) but nevertheless can cause an accelerated expansion (see Section 2.8 later).

Note also the general result that in such models the age of the Universe, t_0, is closely related to the present value of the Hubble parameter, H_0.

2.3 Curved Models: General Properties

After seeing the solutions corresponding to flat models with $\Omega_w = 1$, we now look at some properties of *curved models* with $\Omega_w \neq 1$. We begin by looking at the behaviour of these models at early times.

In (2.1.12) and (2.1.13) the term $(1 - \Omega_{0\mathrm{w}})$ inside the parentheses is negligible with respect to the other term, while

$$\frac{a_0}{a} = 1 + z \gg |\Omega_{0\mathrm{w}}^{-1} - 1|^{1/(1+3w)} \equiv \frac{a_0}{a^*} = 1 + z^*. \tag{2.3.1}$$

During the interval $0 < a \ll a^*$, Equations (2.1.12) and (2.1.13) become, respectively,

$$\left(\frac{\dot{a}}{a_0}\right)^2 \simeq H_0^2\Omega_{0\mathrm{w}}\left(\frac{a_0}{a}\right)^{1+3w} = H_0^2\Omega_0 w(1+z)^{1+3w} \tag{2.3.2}$$

and

$$H^2 \simeq H_0^2\Omega_{0\mathrm{w}}\left(\frac{a_0}{a}\right)^{3(1+w)} = H_0^2\Omega_{0\mathrm{w}}(1+z)^{3(1+w)}, \tag{2.3.3}$$

which are exactly the same as those describing the case $\Omega_w = 1$, as long as one replaces H_0 by $H_0\Omega_w^{1/2}$. In particular, we have

$$H \simeq H_0\Omega_{0\mathrm{w}}^{1/2}(1+z)^{3(1+w)/2} \tag{2.3.4}$$

and

$$t \simeq t_{0\mathrm{w}}\Omega_{0\mathrm{w}}^{-1/2}(1+z)^{-3(1+w)/2}. \tag{2.3.5}$$

At early times, all these models behave in a manner very similar to the Einstein–de Sitter model at times sufficiently close to the Big Bang. In other words, it is usually a good approximation to ignore curvature terms when dealing with models of the very early Universe. The expressions for $\rho(t)$ and $q(t)$ are not modified, because they do not contain explicitly the parameter H_0.

2.3.1 Open models

In models with $\Omega_w < 1$ (*open universes*), the expansion parameter a grows indefinitely with time, as in the Einstein–de Sitter model. From (2.1.12), we see that $\dot{a}$ is never actually zero; supposing that this variable is positive at time t_0, the derivative $\dot{a}$ remains positive forever. The first term inside the square brackets in (2.1.12) is negligible for $a(t) \gg a(t^*) = a^*$, where a^* is given by (2.3.1)

$$a^* = a_0\left(\frac{\Omega_{0\mathrm{w}}}{1-\Omega_{0\mathrm{w}}}\right)^{1/(1+3w)} \tag{2.3.6}$$

(in the case with $w = 0$ this time corresponds to a redshift $z^* = (1-\Omega_0)/\Omega_0 \simeq \Omega_0^{-1}$ if $\Omega \ll 1$); before t^* the approximation mentioned above will be valid, so

$$t^* \simeq t_{0w}\Omega_{0w}^{-1/2}\left(\frac{a^*}{a_0}\right)^{3(1+w)/2} = t_{0w}\Omega_{0w}^{-1/2}\left(\frac{\Omega_{0w}}{1-\Omega_{0w}}\right)^{3(1+w)/2(1+3w)}. \tag{2.3.7}$$

For $t \gg t^*$ one obtains, in the same manner,

$$\dot{a} \simeq a_0H_0\Omega_{0\mathrm{w}}^{1/2}\left(\frac{a_0}{a^*}\right)^{(1+3w)/2} = a_0H_0(1-\Omega_{0\mathrm{w}})^{1/2} \tag{2.3.8}$$

and hence

$$a \simeq a_0H_0(1-\Omega_{0\mathrm{w}})^{1/2}t = a^*\frac{2}{3(1+w)}\frac{t}{t^*} \simeq a^*\frac{t}{t^*}. \tag{2.3.9}$$

One therefore obtains

$$H = \frac{\dot{a}}{a} \simeq t^{-1}, \tag{2.3.10a}$$

$$q \simeq 0, \tag{2.3.10b}$$

$$\rho \simeq \frac{\rho_{0c}\Omega_{0\mathrm{w}}}{[H_0(1-\Omega_{0\mathrm{w}})^{1/2}t]^{3(1+w)}} \simeq \rho(t^*)\left(\frac{t}{t^*}\right)^{-3(1+w)}. \tag{2.3.10c}$$

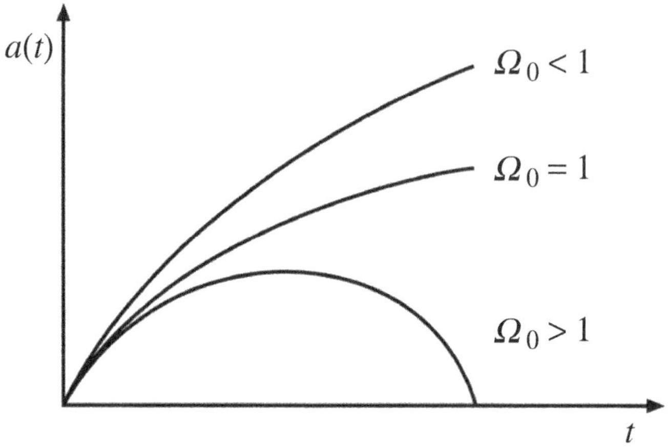

Figure 2.2 Evolution of the expansion parameter $a(t)$ in an open model ($\Omega_0 < 1$), flat or Einstein–de Sitter model ($\Omega_0 = 1$) and closed model ($\Omega_0 > 1$).

It is interesting to note that, if t_0 is taken to coincide with t^*, equation (2.3.6) implies

$$\Omega_{0w}(t^*) = \tfrac{1}{2}; \tag{2.3.11}$$

the parameter $\Omega_{0w}(t)$ passes from a value very close to unity, at $t \ll t^*$, to a value of $\frac{1}{2}$, for $t = t^*$, and to a value closer and closer to zero for $t \gg t^*$.

2.3.2 Closed models

In models with $\Omega_w > 1$ (*closed universes*) there exists a time t_m at which the derivative $\dot{a}$ is zero. From (2.1.12), one can see that

$$a_m \equiv a(t_m) = a_0\left(\frac{\Omega_{0w}}{\Omega_{0w} - 1}\right)^{1/(1+3w)}. \tag{2.3.12}$$

After the time t_m the expansion parameter decreases with a derivative equal in modulus to that holding for $0 \leqslant a \leqslant a_m$: the curve of $a(t)$ is therefore symmetrical around a_m. At $t_f = 2t_m$ there is another singularity in a symmetrical position with respect to the Big Bang, describing a final collapse or Big Crunch.

In Figure 2.2 we show a graph of the evolution of the expansion parameter $a(t)$ for open, flat and closed models.

2.4 Dust Models

Models with $w = 0$ have an exact analytic solution, even for the case where $\Omega \neq 1$ (we gave the solution for $\Omega = 1$ in Section 2.2). In this case, Equation (2.1.12) becomes

$$\left(\frac{\dot{a}}{a_0}\right)^2 = H_0^2\left(\Omega_0\frac{a_0}{a} + 1 - \Omega_0\right). \tag{2.4.1}$$

2.4.1 Open models

For these models Equation (2.4.1) has a solution in the parametric form:

$$a(\psi) = a_0 \frac{\Omega_0}{2(1-\Omega_0)}(\cosh\psi - 1), \tag{2.4.2}$$

$$t(\psi) = \frac{1}{2H_0}\frac{\Omega_0}{(1-\Omega_0)^{3/2}}(\sinh\psi - \psi). \tag{2.4.3}$$

We can obtain an expression for t_0 from the two preceding relations

$$t_0 = \frac{1}{2H_0}\frac{\Omega_0}{(1-\Omega_0)^{3/2}}\left[\frac{2}{\Omega_0}(1-\Omega_0)^{1/2} - \cosh^{-1}\left(\frac{2}{\Omega_0} - 1\right)\right] > \frac{2}{3H_0}. \tag{2.4.4}$$

Equation (2.4.4) has the following approximate form in the limit $\Omega \ll 1$:

$$t_0 \simeq (1 + \Omega_0 \ln \Omega_0)\frac{1}{H_0}. \tag{2.4.5}$$

2.4.2 Closed models

For these models Equation (2.4.1) has a parametric solution in the form of a cycloid:

$$a(\vartheta) = a_0 \frac{\Omega_0}{2(\Omega_0 - 1)}(1 - \cos\vartheta), \tag{2.4.6}$$

$$t(\vartheta) = \frac{1}{2H_0}\frac{\Omega_0}{(\Omega_0 - 1)^{3/2}}(\vartheta - \sin\vartheta). \tag{2.4.7}$$

The expansion parameter $a(t)$ grows in time for $0 \leqslant \vartheta \leqslant \vartheta_m = \pi$. The maximum value of a is

$$a_m = a(\vartheta_m) = a_0 \frac{\Omega}{\Omega - 1}, \tag{2.4.8}$$

which we have obtained previously in (2.3.12), occurring at a time t_m given by

$$t_m = t(\vartheta_m) = \frac{\pi}{2H_0}\frac{\Omega_0}{(\Omega_0 - 1)^{3/2}}. \tag{2.4.9}$$

The curve of $a(t)$ is symmetrical around t_m, as we have explained before. One can obtain an expression for t_0 from Equations (2.4.6) and (2.4.7). The result is

$$t_0 = \frac{1}{2H_0}\frac{\Omega_0}{(\Omega_0 - 1)^{3/2}}\left[\cos^{-1}\left(\frac{2}{\Omega_0} - 1\right) - \frac{2}{\Omega_0}(\Omega_0 - 1)^{1/2}\right] < \frac{2}{3H_0}. \tag{2.4.10}$$

2.4.3 General properties

In the dust models it is possible to calculate analytically in terms of redshift all the various distance measures introduced in Section 1.7. Denote by t the time of emission of a light signal from a source and t_0 the moment of reception of the signal by an observer. We have then, from the definition of the redshift of the source,

$$a(t) = \frac{a_0}{1+z}. \qquad (2.4.11)$$

From the Robertson-Walker metric one obtains

$$f(r) = \int_0^r \frac{\mathrm{d}r'}{(1-Kr'^2)^{1/2}} = \int_t^{t_0} \frac{c\,\mathrm{d}t'}{a(t')} = \int_{a(t)}^{a_0} \frac{c\,\mathrm{d}a'}{a'\dot{a}'}, \qquad (2.4.12)$$

where r is the comoving radial coordinate of the source. From (2.4.11) and (2.1.12) with $w = 0$, Equation (2.4.12) becomes

$$f(r) = \frac{c}{a_0 H_0} \int_{(1+z)^{-1}}^{1} \left[1 - \Omega_0 + \frac{\Omega_0}{x}\right]^{-1/2} \frac{\mathrm{d}x}{x}. \qquad (2.4.13)$$

One can use (2.4.13) to show that, for any value of K (and therefore of Ω_0),

$$r = \frac{2c}{H_0 a_0} \frac{\Omega_0 z + (\Omega_0 - 2)[-1 + (\Omega_0 z + 1)^{1/2}]}{\Omega_0^2 (1+z)}. \qquad (2.4.14)$$

From Equation (1.7.3) of the previous chapter, the *luminosity distance* of a source is then

$$d_{\mathrm{L}} = \frac{2c}{H_0 \Omega_0^2} \{\Omega_0 z + (\Omega_0 - 2)[-1 + (\Omega_0 z + 1)^{1/2}]\}, \qquad (2.4.15)$$

a result sometimes known as the *Mattig formula*. Analogous relationships can be derived for the other important cosmological distances.

Another relation which we can investigate is that between cosmic time t and redshift z. From (2.1.13), for $w = 0$, we easily find that

$$\mathrm{d}t = -\frac{1}{H_0}(1+z)^{-2}(1+\Omega z)^{-1/2}\,\mathrm{d}z. \qquad (2.4.16)$$

The integral of (2.4.16) from the time of emission of a light signal until it is observed at t, where it has a redshift z, is

$$t(z) = \frac{1}{H_0} \int_z^{\infty} (1+z')^{-2}(1+\Omega z')^{-1/2}\,\mathrm{d}z'. \qquad (2.4.17)$$

Thus we can think of redshift z as being a coordinate telling us the cosmic time at which light was emitted from a source; this coordinate runs from infinity if $t = 0$, to zero if $t = t_0$. For $z \gg 1$ and $\Omega_0 z \gg 1$ Equation (2.4.17) becomes

$$t(z) \simeq \frac{2}{3H_0 \Omega_0^{1/2}} z^{-3/2} \simeq \frac{2}{3H_0 \Omega_0^{1/2}} (1+z)^{-3/2}, \qquad (2.4.18)$$

which is a particular case of Equation (2.3.5). We can therefore define a *look-back time* by

$$t_{\rm lb} = t_0 - t(z) = \frac{1}{H_0}\int_0^z (1+z')^{-2}(1+\Omega_0 z')^{-1/2}\,{\rm d}z'. \tag{2.4.19}$$

This represents the time elapsed since the emission of a signal which arrives now, at t_0, with a redshift z. In other words, the time it has taken light to reach us from a source which we observe now at a redshift z.

2.5 Radiative Models

The models with $w = \frac{1}{3}$ also have simple analytic solutions for $\Omega_{\rm r} \neq 1$ (the solution for $\Omega_{\rm r} = 1$ was given in Section 2.2). Equation (2.1.12) can be written in the form

$$\left(\frac{\dot a}{a_0}\right)^2 = H_0^2\left[\Omega_{0\rm r}\left(\frac{a_0}{a}\right)^2 + (1-\Omega_{0\rm r})\right]; \tag{2.5.1}$$

the solution is

$$a(t) = a_0(2H_0\Omega_{0\rm r}^{1/2}t)^{1/2}\left(1 + \frac{1-\Omega_{0\rm r}}{2\Omega_{0\rm r}^{1/2}}H_0 t\right)^{1/2}. \tag{2.5.2}$$

2.5.1 Open models

For $t \gg t_{\rm r}^*$ or, alternatively, $(a \gg a_{\rm r}^*)$, where

$$t_{\rm r}^* = \frac{2}{H_0}\frac{\Omega_{0\rm r}^{1/2}}{1-\Omega_{0\rm r}}, \tag{2.5.3 a}$$

$$a_{\rm r}^* = a_0\left(\frac{\Omega_{0\rm r}}{1-\Omega_{0\rm r}}\right)^{1/2}. \tag{2.5.3 b}$$

Equation (2.5.2) shows that the behaviour of $a(t)$ takes the form of an undecelerated expansion

$$a(t) \simeq a_0(1-\Omega_{0\rm r})^{1/2}H_0 t. \tag{2.5.4}$$

One can also find the present cosmic time by putting $a = a_0$ in this equation:

$$t_0 = \frac{1}{H_0}\frac{1}{\Omega_{0\rm r}^{1/2}+1} > \frac{1}{2H_0}. \tag{2.5.5}$$

2.5.2 Closed models

In this case Equation (2.5.2) shows that there is a maximum value of a at

$$a_{\mathrm{m}} = a_0 \left(\frac{\Omega_{0\mathrm{r}}}{\Omega_{0\mathrm{r}} - 1} \right)^{1/2} \tag{2.5.6}$$

at a time

$$t_{\mathrm{m}} = \frac{1}{H_0} \frac{\Omega_{0\mathrm{r}}^{1/2}}{\Omega_{0\mathrm{r}} - 1}. \tag{2.5.7}$$

The function $a(t)$ is symmetrical around t_{m}. One obtains an expression for t_0 by putting $a = a_0$ in (2.5.2); the result is

$$t_0 = \frac{1}{H_0} \frac{1}{\Omega_{0\mathrm{r}}^{1/2} + 1} < \frac{1}{2H_0}. \tag{2.5.8}$$

There obviously also exists another solution of Equation (2.5.2), say t_0', obtained by reflecting t_0 around t_{m}; at this time $\dot{a}(t_0') < 0$.

2.5.3 General properties

The formula analogous to (2.4.16) is, in this case,

$$\mathrm{d}t = -\frac{1}{H_0}(1+z)^{-2}[1 + \Omega_{0\mathrm{r}} z(2+z)]^{-1/2}\,\mathrm{d}z. \tag{2.5.9}$$

For $z \gg 1$ and $\Omega_{0\mathrm{r}} z \gg 1$, Equation (2.5.9) yields

$$t(z) \simeq \frac{1}{2H_0\Omega_{0\mathrm{r}}^{1/2}} z^{-2} \simeq \frac{1}{2H_0\Omega_{0\mathrm{r}}^{1/2}}(1+z)^{-2}, \tag{2.5.10}$$

which is, again, a particular case of (2.3.5).

2.6 Evolution of the Density Parameter

In most of the expressions derived so far in this chapter, the quantity that appears is Ω_{w0} or, in the special case of $w = 0$, just Ω_0. This is simply because we have chosen to parametrise the solutions with the value of Ω at the time $t = t_0$. However, it is very important to bear in mind that Ω is a function of time in all these models. If we instead wish to calculate the density parameter at an arbitrary redshift z, the relevant expression is

$$\Omega_w(z) = \frac{\rho_w(z)}{[3H^2(z)/8\pi G]}, \tag{2.6.1}$$

where $\rho_w(z)$ is, from (2.1.6),

$$\rho_w(z) = \rho_{0w}(1+z)^{3(1+w)}, \tag{2.6.2}$$

and the Hubble constant $H(z)$ is, from (2.1.13),

$$H^2(z) = H_0^2(1+z)^2[\Omega_{0\mathrm{w}}(1+z)^{1+3w} + (1-\Omega_{0\mathrm{w}})]. \tag{2.6.3}$$

Equation (2.6.1) then becomes

$$\Omega_w(z) = \frac{\Omega_{0\mathrm{w}}(1+z)^{1+3w}}{(1-\Omega_{0\mathrm{w}}) + \Omega_{0\mathrm{w}}(1+z)^{1+3w}}; \tag{2.6.4}$$

this relation looks messy but can be written in the more useful form

$$\Omega_w^{-1}(z) - 1 = \frac{\Omega_{0\mathrm{w}}^{-1} - 1}{(1+z)^{1+3w}}, \tag{2.6.5}$$

which will be useful later on, particularly in Chapter 7. Notice that if $\Omega_{0\mathrm{w}} > 1$, then $\Omega_w(z) > 1$ for likewise, if $\Omega_{0w} < 1$, then $\Omega_w(z) < 1$ for all z; on the other hand, if $\Omega_{0\mathrm{w}} = 1$, then $\Omega_w(z) = 1$ for all time. The reason for this is clear: the expansion cannot change the sign of the curvature parameter K. also worth noting that, as z tends to infinity, i.e. as we move closer and closer to the Big Bang, $\Omega_w(z)$ always tends towards unity.

These results have already been obtained in different forms in the previous parts of this chapter: one can summarise them by saying that any universe with $\Omega_w \neq 1$ behaves like an Einstein–de Sitter model in the vicinity of the Big Bang. We shall come back to this later when we discuss the *flatness problem*, in Chapter 7.

2.7 Cosmological Horizons

Consider the question of finding the set of points capable of sending light signals that could have been received by an observer up to some generic time t. For simplicity, place the observer at the origin of our coordinate system O. The set of points in question can be said to have the possibility of being causally connected with the observer at O at time t. It is clear that any light signal received at O by the time t must have been emitted by a source at some time t' contained in the interval between $t = 0$ and t. The set of points that could have communicated with O in this way must be inside a sphere centred upon O with proper radius

$$R_{\mathrm{H}}(t) = a(t)\int_0^t \frac{c\,\mathrm{d}t'}{a(t')}. \tag{2.7.1}$$

In (2.7.1), the generic distance $c\,\mathrm{d}t'$ travelled by a light ray between t' and $t' + \mathrm{d}t'$ has been multiplied by a factor $a(t)/a(t')$, in the same way as one obtains the relative proper distance between two points at time t. In (2.7.1), if one takes the lower limit of integration to be zero, there is the possibility that the integral diverges because $a(t)$ also tends to zero for small t. In this case the observer at O can, in principle, have received light signals from the whole Universe. If, on the other hand, the integral converges to a finite value with this limit, then the spherical surface with centre O and radius R_{H} is called the *particle horizon* at time t of the observer. In this case, the observer cannot possibly have received light signals, at any time in his history, from sources which are situated at proper distances greater than $R_{\mathrm{H}}(t)$ from him at time t. The particle horizon thus divides the set of all points into two classes: those which can, in principle, have been observed by O (inside the horizon), and those which cannot (outside the horizon). From (2.1.12) and (2.7.1) we obtain

$$R_{\mathrm{H}}(t) = \frac{c}{H_0}\frac{a(t)}{a_0}\int_0^{a(t)} \frac{da'}{a'[\Omega_{0\mathrm{w}}(a_0/a')^{1+3w} + (1-\Omega_{0\mathrm{w}})]^{1/2}}. \tag{2.7.2}$$

The integral in (2.7.2) can be divergent because of contributions near to the Big Bang, when $a(t)$ is tending to zero. At such times, the second term in the square brackets is negligible compared with the first, and one has

$$R_{\mathrm{H}}(t) \simeq \frac{c}{H_0\Omega_{0\mathrm{w}}^{1/2}}\frac{2}{3w+1}\left(\frac{a}{a_0}\right)^{3(1+w)/2}, \tag{2.7.3}$$

which is finite and which also vanishes as $a(t)$ tends to zero. It can also be shown that

$$R_{\mathrm{H}}(t) \simeq 3\frac{1+w}{1+3w}ct. \tag{2.7.4}$$

The solution (2.7.4) is valid exactly in any case if $\Omega_w = 1$; interesting special cases are $R_{\mathrm{H}}(t) = 3ct$ for the flat dust model and $R_{\mathrm{H}}(t) = 2ct$ for a flat radiative model.

For reference, the integral in (2.7.2) can be solved exactly in the case $w = 0$ and $\Omega_0 \neq 1$. The result is

$$R_{\mathrm{H}}(t) = \frac{c}{H_0(1-\Omega_0)^{1/2}}(1+z)^{-1}\cosh^{-1}\left[1 - \frac{2(\Omega_0-1)}{\Omega_0}(1+z)^{-1}\right] \tag{2.7.5 a}$$

and

$$R_{\mathrm{H}}(t) = \frac{c}{H_0(\Omega_0-1)^{1/2}}(1+z)^{-1}\cos^{-1}\left[1 - \frac{2(\Omega_0-1)}{\Omega_0}(1+z)^{-1}\right], \tag{2.7.5 b}$$

in the cases $\Omega_0 < 1$ and $\Omega_0 > 1$, respectively. The previous analysis establishes that there does exist a particle horizon in Friedmann models with equation-of-state parameter $0 \leqslant w \leqslant 1$. Notice, however, that in a pure de Sitter cosmological model, which expands exponentially and lasts forever, there is no particle horizon because the integral (2.7.1) is not finite. We shall return to the nature of these horizons and some problems connected with them in Chapter 7.

We should point out the distinction between the cosmological particle horizon and the *Hubble sphere*, or *speed-of-light sphere*, R_c. The radius of the Hubble sphere, the Hubble radius, is defined to be the distance from O of an object moving with the cosmological expansion at the velocity of light with respect to O. This can be seen very easily to be

$$R_c = c\frac{a}{\dot{a}} = \frac{c}{H}, \tag{2.7.6}$$

by virtue of the Hubble expansion law. One can see that, if $p > -\frac{1}{3}\rho c^2$, the value of R_c coincides, at least to order of magnitude, with the distance to the particle horizon, R_{H}. For example, if $\Omega_w = 1$, we have

$$R_c = \tfrac{3}{2}(1 + w)ct = \tfrac{1}{2}(1 + 3w)R_{\mathrm{H}} \simeq R_{\mathrm{H}}. \tag{2.7.7}$$

One can think of R_c as being the proper distance travelled by light in the characteristic expansion time, or *Hubble time*, of the universe, τ_{H}, where

$$\tau_{\mathrm{H}} \equiv \frac{a}{\dot{a}} = \frac{1}{H}. \tag{2.7.8}$$

The Hubble sphere is, however, not the same as the particle horizon. For one thing, it is possible for objects to be outside an observer's Hubble sphere but inside his particle horizon. It is also the case that, once inside an observer's horizon, a point stays within the horizon forever. This is not the case for the Hubble sphere: objects can be within the Hubble sphere at one time t, outside it sometime later, and, later still, they may enter the sphere again. The key difference is that the particle horizon at time t takes account of the entire past history of the observer up to the time t, while the Hubble radius is defined instantaneously at t. Nevertheless, in some cosmological applications, the Hubble sphere plays an important role which is similar to that of the horizon, and is therefore often called the *effective cosmological horizon*. We shall see the importance of the Hubble sphere when we discuss inflation, and also the physics of the growth of density fluctuations. It also serves as a reminder of the astonishing fact that the Hubble law in the form (1.4.6) is an exact relation no matter how large the distance at which it is applied. Recession velocities greater than the speed of light do occur in these models as when the proper distance is larger than $R_c = c/H_0$.

There is yet another type of horizon, called the *event horizon*, which is a most useful concept in the study of black holes but is usually less relevant in cosmology. The event horizon again divides space into two sets of points, but it refers to the future ability of an observer O to communicate. The event horizon thus separates those points which can emit signals that O can, in principle, receive at some time in the future from those that cannot. The mathematical definition is the same as in (2.7.1) but with the limits of the integral changed to run from t to either $t_{\max}$, which is either t_{f} (the time of the Big Crunch) in a closed model, or $t = \infty$ in a flat or open model. The radius of the event horizon is given by

$$R_{\mathrm{E}}(t) = a(t)\int_t^{t_{\max}} \frac{c\,\mathrm{d}t'}{a(t')}. \tag{2.7.9}$$

The event horizon does not exist in Friedmann models with $-\frac{1}{3} < w < 1$, but does exist in a de Sitter model.

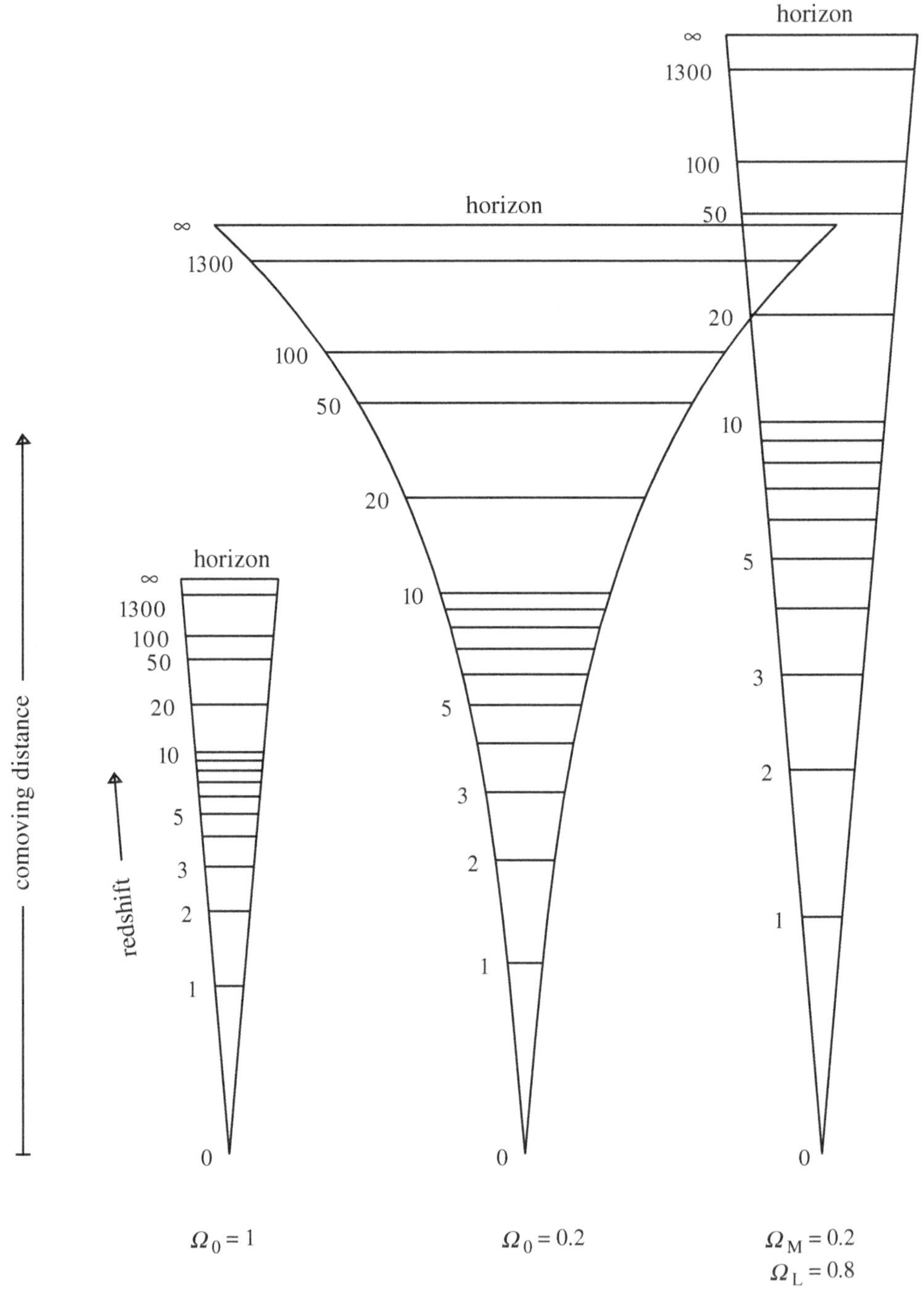

Figure 2.3 Illustration of the behaviour of angular diameters and distances as functions of redshift for cosmological models with and without curvature and cosmological constant terms. From Hamilton (1998).

2.8 Models with a Cosmological Constant

We have already shown how a cosmological constant can be treated as a fluid with equation of state $p = -\rho c^2$, i.e. with $w = -1$. We know, however, that there is at least some non-relativistic matter and some radiation in the Universe, so a model with only a Λ term can not be anything like complete. In mixed models, with more than one type of fluid and/or contributions from a cosmological constant, the equations describing the evolution become more complicated and closed-form solutions much harder to come by. This is not a problem in the era of fast computers, however, as equivalent results to those of single-fluid cases can be solved by numerical integration.

Many of the results we have developed so far in this chapter stem from the expression (2.1.12), which is essentially the Equation (1.13.2) in different variables. The generalisation to the multi-component case is quite straightforward. In cases involving matter, radiation and a cosmological constant, for example, the appropriate form is

$$\left(\frac{\dot{a}}{a_0}\right)^2 = H_0^2\left[\Omega_{0\mathrm{m}}\left(\frac{a}{a_0}\right)+\Omega_{0\mathrm{r}}\left(\frac{a}{a_0}\right)^2+\Omega_{0\Lambda}\left(\frac{a}{a_0}\right)^{-2}+(1-\Omega_{0\mathrm{m}}-\Omega_{0\mathrm{r}}-\Omega_{0\Lambda})\right]. \quad (2.8.1)$$

The simpler forms of this expression, like (2.1.12), are what we have been using to work out such things as the relationship between t_0 and H_0 for given values of Ω_0. In the presence of a cosmological constant there is generally no simple equation relating Ω_0, $\Omega_{0\Lambda}$ and t_0. A closed-form expression is, however, available for the $k = 0$ models containing a cosmological constant and dust mentioned at the end of Chapter 1. In such cases

$$t_0 = \frac{2}{3H_0}\left[\frac{1}{2\sqrt{1-\Omega_0}}\log\frac{1+\sqrt{1-\Omega_0}}{1-\sqrt{1-\Omega_0}}\right]. \quad (2.8.2)$$

Generally speaking, however, one can see that a positive cosmological constant term tends to act in the direction of accelerating the universe and therefore tends to increase the age relative to decelerated models for the same value of H_0. The cosmological constant also changes the relationship between r and z through the form of $f(r)$ shown in Equation (2.4.13). Since $\dot{a}$ must now include a contribution from the $\Omega_{0\Lambda}$ terms in Equation (2.8.1), the value of $f(r)$ for a given redshift z will actually be larger in an accelerated model than in a decelerated example. This has a big effect on the luminosity distance to a given redshift z as well as the volume surveyed as a function of z. This is illustrated dramatically in Figure 2.3. We shall return to these potential observational consequences of a cosmological constant in Chapter 4.

Bibliographic Notes on Chapter 2

Most of the material for this chapter is covered in standard cosmological texts. In particular, see Weinberg (1972), Berry (1989), Narlikar (1993) and Peacock (1999).

Problems

1. For a universe with $k = 0$ and in which $(a/a_0) = (t/t_0)^n$, where $n < 1$, show that the coordinate distance of an object seen at redshift z is
$$r = \frac{ct_0}{(1-n)a_0}[1-(1+z)^{1-1/n}].$$
For $n = \frac{2}{3}$ deduce that the present proper distance to a quasar at redshift $z = 5$ is
$$\frac{2c}{H_0}\tau_0\left(1-\frac{1}{\sqrt{6}}\right),$$
where H_0 is present value of the Hubble constant.

2. Consider a dust model in the limit $\Omega_0 \to 0$. On the one hand, this is an example of an open Friedmann model which has negatively curved spatial sections. On the other hand, being undecelerated and purely kinematic, it ought to be described by special relativity, which is described by the flat metric of Minkowski space. Can these two views be reconciled?

3. By substituting in (2.4.1), show that the parametric open solution given by (2.4.2) and (2.4.3) does indeed solve the Friedmann equation. Repeat the exercise for the closed solution (2.4.6) and (2.4.7).

4. A closed Friedmann universe contains a single perfect fluid with an equation of state of the form $p = w\rho c^2$. Transforming variables to conformal time τ using $dt = a(t)\,d\tau$, show that the variable $y = a^{(1+3w)/2}$ is described by a simple harmonic equation as a function of τ. Hence argue that all closed Friedmann models with a given equation of state have the same conformal lifetime.

5. Calculate the present proper distance to the event horizon in a de Sitter model described by (1.2.14). What is the radius of the Hubble sphere in this case? Is there a particle horizon in this model?

6. A flat matter-dominated (Einstein–de Sitter) universe is populated with galaxies at various proper distances l from an observer at the origin. The distance of these galaxies increases with cosmological proper time in a manner described by the Hubble law. If the galaxies emit light at various times t_e, calculate the locus of points in the l–t_e plane that lie on the observer's past light cone (i.e. those points that emit light at t_e that can be detected at $t = t_0$ by the observer). Show that the maximum proper distance of a galaxy on this locus is $l_{max} = \frac{4}{9}ct_0$.

3

Alternative Cosmologies

Most of this book is devoted to a survey of the standard (Big Bang) cosmology and its consequences for the large-scale structure of the Universe. We nevertheless feel it is important to mention some non-standard cosmologies as illustrations of how different world models can behave. Some of these alternative cosmologies have been important in the past, during the development of modern cosmology as an observational science. Others are more recent speculations about how the Big Bang model may be affected by developments in fundamental physics. Although there are good grounds for believing that the standard cosmology is basically correct, one should never close one's eyes to the possibility that it may turn out to be wrong and that one of the non-standard alternatives may be a better or more complete description of reality. We have not the space, however, to give a panoramic view of all possible alternative cosmologies so we shall concentrate on a few which are of particular historical or contemporary interest and confine ourselves to brief remarks upon them. Those readers not interested in this material may skip this chapter at a first reading.

Before proceeding, we should remind the reader that the fundamentals of the standard Big Bang model are essentially the theory of general relativity, the expanding Universe and the Cosmological Principle. These basic assumptions allow the flexibility to incorporate the models of Einstein, de Sitter and Lemaître characterised by $\Lambda \neq 0$ in Section 1.11 within this standard framework. These models are of historical interest as well as sharing many of the modern 'inflationary' cosmologies constructed using a scalar field whose vacuum energy essentially plays the role of a time-varying cosmological constant. We discuss inflation in more detail in Chapter 7.

3.1 Anisotropic and Inhomogeneous Cosmologies

The Cosmological Principle plays such an important role in the development of the Friedmann models that it is well worth looking at the consequences of relaxing the assumptions of homogeneity and isotropy. One motivation for this is that the Universe is neither homogeneous nor isotropic. In the standard cosmology, however, variations in density are treated as perturbations of a Friedmann model. This means that structure-formation theory is inherently approximate. It would be nice to be able to solve Einstein's equations exactly for lumpy models, but this is extremely difficult except in cases of special symmetry. Indeed, only a few exact anisotropic or inhomogeneous cosmological solutions are known. We shall discuss a few examples here, just to give an idea of the different behaviour one might expect.

3.1.1 The Bianchi models

The first class of non-standard models we discuss are spatially homogeneous but anisotropic. In the Friedmann models the constant time surfaces upon which the matter density is constant are surfaces of constant cosmological proper time. We can give a more general definition of homogeneity by requiring that all comoving observers see essentially the same version of cosmic history. In mathematical terms this means that there must be some symmetry that relates what the Universe looks like as seen by observer A to what is seen in a coordinate system centred on any other observer B. The possible symmetries can be classified into classes usually called the Bianchi types, although there is one peculiar solution of the Einstein equations, called the Kantowski–Sachs solution, that does not fit into this scheme.

The Bianchi classification is based on the construction of spacelike hypersurfaces upon which it is possible to define at least three independent vector fields, ξ_α (α and other Greek indices run from 1 to 3), that satisfy the constraint

$$\xi_{i;j} + \xi_{j;i} = 0. \tag{3.1.1}$$

This is called Killing's equation and the vectors that satisfy it are called Killing vectors. The commutators of the ξ_α are defined by

$$[\xi_\alpha, \xi_\beta] \equiv \xi_\alpha \xi_\beta - \xi_\beta \xi_\alpha = C^{\delta}_{\alpha\beta} \xi_\delta, \tag{3.1.2}$$

where the $C^{\delta}_{\alpha\beta}$ are called structure constants. These are antisymmetric, in the sense that,

$$C^{\delta}_{\alpha\beta} = -C^{\delta}_{\beta\alpha}. \tag{3.1.3}$$

The components of the metric, g_{ij}, describing a Bianchi space are invariant under the isometry generated by infinitesimal translations of the Killing vector fields. In other words, the time-dependence of the metric is the same at all points. The

Table 3.1 The Bianchi types shown in terms of the number of arbitrary constants needed to specify the model on a given constant time surface in vacuum r and with a perfect fluid s.

Bianchi type	group dimension p	vacuum r	fluid s
I	0	1	2
II	3	2	5
VI_0	5	3	7
VII_0	6	4	8
VIII	6	4	8
IX	6	4	8
IV	5	3	7
V	3	1	5
VI_h	6	4	8
VII_h	6	4	8
$\mathrm{VI}_{h=-1/9}$	6	4	7

Einstein equations relate the energy-momentum tensor T_{ij} to the derivatives of g_{ij}, so if the metric is invariant under a given set of operations, then so are the physical properties encoded by T_{ij}.

The set of n Killing vectors will have some n-dimensional group structure, say G_n, that depends on the properties of the structure constants and this is used to classify all spatially homogeneous cosmological models. The most useful form of this classification proceeds as follows. On any particular spacelike hypersurface, the Killing vector basis can be chosen so that the structure constants can be decomposed as

$$C^{\eta}_{\alpha\beta} = \epsilon_{\alpha\beta\gamma} n^{\gamma\eta} + \delta^{\eta}_{\beta} a_{\alpha} - \delta^{\eta}_{\alpha} a_{\beta}, \tag{3.1.4}$$

where $\epsilon_{\alpha\beta\gamma}$ is the total antisymmetric tensor and δ^{β}_{α} is the Kronecker delta. The tensor $n^{\alpha\beta}$ is diagonal with entries, say, n_1, n_2, and n_3. The vector $a_\alpha = (a, 0, 0)$ for some constant a. All the parameters a and n_α can be normalised to be ± 1 or zero. If $an_2n_3 = 0$, then n_2 and n_3 can be set to ± 1 and a is then conventionally taken to be $\sqrt{|h|}$, where h is a parameter used in the classification. The possible combinations of n_1 and a then fix the Bianchi types, which can also be described in terms of the number of arbitrary functions needed to specify the solution in vacuum (r) or in the presence of a perfect fluid (s) as shown in Table 3.1. The 'most general' anisotropic models are therefore those that have the largest number of free functions, or free parameters on each hypersurface.

The Friedmann models form special cases of the Bianchi types. These have G_6 symmetry groups with G_3 subgroups. The flat Friedmann model is a special case of either Bianchi I or Bianchi VII_0, the open Friedmann model is a special case of types V or VII_h and the closed model belongs to type IX.

General solutions of the Einstein equations are only known for some special cases of the Bianchi types, which demonstrates the difficulty of finding meaningful exact solutions in situations of restricted symmetry. There is, however, one very-well-known example which is a useful illustration of the sort of behaviour one can obtain. This solution, called the Kasner solution, belongs to Bianchi type I. The metric in this case has a relatively simply form:

$$ds^2 = c^2\,dt^2 - X_1^2(t)\,dx_1^2 - X_2^2\,dx_2^2 - X_3^2\,dx_3^2. \tag{3.1.5}$$

Substituting this metric into the Einstein Equations (1.2.20) (with $\Lambda = 0$ and a perfect fluid with pressure p and density ρ) yields

$$\frac{\ddot{X}_i}{X_i} - \left(\frac{\dot{X}_i}{X_i}\right)^2 + 3\left(\frac{\dot{X}_i}{X_i}\right)\left(\frac{\dot{a}}{a}\right) = \frac{4\pi G}{c^4}\left(\rho - \frac{p}{c^2}\right), \tag{3.1.6}$$

in which $a^3 = X_1X_2X_3$. Note that this emerges from the diagonal part of the Einstein equations so the summation convention does not apply in Equation (3.1.6). One also obtains

$$\frac{\dot{X}_1\dot{X}_2}{X_1X_2} + \frac{\dot{X}_2\dot{X}_3}{X_2X_3} + \frac{\dot{X}_3\dot{X}_1}{X_3X_1} = \frac{8\pi G}{c^4}\rho. \tag{3.1.7}$$

This is easy to interpret: the spatial sections expand at a rate $\dot{X}_i/X_i$ in each direction. The mean rate of expansion is just

$$\frac{\dot{a}}{a} = \frac{1}{3}\left(\frac{\dot{X}_1}{X_1} + \frac{\dot{X}_2}{X_2} + \frac{\dot{X}_3}{X_3}\right). \tag{3.1.8}$$

In the neighbourhood of an observer at the centre of a coordinate system x_i, fluid particles will move with some velocity u_i. In general,

$$\frac{\partial u_i}{\partial x_j} = \frac{1}{2}\left(\frac{\partial u_i}{\partial x_j} - \frac{\partial u_j}{\partial x_i}\right) + \frac{1}{2}\left(\frac{\partial u_i}{\partial x_j} + \frac{\partial u_j}{\partial x_i}\right) = \omega_{ij} + \theta_{ij}, \tag{3.1.9}$$

where ω_{ij} is the rate of rotation: in more familiar language, the vorticity vector $\omega_i = \epsilon_{ijk}\omega_{jk}$, which is just the curl of u_i. The tensor θ_{ij} can be decomposed into a diagonal part and a trace-free part according to

$$\theta_{ij} = \tfrac{1}{3}\delta_{ij}\theta + \sigma_{ij}, \tag{3.1.10}$$

where $\sigma_{ii} = 0$. In this description θ, σ_{ij} and ω_{ij}, respectively, represent the expansion, shear and rotation of a fluid element.

In the particular case of Bianchi I we have

$$\theta = 3(\dot{a}/a) \tag{3.1.11}$$

and

$$\omega_{ij} = 0. \tag{3.1.12}$$

More complicated Bianchi models have non-zero rotation. We can further rewrite Equation (3.1.6) in the form of evolution equations for

$$\sigma_i = \frac{\dot{X}_i}{X_i} - \frac{\dot{a}}{a}. \tag{3.1.13}$$

In particular we get

$$\dot{\sigma}_i + \theta\sigma_i = 0, \tag{3.1.14}$$

which can be immediately integrated to give

$$\sigma_i = \frac{\Sigma_i}{a^3}, \tag{3.1.15}$$

where the Σ_i are constants such that $\Sigma_1 + \Sigma_2 + \Sigma_3 = 0$. The Kasner solution itself is for a vacuum $p = \rho = 0$, which has a particularly simple behaviour described by $X_i = A_i t^p$, where $p_1 + p_2 + p_3 = p_1^2 + p_2^2 + p_3^2 = 1$. Notice that in general these models possess a shear that decreases with time. They therefore tend to behave more like a Friedmann model as time goes on. Their behaviour as $t \to 0$ is, however, quite complicated and interesting.

There is one other particularly interesting case to mention before we leave this discussion. The mix-master universe of Misner (1968) we mentioned in Chapter 1 is of Bianchi type IX.

3.1.2 Inhomogeneous models

Before the formulation of general relativity and the discovery of the Hubble expansion, which is describable by the Friedmann models founded on Einstein's theory, most astronomers imagined the Universe to be infinite, eternal, static and Euclidean. The distribution of matter within the Universe was likewise assumed to be more or less homogeneous and static. It is worth mentioning at this point that the discovery that galaxies were actually external and comparable in size with the Milky Way was made only a few years or so before Hubble's discovery of the expansion of the Universe.

It is nevertheless noteworthy that, beginning in the last century, there were a number of prominent supporters also of the *hierarchical cosmology*, according to which the material contents of the Universe are distributed in a hierarchical manner reminiscent of the modern concept of a fractal. In such a model, the mean density of matter on a scale r varies with scale as $\rho(r) \propto r^{-\gamma}$, where γ is some constant $\gamma \simeq 2$. In this way the mean density of the Universe tends to zero on larger and larger scales. On the other hand, the velocity induced by the hierarchical fluctuations varies with scale according to $v^2(r) = G\rho(r)r^2 \propto r^{2-\gamma} \simeq$ const. The idea of a fractal Universe still has its adherents today, although the evidence we have from the extreme isotropy of the cosmic microwave background suggests that the Universe is homogeneous and isotropic on scales greater than a few hundred Mpc.

Given the considerable leap in complexity we were forced to take when we dropped one of the two components of the Cosmological Principle, it will come as no surprise that there are few inhomogeneous cosmological models available as exact solutions of the Einstein equations. Moreover, those that do exist tend to be cases of particular symmetry. One of the problems of identifying exact solutions is illustrated by the following metric:

$$ds^2 = \left(1 + \frac{\epsilon}{1 + c^2t^2}\right)^2 c^2\,dt^2 - \left(\frac{\epsilon}{1 + x^2}\right)^2 x^2 - \left(\frac{\epsilon}{1 + y^2}\right)^2 y^2 - \left(\frac{\epsilon}{1 + z^2}\right)^2 z^2, \tag{3.1.16}$$

where ϵ is a small parameter. This looks for all the world like it must describe a small departure from Minkowski space, but it is not. In fact, it is exactly the same as Minkowski space but using a very strange coordinate system.

A notable example of a meaningful exact solution is the Tolman–Bondi solution (Tolman 1934; Bondi 1947) which is spherically symmetric. The metric in this case can be written in the form

$$ds^2 = c^2\,dt^2 - \exp[\lambda(r,t)]\,dr^2 - R^2(r,t)\,d\Omega^2, \tag{3.1.17}$$

in which $d\Omega$ represents the usual collection of angular terms. By working backwards, i.e. substituting the form of this metric back into the Einstein equations, one can show quite easily that

$$\exp[\lambda(r,t)] = \frac{(R')^2}{f^2(r)}, \tag{3.1.18}$$

in which the prime denotes derivative with respect to r and f is one of three undetermined functions in the Tolman–Bondi models. Let us now use $\dot{R}(r,t)$ to denote a partial derivative with respect to t. Again from the Einstein equations we can obtain

$$2\ddot{R}R + R^2 + 1 - f^2 = 0. \tag{3.1.19}$$

This can be integrated to give

$$\dot{R}^2(r,t) = f^2(r) - 1 + \frac{F(r)}{2R(r,t)}, \tag{3.1.20}$$

where $F(r)$ is the second undetermined function. We leave it as an exercise to go further and obtain the third free function.

The Tolman–Bondi solution has been used to understand the passage of photons through inhomogeneous matter distributions such as galaxy clusters, and also to understand some of the possible observational consequences of the kind of fractal inhomogeneity we discussed above (Ribeiro 1992).

3.2 The Steady-State Model

The model of the steady-state universe is now primarily of historical interest. In the past, however, from its original conception by Bondi, Gold and Hoyle in 1948 it was for many years a compelling rival to the Big Bang. Indeed it is ironic that Hoyle, a bitter opponent of the Big Bang, was the man who actually gave that model its name. He meant the term 'Big Bang' to be derogatory, but the term stuck.

The theory of the steady-state universe is based on the *Perfect Cosmological Principle*, according to which the universe must appear identical (at least in some average sense) when viewed from any point, in any direction and at any time. This is clearly a stronger version of the usual Cosmological Principle, which applies to spatial positions only. A particular consequence of this principle is that the Hubble constant really has to be constant in time:

$$\frac{\dot{a}}{a} = H(t) = \text{const.} = H_0; \tag{3.2.1}$$

from this relationship one can immediately deduce that the universe is expanding exponentially:

$$a(t) = a_0 \exp[H_0(t - t_0)]. \tag{3.2.2}$$

It is worth mentioning one immediate conundrum arising from this requirement. Although, as we have seen, it is difficult to measure the Hubble parameter unambiguously, most observations do seem to suggest a value of H_0^{-1}, which is at least within an order of magnitude of the ages of the oldest objects we can see. In a steady-state universe this is a surprise. There is no reason *a priori* why the age of the matter at a particular spatial location should bear any relation at all to the value of H_0^{-1}. The steady-state universe was partly motivated by the fact that, in the 1940s, the 'best' observational estimates of the Hubble constant were very large: $H_0 \simeq 300\ \text{km s}^{-1}\ \text{Mpc}^{-1}$. With this value, the ages of the oldest stars are much larger than H_0^{-1}, which is a powerful argument against the Big Bang. Modern estimates of H_0 are much lower and have blunted most of the force of this argument.

One can demonstrate, starting from the perfect Cosmological Principle, that the curvature parameter K which appears in the Robertson–Walker metric must be zero, and that the spatial sections in this model must therefore be flat. One consequence of Equation (3.2.2) is that, if the Universe is to look the same to all observers at all times, there must be a continuous creation of matter, in such a way that the mean density of particles remains constant. This creation must take place at a rate

$$\frac{3H_0\rho_0}{m_\text{p}} \simeq 10^{-16}h \text{ nucleons cm}^{-3}\text{ year}^{-1}. \tag{3.2.3}$$

It has never been clear exactly how this matter can be created, though it has been suggested that creation events might be responsible for driving active galactic nuclei. Hoyle's idea was to postulate a modification of the Einstein equations to take account of the non-conservation of the energy–momentum tensor through the famous 'C-field', via a term C_{ij}

$$R_{ij} - \tfrac{1}{2}g_{ij}R + C_{ij} = \frac{8\pi G}{c^4}T_{ij}; \tag{3.2.4}$$

substituting the Robertson–Walker metric appropriate for a steady-state model in Equations (3.2.4) one obtains

$$C_{ij} = -\left(8\pi G\frac{p_0}{c^2} + 3H_0^2\right)g_{ij} + 8\pi G\left(\rho_0 + \frac{p}{c^2}\right)U_i U_j. \tag{3.2.5}$$

Hoyle suggested that C_{ij} should be given by

$$C_{ij} = C_{;i;j} \tag{3.2.6}$$

(as usual, the symbol ';' stands for the covariant derivative), and the scalar field C is given by

$$C = -\frac{8\pi G}{H_0}\left(\rho_0 + \frac{p_0}{c^2}\right)t, \tag{3.2.7}$$

with

$$\rho_0 = \frac{3H_0^2}{8\pi G}. \tag{3.2.8}$$

The popularity of the steady-state universe took a nosedive with the discovery of the 3 K cosmic background radiation by Penzias and Wilson (1965), which has a natural explanation only within the framework of the hot Big Bang model. To reconcile the presence of the microwave background radiation with the steady-state theory it would be necessary to postulate the continuous creation not just of matter, but also of photons. Such a hypothesis appears even more unnatural than the creation of matter. An important development was also Sandage's revision of the cosmological distance scale, which brought the ages of astronomical objects into rough agreement with the Hubble timescale, H_0^{-1}. Until recently, the last significant works in defence of the steady-state model were made by Hoyle and Narlikar in the late 1960s. More recently, however, a variant of this model called the 'quasi-steady-state' universe has been proposed. In this scenario, matter is created in chunks of cosmological scale, rather than individually in nucleons. These elaborations remind one of the epicycles used in an attempt to rescue the Earth-centred Solar System model; the steady-state model being advanced nowadays certainly shares none of the compelling simplicity of its predecessor.

Nevertheless, some ideas from the steady-state universe do live on in modern cosmology. In particular, many aspects of the inflationary universe scenario, such as the exponential expansion, are exactly the same as in the steady-state model. However, in the former case, the driving force is not particle creation but rather the vacuum energy of a scalar quantum field with effective potential $V(\Phi) \simeq$ const.

3.3 The Dirac Theory

Dirac (1937, 1974) originated a novel approach to cosmology based on the consideration of dimensionless numbers constructed from fundamental physical quantities. For example, the dimensionless number

$$\frac{e^2}{Gm_\mathrm{p}m_\mathrm{e}} \simeq 0.23 \times 10^{40} \tag{3.3.1}$$

represents the ratio of the Coulomb force and the gravitational force between an electron and a proton;

$$\frac{\hbar c}{Gm_\mathrm{p}^2} \simeq 1.5 \times 10^{38} \tag{3.3.2}$$

is the ratio between the Compton wavelength and the Schwarzschild radius of a proton;

$$\frac{cH_0^{-1}}{(e^2/m_\mathrm{e}c^2)} \simeq 3.7 \times 10^{40} \tag{3.3.3}$$

is roughly the ratio between the cosmological horizon distance (sometimes somewhat inaccurately called the 'radius of the Universe') and the classical electron radius. One must make a distinction between relations of the type (3.3.3) and similar expressions, such as

$$\frac{1}{m_\pi}\left(\frac{\hbar^2 \mathrm{H}_0}{Gc}\right)^{1/3} \simeq \frac{1}{m_\mathrm{e}}\left(\frac{e^4 H_0}{Gc^3}\right)^{1/3} \simeq 1 \tag{3.3.4}$$

(m_π is the pion mass), which are between cosmological and microphysical quantities, and other relations which exist between either two cosmological or two microphysical quantities. For example,

$$\frac{\rho_{0\mathrm{m}}(cH_0^{-1})^3}{m_\mathrm{p}} \simeq 10^{80} = (10^{40})^2 \tag{3.3.5}$$

represents the number of baryons within the cosmological horizon;

$$\rho_{0\mathrm{m}} G H_0^{-2} \simeq 1 \tag{3.3.6}$$

expresses the near-flatness of the Universe; and

$$\left(\frac{k_\mathrm{B}T_{0\mathrm{r}}}{\hbar c}\right)^3 \frac{m_\mathrm{p}}{\rho_{0\mathrm{m}}} \simeq 10^{10} = (10^{40})^{1/4} \tag{3.3.7}$$

represents the ratio between the number densities of photons and baryons. Relations like (3.3.5)–(3.3.7) can be explained within the framework of an adequate cosmological model such as the inflationary universe. The relations (3.3.1)–(3.3.4) cannot be explained in this manner, and must be thought about in some other way.

There seem to be two possibilities: either they are essentially numerical coincidences, which occur because of some special property of the present epoch when we happen to be observing the Universe; or they have some deep physical significance which is yet to be elucidated. Arguments of the first type were advanced by Dicke in the 1960s, who explained that the present value of H_0^{-1} in the Big Bang model must be constrained by the requirement that life must have had time to evolve. This requires at least a main sequence stellar lifetime to have passed. The horizon must therefore be large simply in order for us to have evolved, and the number of baryons it contains must also be large. In the second type of argument a deeper explanation, based on fundamental physics, must be sought of the relations such as Equations (3.3.5) to (3.3.7).

This second approach was adopted by Dirac in numerous writings between 1934 and 1974. His basic assumption was that the large dimensionless numbers that keep appearing in relations between microphysical and cosmological scales are connected by a simple relation in which the only dimensionless coefficients that appear are of order unity. For example, let the first terms in Equations (3.3.1) and (3.3.3) be R_1 and R_2, respectively, so that

$$\frac{R_1}{R_2} = \frac{e^4 H_0}{G m_p m_e^2 c^3} \simeq 1. \tag{3.3.8}$$

If Equation (3.3.8) is valid at any cosmological epoch, given that H_0 varies, then at least one of the relevant physical 'constants' - e, G, m_e, m_p, c - must be time dependent. Dirac proposed two alternatives: either the charge of the electron or the constant of gravitation are variable. For simplicity, let us look at the second of these possibilities. From Equation (3.3.8) we obtain

$$G(t) \propto H(t) = \frac{\dot{a}}{a}, \tag{3.3.9}$$

and from (3.3.6), putting $\rho_m \propto a^{-3}$, we get

$$G(t)a^{-3}(t) \propto H^2(t). \tag{3.3.10}$$

One can eliminate $G(t)$ from Equations (3.3.9) and (3.3.10) leading to

$$\frac{\dot{a}}{a} \propto a^{-3}, \tag{3.3.11}$$

which, integrated, gives

$$a = a_0\left(\frac{t}{t_0}\right)^{1/3} \tag{3.3.12}$$

and, therefore,

$$G(t) = G_0\left(\frac{t}{t_0}\right)^{-1}; \tag{3.3.13}$$

G_0 is the present value of the 'constant' of universal gravitation and t_0 is the age of the Universe. We find that

$$t_0 = \tfrac{1}{3}H_0^{-1} \simeq 3.3 \times 10^9 h^{-1} \text{ years,} \tag{3.3.14}$$

too small compared with the nuclear timescale for stellar evolution which does not depend upon the assumption that G varies with time.

This result is bad news for the Dirac hypothesis. Nevertheless, Dirac's idea has inspired many attempts to construct theories of gravitation with a variable G. The most complete and interesting example is the scalar–tensor theory of Brans and Dicke (1961), which we describe in the next section. It is noteworthy, however, that the large-number coincidences which were the inspiration for Dirac's theory either became of secondary importance or were completely neglected in these alternatives. Nowadays it is generally accepted that the correct interpretation of the large-number coincidences is that due to Dicke, and that they are essentially consequences of the *Weak Anthropic Principle* which we shall discuss later, near the end of Chapter 7.

3.4 Brans–Dicke Theory

The Einstein equations of general relativity can be obtained by applying a variational principle to a Lagrangian of the form

$$L_{\mathrm{GR}} = L + \frac{c^4}{16\pi G}R, \tag{3.4.1}$$

where R is the scalar curvature and L is the Lagrangian action corresponding to the matter. In the Brans–Dicke theory, the appropriate gravitational Lagrangian is instead assumed to be

$$L_{\mathrm{BD}} = L + \frac{c^4}{16\pi}\varphi R - \frac{c^4}{16\pi}\frac{\omega g^{ij}\varphi_{;i}\varphi_{;j}}{\varphi}, \tag{3.4.2}$$

where φ is a scalar field and ω is a dimensionless coupling constant. Comparing Equation (3.4.2) with (3.4.1) shows that the inverse of the field φ plays the role of the gravitational constant G. From (3.4.2) we can derive the relation

$$\Box\varphi \equiv g^{ik}\varphi_{;i;k} = \frac{8\pi}{(3+2\omega)c^4}T^i{}_i, \tag{3.4.3}$$

where T_{ij} is the energy-momentum tensor appropriate for L and, in the place of the Einstein equations, we get

$$R_{ij} - \tfrac{1}{2}g_{ij}R = \frac{8\pi}{c^4\varphi}T_{ij} - \frac{\omega^2}{\varphi^2}(\varphi_{;i}\varphi_{;j} - g_{ij}\varphi_{;k}\varphi_{;}^{k}) - \frac{1}{\varphi}(\varphi_{i;j} - g_{ij}\Box\varphi), \tag{3.4.4}$$

which, after introducing the Robertson–Walker metric to get the cosmological equations, give the following:

$$3\frac{\ddot{a}}{a} = -\frac{8\pi}{(3+2\omega)}\frac{1}{\varphi}\left[(2+\omega)\rho + 3(1+\omega)\frac{p}{c^2}\right] - \omega\left(\frac{\dot{\varphi}}{\varphi}\right)^2 - \frac{\ddot{\varphi}}{\varphi}, \tag{3.4.5}$$

$$\dot{\rho} = -\frac{3\dot{a}}{a}\left(\rho + \frac{p}{c^2}\right), \tag{3.4.6}$$

$$\left(\frac{\dot{a}}{a}\right)^2 + \frac{K}{a^2} = \frac{8\pi\rho}{3\varphi} - \frac{\dot{\varphi}}{\varphi}\frac{\dot{a}}{a} + \frac{\omega}{6}\frac{\dot{\varphi}^2}{\varphi^2}, \tag{3.4.7}$$

$$\frac{\mathrm{d}}{\mathrm{d}t}(\dot{\varphi}a^3) = \frac{8\pi}{(3+2\omega)}\left(\rho - 3\frac{p}{c^2}\right)a^3. \tag{3.4.8}$$

One can also show that, in the framework of a Newtonian approximation, the 'constant' in Newton's law of gravitation is

$$G = \frac{2\omega+4}{2\omega+3}\frac{1}{\varphi}. \tag{3.4.9}$$

The cosmological models which solve Equations (3.4.5)–(3.4.8) depend on the four quantities a_0, $\dot{a}_0$, φ_0 and ρ_0, and the two parameters K (which takes the values 1, 0 or -1) and $\omega > 0$. Recall that the Friedmann models depend only on three initial values and only one parameter K. The set of cosmological solutions to the Brans–Dicke theory therefore forms a family of solutions which is much larger than that of the Friedmann models. We shall not describe these solutions in any detail, though it is perhaps worth mentioning that the homogeneous and isotropic Brans–Dicke solutions also possess a singularity in the past. Just to give one example, however, consider the flat Universe ($K = 0$). The present matter density is given by

$$\rho_{0\mathrm{m}} = \frac{3H_0^2}{8\pi G}\frac{(4+3\omega)(4+2\omega)}{6(1+\omega)^2}, \tag{3.4.10}$$

the age of the Universe by

$$t_{0\mathrm{H}} = \frac{2(1+\omega)}{(4+3\omega)}H_0^{-1}, \tag{3.4.11}$$

and the deceleration parameter by

$$q_0 = \frac{1}{2}\frac{\omega+2}{\omega+1}; \tag{3.4.12}$$

Equations (3.4.10)–(3.4.12) all become identical to the Einstein–de Sitter case for $\omega \to \infty$.

The mysterious relations (3.3.1)–(3.3.7) do not find an explanation in the framework of this theory, which was not formulated with that intention. The situation with respect to the observational implications of this theory is very complicated, given the large set of allowed models. Cosmological considerations (such as the age of the Universe, nucleosynthesis, etc.) do not place strong constraints on the

Brans–Dicke theory. The most important tests of the validity of this theory are those that involve the time-variation of G. There are various relevant observations: the orbital behaviour of Mercury and Venus; historical data about lunar eclipses; properties of fossils; stellar evolution (particularly the Sun); deflection of light by celestial bodies; the perihelion advance of Mercury. These observations together do not rule out the Brans–Dicke theory, but a rough limit on the parameter ω is obtained: $\omega > 500$.

In recent years, interest in the Brans–Dicke theory as an alternative to general relativity has greatly diminished, but there has been a great deal of recent work on the behaviour of certain types of inflationary model which involve a scalar field with essentially the same properties as the Brans–Dicke field φ; these are usually called *extended inflation* models.

3.5 Variable Constants

One of the consequences of Brans–Dicke theory is that the Newtonian gravitational constant changes with time. In recent years this general framework has given rise to suggestions that other fundamental physical quantities may also not be constant. For example, the fine-structure constant α, given in SI units as

$$\alpha = \frac{e^2}{4\pi\epsilon_0 \hbar c}, \tag{3.5.1}$$

may change with time. The presence of e in this expression indicates that the parameter α measures the strength of the electromagnetic interaction. To have this strength change on a cosmological timescale we therefore need to introduce into the Lagrangian a term involving the electromagnetic field. In general the electromagnetic field is described by a tensor of the form

$$F_{\mu\nu} = A_{\nu,\mu} - A_{\mu,\nu}, \tag{3.5.2}$$

where A_μ is the usual vector potential that appears in Maxwell's equations. The appropriate Lagrangian for electromagnetism can be seen to be

$$L_{\rm em} = -\tfrac{1}{4} F^{\mu\nu} F_{\mu\nu}. \tag{3.5.3}$$

One way of building a model in which the coupling to electromagnetism changes is then to use a Lagrangian containing an extra term that couples some scalar field ψ to this in much the same way that the Brans–Dicke theory (3.4.2) couples a scalar field to the metric in order to change the strength of gravity. A possibility is to add a term like $L_{\rm em}\exp(-2\psi)$. In this case the Einstein equations become

$$G_{\mu\nu} = \frac{8\pi G}{c^4}[T^{\rm m}_{\mu\nu} + T^{\psi}_{\mu\nu} + T^{\rm em}_{\mu\nu}\exp(-2\psi)] \tag{3.5.4}$$

leading to changes in the cosmological equations and possible observational consequences in absorption line systems (e.g. Sandvik *et al.* 2002).

However, interpreting this change as a change of α alone is not the only possibility. It is possible to use this general idea also to motivate models in which the speed of light c is also variable. The connection between variable c theories and variable α theories lies in (3.5.1). For example, given a variable α theory it is always possible to redefine units so that c and $\hbar$ are constant and e varies. It is possible therefore to interpret the model described above as a variable c cosmology in which ψ is just some function of c or vice versa. Somewhat surprisingly, it is possible to make such a theory both covariant and Lorentz invariant (Moffat 1993; Magueijo 2000).

3.6 Hoyle–Narlikar (Conformal) Gravity

Another theory of gravitation that has given rise to interesting cosmological models was proposed by Hoyle and Narlikar in 1964; we shall hereafter call this the HN theory. The important difference between HN theory and both general relativity and the Brans–Dicke theory mentioned above is that the latter are *field theories*, while the former is based on the idea of direct interparticle action. *Mach's Principle* suggests the existence of action-at-a-distance by the following argument. The mass of an object m_i according to Mach's Principle is not entirely an intrinsic property of the object, but is due to the background provided by all the other objects in the Universe. Building on some ideas of Dirac at representing electromagnetism in a similar way and exploiting the notion of *conformal invariance*, Hoyle and Narlikar produced a theory of gravitation which, when expressed in the language of field theory, is identical to general relativity.

So what has been gained in this exercise? It seems that this theory provides no new predictions. In fact there are a number of subtle and interesting ways in which this theory differs from general relativity. First, while the Einstein equations have valid solutions for an empty Universe, the HN equations in this case yield an indeterminate solution for the metric g_{ik}. This makes sense in light of Mach's Principle: without a set of background masses against which to measure motion, the concept of a trajectory is meaningless. Second, the sign of the gravitational constant G is only fixed in general relativity by comparing its weak-field limit with Newtonian gravity. There is no *a priori* reason intrinsic to general relativity why G could not be negative. In HN theory, G is always positive. Likewise, there is no space for the cosmological constant Λ in the field equations of HN theory. Finally, we mention that in the HN cosmological solutions, redshift arises from the variation of particle masses with time.

The HN theory is an interesting physically motivated alternative to Einstein's general relativity. While we assume throughout most of this book that GR is the correct law of gravity on cosmological scales, we still feel it is important to stress that there have been no compelling strong-field tests of Einstein's theory. Alternatives like the HN theory have an important role to play in reminding us how different cosmology could be if Einstein's theory turned out to be wrong!

Bibliographic Notes on Chapter 3

A wide-ranging review of alternatives to the Big Bang cosmology may be found in Ellis (1987). In the early 1990s there was an interesting sequence of review articles in *Nature* for and against the standard cosmology: see Arp *et al.* (1990) for the discontents and the riposte by Peebles *et al.* (1991). A nice review of anisotropic and inhomogeneous cosmologies is given by MacCallum (1993).

Problems

1. Prove that the largest possible group for a spatially homogeneous model is six dimensional.
2. What is special about $h = -\frac{1}{9}$ in the Bianchi classification?
3. Investigate the possible behaviour of the singularity as $t \to 0$ in the Kasner solution.
4. Integrate Equation (3.1.8) to identify the third undetermined function in the Tolman-Bondi model and discuss its physical interpretation.
5. Identify the coordinate transformation that turns (3.1.16) into the Minkowski metric.
6. Is there an Olbers Paradox in the steady-state model?

4

Observational Properties of the Universe

4.1 Introduction

Our approach to cosmology so far has been almost entirely theoretical, apart from reference to the observational motivation for the Cosmological Principle which was essential in constructing the Friedmann models. We should now fill in some details on what is known about the bulk properties of our Universe, and how one makes measurements in cosmology. Before doing so, however, we take this opportunity to remind the reader of some simple background material from observational astronomy.

4.1.1 Units

The standard unit of distance in astronomy is the *parsec*, which is defined as the distance at which the deflection of an object's angular position on the sky in the course of the Earth's orbital motion is one second of arc. (Note that, during half an orbit, the angular change is two arcseconds.) Alternatively and equivalently, one parsec is the distance of an object at which the semi-major axis of the Earth's orbit around the Sun subtends an angle of one arcsecond at the object. It turns out that

$$1\ \mathrm{pc} \simeq 3.086 \times 10^{13}\ \mathrm{km} \simeq 3.26\ \text{light years}, \tag{4.1.1}$$

where a light year is the distance travelled by light in a time of one year. A thousand parsecs is called a kiloparsec (kpc) and a million parsecs a megaparsec (Mpc). The typical separation of stars in a galaxy like the Milky Way is of the order of a parsec, while the typical separation of bright galaxies is of the order of an Mpc. The most useful unit for cosmology is therefore the megaparsec. One typically has to use the Hubble law (1.4.6) to estimate extragalactic distances from velocities, since distances are hard to measure directly. There has always been some uncertainty in the value of the Hubble constant H_0, with the result that cosmologists usually still parametrise it in terms of a dimensionless number h, where

$$h = \frac{H_0}{100 \text{ km s}^{-1} \text{ Mpc}^{-1}}. \tag{4.1.2}$$

Using this notation, distances inferred from velocities have units h^{-1} Mpc. We discuss the distance scale further in Section 4.2.

The usual unit of mass is the *solar mass*

$$1M_\odot \simeq 1.99 \times 10^{33} \text{ g}, \tag{4.1.3}$$

and for luminosity L we adopt the *solar luminosity*

$$1L_\odot \simeq 3.9 \times 10^{33} \text{ erg s}^{-1}. \tag{4.1.4}$$

The absolute luminosity L of a source is simply the total energy emitted by the source per unit time, while the apparent luminosity l is the energy received by an observer per unit time per unit area from the source. The latter obviously depends on the distance from the source to the observer. In place of L and l, astronomers frequently use absolute magnitude M and apparent magnitude m. These quantities were defined in Section 1.8, based on a logarithmic scale in which five magnitudes correspond to a factor 100 in luminosity. In fact there are several definitions of apparent magnitude (m_{U}, m_{B}, m_{V}, m_{IR}, etc.) because one often cannot measure the total flux from a source, but only that part which lies within some finite band of wavelengths to which a particular instrument is sensitive. The above examples stand for ultraviolet, blue, visible and infrared, respectively, and are all based on standard filters. The total apparent luminosity of a source, integrated over all wavelengths, is called the bolometric luminosity. In all cases the relationship between apparent magnitude and apparent luminosity is defined in such a way that the apparent magnitudes are the same for stars of spectral type $A0V$.

We shall also, from time to time, have need to use astronomical coordinate systems to describe the location of various objects on the sky. Because we are dealing exclusively with extragalactic objects, we prefer to use galactic coordinates whenever possible. The galactic latitude b is the angle made by a source and the galactic plane; an object in the galactic plane has $b = 0$ and an object vertically above or below the plane has $b = \pm 90°$; the northern galactic pole is defined to be at $b = +90°$ and this pole lies in the northern part of the sky as visible from Earth.

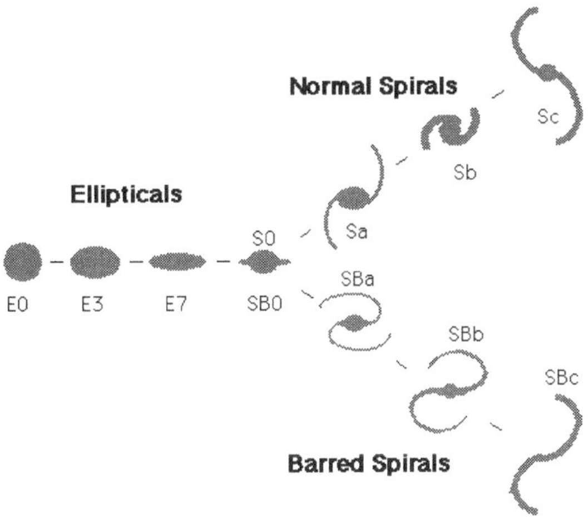

Figure 4.1 The Hubble 'tuning fork' classification of galaxies. The sequence from left to right runs through various types of elliptical galaxies (E), then divides into two branches, corresponding to 'normal' spirals (S0, Sa, Sb, Sc) and barred spirals (SB0, SBa, SBb, SBc). Irregular galaxies are not shown.

Galactic longitude is measured anticlockwise with respect to the galactic meridian, the plane passing through the centre of the galaxy, the Earth and the north and south galactic poles. Standard books on spherical trigonometry explain how to convert l and b coordinates into the usual right ascension α and declination δ.

4.1.2 Galaxies

Observational cosmology is concerned with the distribution of matter on scales much larger than that of individual stars, or even individual galaxies. For many purposes, therefore, we can regard the basic building block of cosmology to be the galaxy. Much of this book is concerned with the problem of understanding galaxy formation and we shall defer a detailed study of galaxies and the way they are distributed until Part 4, where we confront the theories we have described with the observed facts. It is worth, however, describing some of the basic properties of galaxies to give an idea of the richness of structure one can observe.

Galaxies come in three basic types: *spirals*, *ellipticals* and *irregular*. Hubble proposed a morphological classification, or taxonomy, for galaxies in which he envisaged these three types as forming a kind of evolutionary sequence. Although it is now not thought that this evolutionary sequence is correct, Hubble's nomenclature, in which ellipticals are 'early' type and spirals and irregulars 'late', is still commonly used. Figure 4.1 shows Hubble's classification scheme. The elliptical galaxies (E), which account for only around 10% of observed bright galaxies, are elliptical in shape and have no discernible spiral structure. They are usually red in colour, have very little dust and show no sign of active star formation. The luminosity profile of an elliptical galaxy is of the form

$$I(r) = I_0\left(1 + \frac{r}{R}\right)^{-2}, \tag{4.1.5}$$

where I_0 and R are constants and r is the distance from the centre. The scale length R is typically around 1 kpc. The classification of elliptical galaxies into En depends on the ratio of major to minor axes of the ellipse: the integer n is defined by $n \simeq 10(1 - b/a)$, where a and b are the major and minor axes, respectively. Ellipticals show no significant rotational motions and their shape is thought to be sustained by the anisotropic 'thermal' motions of the stars within them. Ellipticals occur preferentially in dense regions, i.e. inside clusters of galaxies.

Spiral galaxies account for more than half the galaxies observed out to 100 Mpc and brighter than $m = 14.5$. Hubble's division into normal (S) and barred (SB) spirals depends on whether the prominent spiral arms emerge directly from the nucleus, or originate at the ends of a luminous bar projecting symmetrically through the nucleus. Spirals often contain copious amounts of dust, and the spiral arms in particular show evidence of ongoing star formation (i.e. lots of young supergiant stars), giving the arms a blue colour. The nucleus of a spiral galaxy resembles an elliptical galaxy in morphology, luminosity profile and colour. Many spirals also demonstrate some kind of 'activity' (non-thermal emission processes). The intensity profile of spiral galaxies (outside the nucleus) does not follow Equation (4.1.4) but can instead be fitted by an exponential form:

$$I(r) = I_0 \exp(-r/R). \tag{4.1.6}$$

The subdivision of S and SB into a, b or c depends on how tightly the spiral arms are wound up. Spirals show ordered rotational motion which can be used to estimate their masses (see Section 4.5).

Lenticular, or S0, galaxies were added later by Hubble to bridge the gap between normal spirals and ellipticals. Around 20% of galaxies we see have this morphology. They are more elongated than elliptical galaxies but have neither bars nor spiral structure. Irregular galaxies have no apparent structure and no rotational symmetry. They are relatively rare, are often faint and small and are consequently very hard to see. The distribution of masses of elliptical galaxies is very broad, extending from 10^5 to $10^{12} M_\odot$, which includes the mass scale of globular star clusters. Small elliptical galaxies appear to be very common: for example, 7 out of 17 galaxies in the Local Group are of this type. Spiral galaxies have a smaller spread in masses, with a typical mass of $10^{11} M_\odot$.

4.1.3 Active galaxies and quasars

Many galaxies, especially spirals, show various types of activity, characterised by non-thermal emission at a wide range of wavelengths from radio to X-ray. A full classification of all the different types of active galaxy is outside the scope of this book, let alone any attempt to explain the bewildering variety of properties they

Figure 4.2 The 'Whirlpool' Galaxy M51, a fine example of a face-on spiral galaxy. Picture courtesy of the National Optical Astronomy Observatory/Association of Universities for Research in Astronomy/National Science Foundation.

possess. One possible explanation is that they are all basically the same kind of 'animal', but we happen to be observing them at different angles and therefore we see radiation from different regions within them. We shall not discuss this idea in detail, however, but merely restrict ourselves to listing the main types. The usual abbreviation for all these phenomena is AGN (active galactic nucleus).

Seyfert galaxies are usually spiral galaxies. They have very little radio emission and no sign of any jets. Seyferts display a strong continuum radiation all the way from the infrared to X-ray parts of the spectrum. They also have emission lines, which may be variable.

Radio galaxies are usually ellipticals. They typically possess two lobes of radio emission and sometimes have a compact core; often they show signs of some kind of 'jet'. The nucleus of these sources tends to have spectral properties similar to Seyfert galaxies.

BL Lac objects have no emission lines, but a strong smooth continuum from radio to X-ray wavelengths. They show dramatic and extremely rapid variability. It is thought that these objects might be explained as the result of looking at a relativistic jet end-on. Relativistic effects might shorten the apparent variability timescale, and the emission lines might be swamped by the jet.

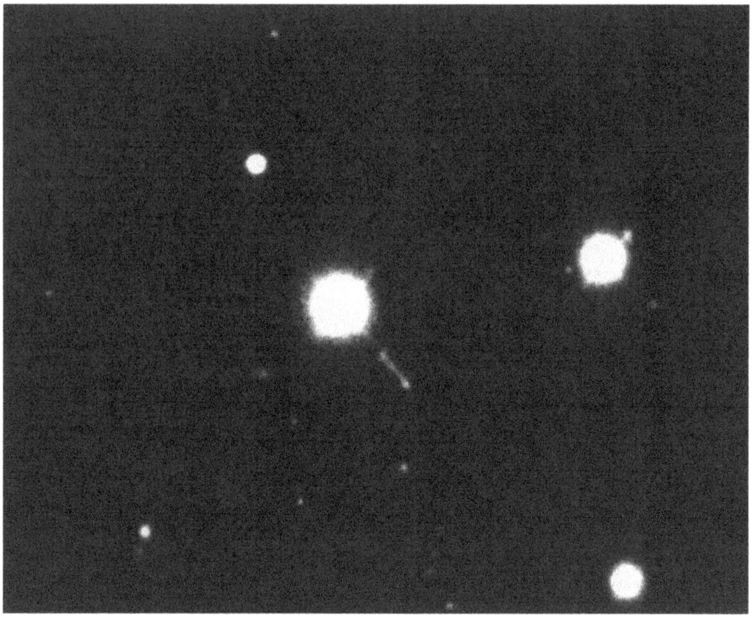

Figure 4.3 The quasar 3C273, seen in optical light, showing a jet of radiating material. Photograph courtesy of the National Optical Astronomy Observatory/Association of Universities for Research in Astronomy/National Science Foundation.

Quasars are point-like objects and are typically at high redshifts. Indeed the current record holder has $z \sim 6$! They are phenomenally luminous at all frequencies. Moreover, they are variable on a timescale of a few hours: this shows that much of their radiant energy must be emitted from within a region smaller than a few light hours across. Such is the energy they emit from a small region that it is thought they might be powered by accretion onto a central black hole. Most quasars are radio-quiet, but some are radio-loud. Long exposures sometimes reveal structure in the form of a jet.

A somewhat milder form of activity is displayed by the *starburst galaxies*, which, as their name suggests, are galaxies undergoing a strong burst of star formation which may be triggered by the interaction of the galaxy with a neighbour.

4.1.4 Galaxy clustering

All self-gravitating systems tend to form clumps, or density concentrations, so one should not be surprised to find that galaxies are not sprinkled randomly throughout space but are clustered. As we shall see in Chapter 16, the way galaxies cluster is approximately hierarchical: many galaxies occur in pairs or small groups which in turn are often clustered into larger associations. Just how large a scale this hierarchy reaches is an important test of theories of structure formation, as we shall see.

Figure 4.4 The Coma cluster of galaxies observed in optical light. Only the central regions are shown; the cluster contains more than a thousand galaxies, most of which are elliptical. Picture courtesy of the National Optical Astronomy Observatory/Association of Universities for Research in Astronomy/National Science Foundation.

Our galaxy, the Milky Way, is a member of a group of around 20 galaxies (most of them small) called the *Local Group*, which also includes the Andromeda spiral M31, and is altogether a few Mpc across. The nearest galaxies to us, the Large and Small Magellanic Clouds, are members of this group. Further away, at a distance of about $10h^{-1}$ Mpc, lies a prominent cluster of galaxies called the Virgo cluster which is pulling the Local Group towards itself. There are several prominent clusters within $100h^{-1}$ Mpc of the Local Group, the most impressive being the Coma cluster which lies about $60h^{-1}$ Mpc away and which contains literally thousands of galaxies. One should stress, however, that it is probably not helpful to think of clusters as discrete entities: all galaxies are clustered to some extent, but most of them reside in small groups with a low density contrast. When one looks at objects like Coma, one is seeing the upper extreme of the distribution of cluster sizes.

Nevertheless, an important part of the analysis of galaxy clustering is played by the study of the richest clusters. George Abell catalogued the most prominent clusters according to their apparent richness and estimated distance in the 1950s. The manner in which he did this was somewhat subjective and, as we shall discuss in Chapter 16, the methods he used to identify 'Abell' clusters may have introduced some systematic errors. Nevertheless, his catalogue is still used today for studies of large-scale structure. Rich clusters of galaxies also have other uses. These objects are so dense that they are probably gravitationally fully collapsed systems and one can therefore use statistical mechan-

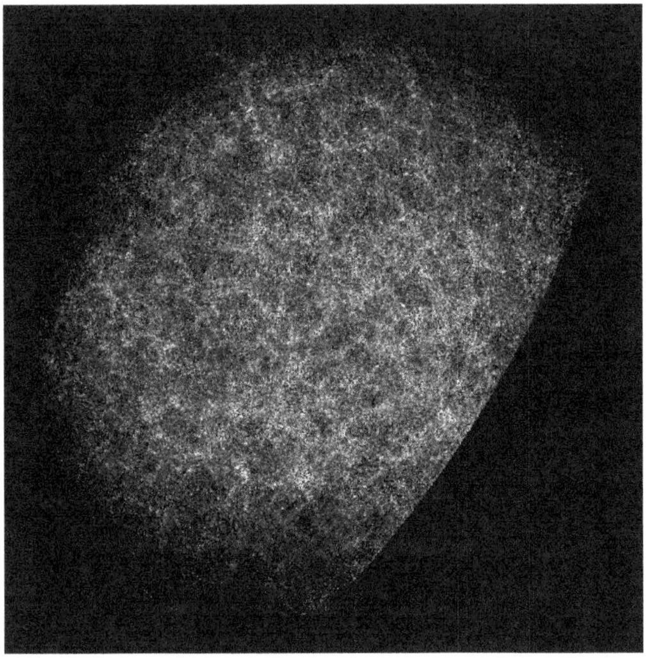

Figure 4.5 The Lick map showing a region of the northern galactic sky. A strong visual impression of 'bubbly' and/or 'filamentary' pattern is revealed. Picture courtesy of Ed Groth.

ics to estimate their mass (see Section 4.5). Moreover, they are also very bright in the X-ray part of the spectrum because they contain large amounts of hot, ionised gas. X-ray observations can therefore be used to measure the relative contributions to the total cluster mass of individual galaxies and hot gas, as well as any unseen component of dark matter. Maps of the general pattern of clustering on the sky require systematic surveys of galaxies with some well-defined selection criterion (usually a strict apparent magnitude limit). Usually such surveys avoid regions of the sky close to the galactic plane, say with galactic latitude $b < 20°$, because of the observational difficulties posed by interstellar dust within our Galaxy. The first survey of galaxy positions was due to Shapley and Ames (1932) which catalogued 1250 galaxies with $m < 13$. This was the first strong indicator of galaxy clustering. Later, Zwicky accumulated a sample of 5000 galaxies with $m < 15$ using the Palomar Sky Survey. Enormous strides were then taken by Shane and Wirtanen (1967), who created the famous Lick map of galaxies. This shows around a million galaxies with $m < 19$ and covers most of the sky. Figure 4.5 shows clear evidence of clustering in the form of filamentary patterns, large clusters and regions of very low density. The Lick map was compiled using relatively primitive eyeball techniques. More recent surveys using automatic plate-measuring machines, such as the APM and COSMOS, have made the acquisition of large quantities of data rather less problematic. The APM catalogue, for example, contains about two million galaxies (Maddox *et al.* 1990). Important though these sky surveys are, because of the sheer number of galaxies they contain, they do not reveal directly the positions of galaxies in three-dimensional space, but only in two-dimensional projection on the sky. No distance information is present in sky catalogues, except in the statistical sense that the fainter galaxies will, on average, be further away than the bright ones. The third dimension can at least be estimated by using the galaxy redshift z. This, however, requires not just an image of the galaxy but a spectrum. Systematic surveys of the redshifts of galaxies identified on sky survey plates more or less began in the 1980s with the Harvard–Smithsonian Center for Astrophysics (CfA) survey, which used the Zwicky catalogue as its 'parent' (de Lapparent *et al.* 1986). This resulted in maps of the redshifts of several thousand galaxies in various 'slices' on the sky. Improvements in instrumentation technology have led to a revolution in the field of 'cosmography', i.e. mapping the distribution of galaxies in our Universe. For example, a large-scale map of the galaxy distribution was obtained by the QDOT (Queen Mary, Durham, Oxford and Toronto) team using not optical galaxies, but galaxies detected by the IRAS satellite through their infrared radiation. The survey was subsequently expanded by a factor of six and, now complete, contains more than 10 000 galaxies. As far as optical surveys are concerned the great step forward has been the advent of multi-fibre spectroscopic devices on wide-field telescopes, enabling redshifts to be obtain of several hundred galaxies in a single pointing of a telescope. The first large survey of this type, the Las Campanas Redshift Survey, contained about 25 000 galaxies; the catalogue was published in 1996. A survey of around a quarter of a million galaxies, using the APM survey as its parent and exploiting the 'two-degree field' (2dF) on the Anglo-Australian telescope, is nearing completion by a British–Australian consortium. While in the USA the Sloan Digital Sky Survey aims eventually to measure a million galaxy redshifts. The picture that emerges is a fascinating one. The galaxy distribution is characterised by filaments, sheets and clusters. Clusters are themselves grouped into superclusters, such as the Virgo supercluster and the so-called Shapley concentration. In between these structures there are large regions almost devoid of galaxies. These are usually called *voids*. There are two important tasks for modern cosmology, connected with the way in which galaxies and clusters are distributed throughout space. The first is to quantify, using appropriate statistical tools, the level of present clustering. The second is then to explain this clustering using a theory for the evolution of structure within expanding universe models. Part 3 of this book will be devoted to the standard theory for structure formation and Part 4 to the various constraints placed on these theories by detailed statistical analysis of galaxy clustering and other cosmological observations.

4.2 The Hubble Constant

As we have explained, the Hubble law is implicit in the requirement that the Universe is homogeneous and isotropic. There is therefore a strong theoretical motivation for it stemming from the Cosmological Principle. In fact, the Hubble

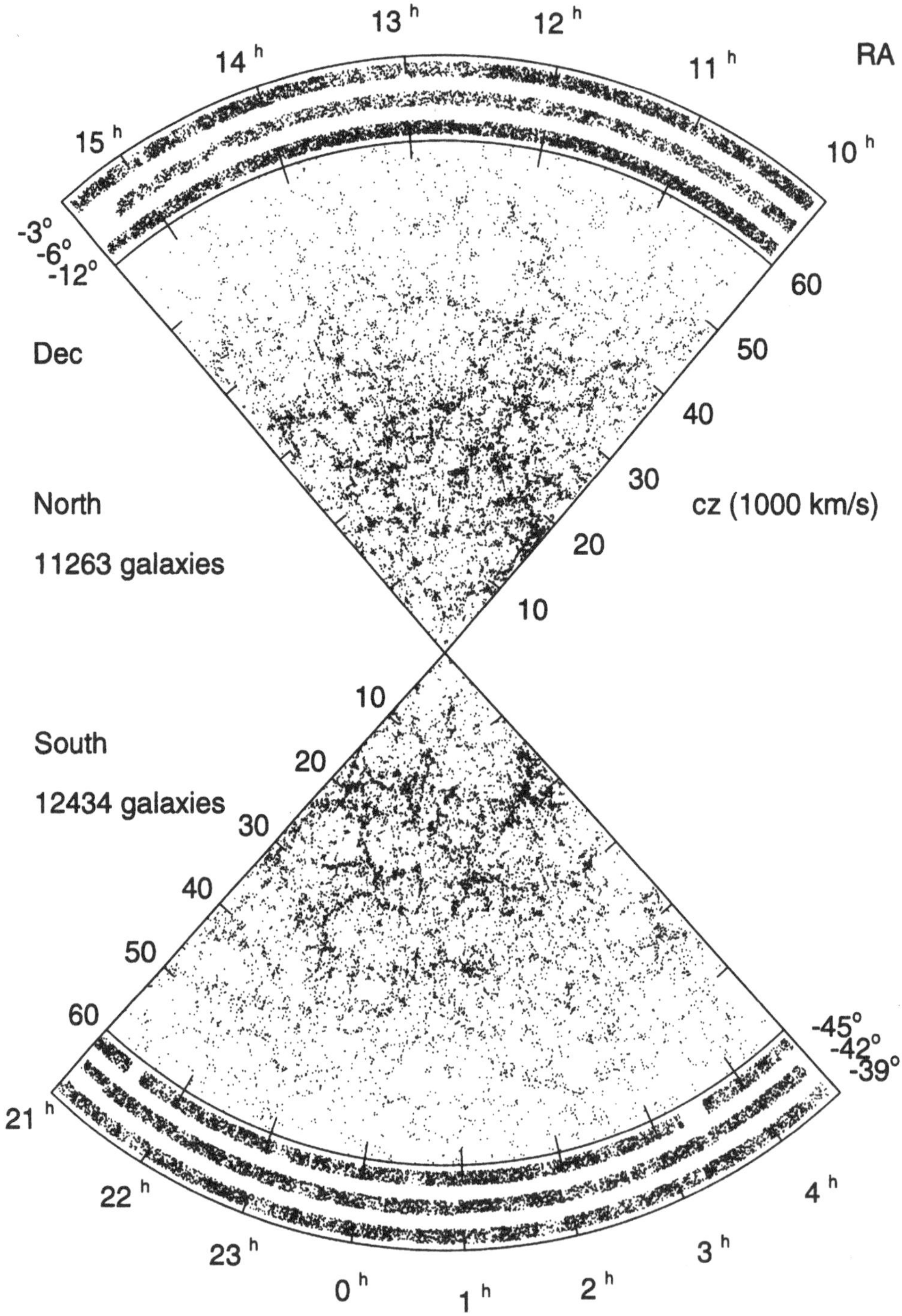

Figure 4.6 The Las Campanas Redshift Survey. Picture courtesy of Bob Kirschner.

expansion was first discovered observationally by Slipher but he did not make the bold interpretation of his data that Hubble did. After many years of painstaking observations, Hubble (1929) formulated his law in the form that galaxies seem to be receding with a velocity v proportional to their distance d from the observer:

$$v = H_0 d. \tag{4.2.1}$$

This relation is called the *Hubble law* and the constant of proportionality H_0 is called the *Hubble constant*. The numerical value of H_0 is most conveniently expressed in units of km s^{-1} for the velocity and Mpc for the distance, i.e. in km s^{-1} Mpc^{-1}. As we have mentioned before, and shall discuss in much detail soon, H_0 is very difficult to measure accurately. Until recently there was an uncertainty of about a factor of two in H_0. Given the scale of the possible error, it is useful to introduce the dimensionless parameter h defined in (4.1.2).

We should now make some comments about the limits of the validity of Equation (4.2.1). For a start, the distance d must be sufficiently large that the recession velocity deduced from (4.2.1) is much larger than the radial component of the peculiar velocities. This can be up to 1000 km s^{-1} for galaxies inside clusters; this places the requirement that $d \gg 10h^{-1}$ Mpc. In terms of redshift this means that $z \gg 10^{-2}$. On the other hand, the distance should not be so large that Equation (4.2.1) implies a recession velocity greater than the velocity of light. In fact Equation (4.2.1) is true if d is the proper distance of the galaxy, but we cannot measure this directly and one has to use measures such as the luminosity distance for which Equation (4.2.1) is no longer valid. Roughly speaking one should therefore only use this equation for $d \ll 300h^{-1}$ Mpc (or $z \ll 10^{-1}$). From Section 1.5 it can be shown that the distance d of a galaxy with redshift in the range $10^{-2} \leqslant z \leqslant 10^{-1}$ is given, to a good approximation, by

$$d \simeq \frac{c}{H_0} z \simeq 3000h^{-1} z \text{ Mpc}. \tag{4.2.2}$$

This equation should be thought of as the first approximation to the formula for the luminosity distance as a function of redshift for Friedmann models:

$$d_{\mathrm{L}} = \frac{c}{H_0}\frac{1}{q_0^2}\{q_0 z + (q_0 - 1)[-1 + (2q_0 z + 1)^{1/2}]\} \simeq \frac{c}{H_0}[z + \tfrac{1}{2}(1 - q_0)z^2], \tag{4.2.3}$$

which one can prove quite easily starting from Equation (1.7.3) (see also Equation (2.4.15)).

As we have mentioned, Equation (4.2.1) can be derived from the assumption that the Universe is homogeneous and isotropic, i.e. that the Cosmological Principle applies. All the relations one can use to demonstrate this property from an observational point of view, such as the m–z (magnitude–redshift) and N–z relations, obviously contain the parameter H_0 explicitly.

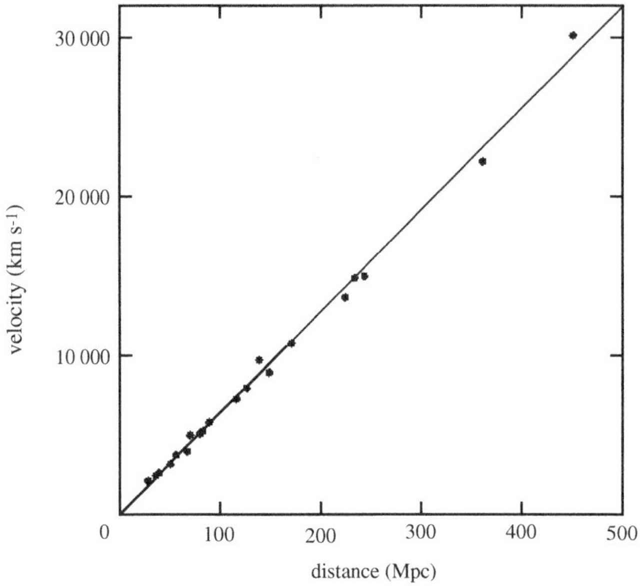

Figure 4.7 The Hubble diagram showing the correlation between redshift (y-axis) and a distance indicator based on the first-ranked cluster elliptical (x-axis). Adapted from Sandage (1972).

As we have seen, H_0 is the first of the important parameters one needs to know in order to construct a useful cosmological model. Knowledge of it would establish three quantities:

1. the *distance scale* of the present cosmological horizon

$$l_{0\mathrm{H}} \simeq \frac{c}{H_0} \simeq 3000h^{-1}\ \mathrm{Mpc}; \tag{4.2.4}$$

2. the characteristic *timescale* for the expansion of the Universe

$$t_{0\mathrm{H}} \simeq \frac{1}{H_0} \simeq 0.98 \times 10^{10} h^{-1}\ \text{years} \simeq 3 \times 10^{17} h^{-1}\ \mathrm{s}; \tag{4.2.5}$$

and

3. the *density scale* required to close the universe

$$\rho_{0\mathrm{c}} = \frac{3H_0^2}{8\pi G} \simeq 1.9 \times 10^{-29} h^2\ \mathrm{g\ cm^{-3}}, \tag{4.2.6}$$

where $\rho_{0\mathrm{c}}$ is the present value of the *critical density*.

The significance of these quantities was explained in Chapter 2.

4.3 The Distance Ladder

The value of H_0 found by Hubble in 1929 was around 500 km s^{-1} Mpc^{-1}, much larger than the values currently accepted. This discrepancy was due to errors in the calibration of distance indicators that he used, which were only corrected many years later. In the 1950s, Baade derived a value of H_0 of order 250 km s^{-1} Mpc^{-1}, but this was also affected by a calibration error. A later recalibration by Sandage in 1958 brought the value down to between 50 and 100 km s^{-1} Mpc^{-1}; present observational estimates still lie in this range. This demonstrates the truth of the comment we made above: Hubble's 'constant' is not actually constant because it has changed by a factor of 10 in only 50 years! Joking apart, the term 'constant' was never intended to mean constant in time, but constant in the direction in which one observes the recession of a galaxy. As far as time is concerned, the Hubble constant changes in a period of order H^{-1}.

One simple way to estimate the Hubble constant is to determine the absolute luminosity of a distant source and to measure its apparent luminosity l. From these two quantities one can calculate its luminosity distance

$$d_{\mathrm{L}} = \left(\frac{L}{4\pi l}\right)^{1/2}, \tag{4.3.1}$$

which, together with the redshift z which one can measure via spectroscopic observations of the source, provides an estimate of the Hubble constant through Equation (4.2.3) (in the appropriate interval of z). The main difficulty with this approach is to determine L. The usual approach, which is the same as that developed by Hubble, is to construct a sort of *distance ladder*: relative distance measures are used to establish each 'rung' of the ladder and calibrating these measures against each other allows one to measure distances up to the top of the ladder. A modern analysis might use several rungs, based on different distance measures, in the following manner.

First, one exploits local kinematic distance measures to establish the length scale within the galaxy. Kinematic methods do not rely upon knowledge of the absolute luminosity of a source. Nearby distances can be derived using the *trigonometric parallax* ϖ of a star, i.e. the change in angular position of a star on the sky in the course of a year due to the Earth's motion in space. Measuring ϖ in arcseconds is convenient here because the distance in parsecs is then just $d = \varpi^{-1}$, as we mentioned in Section 4.1. Until recently this direct technique was limited to distances of order 30 pc or so, but the astrometric satellite Hipparcos has established a distance scale based on parallax to kiloparsec scales.

The *secular parallax* of nearby stars is due to the motion of the Sun with respect to them. For stellar binaries one can derive distances using the *dynamical parallax*, based on measurements of the angular size of the semi-major axis of the orbital ellipse, and other orbital elements of the binary system. Another method is based on the properties of a *moving cluster* of stars. Such a cluster is a group of stars which move across the Galaxy with the same speed and parallel trajectories; a perspective effect makes these stars appear to converge to a point on the sky. The position of this point and the proper motion of the stars lead one to the distance. This method can be used on scales up to a few hundred parsecs; the Hyades cluster is a good example of a suitable cluster. With the method of *statistical parallax* one can derive distances of order 500 pc or so; this technique is based on the statistical analysis of the proper motions and radial velocities of a group of stars. Taken together, such kinematic methods allow us to establish distances up to the scale of a few hundred parsecs, much smaller even than the scale of our Galaxy.

Once one has determined the distances of nearby stars with a kinematic method, one can then calculate their absolute luminosities from their apparent luminosities and their (known) distances. In this way it was learned that most stars, the so-called *main sequence stars*, follow a strict relationship between spectral type (an indicator of surface temperature) and absolute luminosity: this is usually visualised in the form of the HR (Hertzsprung–Russell) diagram. Using the properties of this diagram one can measure the distances of main sequence stars of known apparent luminosity and spectral type. With this method, one can measure distances up to around 30 kpc.

Another important class of distance indicators contains variables stars of various kinds, including *RR Lyrae* and *Classical Cepheids*. The RR Lyrae all have a similar (mean) absolute luminosity; a simple measurement of the apparent luminosity suffices to provide a distance estimate for this type of star. These stars are typically rather bright, so this can extend the distance ladder to around 300 kpc. The classical Cepheids are also bright variable stars which have a very tight relationship between the period of variation P and their absolute luminosity: $\log P \propto \log L$. The measurement of P for a distant Cepheid thus allows one to estimate its distance. These stars are so bright that they can be seen in galaxies outside our own and they extend the distance scale to around 4 Mpc. Errors in the Cepheid distance scale, due to interstellar absorption, galactic rotation and, above all, a confusion between Cepheids and another type of variable star, called W Virginis variables, were responsible for Hubble's large original value for H_0. Other distance indicators based on *novae*, *blue supergiants* and *red supergiants* allow the ladder to be extended slightly to around 10 Mpc. Collectively, these methods are given the name *primary distance indicators.*

The *secondary distance indicators* include *HII regions* (large clouds of ionised hydrogen surrounding very hot stars) and *globular clusters* (clusters of around 10^5–10^7 stars). The former of these has a diameter, and the latter an absolute luminosity, which has a small scatter around the mean. With such indicators one can extend the distance ladder out to about 100 Mpc.

The *tertiary distance indicators* include *brightest cluster galaxies* and *supernovae.* Clusters of galaxies can contain up to about a thousand galaxies. One finds that the brightest galaxy in a rich cluster has a small dispersion around the mean value (various authors have also used the third, fifth or tenth brightest cluster galaxy as a distance indicator). With the brightest galaxies one can reach distances of several hundred Mpc. Supernovae are stars that explode, producing a luminosity roughly equal to that of an entire galaxy. These stars are therefore easily seen in distant galaxies, but the various indicators that use them are not too precise.

More recently, much attention has been paid to observed correlations of intrinsic properties of galaxies themselves as distance indicators. In spiral galaxies, one can use the empirical *Tully-Fisher relationship*:

$$L \propto V_c^{\alpha}, \tag{4.3.2}$$

where L is the absolute luminosity of the galaxy and V_c is the circular rotation velocity (most massive spirals have rotation curves which are constant with radial distance from the centre). The index $\alpha \sim 3$, but depends on the waveband within which L is measured. The correlation is so tight that the measurement of V_c allows the luminosity to be determined to an accuracy of about 40%. Since the apparent flux can be measured accurately, and this depends on the square of the distance to the galaxy, the resulting distance error is about 20%. This can be reduced further by applying the method to a number of spirals in the same cluster.

The situation is somewhat more complicated for elliptical galaxies because the correlation involves three parameters: the characteristic size of the galaxy R; its surface brightness Σ; and the central velocity dispersion σ. (Recall that elliptical galaxies do not have ordered motions, but random ones characterised by a dispersion rather than a mean value.) These three parameters are correlated in such a way that they occupy the so-called *fundamental plane* defined by a relation of the form

$$\log R = A \log \sigma - B \log \Sigma + C, \tag{4.3.3}$$

where C is a constant. Before the fundamental plane was established there were attempts to find relations of the form (4.3.2), such as the Faber–Jackson relation,

$$L \propto \sigma^{\alpha}, \tag{4.3.4}$$

and the D_n-σ relation

$$D_n \propto \sigma^{1.2}, \tag{4.3.5}$$

where D_n is the radius within which the mean surface brightness of the galaxy image exceeds a certain threshold value. The problem with these two-parameter correlations is that they suppress one variable in the relation (4.3.3). The Faber–Jackson relation does not take account of varying Σ and consequently has a large scatter. On the other hand, the relation (4.3.5) is close to an edge-on view of the fundamental plane and is almost as good as (4.3.2). The value of α needed to fit the objections in this case is $\alpha \sim 4$. The use of these distance measures, together with redshift, to map the local peculiar velocity field is described in Section 4.6 and in Chapter 18.

So there seems to be no shortage of techniques for measuring H_0. Why is it then that observational limits constrain H_0 so poorly, as in Equation (4.2.2)? One problem is that a small error in one 'rung' of the distance ladder also affects higher levels of the ladder in a cumulative way. At each level there are actually many corrections to be made, some of them well known, others not. Some such corrections are as follows.

Galactic rotation: the Sun rotates around the galactic centre at a distance of around 10 kpc and with a velocity around 215 km s^{-1}. This motion can produce spurious systematic shifts towards the red or the violet in observed spectra.

Aperture effects: it is necessary to refer all the measurements regarding galaxies to a standard telescope aperture. At different distances the aperture may include different fractions of the galaxy.

K**-correction:** the redshift distorts the observed spectrum of a source in the sense that the luminosity observed at a certain frequency was actually emitted at a higher frequency. To correct this, one needs to know the true spectrum of the source.

Absorption: our Galaxy absorbs a certain fraction of the light coming to it from an extragalactic source. In fact the intensity of light received at the Earth varies as $\exp(-\lambda \operatorname{cosec} b)$, where λ is a positive constant and b is the angle between the line of sight and the galactic plane, i.e. the galactic latitude.

Malmquist bias: there are various versions of this effect, which is basically due to the fact that the properties of samples of astronomical objects limited by apparent luminosity (i.e. containing all the sources brighter than a certain apparent flux limit) are different from the properties of samples limited in distance because the objects in distant regions will have to be systematically brighter in order to get into the sample.

Scott effect: there is a correlation between the luminosity of the brightest galaxy in a cluster and the richness (i.e. number of galaxies) of the cluster. At large distances one tends to see only the richest clusters, which biases the brightest galaxy statistics.

Baunt–Morgan effect: in fact, clusters are divided into at least five classes in each of which the luminosity of the brightest galaxy is different from the others.

Shear: there is an apparent rotation in the Local Supercluster, as well as of the Local Group and the Virgo cluster.

Galactic evolution: the luminosity of the most luminous cluster galaxies is a function of time and, therefore, of the distance between the galaxy and us. The main reason for this is that the stellar populations of such galaxies are modified as the central cluster galaxy swallows smaller galaxies in its vicinity in a sort of 'cannibalism'.

Given this large number of uncertain corrections, it is perhaps not surprising that we are not yet in a position to determine H_0 with any great precision. We should mention at this point, however, that some methods have recently been proposed to determine the distance scale directly, without the need for a ladder. One of them is the Sunyaev–Zel'dovich effect, which we discuss in Section 17.7. The Hubble Space Telescope (HST) is able to image stars directly in galaxies within the Virgo cluster of galaxies, an ability which bypasses the main sources of uncertainty in the calibration of the traditional distance ladder approaches. This 'key' project is now more-or-less complete, and has produced a value of $h \simeq 0.7$ with an error of about 10%.

4.4 The Age of the Universe

We now turn to the determination of the characteristic timescale for the evolution of the Universe with the ultimate aim of determining t_0, the time elapsed from the Big Bang until now. The quantity we call the Hubble time is defined in Section 2.7, and is simply the reciprocal of the Hubble constant. It is interesting to note - we shall demonstrate this later - that this timescale is in rough order-of-magnitude agreement with the ages of stars and galaxies and of the nuclear timescale obtained from the radioactive decay of long-lived isotopes.

4.4.1 Theory

In a matter-dominated Friedmann model, the age of the Universe is given to a good approximation by

$$t_0 = F(\Omega_0)H_0^{-1} \simeq 0.98 \times 10^{10} F(\Omega_0) h^{-1} \text{ years}, \tag{4.4.1}$$

where, as a reminder, the density parameter Ω_0 is the ratio between the present total density of the Universe ρ_0 and the critical density for closure ρ_{0c},

$$\Omega_0 = \frac{\rho_0}{\rho_{0c}} = \frac{8\pi G \rho_0}{3H_0^2}, \tag{4.4.2}$$

and the function $F(\Omega_0)$ is given by

$$F(\Omega_0) = \frac{\Omega_0}{2}(\Omega_0 - 1)^{-3/2} \cos^{-1}\left(\frac{2}{\Omega_0} - 1\right) - (\Omega_0 - 1)^{-1}, \tag{4.4.3 a}$$

$$F(\Omega_0) = \tfrac{2}{3}, \tag{4.4.3 b}$$

$$F(\Omega_0) = (1 - \Omega_0)^{-1} - \frac{\Omega_0}{2}(1 - \Omega_0)^{-3/2} \cosh^{-1}\left(\frac{2}{\Omega_0} - 1\right), \tag{4.4.3 c}$$

in the cases $\Omega_0 > 1$, $\Omega_0 = 1$ and $\Omega_0 < 1$. These results can be compared with Equations (2.4.10), (2.2.6 *e*) and (2.4.3), respectively. The results (4.4.3 *a*) and (4.4.3 *c*) are well approximated by the relations

$$F(\Omega_0) \simeq \tfrac{1}{2}\pi \Omega_0^{-1/2} \quad \text{for } \Omega_0 \gg 1, \tag{4.4.4 a}$$

$$F(\Omega_0) \simeq 1 + \Omega_0 \ln \Omega_0 \quad \text{for } \Omega_0 \ll 1. \tag{4.4.4 b}$$

Some illustrative values are $F = 1, 0.90, 0.67, 0.5$ and 0 for $\Omega_0 = 0, 0.1, 1, 10$ and ∞, respectively; for values of Ω_0 which are reasonably in accord with observations, as we shall discuss shortly, the age is always of order $1/H_0$.

As we shall see in the next section, the density parameter Ω_0 is also extremely uncertain. A (conservative) interval for Ω_0 is

$$0.01 < \Omega_0 < 2, \tag{4.4.5}$$

from which the Equations (4.4.1) and (4.4.3) give

$$t_{0H} \simeq (6.5\text{–}10) \times 10^9 h^{-1} \text{ years.} \tag{4.4.6}$$

The age of the Universe as deduced from stellar ages (see below) is probably in the range $1.4\text{–}1.6 \times 10^{10}$ years. This result places severe constraints on the Hubble constant through Equation (4.4.1): universes with $\Omega_0 \simeq 1$ are only compatible with these age estimates if $h \simeq 0.5$ or less, a value which is already at the bottom of the allowed range of estimates. This problem is less severe if $\Omega_0 \simeq 0.1$; in this case we need an $h \simeq 0.6\text{–}0.8$. Note, however, that in models with a cosmological constant term Λ, the universe can be accelerating so that $F(\Omega_0, \Lambda) > 1$ in some cases.

4.4.2 Stellar and galactic ages

The age of a stellar population can be deduced from various relationships between their observed properties and the predictions of models of stellar evolution. In this field, one pays great attention to stars belonging to globular clusters because of the good evidence that the stars in a given globular cluster all have the same age and differ only in their masses. Less massive stars evolve very slowly and look very much as they did at the moment of their 'birth' (when hydrogen burning began in their cores). These stars are situated predominantly on the main sequence in the HR diagram. On the other hand, the most massive stars evolve very rapidly and, at a certain point, leave the main sequence and move towards the region of the HR diagram occupied by red giants; the time when they do this is called the 'turnoff' point and it is a function of the mass of the star. The age of the cluster t_c is taken to be the age of those stars that have just left the main sequence for the red-giant branch. Estimates of such ages are prone to an error of about 10% because the red-giant phase of stellar evolution lasts around 10% of the main sequence lifetime. The theory of stellar evolution applied to this problem generally gives a value of around $1.3\text{–}1.4 \times 10^{10}$ years for the age of globular clusters, though much higher ages have appeared in the literature. Given that the time for the formation of galaxies is probably in the range $1\text{–}2 \times 10^9$ years, one should conclude that the age of the Universe is probably around

$$t_0 \simeq 1.4\text{–}1.6 \times 10^{10} \text{ years.} \tag{4.4.7}$$

4.4.3 Nucleocosmochronology

The term 'nucleocosmochronology' is given to attempts to estimate the age of the Universe by means of the relative abundances of long-lived radioactive nuclei and their decay products. Most long-lived radioactive nuclei are synthesised in the so-called r-process reactions involving the rapid absorption of neutrons by heavy nuclei such as iron. Such processes are generally thought to occur in supernovae explosions. Given that the stars that become supernovae are very short lived (of order 10^7 years), nucleocosmochronology is a good way to determine the time at which stars and galaxies were formed. If the origin of our Galaxy was at $t \simeq 0$, at which time there occurred an era of nucleosynthesis of heavy elements lasting for some time T, and this was followed by a time Δ in which the Solar System became isolated from the rest of the galaxy, and after which there was a period t_s corresponding to the age of the Solar System, then the age estimate of the Universe one would produce is $t_n = T + \Delta + t_s$.

The age of the Solar System can be deduced in the following way. The isotope ^{235}U decays into ^{207}Pb with a mean lifetime $\tau_{235} = 10^9$ years; ^{238}U produces ^{206}Pb with $\tau_{238} = 6.3 \times 10^9$ years; the isotope ^{204}Pb does not have radioactive progenitors. Let us indicate the abundances of each of these elements by their atomic symbols and the suffices 'i' and '0' to denote the initial and present time, respectively. We have

$$^{235}\mathrm{U_i} + {}^{207}\mathrm{Pb_i} = {}^{235}\mathrm{U_0} + {}^{207}\mathrm{Pb_0} = {}^{235}\mathrm{U_0}\exp\left(\frac{t_s}{\tau_{235}}\right) + {}^{207}\mathrm{Pb_i}, \tag{4.4.8}$$

$$^{238}\mathrm{U_i} + {}^{206}\mathrm{Pb_i} = {}^{238}\mathrm{U_0} + {}^{206}\mathrm{Pb_0} = {}^{238}\mathrm{U_0}\exp\left(\frac{t_s}{\tau_{238}}\right) + {}^{206}\mathrm{Pb_i}, \tag{4.4.9}$$

from which, dividing by the abundance of $^{204}\mathrm{Pb_0} = {}^{204}\mathrm{Pb_i}$, we obtain

$$R_{207} \equiv \frac{^{207}\mathrm{Pb_0}}{^{204}\mathrm{Pb_0}} = \frac{^{207}\mathrm{Pb_i}}{^{204}\mathrm{Pb_0}} + \frac{^{235}\mathrm{U_0}}{^{204}\mathrm{Pb_0}}\left[\exp\left(\frac{t_s}{\tau_{235}}\right) - 1\right], \tag{4.4.10}$$

$$R_{206} \equiv \frac{^{206}\mathrm{Pb_0}}{^{204}\mathrm{Pb_0}} = \frac{^{206}\mathrm{Pb_i}}{^{204}\mathrm{Pb_0}} + \frac{^{238}\mathrm{U_0}}{^{204}\mathrm{Pb_0}}\left[\exp\left(\frac{t_s}{\tau_{238}}\right) - 1\right]. \tag{4.4.11}$$

Measuring R_{207} and R_{206} in two different places, for example in two meteorites, which we indicate with 'I' and 'II', one can easily get

$$\frac{R_{207,\mathrm{I}} - R_{207,\mathrm{II}}}{R_{206,\mathrm{I}} - R_{206,\mathrm{II}}} = \frac{^{235}\mathrm{U_0}\exp(t_s/\tau_{235}) - 1}{^{238}\mathrm{U_0}\exp(t_s/\tau_{238}) - 1}, \tag{4.4.12}$$

from which one can recover t_s. In this way one finds an age for the Solar System of order 4.6×10^9 years. Analogous results can be obtained with other radioactive nuclei such as ^{87}Rb, which decays into ^{87}Sr with $\tau_{87} = 6.6 \times 10^{10}$ years.

By analogous reasoning to that above, one finds that $T + t_s \simeq (0.6\text{–}1.5) \times 10^{10}$ years and that $\Delta \simeq (1\text{–}2) \times 10^8$ years $\ll T + t_s$, from which the age of the Universe must be

$$t_n \simeq (0.6\text{–}1.5) \times 10^{10} \text{ years}. \tag{4.4.13}$$

It is worth remarking that the time deduced for the isolation of the Solar System Δ is of the same order as the interval between successive passages of a spiral arm through a given location in a galaxy.

In summary, we can see that the theoretical age of the Universe t_0, the ages of globular clusters t_c and the nuclear timescale t_n are all in rough agreement with each other. This does not necessarily mean that the Universe was 'born' at a time t_0 in the past, in the sense that it must have been created with a singularity at $t = 0$. Some ways of avoiding this kind of 'creation' are discussed in Chapter 6.

4.5 The Density of the Universe

Let us now give some approximate estimates of the total energy density of the Universe. We shall see that this is also uncertain by a large factor. More sophisticated methods for measuring the density parameter are discussed in Chapter 18.

4.5.1 Contributions to the density parameter

The evolution of the Universe depends not only on the total density ρ but also on the individual contributions from the various components present (baryonic matter, photons, neutrinos). Let us denote the contribution of ith component to the present density by

$$\Omega_i = \frac{\rho_{0i}}{\rho_{0c}}. \tag{4.5.1}$$

For this section only we drop the zero suffix on Ω that indicates the present value of this parameter. All quantities in this section are at the present time, so it should do no harm to simplify the notation. We shall estimate the contribution Ω_g from the mass concentrated in galaxies a little later. Within a considerable uncertainty we have

$$\Omega_g = \frac{\rho_{0g}}{\rho_{0c}} \simeq 0.03. \tag{4.5.2}$$

There may, of course, be a contribution from matter which is not contained in galaxies, but is present, for example, in clusters of galaxies. The size of this contribution is even more uncertain. We shall see later that a reasonable estimate for the total amount of mass contributing to the gravitational dynamics of large-scale objects is around

$$\Omega_{dyn} \simeq 0.2\text{–}0.4. \tag{4.5.3}$$

The discrepancy between the two values of Ω given by Equations (4.5.2) and (4.5.3) is attributed to the presence of non-luminous matter, called *dark matter*, which may play an important role in structure formation, as we shall see in Section 4.6 and, in much more detail, later on.

As well as matter, the Universe is filled with a thermal radiation background, called the *cosmic microwave background* (CMB) radiation. This was discovered in 1965, and we shall discuss it later in Section 4.9 and Chapter 17. The radiation has a thermal spectrum and a well-defined temperature of $T_{0r} = 2.726 \pm 0.005$ K. The mass density corresponding to this radiation background is

$$\rho_{0r} = \frac{\sigma_r T_{0r}^4}{c^2} \simeq 4.8 \times 10^{-34} \text{ g cm}^{-3} \tag{4.5.4}$$

($\sigma_r = \pi^2 k_B^4/15\hbar^3 c^3$ is the so-called black-body constant; the Stefan–Boltzmann constant is just $\sigma c/4$), so that the corresponding density parameter is

$$\Omega_r \simeq 2.3 \times 10^{-5} h^{-2}. \tag{4.5.5}$$

As we shall see in Section 8.5, there is also expected to be a contribution to Ω from a cosmological neutrino background which, if the neutrinos are massless, yields

$$\rho_{0\nu} \simeq N_\nu \times 10^{-34} \text{ g cm}^{-3}, \tag{4.5.6}$$

where N_ν indicates the number of massless neutrino species ($N_\nu \simeq 3$, according to modern particle physics experiments). The resulting $\rho_{0\nu}$ is comparable with ρ_{0r} expressed by (4.5.4). If the neutrinos have mean mass of order 10 eV, as used to be thought in the 1980s, then

$$\rho_{0\nu} \simeq 1.9 N_\nu \frac{\langle m_\nu \rangle}{10 \text{ eV}} 10^{-30} \text{ g cm}^{-3}, \tag{4.5.7}$$

corresponding to

$$\Omega_\nu \simeq 0.1 N_\nu \frac{\langle m_\nu \rangle}{10 \text{ eV}} h^{-2}, \tag{4.5.8}$$

which is much larger than that implied by Equation (4.5.2); if neutrinos have a mass of this order, then they would dominate the density of the Universe. However, more recent experimental measurements of neutrino oscillations suggest they have a much smaller mass than this, much less than one electronvolt. Such light neutrinos have some effect on cosmic evolution, but they do not dominate.

As far as the contribution to Ω from relativistic particles in general is concerned, there is a good argument, which we shall explain in Section 11.7, why such particles should not dominate the matter component. If this were the case, then fluctuations would not be able to grow in order to generate galaxies and large-scale structure by the present epoch.

Upper and lower limits on the contribution Ω_b from baryonic material can be obtained by comparing the observed abundances of light elements (deuterium, ^{3}He, ^{4}He and ^{7}Li) with the predictions of primordial nucleosynthesis computations. The latest results, described in more detail in Chapter 8, give

$$\Omega_b \sim 0.02 h^{-2}; \tag{4.5.9}$$

if we allow the historical lower limit for the Hubble constant, $h \simeq 0.5$, then the largest allowed upper limit on Ω_b becomes 0.08 and, if $h \simeq 1$, the lower limit is just 0.01. For small h it is therefore clear that Ω_b may be compatible with Ω_g, but not with Ω_{dyn}.

4.5.2 Galaxies

Let us now explain in a little more detail how we arrive at the estimate Ω_g given in Equation (4.5.2). We proceed by calculating the mean luminosity per unit volume produced by galaxies, together with the mean value of M/L, the mass-to-light ratio, of the galaxies. Thus,

$$\rho_{0g} = \mathcal{L}_g \left\langle \frac{M}{L} \right\rangle. \tag{4.5.10}$$

The value $\mathcal{L}_g$ can be obtained from the *luminosity function* of the galaxies, $\Phi(L)$. This function is defined such that the number of galaxies per unit volume with luminosity in the range L to $L + dL$ is given by

$$dN = \Phi(L)\, dL. \tag{4.5.11}$$

Thus,

$$\mathcal{L}_g = \int_0^\infty \Phi(L) L\, dL. \tag{4.5.12}$$

The best fit to the observed properties of galaxies is afforded by the *Schechter function*

$$\Phi(L) = \frac{\Phi_*}{L_*}\left(\frac{L}{L_*}\right)^{-\alpha} \exp\left(-\frac{L}{L_*}\right), \tag{4.5.13}$$

where the parameters are, approximately, $\Phi_* \simeq 10^{-2}h^3\ \mathrm{Mpc}^{-3}$, $L_* \simeq 10^{10}h^{-2}L_\odot$ and $\alpha \simeq 1$. The value of $\mathcal{L}_g$ that results is therefore

$$\mathcal{L}_g \simeq 3.3 \times 10^8 h L_\odot\ \mathrm{Mpc}^{-3}. \tag{4.5.14}$$

To derive the mass-to-light ratio M/L we must somehow measure the value of M. One can calculate the mass of a spiral galaxy if one knows the behaviour of the orbital rotation velocity of stars with distance from the centre of the galaxy, the *rotation curve*. One compares the observed curve with a theoretical model in which the rotation curve is produced by a distribution of gravitating material. There is strong evidence from 21 cm radio and optical observations that the rotation curves of spiral galaxies remains flat well outside the region in which most of the luminous material resides. This demonstrates that spiral galaxies possess large 'haloes' of dark matter, concerning the nature of which there is a huge debate. Some of the possibilities are neutral hydrogen gas, white dwarfs, massive planets, black holes, massive neutrinos and exotic particles, like for instance photinos. The mass of these haloes is thought to be between 3 and 10 times the mass of the luminous component of the galaxy.

Elliptical and S0 galaxies do not have such ordered orbital motions as spiral galaxies, so one cannot use rotation curves. One uses instead the virial theorem:

$$2E_k + U = 0, \tag{4.5.15}$$

where the mean kinetic energy E_K is estimated from the velocity dispersion of the stars and the potential energy U is estimated from the size and shape of the galaxy. The typical value of M/L one obtains is

$$\left\langle \frac{M}{L} \right\rangle \simeq 30h \frac{M_\odot}{L_\odot}, \tag{4.5.16}$$

for which

$$\rho_{0g} \simeq 6 \times 10^{-31} h^2 \text{ g cm}^{-3}, \tag{4.5.17}$$

corresponding to

$$\Omega_g = \frac{\rho_{0g}}{\rho_{0c}} \simeq 0.03. \tag{4.5.18}$$

This should probably be regarded as a lower limit on the contribution due to galaxies because it refers only to the luminous part and does not take account of the full extent of the dark haloes.

4.5.3 Clusters of galaxies

Using the virial theorem we can also estimate the mass of groups and clusters of galaxies. This method is particularly useful for rich clusters of galaxies like the Coma and Virgo clusters. The kinetic energy can be estimated from the velocity dispersion of the galaxies in the cluster

$$E_K \simeq \tfrac{3}{2} M_{cl} \langle v_r^2 \rangle; \tag{4.5.19}$$

M_{cl} is the total mass of the cluster and $\langle v_r^2 \rangle^{1/2}$ is the line-of-sight velocity dispersion of the galaxies. The potential energy is given by

$$U \simeq -\frac{GM_{cl}^2}{R_{cl}}, \tag{4.5.20}$$

where R_{cl} is the radius of the cluster which can be estimated from a model of its density profile. One typically obtains from this type of analysis values of order

$$M_{cl} \simeq 10^{15} h^{-1} M_\odot. \tag{4.5.21}$$

A more sophisticated approach involves more detailed modelling of the velocities within the cluster:

$$M(r) = -\frac{r\sigma^2(r)}{G}\left[\frac{\mathrm{d}\log\rho}{\mathrm{d}\log r} + \frac{\mathrm{d}\log\sigma_r^2}{\mathrm{d}\log r} + 2\beta\right]. \tag{4.5.22}$$

This gives the mass contained within a radius r in terms of the density profile $\rho(r)$ and the two independent velocity dispersions in the radial and tangential directions σ_r^2 and σ_t^2; the quantity

$$\beta = 1 - \frac{\sigma_t^2}{\sigma_r^2} \tag{4.5.23}$$

is a measure of the anisotropy of the radial velocity dispersion. In order to use this equation, one needs to know the profile of galaxies and velocity dispersion as a function of radius from the centre of the cluster. In reality, one can only measure the projected versions of these quantities, so the problem is formally indeterminate. One can, however, use a modelling procedure to perform an inversion of the projected profiles. For the Coma cluster, the result is a total dynamically inferred mass within an Abell radius of

$$M_{\text{tot}} \simeq 6.8 \times 10^{14} h^{-1} M_{\odot}, \tag{4.5.24}$$

which corresponds to a value of $M/L \simeq 320h$. Galaxies themselves therefore contribute only about 15% of the mass of the Coma cluster.

This value can be compared with two alternative determinations of cluster masses. One of these takes account of the fact that rich clusters of galaxies are permeated by a tenuous gaseous atmosphere of X-ray emitting gas. Since the temperature and density profiles of the gas can be obtained with X-ray telescopes such as ROSAT and data on the X-ray spectrum of these objects is also often available, one can break the indeterminacy of the modelling method. The X-ray data also have the advantage that they are not susceptible to Poisson errors coming from the relatively small number of galaxies that exist at a given radius. Assuming the cluster is spherically symmetric and considering only the gaseous component, for simplicity, the equation of hydrostatic equilibrium becomes

$$M(r) = -\frac{k_{\text{B}} T(r) r}{G \mu m_{\text{p}}} \left[\frac{\mathrm{d} \log \rho}{\mathrm{d} \log r} + \frac{\mathrm{d} \log T}{\mathrm{d} \log r} \right]; \tag{4.5.25}$$

μ is the mean molecular weight of the gas. The procedure adopted is generally to use trial functions for $M(r)$ in order to obtain consistency with $T(r)$ and the spectrum data.

Good X-ray data from ROSAT have been used to model the gas distribution in the Coma cluster (Briel *et al.* 1992) with the result that

$$M_{\text{gas}} \simeq 5.5 \times 10^{13} h^{-5/2} M_{\odot} \tag{4.5.26}$$

for the mass inside the Abell radius. The gas contributes more than the galaxies, but is still less than the total mass.

The third method for obtaining cluster masses is to use gravitational lensing. We discuss this later, in Chapter 19. Generally speaking, all three of these methods give cluster masses of the same order of magnitude, although they do not agree in all details.

Given that there are approximately 4×10^3 large clusters of galaxies within a distance of $6 \times 10^2 h^{-1}$ Mpc from the Local Group, the density of matter produced by such clusters is roughly

$$\rho_{0\text{cl}} \simeq 4 \times 10^{-31} h^2 \text{ g cm}^{-3}, \tag{4.5.27}$$

which is of the same order as $\rho_{0\text{g}}$ given by Equation (4.5.17). The reason for this is not that virtually all galaxies reside in such clusters, which they certainly do not, but that the ratio M/L for the matter in clusters is much higher than that for individual galaxies. In fact this ratio is of order $300M_\odot/L_\odot$, roughly a factor of ten greater than that of galaxies. This discrepancy is the origin of the so-called 'hidden mass problem' in galaxy clusters, namely that there seems to be matter there in some unknown form.

If the value of M/L for galaxies were to be reconciled with the galactic value, one would have to have systematically overestimated the virial mass of the cluster. This might happen if the cluster were not gravitationally bound and virialised, but instead were still freely expanding with the background cosmology. In such a case we would have

$$2E_{\mathrm{k}} + U > 0 \tag{4.5.28}$$

and, therefore, a smaller total mass. However, we would expect the cluster to disperse on a characteristic timescale $t_{\mathrm{c}} \simeq l_{\mathrm{c}}/\langle v^2\rangle^{1/2}$, where l_{c} is a representative length scale for the cluster and $\langle v^2\rangle^{1/2}$ is the root-mean-square peculiar velocity of the galaxies in the cluster; for the Coma cluster $t_{\mathrm{c}} \simeq 1/16H_0$ and it is generally the case that t_{c} for clusters is much less than a Hubble time. If the clusters we observe were formed in a continuous fashion during the expansion of the Universe, many such clusters must have already dispersed in this way. The space between clusters should therefore contain galaxies of the type usually found in clusters, i.e. elliptical and lenticular galaxies, and they might be expected to have large peculiar motions. One observes, however, that 'field' galaxies are usually spirals and they do not have particularly large peculiar velocities. It seems reasonable therefore to conclude that clusters must be bound objects.

In light of this, it is necessary to postulate the existence of some component of dark matter (matter with a large value of M/L) to explain the virial masses of galaxy clusters. It is known from X-ray observations of clusters that a large fraction of the mass is in the form of hot gas. In particular, an analysis by White *et al.* (1993b) of the ubiquitous Coma cluster, in conjunction with Equation (4.5.9), indicates that, if the ratio of baryonic matter to total gravitating matter in Coma is representative of the global ratio, then one can constrain Ω to be

$$\Omega \leqslant \frac{0.15h^{-1/2}}{1 + 0.55h^{3/2}}, \tag{4.5.29}$$

which is less than unity for most sensible values of h. It seems, however, that this hot gas component is not sufficient to explain the dynamical mass; another component is needed. This component is probably collisionless and could in principle be in the form of cometary or asteroidal material, large planets (Jupiter-like objects), low-mass stars (brown dwarfs), or even black holes. There are problems, however, in reconciling the value of Ω_{dyn} with nucleosynthesis predictions if all the cluster mass were baryonic. A favoured option is that at least some of this material is in the form of weakly interacting non-baryonic particles (photinos, axions, neutrinos, etc.) left over after the Big Bang. It is even possible, as we shall explain in Section 4.7, that these particles actually constitute the dominant contribution to Ω globally, not just in cluster cores. This is an attractive notion because, as we shall see, a universe with $\Omega \simeq 1$ dominated by non-baryonic matter has some advantages when it comes to explaining the formation of galaxies and large-scale structure. The existence of such a high density of non-baryonic matter would not contradict nucleosynthesis because the weakly interacting matter would not be involved in nuclear reactions in the early Universe. Modern inflationary cosmologies also favour $\Omega_0 \simeq 1$ for theoretical reasons and it is often argued that if the Universe turned out to have $\Omega \simeq 1$, this could be construed as evidence for inflation. There is not much evidence that $\Omega_0 \sim 1$, but we can say that it is (probably) at least $\Omega_0 \simeq 0.2$.

4.6 Deviations from the Hubble Expansion

In the previous section we showed how one can use virial arguments relating velocities to gravitating mass in order to estimate masses from velocity data. The logical extension of this type of argument is to attempt to explain the peculiar motions of galaxies with respect to the Hubble expansion as being due to the cosmological distribution of mass. This idea is of great current interest but the arguments are more technical than we can accommodate in this introductory section; details are given in Chapter 18. We can nevertheless introduce some of the ideas here to whet the reader's appetite.

The (radial) peculiar velocity of a galaxy is defined to be the difference between the galaxy's total measured radial velocity v_r (obtained from the redshift) and the expected Hubble recession velocity for a galaxy at distance d from the observer:

$$v_p = v_r - H_0 d. \tag{4.6.1}$$

Obviously, knowledge of v_p requires both the redshift and an independent measurement of distance to the galaxy. The latter is not easy to acquire, so the construction of catalogues of peculiar motions is not a simple task. Nevertheless, some properties of the local flow pattern of galaxies are known. The motion of our Local Group of galaxies towards the Virgo cluster has been known for some time to be $v \simeq 250 \pm 50$ km s^{-1} and, as we shall see in Section 4.8, it is possible to estimate our velocity with respect to the reference frame in which the cosmic microwave background is at rest: $v \simeq 550 \pm 40$ km s^{-1} in a direction $\alpha = 10.7 \pm 0.3$ h and $\delta = -22 \pm 5°$, 44° away from the Virgo cluster. For reasons we shall explain later, one expects the resultant velocity of the Local Group to lie in the same direction as the net gravitational acceleration on it produced by the distribution of matter around it. Clearly then, our velocity with respect to the microwave background is not explained by the action of the Virgo cluster. In fact, studies of galaxy-peculiar motions show that the peculiar flow of galaxies is actually coherent over a large scale. A region of radius $50h^{-1}$ Mpc centred on the Local Group seems to be moving *en masse* in a direction corresponding to the Hydra and Centaurus clusters with a velocity of $v \simeq 600$ km s^{-1}. It was thought that this bulk flow was due to the action of a huge concentration of mass at a distance of order $50h^{-1}$ Mpc from the Local Group, called the *Great Attractor*, but it is now generally accepted that the pull is not due to a single mass but to the concerted effort of a large number of clusters.

So how can the observed peculiar motions tell us about the distribution of mass and, in particular, the total density? The arguments rely on the theory of gravitational instability which we shall explain later, but a qualitative example can be given here based on the motion of the Local Group with respect to the Virgo cluster. One takes this motion to be the result of 'infall', which can be modelled by a simple linear model in which a 'shell' of galaxies containing the Local Group falls symmetrically onto the Virgo cluster, which is assumed to be spherical. If the density of galaxies in the Virgo cluster is a factor $(1 + \Delta_g)$ higher than the cosmological average, the infall velocity is v_{LG}, and the Virgocentric distance of the Local Group is r_{LG}, then one can estimate

$$\Omega_{dyn} \simeq \Delta_g^{-1.7} \left(\frac{3v_{LG}}{H_0 r_{LG}} \right)^{1.7}. \tag{4.6.2}$$

This type of argument leads one to a value of Ω_{dyn} which is consistent with that obtained from virial arguments in clusters, i.e. $\Omega_{dyn} \simeq 0.2$–0.4. More recent analyses using data covering much larger scales give results apparently consistent with $\Omega_{dyn} = 1$ though with a great uncertainty.

One of the problems with analyses of this type is that one has to estimate the density fluctuation Δ_g producing the peculiar motion. In the example this is estimated as the excess density of galaxies inside the cluster compared with the 'field'. Given that much of the mass one detects is dark, there is no reason *a priori* why the fluctuation in mass density Δ_m has to be the same as the fluctuation in number density of galaxies Δ_g. If these differ by a factor b, then, according to Equation (4.6.2), one's estimate of Ω_{dyn} is wrong by a factor $\simeq b^{1.7}$. The idea that galaxies might not trace the mass is usually called *biased galaxy formation* and it considerably complicates the analysis of galaxy clustering and peculiar motion studies; we discuss bias in detail in Section 14.8. Note that a value of $b \simeq 2$ can reconcile the Virgocentric flow with $\Omega = 1$.

A more accurate determination of the anisotropy of the Hubble expansion on large scales allows the construction of a map of the peculiar velocity field, which, as we shall see in Chapter 18, is an important goal of modern observational cosmology. It is hoped that such a map will allow an accurate determination of the distribution of matter in the Universe, even if galaxies are biased tracers of the mass. The reason for this optimism is that all matter components exert gravity and react to it, not just the component of luminous matter which appears in galaxies. Regardless of how a galaxy forms and what it is made of, its motion is due to the action of all the gravitating mass around it. Modern theoretical developments, as well as new observational techniques for measuring distances to galaxies, give good grounds for believing that this is a reasonable task.

We should also take this opportunity to make some more formal comments about the nature of deviations from the Hubble flow in the context of the Cosmological Principle. Deviations of the type (4.6.1) can be regarded as being due to an anisotropic expansion such that the velocity of a distant galaxy is

$$v_\alpha = H_\alpha{}^\beta d_\beta \tag{4.6.3}$$

with respect to a coordinate origin at our Galaxy. We discussed this in the context of globally anisotropic models in Chapter 3. The tensor $H_{\alpha\beta}$ is called the Hubble tensor and can be written in the form

$$H_{\alpha\beta} = H\delta_{\alpha\beta} + \omega_{\alpha\beta} + \sigma_{\alpha\beta}, \tag{4.6.4}$$

where $\delta_{\alpha\beta}$ is the Kronecker symbol, $\omega_{\alpha\beta}$ is an antisymmetric tensor which represents a rotation ($\omega_{\alpha\beta} = -\omega_{\beta\alpha}$), and $\sigma_{\alpha\beta}$ is a symmetric traceless tensor which represents shear ($\sigma_{\alpha\beta} = \sigma_{\beta\alpha}$; $\sigma_{\alpha\alpha} = 0$). The constant H is the familiar Hubble constant.

The only observable quantity is the line-of-sight velocity v_r

$$v_\mathrm{r} = \frac{d_\alpha v^\alpha}{d} = Hd + \sigma_{\alpha\beta} n^\alpha n^\beta d, \tag{4.6.5}$$

where the n_α are the direction cosines of a distant galaxy at $\boldsymbol{d}$. It is found that the contribution to the shear $\sigma_{\alpha\beta}$ from massive distant clusters is of the order of 10%. In fact, by considering a large-redshift sample of distant clusters, one can find a coordinate system in which $\sigma_{\alpha\beta}$ is diagonal; in this system one finds that

$$|\sigma_{\alpha\alpha}| < 0.1H. \tag{4.6.6}$$

This provides some evidence for the Cosmological Principle.

4.7 Classical Cosmology

In the early days of observational cosmology, much emphasis was placed on the geometrical properties of expanding-universe models as tools for estimating parameters of the cosmological models. Indeed, famous articles by Sandage (1968, 1970) called 'Cosmology: the search for two numbers' reduced all cosmology to the task of determining H_0 and q_0, the deceleration parameter. Remember that, at a generic time t the deceleration parameter is defined by

$$q = -\frac{\ddot{a}a}{\dot{a}^2}; \tag{4.7.1}$$

as usual, the zero suffix means that q_0 is defined at the present time. Matter-dominated models with vanishing Λ have

$$q_0 = \tfrac{1}{2}\Omega_0, \tag{4.7.2}$$

so the parameters q_0 and Ω_0 are essentially equivalent. If there is a cosmological constant contributing towards the spatial curvature, however, we have the general relation

$$q_0 = \tfrac{1}{2}\Omega_0 - \Omega_\Lambda. \tag{4.7.3}$$

In the case where $\Omega_\Lambda + \Omega_0 = 1$ ($\kappa = 0$) we have $q_0 < 0$ for $\Omega_0 < \frac{2}{3}$.

The parameters H_0 and q_0 thus furnish a general description of the expansion of a cosmological model: these are Sandage's famous 'two numbers'. Their importance is demonstrated in standard cosmology textbooks (Weinberg 1972; Peebles 1993; Narlikar 1993; Peacock 1999), which show how the various observational relationships, such as the angular diameter-redshift and apparent magnitude-redshift relations for standard sources, can be expressed in simple forms using these parameters and the Robertson-Walker metric. In the standard Friedmann-Robertson-Walker models, the apparent flux density and angular size of a standard light source or standard rod depend in a relatively simple way on q_0 (Hoyle 1959; Sandage 1961, 1968, 1970, 1988; Weinberg 1972), but the relationships are more complex if the cosmological constant term is included (e.g. Charlton and Turner 1987).

During the 1960s and early 1970s, a tremendous effort was made to determine the deceleration parameter q_0 from the magnitude-redshift diagram. For a while, the preferred value was $q_0 \simeq 1$ (Sandage 1968) but eventually the effort died away when it was realised that evolutionary effects dominated the observations; no adequate theory of galaxy evolution is available that could enable one to determine the true value of q_0 from the observations. To a large extent this is the state of play now, although the use of the angular size-redshift and, in particular, the magnitude-redshift relation for Type Ia supernovae have seen something of a renaissance of this method. We shall therefore discuss only the recent developments in the subsequent sections.

4.7.1 Standard candles

The fundamental property required here is the *luminosity distance* of a source, which, for models with $p = \Lambda = 0$, is given by

$$d_{\mathrm{L}}(z) = \frac{c}{H_0 q_0^2}[q_0 z + (q_0 - 1)(\sqrt{2q_0 z + 1} - 1)]; \tag{4.7.4}$$

this relationship is simply defined in terms of the intrinsic luminosity of the source L and the flux l received by an observer using the Euclidean relation

$$d_{\mathrm{L}} = \left(\frac{L}{4\pi l}\right)^{1/2}. \tag{4.7.5}$$

One usually seeks to exploit this dependence by plotting the so-called 'Hubble diagram' of apparent magnitude against redshift for objects of known intrinsic luminosity: this boils down to plotting $\log l$ against z, hence the dependence on d_{L}.

The problem with exploiting such relations to prove the value of q_0 directly is that one needs to have a standard 'candle': an object of known intrinsic luminosity. The dearth of classes of object suitable for this task is, of course, one of the reasons why the Hubble constant is so poorly known locally. If it were not for recent developments based on one particular type of object – Type Ia supernovae – we would have been inclined to have omitted this section entirely. As it is now,

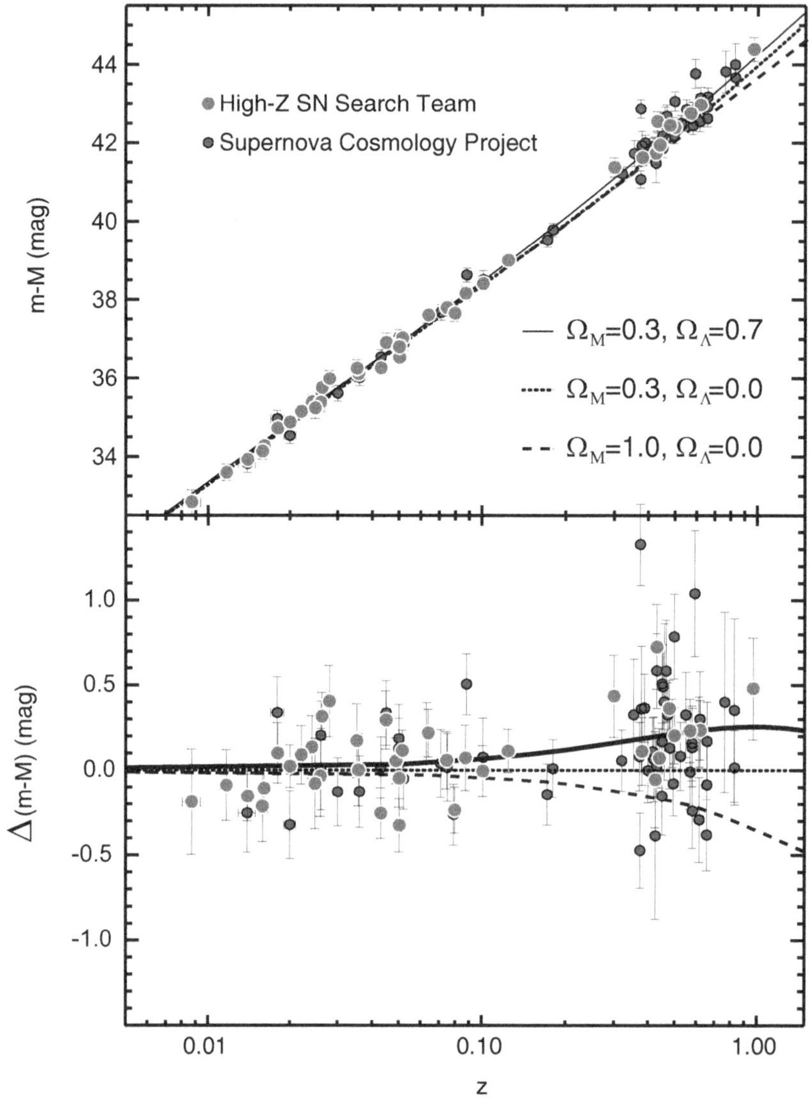

Figure 4.8 The magnitude-redshift diagram for high-redshift supernovae measured by two independent groups. The data show a preference for models with a contribution from Λ. Picture courtesy of Bob Kirschner.

we consider that these sources offer the most exciting prospects for classical cosmology within the next few years.

The homogeneity and extremely high luminosity of the peak magnitudes of Type Ia supernovae, along with physical arguments as to why they should be standard sources, have made these attractive objects for observational cosmologists in recent years (e.g. Branch and Tammann 1992), though the use of supernovae has been discussed before, for example, by Sandage (1961). The current progress stems from the realisation that these objects are not in fact identical, but form a family which can nevertheless be mapped onto a standard object by using independent observations. Correlations between peak magnitude and the shape of the light curve (Hamuy *et al.* 1995; Riess *et al.* 1995) or spectral features (Nugent *et al.* 1995) have reduced the systematic variations in peak brightness to about two-tenths of a magnitude. The great advantages of these objects are

1. because their behaviour depends only on the local physics, they are expected to be independent of environment and evolution and so are good candidates for standard candles, and
2. that they are bright enough to be seen at quite high redshifts, where the dependence on cosmological parameters (4.7.4) is appreciable.

Two teams are pursuing the goal of measuring cosmological parameters using Type Ia supernovae. Originally, results seemed to suggest a measurement of positive q_0, but more recently it has become apparent that the high-redshift supernovae may be fainter, i.e. be at larger luminosity distance, for a given z than is compatible with $q_0 > 0$. If these measurements are being interpreted correctly, and there is as yet no reason to believe they are not, this is compelling evidence for a cosmological constant.

4.7.2 Angular sizes

The angle subtended by a standard metric 'rod' behaves in an interesting fashion as its distance from the observer is increased in standard cosmologies. It first decreases, as expected, then reaches a minimum after which it increases again (Sandage 1961). The position of the minimum depends upon q_0 (Ellis and Tivon 1985; Janis 1986). This somewhat paradoxical behaviour can be more easily understood by remembering that the light from very-high-redshift objects was emitted a long time ago when the proper distance to the object would have been much smaller than it is at the present epoch. Given appropriate dynamics, therefore, it is quite possible that distant objects appear larger than nearby ones with the same physical size.

For models with $\Lambda = 0$ the relationship between angular diameter θ and redshift z for objects moving with the Hubble expansion and with a fixed metric diameter d is simply

$$\theta = d\frac{(1+z)^2}{d_{\mathrm{L}}(z)}, \tag{4.7.6}$$

where $D_{\mathrm{L}}(z)$ is the *luminosity distance* given by Equation (4.7.4).

As with the standard candles, astronomers are generally not equipped with standard sources they are able to place at arbitrarily large distances. To try to use this method, one must select galaxies or other sources and hope that the intrinsic properties of the objects selected do not change with their distance from the observer. Because light travels with a finite speed, more distant objects emitted their light further in the past than nearby objects. Lacking an explicit theory of

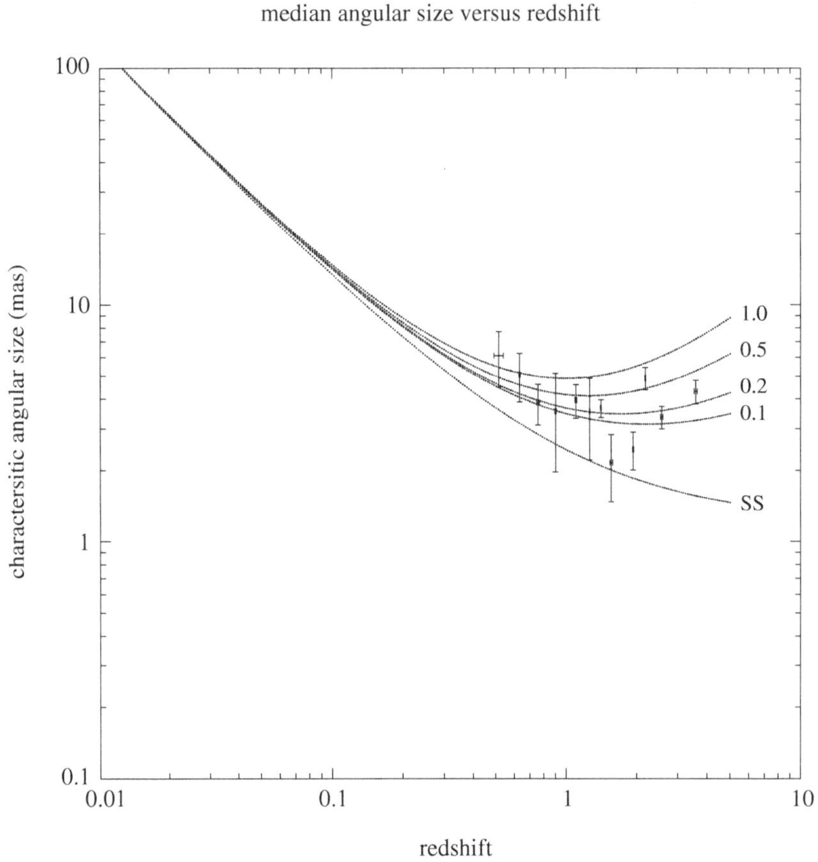

Figure 4.9 Angular diameter versus redshift for 145 radio sources. From Gurvits *et al.* (1999). Picture courtesy of Leonid Gurvits.

source evolution, one must assume the source properties do not vary with cosmological time. Since there is overwhelming evidence for strong evolution with time in almost all classes of astronomical object, the prospects for using this method are highly limited.

An example is the attempt by Kellermann (1993) to resurrect this technique by applying it to compact radio sources. These sources are much smaller than the extended radio sources discussed in previous studies, so one might therefore expect them to be less influenced by, for example, the evolution of the cosmological density. Kellermann originally found a minimum in the angular-size versus distance relationship, but a subsequent analysis by Gurvits *et al.* (1999) found a larger scatter in the data. We must therefore conclude that the evidence from the angular size data is not particularly compelling. Indeed, it is not at all obvious that there are any 'standard metre sticks' in sight that will be visible at high redshift and also will have well-understood evolutionary properties that could lead to a change in this situation. It is wise not to be too optimistic about this method yielding decisive results, although it is possible that angular size estimates of clusters of sources, or measurements of angular separation of similar objects, could eventually give the statistical data needed for this test.

4.7.3 Number-counts

An alternative approach is not to look at the properties of objects themselves but to try to account for the cumulative number of objects one sees in samples that probe larger and larger distances. A first application of this idea was by Hubble (1929); see also Sandage (1961). By making models for the evolution of the galaxy luminosity function one can predict how many sources one should see above an apparent magnitude limit and as a function of redshift. If one accounts for evolution of the intrinsic properties of the sources correctly, then any residual dependence on redshift is due to the volume of space encompassed by a given interval in redshift; this depends quite strongly on Ω_0. The considerable evolution seen in optical galaxies, even at moderately low redshifts, as well as the large K-corrections and uncertainties in the present-day luminosity function, renders this type of analysis prone to all kinds of systematic uncertainties. One of the major problems here is that one does not have complete information about the redshift distribution of galaxies appearing in the counts. Without that information, one does not really know whether one is seeing intrinsically fainter galaxies relatively nearby, or relatively bright galaxies further away. This uncertainty makes any conclusions dependent upon the model of evolution assumed.

Controversies are rife in the history of this field. A famous application of this approach by Loh and Spillar (1986) yielded a value $\Omega_0 = 1^{+0.7}_{-0.5}$. This is, of course, consistent with unity but cannot be taken as compelling evidence. A slightly later analysis of these data by Cowie (1988) showed how, with slightly different assumptions, one can reconcile the data with a much smaller value of Ω_0. Further criticisms of the Loh–Spillar analysis have been lodged by other authors (Bahcall and Tremaine 1988; Caditz and Petrosian 1989). Such is the level and apparent complexity of the evolution in the stellar populations of galaxies over the relevant timescale that we feel that it will be a long time before we understand what is going on well enough to even try to disentangle the cosmological and evolutionary aspects of these data. There has been significant progress, however, with number-counts of faint galaxies, beginning in the late 1980s (Tyson and Seitzer 1988; Tyson 1988) and culminating with the famous 'deep field' image taken with the Hubble Space Telescope, which is shown in Figure 4.10. The 'state-of-the-art' analysis of number-counts (Metcalfe *et al.* 2001) is shown in Figure 4.11, which displays the very faint number-counts from the HST in two wavelength bands, together with ground-based observations from other surveys. The implications of these results for cosmological models are unlikely to be resolved unless and until there are major advances in the theory of galactic evolution.

Figure 4.10 Part of the HST deep field image, showing images of galaxies down to limiting visual magnitude of about 28.5 in blue light. By extrapolating the local luminosity function of galaxies, one concludes that a large proportion of the galaxies at the faint limit have $z > 2$. Picture courtesy of the Space Telescope Science Institute.

4.7.4 Summary

The problem with most of these tests is that, if the Big Bang is correct, objects at high redshift are younger than those nearby. One should therefore expect to see evolutionary changes in the properties of galaxies, and any attempt to define a standard 'rod' or 'candle' to probe the geometry will be very prone to such evolution. Indeed, as we shall see, many of these tests require considerable evolution in order to reconcile the observed behaviour with that expected in the standard models. It is worth mentioning these problems at this point in order to introduce the idea of evolution in galaxy properties, which we shall return to in Section 19.4.

Direct observations of gravitational lensing may prove to be a more robust diagnostic of spatial curvature and hence of the cosmological model. The statistics of the frequency of occurrence of multiply lensed quasars can, in principle, be used to measure q_0. This method is in its infancy at the moment, however, and no strong constraint on the spatial geometry has yet emerged; see Chapter 20 for more details of this.

4.8 The Cosmic Microwave Background

The discovery of the microwave background by Penzias and Wilson in 1965, for which they later won the Nobel Prize, provided one of the most impor-

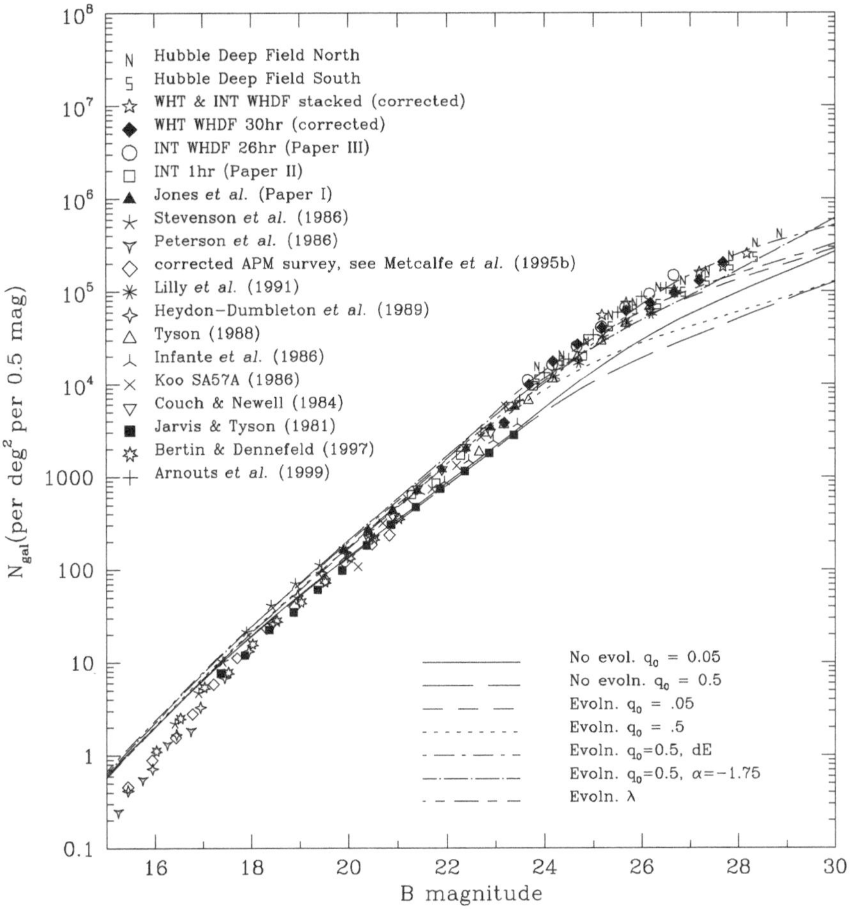

Figure 4.11 Compilation of number-count data in the B (blue) band, from Metcalfe *et al.* (2001). Picture courtesy of Tom Shanks.

tant pieces of evidence for the hot Big Bang model. In fact this discovery was entirely serendipitous. Penzias and Wilson were radio engineers investigating the properties of atmospheric noise in connection with the Telstar communication satellite project. They found an apparently uniform background 'hiss' at microwave frequencies which could not be explained by instrumental noise or by any known radio sources. After careful investigations they admitted the possible explanation that they had discovered a thermal radiation background such as that expected to be left as a relic of the primordial fireball phase. In fact, the existence of a radiation background of roughly the same properties as that observed was predicted by George Gamow in the mid-1940s, but this prediction was not known to Penzias and Wilson. A group of theorists at Princeton University, including Dicke and Peebles, soon saw the possible interpretation of the background 'hiss' as relic radiation, and their paper (Dicke *et al.* 1965) was published alongside the Penzias and Wilson (1965) paper in the *Astrophysical Journal.*

The cosmic microwave background is a source of enormous observational and theoretical interest at the present time, so we have devoted the whole of Chapter 17 to it. For the present we shall merely mention two important properties.

First, the CMB radiation possesses a near-perfect *black-body spectrum*. The theoretical ramifications of this result are discussed in Chapter 9 and Section 19.3; the latest spectral data are also shown later, in Figure 9.1. At the time of its discovery the CMB was known to have an approximately thermal spectrum, but other explanations were possible. Advocates of the steady state proposed that one was merely observing starlight reprocessed by dust and models were constructed which accounted for the observations reasonably well. In the past 30 years, however, continually more sophisticated experimental techniques have been directed at the measurement of the CMB spectrum, exploiting ground-based antennae, rockets, balloons and, most recently and effectively, the COBE satellite. The COBE satellite had an enormous advantage over previous experiments: it was able to avoid atmospheric absorption, which plays havoc with ground-based experiments at microwave and submillimetric frequencies. The spectrum supplied by COBE reveals just how close to an ideal black body the radiation background is; the temperature of the CMB is now known to be 2.725 ± 0.002 K. Attempts to account for this in a steady-state model by non-thermal processes are entirely contrived. The CMB radiation really is good evidence that the Big Bang model is correct.

The second important property of the CMB radiation is its *isotropy* or, rather, its small anisotropy. The temperature anisotropy is usually expressed in terms of the quantity

$$\frac{\Delta T}{T}(\theta, \phi) = \frac{T(\theta, \phi) - T_0}{T_0}, \tag{4.8.1}$$

which gives the temperature fluctuation as a fraction of the mean temperature T_0 as a function of angular position on the sky. Penzias and Wilson (1965) were only able to give rough constraints on the departure of the sky temperature of the CMB from isotropy. Theorists soon realised, however, that if the CMB actually did originate in the early stages of a Big Bang, it should bear the imprint of various physical processes both during and after its production. However, attempts to detect variations in the temperature of the CMB on the sky have, until recently (with the exception of the dipole anisotropy; see below), been unsuccessful. The observed level of isotropy of the cosmic microwave background radiation is important because:

1. it provides strong evidence for the large-scale isotropy of the Universe;
2. it excludes any model in which the radiation has a galactic origin or is produced by a random distribution of sources, also on the grounds of its near-perfect black-body spectrum; and
3. it can provide important information on the origin, nature and evolution of density fluctuations which are thought to give rise to galaxies and large-scale structures in the Universe.

Let us mention some of the possible sources of anisotropy here, though we shall return to the CMB in much more detail in Chapter 17. First, there is known to be a *dipole anisotropy* (a variation on a scale of 180°)

$$T(\vartheta) = T_0\left(1 + \frac{\Delta T_D}{T_0}\cos\vartheta\right), \tag{4.8.2}$$

which is due to the motion of the observer through a reference frame in which the CMB is 'at rest', meaning the frame in which the CMB appears isotropic; notice that there is no dependence upon ϕ in this expression. The amplitude and direction of the dipole anisotropy have been known for some time: the amplitude is around $\Delta T_D/T_0 \simeq 10^{-3} \simeq v/c$, where v is the velocity of the observer. After subtracting the Earth's motion around the Sun, and the Sun's motion around the galactic centre, this observation can be used to determine the velocity of our Galaxy with respect to this 'cosmic reference frame'. The result is a rather large velocity of $v \simeq 600\ \mathrm{km\ s^{-1}}$ in the direction of the constellations of Hydra-Centaurus ($l = 268°$, $b = 27°$). This velocity can be used in an ingenious determination of Ω_0, as we describe later in Chapter 18.

On smaller scales, from the quadrupole (90°) down to a few arcseconds, there are various possible sources of anisotropy as follows.

1. If there are inhomogeneities in the distribution of matter on the surface of last scattering, described in Section 9.5, these can produce anisotropies by the redshift or blueshift of photons from regions of different gravitational potential, the *Sachs-Wolfe effect* (Sachs and Wolfe (1967)).
2. If material on the last scattering surface is moving, then it will induce temperature fluctuations by the *Doppler effect* (material moving towards the observer will be blueshifted, that moving away will be redshifted).
3. The coupling between matter and radiation at last scattering may mean that dense regions are actually intrinsically hotter than underdense regions.
4. An inhomogeneous distribution of material between the observer and the last scattering surface may induce anisotropy by inverse Compton scattering of CMB photons by free electrons in a hot intergalactic plasma (the *Sunyaev-Zel'dovich effect* (Sunyaev and Zel'dovich 1969); see Section 17.7 for the possible use of this effect in determining H_0).
5. Photons travelling through a time-varying gravitational potential field also suffer an effect similar to (i) (usually called the *Rees-Sciama effect* (Rees and Sciama 1968), but actually it is simply a version of the Sachs-Wolfe phenomenon).

As we shall see in Chapter 17, the COBE satellite has recently detected anisotropy on the scale of a few degrees up to the quadrupole. This detection, with an amplitude of $\Delta T/T \simeq 10^{-5}$, has been independently confirmed by an experiment on Tenerife. The characteristics of this signal are consistent with it being due to the Sachs-Wolfe effect (i). If the primordial fluctuations giving rise to this effect are indeed the seeds of galaxies and clusters, then this observation has profound implications for theories of galaxy and cluster formation. Attempts are currently being made to measure the anisotropy on smaller scales than this.

The balloon-borne experiments MAXIMA and Boomerang have mapped the small-scale structure of the cosmic microwave background over small patches of the sky. Soon, the US satellite MAP (Microwave Anisotropy Probe) will map the whole sky and around 2007 a European mission called the Planck Surveyor will do likewise with even higher resolution. As we shall see in Chapter 17, angular scales of a degree or less are a sensitive diagnostic of the form of fluctuations present in the early Universe as well as the geometry of the background Universe.

Bibliographic Notes on Chapter 4

More detailed discussions of galaxy properties can be found in Binney and Tremaine (1987) and Binney and Merrifield (1998). For historical interest, Zwicky (1952) is also worth consulting, as is Faber and Gallagher (1979).

Historically important papers on the development of cosmography are Abell (1958); Bahcall (1988); Rood (1988); Shane and Wirtanen (1967); Shapley and Ames (1932) and Zwicky *et al.* (1961–1968). The classic reference on the expansion of the Universe is Hubble (1929), but readers should be aware that much of the data upon which Hubble based his arguments were obtained by Slipher (1914). Rowan-Robinson (1985) gives a detailed overview of the distance ladder; a more recent paper is that by Fukugita *et al.* (1992). Interesting sources on the density parameter are Peebles (1986), Trimble (1987) and Sciama (1993). Arguments in favour of a Universe with $\Omega_0 < 1$ can be found in Coles and Ellis (1994, 1997).

Problems

1. Show that the Hubble profile of surface brightness (4.1.5) leads to an infinite total luminosity, while the law
$$I = I_0 \exp[-(r/a)^{1/4}],$$
with a a constant, does not. In the second case, estimate (in units of a) the value of r that encloses half the total light and compare your answer for an exponential disc (4.1.6).

2. The half-life of Uranium-235 is 0.7×10^9 years, while that of Uranium-238 is 4.5×10^9 years. A rock has an observed abundance ratio
$$\left[\frac{^{235}\mathrm{U}}{^{238}\mathrm{U}}\right] = 0.00723,$$
while these isotopes are thought to be produced in supernovae explosions with a relative abundance of 1.71. Assuming all the material in the rock was produced in a single supernova event, estimate the time that has elapsed since this event took place.

3. Calculate the rotation curve, $v(R)$, for test particles in circular orbits of radius R: (a) around a point mass M; (b) inside a rotating spherical cloud with uniform density; and (c) inside a spherical halo with density $\rho(r) \propto 1/r^2$.

4. The Tully–Fisher relation (4.3.2) usually has an index $\alpha \simeq 3$. Show that, in a simple model of a galaxy in which stars undergo circular motions in a disc of constant thickness and in which the mass-to-light ratio is constant, a value $\alpha = 4$ would be expected.

5. Assuming all elliptical galaxies have the same central surface brightness and that they are in virial equilibrium, derive the Faber–Jackson relation (4.3.4).

6. Assume that the mass-to-light ratio, M/L, for the Galaxy is, and always has been, 10 in solar units. What is the maximum fraction of the total mass that could have been burnt into helium from hydrogen over 10^{10} years? (The mass deficit for the reaction $4\mathrm{H} \rightarrow {}^4\mathrm{He}$ is 0.7%.)

7. If the luminosity function of galaxies is given by the Schechter function (4.5.13), show that when $\alpha = 1.5$ the total luminosity of all galaxies is approximately $1.77\Phi_* L_*$. The volume through which a galaxy of luminosity L can be seen above a fixed magnitude limit is proportional to $L^{3/2}$. Hence show that in a magnitude-limited survey of galaxies with a luminosity function of this form, about half will have luminosity exceeding about $0.7L_*$ but less than about 5% will have luminosity greater than $3L_*$.

8. Prove the virial theorem (4.5.15) for a system of self-gravitating masses in statistical equilibrium.

PART 2

The Hot Big Bang Model

5

Thermal History of the Hot Big Bang Model

5.1 The Standard Hot Big Bang

The *hot Big Bang* is the name usually given to the standard cosmological model: a homogeneous, isotropic universe whose evolution is governed by the Friedmann equations obtained from general relativity (with or without a cosmological constant), whose main constituents can be described by matter and radiation fluids, and whose kinematic properties (i.e. the Hubble constant) match those we observe in the real Universe. It is further assumed that the radiation component of the energy density is of cosmological origin: this is why the term 'hot' is given to the model. Of course, our real Universe is not exactly homogeneous and isotropic, so this model is to some extent an abstraction. However, as we shall see later, this standard model does provide us with a framework within which we can study the emergence of structures like the observed galaxies and clusters of galaxies from small fluctuations in the density of the early Universe. In this chapter, we give a brief overview of the evolution the basic physical properties of this model; more detailed treatment will be deferred to Chapters 8 and 9.

As we have already seen in Chapter 4, the present-day matter density is

$$\rho_{0\mathrm{m}} = \rho_{0\mathrm{c}}\Omega_{0\mathrm{m}} \simeq 1.9 \times 10^{-29}\Omega_{0\mathrm{m}}h^2 \text{ g cm}^{-3}. \tag{5.1.1}$$

In the following, as in Chapter 4, we shall drop one of the subscripts and use Ω_0 to quantify the density of non-relativistic matter. Observations tell us that

Ω_0 is somewhere in the range $0.01 < \Omega_0 < 2$. The luminous material in galaxies and clusters is primarily hydrogen and a small part of helium. Cosmological nucleosynthesis provides an explanation for the relative abundances of these, and other, light elements: see Chapter 8. As we have seen, however, the Universe is probably dominated by unseen *dark matter*, whose nature is yet to be clarified.

The energy-density contributed by the radiation background at 2.73 K is

$$\rho_{0\mathrm{r}} = \frac{\sigma_\mathrm{r} T_{0\mathrm{r}}^4}{c^2} \simeq 4.8 \times 10^{-34}\ \mathrm{g\ cm^{-3}}, \tag{5.1.2}$$

where σ_r is the radiation density constant. We discussed this before, in Chapter 4. The standard model also predicts the existence of a cosmological background of neutrinos, which we discuss more fully in Chapter 8, with an energy density

$$\rho_{0\nu} \simeq N_\nu \times 10^{-34}\ \mathrm{g\ cm^{-3}}; \tag{5.1.3}$$

N_ν is the number of neutrino species, which is now known from particle physics experiments at LEP/CERN to be very close to $N_\nu = 3$. Equation (5.1.3) applies if the neutrinos are massless, which we shall assume to be the case in this chapter; the idea that they might have a mass of order $\langle m_\nu \rangle \simeq 10$ eV would have important implications for cosmology, as we shall discuss in Chapters 8 and 13. If the neutrinos are massless, then their contribution to the density parameter is $\Omega_{0\nu} \simeq \Omega_{0\mathrm{r}} \simeq 10^{-5} h^{-2}$.

From the point of view of the Friedmann models, the real Universe is well approximated as a *dust* or *matter-dominated* model, with total energy density

$$\rho_0 = \rho_{0\mathrm{m}} + \rho_{0\mathrm{r}} + \rho_{0\nu} \simeq \rho_{0\mathrm{m}}, \tag{5.1.4}$$

and pressure

$$p_0 = p_{0\mathrm{m}} + p_{0\mathrm{r}} + p_{0\nu} \simeq \rho_{0\mathrm{m}} \frac{k_\mathrm{B} T_{0\mathrm{m}}}{m_\mathrm{p}} + \tfrac{1}{3}\rho_{0\mathrm{r}} c^2 \simeq \rho_{0\mathrm{r}} c^2 \ll \rho_0 c^2, \tag{5.1.5}$$

where $T_{0\mathrm{m}}$ is the present temperature of the intergalactic gas (assumed to be hydrogen) and m_p is the proton mass. This temperature is different from the temperature of the radiative component, $T_{0\mathrm{r}}$, because matter and radiation are completely decoupled from each other at the present epoch. In fact the neutrino component is also decoupled from the other two (matter and photons). Matter and radiation are decoupled because the characteristic timescale for collisions between photons and neutral hydrogen atoms, $\tau_{0\mathrm{c}} = m_\mathrm{p}/(\rho_{0\mathrm{m}} \sigma_\mathrm{H} c)$, where σ_H is the scattering cross-section of a hydrogen atom, is much larger than the characteristic time for the expansion of the Universe: $\tau_\mathrm{H} \equiv (a/\dot{a})_0 = H_0^{-1}$.

An important quantity is the ratio, η_0, between the present mean number-density of nucleons (or baryons), $n_{0\mathrm{b}}$, and the corresponding quantity for photons, $n_{0\gamma}$. The present density in baryons is

$$n_{0\mathrm{b}} = \frac{\rho_{0\mathrm{m}}}{m_\mathrm{p}} \simeq 1.12 \times 10^{-5} \Omega_{0\mathrm{b}} h^2\ \mathrm{cm^{-3}}, \tag{5.1.6}$$

while the corresponding number for the photons is obtained by integrating over a Planck spectrum at a temperature of $T_{0r} = 2.73$ K:

$$n_{0\gamma} = \left(\frac{k_B T_{0r}}{\hbar c}\right)^3 \int_0^\infty \frac{8\pi x^2\,dx}{e^x - 1} = 2\frac{\zeta(3)}{\pi^2}\left(\frac{k_B T_{0r}}{\hbar c}\right)^3 \simeq 420\ \text{cm}^{-3}; \tag{5.1.7}$$

the quantity $\zeta(3) \simeq 1.202$, where ζ is the Riemann zeta function which crops up in the integral over the black-body spectrum. We therefore have

$$\eta_0^{-1} = \frac{n_{0\gamma}}{n_{0b}} \simeq 3.75 \times 10^7 (\Omega_{0b} h^2)^{-1}; \tag{5.1.8}$$

we prefer to give the value η_0^{-1} rather than η_0 because, as we shall see, η_0^{-1} practically coincides with the *entropy per baryon*, σ_{0r}, which will figure prominently later on. The fact that η_0^{-1} is so large is of particular importance in the analysis of the standard model; we shall return to it later.

5.2 Recombination and Decoupling

During the period in which matter and radiation are decoupled, the matter temperature, T_m, and the radiation temperature, T_r, evolve independently of each other. If the gas component expands adiabatically, and is assumed to consist only of hydrogen, standard thermodynamics gives us

$$d\left[\left(\rho_m c^2 + \tfrac{3}{2}\rho_m \frac{k_B T_m}{m_p}\right) a^3\right] = -\rho_m \frac{k_B T_m}{m_p}\, da^3. \tag{5.2.1}$$

Given that $\rho_m a^3$ is constant, because of mass conservation, Equation (5.2.1) leads to

$$T_m = T_{0m}\left(\frac{a_0}{a}\right)^2 = T_{0m}(1+z)^2, \tag{5.2.2}$$

which is nothing other than the usual relation $TV^{\gamma-1} = $ const. for a monatomic gas ($\gamma = \frac{5}{3}$). For a gas of photons, we use the relationship between the energy-density and temperature of a black body,

$$\rho_r c^2 = \sigma_r T_r^4, \tag{5.2.3}$$

to find that

$$T_r = T_{0r}\frac{a_0}{a} = T_{0r}(1+z). \tag{5.2.4}$$

If σ_c, the collision cross-section between photons and atoms, is constant, then the collision time τ_c simply scales as the inverse of the number-density of atoms and therefore decreases with redshift much more rapidly than the characteristic timescale for the expansion τ_H: for example, in a flat universe,

$$\tau_c \propto \rho_m^{-1} \propto (1+z)^{-3}, \tag{5.2.5}$$

$$\tau_H = \left(\frac{\dot{a}}{a}\right)^{-1} \propto (1+z)^{-3/2}, \tag{5.2.6}$$

where we have assumed matter domination to calculate τ_H; if the Universe were radiation dominated, this reasoning would still hold good. In fact, the cross-section for scattering of electrons by atoms does not behave as simply as this with z. The main mechanism by which photons interact with matter is Thomson scattering by electrons, but photons of sufficient energy can also be absorbed by the atom, resulting in photo-ionisation. The ions thus produced may then recombine, with the usual cascades producing the Lyman and Balmer series. Photons of exactly the right wavelength can also cause upward transitions, leading to absorption lines. However, in the cosmological situation we are interested in, it suffices to take Thomson scattering by electrons as the dominant mechanism. As we shall see, as the photon energies increase to the energies relevant for the other processes mentioned here, the plasma becomes fully ionised and Thomson scattering is then indeed the dominant interaction between the matter and radiation. In any event, there clearly exists a time, say t_d, before which scattering occurs on a timescale much less than the expansion timescale, resulting in a tight coupling between matter and radiation. After t_d, a process of *decoupling* occurs and, for $t \gg t_d$, matter and radiation effectively evolve separately. As we shall see in Chapter 9, this process is not instantaneous and actually continues over a relatively large range of t (or z). Before decoupling, at $t = t_d$, matter and radiation are held in equilibrium with each other at the same temperature, and T varies with z in a manner intermediate between (5.2.2) and (5.2.4), which we can represent by Equation (5.3.3) below. At very high T (high z), the equilibrium state for the matter component has a very high state of ionisation. As T decreases, the fraction of atoms which are ionised (the degree of ionisation) falls. There exists therefore a time, say t_{rec}, before which the matter is fully ionised, and after which the ionisation is very small. This transition is usually called *recombination*, although it would be more accurate to call it simply *combination*. Recombination is also a relatively gradual process so it does not occur at a single definite $t = t_{rec}$. Notice, however, that in general $t_d \geqslant t_{rec}$. We discuss recombination and decoupling in the context of realistic cosmological models in Section 5.4 and in Chapter 9.

5.3 Matter–Radiation Equivalence

Another important timescale in the thermal history of the Universe is that of *matter-radiation equivalence*, say $t = t_{eq}$, which we take to occur at $z_{eq} = z(t_{eq})$. Remember that the matter density evolves according to

$$\rho_m = \rho_{0m}(1+z)^3, \tag{5.3.1}$$

while the density of radiation follows

$$\rho_r = \rho_{0r}(1+z)^4, \tag{5.3.2}$$

in the period after decoupling, and

$$\rho_r \propto T^4 \propto (1+z)^{4+\epsilon(z)} \tag{5.3.3}$$

before decoupling; in the relation (5.3.3), $0 < \epsilon(z) < 4$ is a term included to take account of the evolution of $T(z)$ in this regime. It turns out that $\epsilon(z)$ is actually very small, for reasons we shall discuss later.

Matter–radiation equivalence occurs when the densities (5.3.1) and (5.3.3) are equal. Of course, if there are other components of the fluid which are relativistic at interesting redshifts, then they should, strictly speaking, be included in the definition of this timescale. In general, if there are several relativistic components, labelled i, each contributing a fraction $\Omega_{0r,i}$ of the present critical density, then the *total* relativistic contribution dominates for

$$1 + z > 1 + z_{eq} = \frac{\Omega_0}{\sum_i \Omega_{0r,i}} = \frac{\Omega}{\Omega_{0r,tot}}, \tag{5.3.4}$$

where Ω_0 is the density parameter for the non-relativistic material. We have assumed $\epsilon = 0$ in Equation (5.3.4). If we neglect the contribution to the sum in (5.3.4) due to relativistic particles other than photons, we find $z_{eq} \simeq 4.3 \times 10^4 \Omega_0 h^2$.

5.4 Thermal History of the Universe

Before decoupling at $t = t_d$, matter and radiation are tightly coupled. This is ultimately due to the fact that, before recombination, the matter component is fully ionised and the relevant photon scattering cross-section is therefore the Thomson scattering cross-section σ_T, which is much larger than that presented by a neutral atom of hydrogen. As we have explained, this guarantees that the radiative component (photons) and the matter component (the electron–proton plasma) have the same temperature T. Let us now investigate the behaviour of this temperature in more detail.

The appropriate expression governing the adiabatic expansion of a gas of matter and radiation is

$$\mathrm{d}\left[\left(\rho_m c^2 + \frac{3\rho_m k_B T}{2m_p} + \sigma_r T^4\right)a^3\right] = -\left(\frac{\rho_m k_B T}{m_p} + \frac{\sigma_r T^4}{3}\right)\mathrm{d}a^3, \tag{5.4.1}$$

in which we assume the matter component has the equation of state of a perfect gas:

$$p = \frac{\rho_m k_B T}{m_p}. \tag{5.4.2}$$

Recall that $\rho_m a^3 = \text{const.}$, and introduce the dimensionless constant

$$\sigma_{rad} = \frac{4 m_p \sigma_r T^3}{3 k_B \rho_m}; \tag{5.4.3}$$

the physical significance of σ_{rad} will become apparent shortly. From (5.4.1) we have

$$\frac{\mathrm{d}T}{T} = -\frac{1 + \sigma_{rad}}{\frac{1}{2} + \sigma_{rad}}\frac{\mathrm{d}a}{a}, \tag{5.4.4}$$

which, unfortunately, cannot be integrated analytically, because $\sigma_{\rm rad}(T)$ depends on the unknown function $T(a)$. It is easy to see that $\sigma_{\rm rad}(T)$ does not depend on a after decoupling if we interpret T as the temperature of the radiation. The value of $\sigma_{\rm rad}$ must therefore coincide with its present value $\sigma_{\rm rad}(t = t_0)$, which can be calculated in terms of the present density of the Universe, $\rho_{0\rm m}$, and the present radiation temperature, $T_{0\rm r}$:

$$\sigma_{\rm rad}(t = t_0) = \frac{4m_{\rm p}\sigma_{\rm r}T_{0\rm r}^3}{3k_{\rm B}\rho_{0\rm m}} \simeq 3.6\eta_0^{-1} \simeq 1.35 \times 10^8(\Omega_{0\rm b}h^2)^{-1}, \tag{5.4.5}$$

which is a very large number given the known bounds on the parameters $\Omega_{0\rm b}$ and h.

The Equation (5.4.4) is valid also at $t = t_{\rm d}$. In a short interval of time at $t_{\rm d}$, we can make use of the fact that $\sigma_{\rm rad}(t) \simeq \sigma_{\rm rad}(t_{\rm d}) = \sigma_{\rm rad}(t_0) \gg 1$, thus obtaining

$$\frac{{\rm d}T}{T} \simeq -\frac{{\rm d}a}{a}, \tag{5.4.6}$$

which, upon integration, leads to Equation (5.2.4). This shows that we indeed expect $\epsilon \simeq 0$; it is virtually guaranteed by the very high actual value of $\sigma_{0\rm r}$.

At higher temperatures, the matter component also becomes relativistic and therefore assumes the equation of state $p = \frac{1}{3}\rho c^2$. In this regime the behaviour of T is very closely represented by Equation (5.2.4). The reason for this is as follows.

Suppose the temperature of the Universe exceeds a value T_p, such that

$$k_{\rm B}T_p \simeq 2mc^2, \tag{5.4.7}$$

where p is a particle with mass m (for example an electron). In this situation the creation-annihilation reaction

$$\gamma + \gamma' \rightleftharpoons {\rm e}^+ + {\rm e}^- \tag{5.4.8}$$

has an equilibrium which lies to the right. A significant number of electron-positron ($\rm e^+$-$\rm e^-$) pairs are therefore created. At higher temperatures still, even more particle species might be created, of higher and higher masses.

The era contained between the two temperatures $T_{\rm e}$ ($\simeq 5\times10^9$ K) and T_π, where e and π are the electron and pion, respectively, is called the *lepton era* because, as besides the radiative fluid of photons and neutrinos, the background of leptons $\rm e^+$, $\rm e^-$, μ^+, μ^- and τ^+ and τ^- dominates the energy density. The brief interval with 200–300 MeV $> k_{\rm B}T > T_\pi \simeq 130$ MeV is called the *hadron era*, because as well as photons, neutrinos and leptons, we now also have hadrons (π_0, π^+, π^-, p, p̄, n, n̄, etc.); they do not, however, dominate the energy density. For $k_{\rm B}T > 200$–300 MeV, the hadrons are separated into their component quarks. We shall discuss these phases in some detail in Chapter 8. There are so many relativistic particle species at such high energies, however, that for the moment it suffices to say that it is a good approximation to take the relativistic equation of state $p = \frac{1}{3}\rho c^2$ and $\rho c^2 = A\sigma T^4$ appropriate for pure radiation, which gives the Equation (5.2.4) exactly, but in which the constant A describes the fact that there are many different relativistic particles in addition to the photons.

5.5 Radiation Entropy per Baryon

As we have seen in Section 5.4, the high value of $\sigma_{\rm rad}$ guarantees that the temperature and density of the radiation, to a very good approximation, evolve as in a pure radiation universe. The quantity $\sigma_{\rm rad}$ is actually related to the ratio between the entropy of the radiation per unit volume,

$$s_{\rm r} = \frac{\rho_{\rm r}c^2 + p_{\rm r}}{T} = \frac{4}{3}\frac{\rho_{\rm r}c^2}{T} = \frac{4}{3}\sigma_{\rm r}T^3, \tag{5.5.1}$$

and the number-density of baryons,

$$n_{\rm b} = \frac{\rho_{\rm m}}{m_{\rm p}}, \tag{5.5.2}$$

written in dimensionless form by dividing by Boltzmann's constant:

$$\sigma_{\rm rad} = \frac{s_{\rm r}}{k_{\rm B}n_{\rm b}}. \tag{5.5.3}$$

The quantity $\sigma_{\rm rad}^{-1}$ is proportional to the ratio η between the number-density of baryons and that of photons. From Equations (5.1.8) and (5.2.3) we get

$$\sigma_{\rm rad} = 3.6\eta^{-1}. \tag{5.5.4}$$

The quantity $\sigma_{\rm rad}$ is also proportional to the ratio of the heat capacity per unit volume of the radiation, $\rho_{\rm r}c_{\rm r}$, and that of the matter, $\rho_{\rm m}c_{\rm m}$. In fact, for the radiation,

$$\rho_{\rm r}c_{\rm r} = \frac{\partial(\rho_{\rm r}c^2)}{\partial T} = \frac{\partial(\sigma_{\rm r}T^4)}{\partial T} = 4\sigma_{\rm r}T^3, \tag{5.5.5}$$

and for the matter,

$$\rho_{\rm m}c_{\rm m} = \frac{\partial(3\rho_{\rm m}k_{\rm B}T/2m_{\rm p})}{\partial T} = \frac{3}{2}\frac{\rho_{\rm m}}{m_{\rm p}}k_{\rm B}, \tag{5.5.6}$$

from which

$$\frac{\rho_{\rm r}c_{\rm r}}{\rho_{\rm m}c_{\rm m}} = 2\sigma_{\rm rad}; \tag{5.5.7}$$

the high value of this ratio makes sure that the coupled matter–radiation fluid follows the cooling law for pure radiation to a very good approximation.

The quantity $\sigma_{\rm rad}$ is also (and finally) related to the scale of primordial *baryon-antibaryon asymmetry* present in the early Universe. Let us indicate by $n_{\rm b}$ and $n_{\bar{b}}$ the baryon and antibaryon number density, respectively. The quantity $(n_{\rm b} - n_{\bar{b}})a^3$ remains constant during the expansion of the Universe because baryon number is a conserved quantity. In fact, one does not observe a significant presence of antibaryons, so the relevant quantity is just $n_{0\rm b}a_0^3$. (If there were significant quantities of antibaryons, annihilation events would lead to a much greater background of gamma rays than is observed.) In the epoch following $T_{\rm GUT} \simeq 10^{15}$ GeV, which we will discuss in Chapter 7, we have

$$n_{\rm b} \simeq n_{\bar{b}} \simeq n_\gamma \propto T^3, \tag{5.5.8}$$

from which the baryon–antibaryon asymmetry is expected to be

$$\frac{n_{\rm b} - n_{\bar{b}}}{n_{\rm b} + n_{\bar{b}}} \simeq \frac{n_{\rm b} - n_{\bar{b}}}{2n_\gamma} \simeq \frac{n_{0\rm b}}{2n_{0\gamma}}. \tag{5.5.9}$$

The baryon–antibaryon asymmetry is very small, of the order of $\sigma_{\rm rad}^{-1}$, so that for every, say, 10^9 antibaryons there will be $10^9 + 1$ baryons. The reason for this asymmetry, and why it is so small, is therefore the same as the reason why the value of $\sigma_{\rm rad}$ is large. Developments in the theory of elementary particles have led to some suggestions as to how cosmological *baryosynthesis* might occur; we shall discuss them in some detail in Chapter 7.

5.6 Timescales in the Standard Model

In the standard model, after the lepton era, the Friedmann Equation (1.12.6) becomes

$$\left(\frac{\dot{a}}{a_0}\right)^2 = H_0^2\left[\Omega_0\frac{a_0}{a} + \Omega_{0\rm r}K_0\left(\frac{a_0}{a}\right)^2 + (1 - \Omega_0 - \Omega_{0\rm r})\right], \tag{5.6.1}$$

where, as usual, the suffix '0' refers to the present epoch. The last bracket neglects contributions from relativistic particles which are small at the present time. Jumping the gun slightly (see Chapter 8 for details), we have replaced the purely radiation contribution $\Omega_{\rm r}$ by $K_0\Omega_{\rm r}$ to take account of the contribution of light neutrinos to the relativistic part of the fluid; that is to say, the sum over i in Equation (5.3.4) now includes both photons and neutrinos. We shall see later, in Chapter 8, that

$$K_0 = 1 + \tfrac{7}{8}(\tfrac{4}{11})^{4/3}N_\nu \simeq 1 + 0.227N_\nu, \tag{5.6.2}$$

with N_ν the number of types of light neutrino; $K_0 \simeq 1.68$ if $N_\nu = 3$. The second part of K_0 derives from the neutrinos, and differs from the photon contribution because they are fermions. The matter component is simply written Ω_0 in Equation (5.6.1).

In light of Section 5.3, we can now calculate the equivalence redshift, $z_{\rm eq}$, at which $\rho_{\rm m} = K_0\rho_{\rm r} = \rho_{\rm eq}$. The result is

$$\rho_{\rm eq} = \rho_{\rm m}(z_{\rm eq}) = \rho_{0\rm c}\Omega_0(1 + z_{\rm eq})^3 = K_0\rho_{\rm r}(z_{\rm eq}) = K_0\rho_{0\rm r}(1 + z_{\rm eq})^4, \tag{5.6.3}$$

from which we obtain

$$1 + z_{\rm eq} = \frac{\rho_{0\rm c}\Omega_0}{K_0\rho_{0\rm r}} = \Omega_{0\rm r}^{-1}K_0^{-1}\Omega_0 \simeq 2.6 \times 10^4\Omega_0 h^2 \tag{5.6.4}$$

if $N_\nu = 3$. In and before the lepton era, Equation (5.6.1) is replaced by

$$\left(\frac{\dot{a}}{a_0}\right)^2 = H_0^2\left[\Omega_0\frac{a_0}{a} + \Omega_{0\mathrm{r}}K_\mathrm{c}\left(\frac{a_0}{a}\right)^2 + (1-\Omega_0)\right] \simeq H_0^2\Omega_\mathrm{r}K_\mathrm{c}\left(\frac{a_0}{a}\right)^2; \tag{5.6.5}$$

the approximation on the right-hand side holds for $z = a_0/a \gg z_{\mathrm{eq}} \gg 1$. The factor $K_\mathrm{c}(z)$ takes account of the creation of pairs of higher and higher mass, as we discussed in Section 5.4. As we shall see in Chapter 8, K_c is not expected to be much bigger than K_0. A good approximation for the period following the lepton era and before decoupling is therefore obtained by using Equation (5.6.5) with $K_\mathrm{c}(z) \simeq K_0$:

$$\left(\frac{\dot{a}}{a_0}\right)^2 \simeq H_0^2\Omega_{0\mathrm{r}}K_0\left(\frac{a_0}{a}\right)^2. \tag{5.6.6}$$

For redshifts $z \gg (\Omega_{0\mathrm{r}}K_0)^{-1} \simeq z_{\mathrm{eq}}$ this equation gives

$$t(z) \simeq \frac{1}{2H_0\Omega_{0\mathrm{r}}^{1/2}K_0^{1/2}}(1+z)^{-2} \simeq 3.2\times 10^{19}K_0^{-1/2}(1+z)^{-2}\ \mathrm{s}. \tag{5.6.7}$$

Extrapolating Equation (5.6.7) to z_{eq} (where in fact it is only marginally valid), one obtains

$$t_{\mathrm{eq}} = t(z_{\mathrm{eq}}) \simeq 10^4(\Omega_0 h^2)^{-2}\ \mathrm{years}. \tag{5.6.8}$$

At much later times, in the interval between $z \ll z_{\mathrm{eq}}$ and $1 + z \gg \Omega_0^{-1}$, Equation (5.6.1) is well approximated by

$$\left(\frac{\dot{a}}{a_0}\right)^2 \simeq H_0^2\Omega_0\frac{a_0}{a}. \tag{5.6.9}$$

In this period it is a good approximation to use Equation (2.4.8), from which we get

$$t(z) - t_{\mathrm{eq}} \simeq \frac{2}{3H_0\Omega_0^{1/2}}[(1+z)^{-3/2} - (1+z_{\mathrm{eq}})^{-3/2}]. \tag{5.6.10}$$

For $t \gg t_{\mathrm{eq}}$, and therefore for $z \ll z_{\mathrm{eq}}$, Equation (5.6.10) can be written

$$t(z) \simeq \frac{2}{3H_0\Omega_0^{1/2}}(1+z)^{-3/2} \simeq 2.1\times 10^{17}\Omega_0^{-1/2}h^{-1}(1+z)^{-3/2}\ \mathrm{s}. \tag{5.6.11}$$

If the recombination redshift, z_{rec}, is of order 10^3, which we shall argue is indeed the case in Chapter 10, it will be lower than that of matter–radiation equivalence as long as $\Omega_0 h^2 > 0.04$. The previous expression gives the recombination time as

$$t_{\mathrm{rec}} = t(z_{\mathrm{rec}}) \simeq 3\times 10^5\ \mathrm{years}. \tag{5.6.12}$$

The age of the Universe, t_0, can be obtained by integrating Equation (5.6.1) from the Big Bang ($t = 0$) to the present epoch. This integral can be divided into two contributions: from the Big Bang until t_{eq}, and from t_{eq} to t_0. Given that $z_{eq} \gg 1$ and, therefore, that $t_{eq} \ll t_0$, the former contribution is negligible compared with the second. It is therefore a good approximation to calculate t_0 by putting $\Omega_r = 0$ in Equation (5.6.1) and taking the lower limit of integration to be $t = 0$. One will thus obtain the values derived in Section 2.4 for the age of a matter-dominated universe.

Bibliographic Notes on Chapter 5

The material in this chapter is very well established. The main results are discussed in Peebles (1971, 1993) and Weinberg (1972). A nice review article of retrospective interest is given by Harrison (1973).

Problems

1. The result (5.1.7) is obtained for photons by integrating over the Planck distribution appropriate for bosons. In the case of neutrinos (or other fermions), show that the number-density in thermal equilibrium at a temperature $T_{0\nu}$ is

$$n_{0\nu} = 3\frac{\zeta(3)}{2\pi^2}\left(\frac{k_B T_{0r}}{\hbar c}\right)^3.$$

2. The Friedmann Equation (5.6.1) describing the evolution of a Universe containing only non-relativistic matter and photons can be written

$$\left(\frac{\dot{a}}{a_0}\right)^2 = H_0^2\left[\Omega_0\frac{a_0}{a} + \Omega_{0r}\left(\frac{a_0}{a}\right)^2 + (1 - \Omega_0 - \Omega_{0r})\right].$$

Show that for any choice of $\Omega_0 < 1$ there is a value of Ω_{0r} that makes the right-hand side a perfect square of a function of a. Obtain an exact solution for $a(t)$ in such a case.

3. Show that, in a flat radiation-dominated Universe, the radiation temperature varies with time t as

$$T = At^{-1/2},$$

and obtain an expression for A in terms of physical quantities. Use your result to estimate the temperature at $t = 1$ second after the Big Bang.

6

The Very Early Universe

6.1 The Big Bang Singularity

As we explained in Chapter 2, all homogeneous and isotropic cosmological models containing perfect fluids of equation of state $p = w\rho c^2$, with $0 \leqslant w \leqslant 1$, possess a singularity at $t = 0$ where the density diverges and the proper distance between any two points tends to zero. This singularity is called the Big Bang. Its existence is a direct consequence of four things: (i) the Cosmological Principle; (ii) the Einstein equations in the absence of a cosmological constant; (iii) the expansion of the Universe (in other words, $(\dot{a}/a)_0 = H_0 > 0$); and (iv) the assumed form of the equation of state.

It it clear that the Big Bang might well just be a consequence of extrapolating deductions based on the theory of general relativity into a situation where this theory is no longer valid. Indeed, Einstein (1950) himself wrote:

> The theory is based on a separation of the concepts of the gravitational field and matter. While this may be a valid approximation for weak fields, it may presumably be quite inadequate for very high densities of matter. One may not therefore assume the validity of the equations for very high densities and it is just possible that in a unified theory there would be no such singularity.

We clearly need new laws of physics to describe the behaviour of matter in the vicinity of the Big Bang, when the density and temperature are much higher than can be achieved in laboratory experiments. In particular, any theory of matter under such extreme conditions must take account of quantum effects on a cosmological scale. The name given to the theory of gravity that replaces general relativity at ultra-high energies by taking these effects into account is *quantum gravity*. We are, however, a very long way from being able to construct a satisfactory theory to accomplish this. It seems likely, however, that in a complete theory of quantum gravity, the cosmological singularity would not exist. In other words, the existence of a singularity in cosmological models based on the classical theory of general relativity is probably just due to the incompleteness of the theory. Moreover, there are ways of avoiding the singularity even without appealing to explicitly quantum-gravitational effects and remaining inside Einstein's theory of gravity.

Firstly, one could try to avoid the singularity by proposing an equation of state for matter in the very early Universe that is different to the usual perfect fluid with $p/\rho > -\frac{1}{3}$. Let us begin by writing down Equation (1.10.3):

$$\ddot{a} = -\tfrac{4}{3}\pi G\left(\rho + 3\frac{p}{c^2}\right)a. \tag{6.1.1}$$

Recall that, if we have a perfect fluid satisfying

$$p < -\tfrac{1}{3}\rho c^2, \tag{6.1.2}$$

then the argument we gave in Section 2.1 based on the concavity of $a(t)$ is no longer valid and the singularity can be avoided. Fluids with $w < -\frac{1}{3}$ in this way are said to violate the *strong energy condition*. There are various ways in which this condition might indeed be violated. For example, suppose we describe the contents of the Universe as an *imperfect fluid*, that is one in which viscosity and thermal conductivity are not negligible. The energy momentum tensor of such a fluid is no longer of the form (1.9.2); it must contain dependences on the coefficient of shear viscosity η, the coefficient of bulk viscosity ζ, and the thermal conductivity χ. The physical significance of the first two of these coefficients can be recognised by looking at the equation of motion (Euler equation) for a non-relativistic fluid neglecting self-gravity:

$$\rho\left[\frac{\partial \boldsymbol{v}}{\partial t} + (\boldsymbol{v}\cdot\nabla)\boldsymbol{v}\right] = -\nabla p + \eta\nabla^2\boldsymbol{v} + (\zeta + \tfrac{1}{3}\eta)\nabla(\nabla\cdot\boldsymbol{v}). \tag{6.1.3}$$

One can demonstrate that in a Robertson-Walker metric the terms in η and χ must be zero because of homogeneity and isotropy: there can be no gradients in pressure or temperature. The terms in the bulk viscosity, however, need not be zero: their effect upon the Friedmann equations is to replace the pressure p by an 'effective' pressure p^*:

$$p \to p^* = p - 3\zeta\frac{\dot{a}}{a}, \tag{6.1.4}$$

for which the energy-momentum tensor becomes

$$T_{ij} = -\left(p - 3\zeta\frac{\dot{a}}{a}\right)g_{ij} + \left(p - 3\zeta\frac{\dot{a}}{a} + \rho c^2\right)U_iU_j. \tag{6.1.5}$$

The resulting Equation (6.1.1) does not change in form, but one must replace p by p^*. Generally speaking, the bulk viscosity is expected to be negligible in non-relativistic fluids as well as ultra-relativistic ones. It need not be small in the intermediate regime, such as one obtains if there is a mixture of relativistic and non-relativistic fluids. With an appropriate expression for ζ (for example $\zeta = \alpha^* \rho$, with $\alpha^* = \text{const.} > 0$, or $\zeta = \text{const.} > 0$), one can obtain homogeneous and isotropic solutions to the Einstein equations that do not possess a singularity. In general, however, ζ has to be very small but non-zero; it is not trivial to come up with satisfactory models in which bulk viscosity is responsible for the absence of a singularity.

The Big Bang does not exist in many models with a non-zero cosmological constant, $\Lambda > 0$. As we shall see, the present value of Λ can be roughly bounded observationally

$$|\Lambda| < \left(\frac{H_0}{c}\right)^2 \simeq 10^{-55}\ \text{cm}^{-2}, \tag{6.1.6}$$

which is very small. The effect of such a cosmological constant at very early times would be very small indeed, since its dynamical importance increases with time. A more realistic option is to interpret the cosmological constant as an effective quantity related to the vacuum energy density of a quantum field; this can be a dynamical quantity and may therefore have been more important in the past than a true cosmological constant. For example, as we shall see in Chapter 7 when we discuss inflation, it is possible that the dynamics of the very early Universe is dominated by a homogeneous and isotropic *scalar quantum field* whose evolution is governed by the effective classical Lagrangian

$$L_\Phi = \tfrac{1}{2}\dot{\Phi}^2 - V(\Phi), \tag{6.1.7}$$

where the first term is 'kinetic' and the second is the 'effective potential'. To simplify Equation (6.1.7) and the following expressions, we have now adopted units in which $c = \hbar = 1$. The energy-momentum tensor for such a field is

$$T_{in}(\Phi) = -p_\Phi g_{ij} + (p_\Phi + \rho_\Phi c^2)U_i U_j, \tag{6.1.8}$$

where the 'energy-density' $\rho_\Phi c^2$ and the 'pressure' p_Φ are to be interpreted as effective quantities (the scalar field is *not* a fluid), and are given by

$$\rho_\Phi c^2 = \tfrac{1}{2}\dot{\Phi}^2 + V(\Phi), \tag{6.1.9 a}$$

$$p_\Phi = \tfrac{1}{2}\dot{\Phi}^2 - V(\Phi). \tag{6.1.9 b}$$

In particular, if the kinetic term is negligible with respect to the potential term, the effective equation of state for the field becomes

$$p_\Phi \simeq -\rho_\Phi c^2. \tag{6.1.10}$$

The scalar field can therefore be regarded as behaving like a fluid with an equation-of-state parameter $w = -1$ (thus violating the strong energy condition) or as an effective cosmological constant

$$\Lambda = \frac{8\pi G}{c^2}\rho_\Phi. \tag{6.1.11}$$

The density ρ_Φ is zero or at least negligible today, but could have been the dominant dynamical factor in certain phases of the evolution of the Universe. It may also have been important in driving an epoch of *inflation*; see Chapter 7.

Whether the singularity is avoidable or not remains an open question, as does the question of what happens to the Universe for $t < 0$. It is reasonable to call this question the *problem of the origin of the Universe*: it is one of the big gaps in cosmological knowledge; some comments about the possible physics of the creation of the Universe are discussed in Sections 6.4 and 6.5.

6.2 The Planck Time

We have already mentioned that the theory of general relativity should be modified in situations where the density tends to infinity, in order to take account of quantum effects on the scale of the cosmological horizon. In fact, Einstein himself believed that his theory was incomplete in this sense and would have to be modified in some way. When do we expect quantum corrections to become significant? Of course, in the absence of a complete theory (or indeed *any* theory) of quantum gravity, it is impossible to give a precise answer to this question. On the other hand, one can make fairly convincing general arguments that yield estimates of the timescales and energy scales where we expect quantum gravitational effects to be large and where we should therefore distrust calculations based only upon the classical theory of general relativity. As we shall now explain, the limit of validity of Einstein's theory in the Friedmann models is fixed by the *Planck time* which is of the order of 10^{-43} s after the Big Bang.

The Planck time t_P is the time for which quantum fluctuations persist on the scale of the Planck length $l_P \simeq ct_P$. From these two scales one can construct a Planck mass, $m_P \simeq \rho_P l_P^3$, where the Planck density ρ_P is of the order of $\rho_P \simeq (Gt_P^2)^{-1}$ (from the Friedmann equations). Starting from the Heisenberg uncertainty principle, in the form

$$\Delta E \Delta t \simeq \hbar, \tag{6.2.1}$$

we see that, on dimensional grounds,

$$\Delta E \Delta t \simeq m_P c^2 t_P \simeq \rho_P (ct_P)^3 c^2 t_P \simeq \frac{c^5 t_P^4}{G t_P^2} \simeq \hbar, \tag{6.2.2}$$

from which

$$t_P \simeq \left(\frac{\hbar G}{c^5}\right)^{1/2} \simeq 10^{-43}\ \text{s}. \tag{6.2.3}$$

Other quantities related to the Planck time are the *Planck length*,

$$l_{\mathrm{P}} \simeq c t_{\mathrm{P}} \simeq \left(\frac{G\hbar}{c^3}\right)^{1/2} \simeq 1.7 \times 10^{-33}\ \mathrm{cm}, \tag{6.2.4}$$

which represents the order of magnitude of the cosmological horizon at $t = t_{\mathrm{P}}$; the *Planck density*

$$\rho_{\mathrm{P}} \simeq \frac{1}{G t_{\mathrm{P}}^2} \simeq \frac{c^5}{G^2\hbar} \simeq 4 \times 10^{93}\ \mathrm{g\ cm^{-3}}; \tag{6.2.5}$$

the *Planck mass* (roughly speaking the mass inside the horizon at t_{P})

$$m_{\mathrm{P}} \simeq \rho_{\mathrm{P}} l_{\mathrm{P}}^3 \simeq \left(\frac{\hbar c}{G}\right)^{1/2} \simeq 2.5 \times 10^{-5}\ \mathrm{g}. \tag{6.2.6}$$

Let us also define an effective number-density at t_{P} by

$$n_{\mathrm{P}} \simeq l_{\mathrm{P}}^{-3} \simeq \frac{\rho_{\mathrm{P}}}{m_{\mathrm{P}}} \simeq \left(\frac{c^3}{G\hbar}\right)^{3/2} \simeq 10^{98}\ \mathrm{cm^{-3}}, \tag{6.2.7}$$

a *Planck energy*

$$E_{\mathrm{P}} \simeq m_{\mathrm{P}} c^2 \simeq \left(\frac{\hbar c^5}{G}\right)^{1/2} \simeq 1.2 \times 10^{19}\ \mathrm{GeV}, \tag{6.2.8}$$

and a *Planck temperature*

$$T_{\mathrm{P}} \simeq \frac{E_{\mathrm{P}}}{k_{\mathrm{B}}} \simeq \left(\frac{\hbar c^5}{G}\right)^{1/2} k_{\mathrm{B}}^{-1} \simeq 1.4 \times 10^{32}\ \mathrm{K}. \tag{6.2.9}$$

The last relation can also be found by putting

$$\rho_{\mathrm{P}} c^2 \simeq \sigma T_{\mathrm{P}}^4. \tag{6.2.10}$$

The dimensionless entropy inside the horizon at the Planck time takes the value

$$\sigma_{\mathrm{P}} \simeq \frac{\rho_{\mathrm{P}} c^2 l_{\mathrm{P}}^3}{k_{\mathrm{B}} T_{\mathrm{P}}} \simeq 1, \tag{6.2.11}$$

which reinforces the point that there is, on average, one 'particle' of Planck mass inside the horizon at the Planck time. It is important to note that all these quantities related to the Planck time can be derived purely on dimensional grounds from the fundamental physical constants c, G, k_{B} and $\hbar$.

6.3 The Planck Era

In order to understand the physical significance of the Planck time, it is useful to derive t_{P} in the following manner, which ultimately coincides with the derivation we gave above. Let us define the *Compton time* for a body of mass m (or of energy mc^2) to be

$$t_{\rm C} = \frac{\hbar}{mc^2}; \tag{6.3.1}$$

this quantity represents the time for which it is permissible to violate conservation of energy by an amount $\Delta E \simeq mc^2$, as deduced from the uncertainty principle. For example one can create a pair of virtual particles of mass m for a time of order $t_{\rm C}$. Let us also define the Compton radius of a body of mass m to be

$$l_{\rm C} = ct_{\rm C} = \frac{\hbar}{mc}. \tag{6.3.2}$$

Obviously $t_{\rm C}$ and $l_{\rm C}$ both decrease as m increases. These scales are indicative of quantum physics.

On the other hand the *Schwarzschild radius* of a body of mass m is

$$l_{\rm S} = \frac{2Gm}{c^2}; \tag{6.3.3}$$

this represents, to order of magnitude, the radius which a body of mass m must have so that its rest-mass energy mc^2 is equal to its internal gravitational potential energy $U \simeq Gm^2/l_{\rm S}$. General relativity leads to the conclusion that any particle (even a photon) cannot escape from a region of radius $l_{\rm S}$ around a body of mass m; in other words, speaking purely in terms of classical mechanics, the escape velocity from a body of mass m and radius $l_{\rm S}$ is equal to the velocity of light: $c^2/2 = Gm/l_{\rm S}$. Notice, however, that in the latter expression we have taken the 'kinetic energy' per unit mass of a photon to be $c^2/2$ as if it were a non-relativistic material. It is curious that the correct result emerges with these approximations. One can similarly define a Schwarzschild time to be the quantity

$$t_{\rm S} = \frac{l_{\rm S}}{c} = \frac{2Gm}{c^3}; \tag{6.3.4}$$

this is simply the time taken by light to travel a proper distance $l_{\rm S}$. A body of mass m and radius $l_{\rm S}$ has a free-fall collapse time $t_{\rm ff} \simeq (G\rho)^{-1/2}$, where $\rho \simeq m/l_{\rm S}^3$, which is of order $t_{\rm S}$. Notice that $t_{\rm S}$ and $l_{\rm S}$ both increase, as m increases.

One can easily verify that for a mass equal to the Planck mass, the Compton and Schwarzschild times are equal to each other, and to the Planck time. Likewise, the relevant length scales are all equal. For masses $m > m_{\rm P}$, that is to say *macroscopic bodies*, we have $t_{\rm C} < t_{\rm S}$ and $l_{\rm C} < l_{\rm S}$: quantum corrections are expected to be negligible in the description of the gravitational interactions between different parts of the body. Here we can describe the self-gravity of the body using general relativity or even, to a good approximation, Newtonian theory. On the other hand, for bodies with $m < m_{\rm P}$, i.e. microscopic entities such as elementary particles, we have $t_{\rm C} > t_{\rm S}$ and $l_{\rm C} > l_{\rm S}$: quantum corrections will be important in a description of their self-gravity. In the latter case, one must use a theory of quantum gravity in place of general relativity or Newtonian gravity.

At the cosmological level, the Planck time represents the moment before which the characteristic timescale of the expansion $\tau_H \sim t$ is such that the cosmological horizon, given roughly by l_P, contains only one particle (see above) having $l_C \geqslant l_S$. On the same grounds, as above, we are therefore required to take into account quantum effects on the scale of the cosmological horizon.

It is also interesting to note the relationship between the Planck quantities given in Section 6.2 to known thermodynamical properties of black holes (Thorne *et al.* 1986). According to theory, a black hole of mass M, due to quantum effects, emits radiation like a black body called Hawking radiation. The typical energy of photons emitted by the black hole is of order $\epsilon \simeq k_B T$, where T is the black-body temperature given by the relation

$$T = \frac{\hbar c^3}{4\pi k_B G M} \simeq 10^{-7}\left(\frac{M}{M_\odot}\right)^{-1} \text{ K.} \tag{6.3.5}$$

The time needed for such a black hole to completely evaporate, i.e. to lose all its rest-mass energy Mc^2 through such radiation, is of the order of

$$\tau \simeq \frac{G^2 M^3}{\hbar c^4} \simeq 10^{10}\left(\frac{M}{10^{15}\text{ g}}\right)^3 \text{ years.} \tag{6.3.6}$$

It is easy to verify that, if one extrapolates these formulae to the Planck mass m_P, the result is that $\epsilon(m_P) \simeq m_P c^2$ and $\tau(m_P) \simeq t_P$. A black hole of mass m_P therefore evaporates in a single Planck time t_P by the emission of one quantum particle of energy E_P.

These considerations show that quantum-gravitational effects are expected to be important not only at a cosmological level at the Planck time, but also continuously on a microscopic scale for processes operating over distances of order l_P and times of order t_P. In particular, the components of a space–time metric g_{ik} will suffer fluctuations of order $|\Delta g_{ik}/g_{ik}| \simeq l_P/l \simeq t_P/t$ on a spatial scale l and a temporal scale t. At the Planck time, the fluctuations are of order unity on the spatial scale of the horizon, which is l_P, and on the timescale of the expansion, which is t_P. One could imagine the Universe at very early times might behave like a collection of black holes of mass m_P, continually evaporating and recollapsing in a Planck time. This picture is very different from the idealised, perfect-fluid universe described by the Friedmann equations, and it would not be surprising if deductions from these equations, such as the existence of a singularity were found to be invalid in a full quantum description.

Before moving on to quantum gravity itself, let us return for a moment to the comments we made above about the creation of virtual particles. From the quantum point of view, a field must be thought of as a flux of virtual pairs of particles that are continually created and annihilated. As we explained above, the time for which a virtual particle of mass m can exist is of order the Compton time t_C, and the distance it moves before being annihilated is therefore the Compton length, l_C.

In an electrostatic field the two (virtual) particles, being charged, can be separated by the action of the field because their electrical charges will be opposite. If the separation achieved is of order l_C, there is a certain probability that the pair will not annihilate. In a very intense electrical field, one can therefore achieve a net creation of pairs. From an energetic point of view, the rest-mass energy of the pair $\Delta E \simeq 2mc^2$ will be compensated by a loss of energy of the electric field, which will tend to be dissipated by the creation of particles. Such an effect has been described theoretically, and can be observed experimentally in the vicinity of highly charged, unstable nuclei.

A similar effect can occur in an intense, non-uniform gravitational field. One creates a pair of particles (similar to the process by which black holes radiate particles). In this case, separation of the particles does not occur because of opposite charges (the gravitational 'charge', which is the mass, is always positive), but because the field is not uniform. One finds that the creation of particles in this way can be very important, for example, if the gravitational field varies strongly in time, as is the case in the early stages of the expansion of the Universe, above all if the expansion is anisotropic. Some have suggested that such particle creation processes might be responsible for the origin of the high entropy of the Universe. The creation of pairs will also tend to isotropise the expansion.

6.4 Quantum Cosmology

We have explained already that there is no satisfactory theory of quantum gravity, and hence no credible formulation of quantum cosmology. The attempt to find such a theory is technically extremely complex and somewhat removed from the main thrust of this book, so here is not the place for a detailed review of the field. What we shall do, however, is to point out aspects of the general formulation of quantum cosmology to give a flavour of this controversial subject, and to give some idea where the difficulties lie. The reader is referred to the reference list for more technical details.

The central concept in quantum mechanics is that of the *wavefunction*. To give the simple example of a single-particle system, one looks at $\psi(\boldsymbol{x}, t)$. Although the interpretation of ψ is by no means simple, it is generally accepted that the square of the modulus of ψ (for ψ will in general be a complex function) determines the *probability* of finding the particle at position $\boldsymbol{x}$ at time t. One popular formulation of quantum theory involves the concept of a 'sum over histories'. In this formulation, the probability of the particle ending up at $\boldsymbol{x}$ (at some time t) is given by an integral over all possible paths leading to that space–time location, weighted by a function depending on the *action*, $S(\boldsymbol{x}, t)$, along the path. Each path, or history, will be a function $\boldsymbol{x}(t)$, so that $\boldsymbol{x}$ specifies the intersection of a given history with a time-like surface labelled by t. In fact, one takes

$$\psi(\boldsymbol{x}, t) \propto \int \mathrm{d}\boldsymbol{x}\, \mathrm{d}t\, \exp[\mathrm{i}S(\boldsymbol{x}', t')], \tag{6.4.1}$$

where the integration is with respect to an appropriate measure on the space of all possible histories. The upper limit of integration will be the point in space–time given by $(\boldsymbol{x}, t)$, and the lower limit will depend on the initial state of the system. The action describes the forces to which the particle is subjected.

This 'sum-over-histories' formalism is the one which appears the most promising for the study of quantum gravity. Let us illustrate some of the ideas by looking at quantum cosmology. To make any progress here one has to make some simplifying assumptions. First, we assume that the Universe is finite and closed: the relevant integrals appear to be undefined in an open universe. We also have to assume that the spatial topology of the Universe is fixed; recall that the topology is not determined in general relativity. We also assume that the relevant 'action' for gravity is the action of general relativity we discussed briefly in Chapters 1 and 3, which we here write as S_E.

In fact, as an aside, we should mention that this is one of the big deficiencies in quantum gravity. There is no choice for the action of space-time coupled to matter fields which yields a satisfactory quantum field theory judged by the usual local standards of renormalisability and so on. There is no reason why the Einstein action S_E should keep its form as one moves to higher and higher energies. For example, it has been suggested that the Lagrangian for general relativity might pick up terms of higher order in the Ricci scalar R, beyond the familiar $L \propto R$. Indeed, second-order Lagrangian theories with $L = -R/(16\pi G) + \alpha R^2$ have proved to be of considerable theoretical interest because they can be shown to be conformally equivalent to general relativity with the addition of a scalar field. Such a theory could well lead to inflation (see Chapter 7 below), but would also violate the conditions necessary for the existence of a singularity. Some alternative cosmological scenarios based on modified gravitational Lagrangians have been discussed in Chapter 3. Since, however, we have no good reason in this context to choose one action above any other, we shall proceed assuming that the classical Einstein action is the appropriate one to take.

To have any hope of formulating cosmology in a quantum manner, we have to first think of the appropriate analogue to a 'history'. Let us simplify this even further by dealing with an empty universe, i.e. one in which there are no matter or radiation fields. It is perhaps most sensible to think of trying to determine a wavefunction for the configuration of the Universe at a particular time, and in general relativity the configuration of such a Universe will be simply given by the 3-geometry of a space-like hypersurface. Let this geometry be described by a 3-metric $h_{\mu\nu}(\boldsymbol{x})$. In this case, the corresponding quantity to a history $\boldsymbol{x}(t)$ is just a (Lorentzian) 4-geometry, specified by a 4-metric g_{ij}, which induces the 3-geometry $h_{\mu\nu}$ on its boundary. In general relativity, the action depends explicitly on the 4-metric g_{ij} so it is clear that, when we construct an integral by analogy with (6.4.1), the space over which it is taken is some space of allowed 4-geometries. The required wavefunction will then be a function of $h_{\mu\nu}(\boldsymbol{x})$ and will be given by an integral of the form

$$\Psi[h_{\mu\nu}(\boldsymbol{x})] = \int \mathrm{d}g_{ij} \exp[\mathrm{i}S(g_{ij})]. \tag{6.4.2}$$

The wavefunction Ψ is therefore defined over the space of all possible 3-geometries consistent with our initial assumptions (i.e. closed and with a fixed topology). Such a space is usually called a *superspace*. To include matter in this formulation then one would have to write $\Psi[h_{\mu\nu}, \Phi]$, where Φ labels the matter field at $\boldsymbol{x}$. Notice that, unlike (6.4.1), there is no need for an explicit time labelling beside $h_{\mu\nu}$ in the argument of Φ: a generic 3-geometry will actually only fit into a generic 4-geometry in at most only one place, so $h_{\mu\nu}$ carries its own labelling of time. The integral is taken over appropriate 4-geometries consistent with the 3-geometry $h_{\mu\nu}$. The usual quantum-mechanical wavefunction ψ evolves according to a Schrödinger equation; our 'wavefunction of the Universe', Ψ, evolves according to a similar equation called the *Wheeler–de Witt equation* (de Witt 1967). It is the determination of what constitutes the appropriate set of histories over which to integrate that is the crux of the problem and it is easy to see that this is nothing other than the problem of initial conditions in quantum cosmology (by analogy with the single-particle problem discussed above). This problem is far from solved.

One suggestion, by Hartle and Hawking (1983), is that the sum on the right-hand side of (6.4.2) is over compact Euclidean 4-geometries. This essentially involves making the change $t \to -i\tau$ with respect to the usual Lorentzian calculations. In this case the 4-geometries have no boundary and this is often called the *no-boundary conjecture*. Amongst other advantages, the relevant Euclidean integrals can be made to converge in a way in which the Lorentzian ones apparently cannot. Other choices of initial condition have, however, been proposed. Vilenkin (1984, 1986), amongst others, has proposed a model wherein the Universe undergoes a sort of quantum-tunnelling from a vacuum state. This corresponds to a definite creation, whereas the Hawking proposal has no 'creation' in the usual sense of the word. It remains to be seen which, if any, of these formulations is correct.

6.5 String Cosmology

Recent years have seen a radically different approach to the problem of quantum gravity which has led to a different idea of the possible structure of a quantum gravity theory. One of the most exciting ideas is that the fundamental entities upon which quantum operations must be performed are not point-like but are one dimensional. Such objects are usually known as strings, or more often superstrings (because they are usually discussed within so-called supersymmetric theories that unite fermions and bosons; see Chapter 8). Many physicists feel that string theory holds the key to the unification of all four forces of nature (gravity included) in a single over-arching theory of everything. Such a theory does not yet exist, but there is much interest in what its possible consequences might be.

String cosmology entered the doldrums in the early 1990s after a period of initial excitement. However it has since seen a resurgence largely because of the realisation that string theories can be thought of in terms of a more general class of theories known as M-theories. It is a property of all these structures that in order to be mathematically consistent they must be defined in space–times having more dimensions than the $(3+1)$-dimensional one with which we are familiar. One of the consequences of such theories is that fundamental constants 'live' in the higher-dimensional space and can vary in the 4-dimensional subspace we inhabit.

They therefore lead naturally to models like those we discussed in Chapter 3 in which fundamental constants may vary with time.

The idea that there may be more than four space-time dimensions is not itself new. Kaluza (1921) and Klein (1926) examined a (4 + 1)-dimensional model which furnished an intriguing geometrical unification of gravity and electromagnetism. In the Kaluza–Klein theory the extra space dimension was compactified on a scale of order the Planck length, i.e. wrapped up so small as to be unobservable.

String theories have to hide many dimensions, not just one, and until recently it was assumed that they would all have to be compactified. However, a radically new idea is called the braneworld scenario in which at least one of the extra dimensions might be large. In this picture we are constrained to live on a three-dimensional *brane* inside a higher-dimensional space called the *bulk*. Gravity is free to propagate in the bulk, and the gravity we see on the brane is a kind of projection of this higher-dimensional force. In the simplest braneworld model, known as the Randall–Sundrum model (Randall and Sundrum 1999) this theory results in a modification of the low-energy form of gravity such that the Newtonian potential becomes

$$V(r) = \frac{GM_1M_2}{r^2}\left(1 + \frac{1}{r^2k^2}\right), \tag{6.5.1}$$

where k corresponds to a very small length scale, of order the Planck length. At higher energies, however, there are interesting effects. In a particular case of the Randall–Sundrum model the high-energy behaviour of the Friedmann equation is modified:

$$\left(\frac{\dot{a}}{a}\right)^2 = \frac{8\pi G}{3}\left(\rho + \frac{\rho^2}{2\lambda}\right), \tag{6.5.2}$$

where λ is the tension in the brane.

A further development of the braneworld scenario is the notion that what we think of as the Big Bang singularity may in fact be the result of a collision between two branes. This has been dubbed the *Ekpyrotic universe*. Interest in this model stems from the fact that the impact of two branes may lead to effects that appear to be acausal when viewed from one of them. It remains to be seen whether this model can be developed to the point where it stands as a rival to the Big Bang.

Bibliographic Notes on Chapter 6

Two classic compilations on fundamental gravity theory are Hawking and Israel (1979, 1987). Duff and Isham (1982) is also full of interesting thoughts on quantum gravity, and Hartle (1988) is readable as well as authoritative.

Problems

1. In natural units $\hbar = c = 1$. Show that in such a system all energies, lengths and times can be expressed in terms of the Planck mass m_P.

2. Show that, in natural units, an energy density may be expressed as the fourth power of a mass. If the vacuum energy contributed by a cosmological constant is now of order the critical density, what is the mass to which this density corresponds?

3. Obtain a formula relating the Hawking temperature (6.3.5) to the radius of the event horizon of a Black Hole. In a de Sitter universe the scale factor increases exponentially with time such that $\dot{a}/a = H$ is constant. Show that in this model there is an event horizon with radius c/H. Assuming the Hawking formula also works for this radius, calculate the temperature of the event horizon in de Sitter space. How do you interpret this radiation?

4. Find a solution of Equation (6.5.2) with λ constant.

7

Phase Transitions and Inflation

7.1 The Hot Big Bang

We shall see in the next chapter that, if cosmological nucleosynthesis is the correct explanation for the observed light-element abundances, the Universe must have been through a phase in which its temperature was greater than $T \simeq 10^{12}$ K. In this chapter we shall explore some of the consequences for the Universe of phases of much higher temperature than this. Roughly speaking, we can define 'Hot' Big Bang models to be those in which the temperature increases as one approaches $t = t_0$. We assume that, after the Planck time, the temperature follows the law:

$$T(t) \simeq T_{\mathrm{P}} \frac{a(t_{\mathrm{P}})}{a(t)}; \tag{7.1.1}$$

we shall give detailed justification for this hypothesis later on.

Travelling backward in time, so that the temperature increases towards T_{P}, the particles making up the contents of the present Universe will all become relativistic and all the interactions between them assume the character of a long-range force such as electromagnetism. One can apply the model of a perfect ultra-relativistic gas of non-degenerate (i.e. with chemical potential $\mu = 0$) particles in thermal equilibrium during this stage. The equilibrium distribution of a particle species i depends on whether it is a fermionfermions or a boson and upon how many spin or helicity states the particle possesses, g_i. The quantity g_i is also sometimes called the statistical weight of the species i. The number-density of particles can be written

$$n_i(T) = g_i\left(\frac{k_B T}{\hbar c}\right)^3 \int_0^\infty \frac{x^2\,dx}{e^x \pm 1} = \begin{pmatrix}3/4\\1\end{pmatrix}\frac{g_i}{2}\frac{2}{\pi^2}\zeta(3)\left(\frac{k_B T}{\hbar c}\right)^3, \tag{7.1.2}$$

where the integrand includes a '+' sign for fermions and a '−' sign for bosons producing a factor of $\frac{3}{4}$ or 1 in these respective cases. In (7.1.2) ζ is the Riemann zeta function which crops up in the integral; $\zeta(3) \simeq 1.202$. Similarly, the energy density of the particles is

$$\rho_i(T)c^2 = \frac{g_i k_B^4 T^4}{2\pi^2\hbar^3 c^3}\int_0^\infty \frac{x^3\,dx}{\exp(x) \pm 1} = \begin{pmatrix}7/8\\1\end{pmatrix}\frac{g_i}{2}\sigma_r T^4, \tag{7.1.3}$$

in which we have used the definition of the radiation density constant σ_r. The total energy density is therefore given by

$$\rho(T)c^2 = \left(\sum_B g_{iB} + \frac{7}{8}\sum_F g_{iF}\right)\frac{\sigma_r T^4}{2} = g^*(T)\frac{\sigma_r T^4}{2}, \tag{7.1.4}$$

in which B stands for bosons and F for fermions; the sums are taken over all the bosons and fermions with their respective statistical weights g_{iB} and g_{iF}. The quantity $g^*(T)$ is called the *effective number of degrees of freedom*. To obtain the total density of the Universe one must add the contribution $\rho_d(T)$, coming from those particles which are no longer in thermal equilibrium (i.e. those which have decoupled from the other particles, such as neutrinos after their decoupling) and the contribution $\rho_{nr}(T)$, coming from those particles which are still coupled but no longer relativistic, as is the case for the matter component in the plasma era. There may also be a component $\rho_{nt}(T)$ due to particles which are never in thermal equilibrium with the radiation (e.g. axions). As we shall see, for the period which interests us in this chapter, the contributions $\rho_d(T)$, $\rho_{nr}(T)$ and $\rho_{nt}(T)$ are generally negligible compared with $\rho(T)$.

The number-densities corresponding to each degree of freedom (spin state) of a boson, n_B, and fermionfermions, n_F, are

$$n_B = \tfrac{4}{3}n_F = \frac{\zeta(3)}{\pi^2}\left(\frac{k_B T}{\hbar c}\right)^3 \simeq \frac{\rho_B c^2}{3k_B T} \simeq \frac{\rho_F c^2}{3k_B T}. \tag{7.1.5}$$

As we shall see later, $g^*(T) < 200$ or so. This means that the average separation of the particles is

$$\bar{d} \simeq [g^*(T)n_B]^{-1/3} \simeq n_B^{-1/3} \simeq \frac{\hbar c}{k_B T}, \tag{7.1.6}$$

so that $\bar{d}$ practically coincides with the 'thermal wavelength' of the particles, $\hbar c/k_B T$, which is in some sense analogous to the Compton radius.

The cross-section of all the particles is, in the asymptotic limit $T \to T_P$,

$$\sigma_a \simeq \alpha^2\left(\frac{\hbar c}{k_B T}\right)^2, \tag{7.1.7}$$

with α of the order of 1/50, so that the collision time is

$$\tau_{\text{coll}} \simeq \frac{1}{n\sigma_{\text{a}}c} \simeq \frac{\hbar}{g^*(T)\alpha^2 k_{\text{B}}T}. \tag{7.1.8}$$

This time is to be compared with the expansion timescale $\tau_{\text{H}} = a/\dot{a}$:

$$\tau_{\text{H}} = 2t \simeq \left(\frac{3}{32\pi G\rho}\right)^{1/2} \simeq \frac{0.3\hbar T_{\text{P}}}{g^*(T)^{1/2}k_{\text{B}}T^2} \simeq \frac{2.42\times 10^{-6}}{g^*(T)^{1/2}}\left(\frac{T}{1\ \text{GeV}}\right)^{-2}\ \text{s} \tag{7.1.9}$$

(note that 1 GeV $\simeq 1.16\times 10^{13}$ K). We therefore have

$$\frac{\tau_{\text{coll}}}{\tau_{\text{H}}} \simeq \frac{1}{g^*(T)^{1/2}\alpha^2}\frac{T}{T_{\text{P}}} \ll 1. \tag{7.1.10}$$

The hypothesis of thermal equilibrium is consequently well founded.

One can easily verify that the assumption that the particles behave like a perfect gas is also valid. Given that asymptotically all the interactions are, so to speak, equivalent to electromagnetism with the same coupling constant, one can verify this hypothesis for two electrons: the ratio r between the kinetic energy, $E_{\text{c}} \simeq k_{\text{B}}T$, and the Coulomb energy, $E_{\text{p}} \simeq e^2/\bar{d}$ (using electrostatic units), is, from Equation (7.1.4),

$$r \simeq \frac{\bar{d}k_{\text{B}}T}{e^2} \simeq \frac{\hbar c}{e^2} \simeq 137 \gg 1: \tag{7.1.11}$$

to a good approximation r is the inverse of the fine-structure constant.

In Equations (7.1.4) and (7.1.5) there is an implicit hypothesis that the particles are not degenerate. We shall see in the next chapter that this hypothesis, at least for certain particles, is held to be the case for reasonably convincing reasons.

7.2 Fundamental Interactions

The evolution of the first phases of the hot Big Bang depends essentially on the physics of elementary particles and the theories that describe it. For this reason in this section we will make some comments on interactions between particles. It is known that there are four types of fundamental interactions: electromagnetic, weak nuclear, strong nuclear and gravitational. As far as the first three of these are concerned, quantum describes them in terms of the exchange of bosonic particles which play the role of force carriers.

The *electromagnetic interactions* are described classically by Maxwell's equations and in the quantum regime by *quantum electrodynamics* (QED). These forces are mediated by the photon, a massless boson: this implies that they have a long range. The coupling constant, a quantity which, roughly speaking, measures the strength of the interaction, is given by $g_{\text{QED}} = e^2/\hbar c \simeq 1/137$. From the point of view of group theory the Lagrangian describing electromagnetic interactions is invariant under the group of gauge transformations denoted U(1) (by gauge transformation we mean a transformation of local symmetry, i.e. depending upon space-time position).

In the standard model of particle physics the fundamental interacting objects are all fermions, either quarks or leptons. There are three families of leptons in this model, each consisting of a charged lepton and an associated neutrino. The charged leptons include the electron e^- but there are also μ^- and τ^- particles. Each of these has an accompanying neutrino designated ν_e, ν_μ and ν_τ. The leptons are therefore arranged in three families, each of which contains a pair of related particles. There are also antiparticles of the charged leptons (i.e. e^+, μ^+, τ^+), and antineutrinos of each type ($\bar{\nu}_e$, $\bar{\nu}_\mu$, $\bar{\nu}_\tau$).

The other fundamental fermions are the quarks, which also occur in three families of pairs mirroring the leptons. Quarks have fractional electronic charge but also possess a property known as *colour* which plays a role in strong nuclear reactions. The six quarks are denoted 'up' (u), 'down' (d), 'strange' (s), 'charmed' (c), 'bottom' (b) and 'top' (t). Each quark comes in three different colours, red, green and blue: these are denoted u_r, u_g, u_b and so on. The three families are arranged as (u, d), (s, c) and (b, t). There are also antiquarks for each quark (i.e. $\bar{u}$, $\bar{d}$) possessing opposite electrical charge and also opposite colour. Note that 'anti-up' is not the same thing as 'down'!

Basic properties of the quarks and leptons are shown in Appendix A.

The *weak nuclear interactions* involve all particles, but are generally of most interest when they involve the leptons. These interactions are of short range because the bosons that mediate the weak nuclear force (called W^+, W^- and Z_0) have masses $m_W \simeq 80$ GeV and $m_{Z_0} \simeq 90$ GeV. It is the mass of this boson that makes the weak interactions short range. The weak interactions can be described from a theoretical point of view by a theory developed by Glashow, Salam and Weinberg around 1970. According to this theory, the electromagnetic and weak interactions are different aspects of a single force (the *electroweak force*) which, for energies greater than $E_{EW} \simeq 10^2$ GeV, is described by a Lagrangian which is invariant under the group of gauge transformations denoted $SU(2)\times U(1)$. At energies above E_{EW} the leptons do not have mass and their electroweak interactions are mediated by four massless bosons (W_1, W_2, W_3, B), called the intermediate vector bosons, with a coupling constant of order g_{QED}. At energies lower than E_{EW} the symmetry given by the $SU(2) \times U(1)$ transformation group is spontaneously broken; the consequence of this is that the leptons (except perhaps the neutrinos) and the three bosons acquire masses (W^+, W^- and Z_0 can be thought of as 'mixtures' of quantum states corresponding to the W_1, W_2, W_3 and B). The only symmetry that remains is, then, the U(1) symmetry of electromagnetism.

The *strong nuclear interactions* involve above all the so-called hadrons. These are composite particles made of quarks, and are themselves divided into two classes: baryons and mesons. Baryons consist of combinations of three quarks of different colours (one red, one green, one blue) in such a way that they are colourless. Mesons are combinations of a quark and an anti-quark and are also colourless. Familiar examples of hadrons are p, $\bar{p}$, n, and $\bar{n}$, while the most relevant mesons are the pions π^+, π^-, π^0. All hadronic states are described from a quantum point of view by a theory called *quantum chromodynamics* (QCD). This theory was developed at a similar time to the theory which unifies the electromagnetic and weak interactions and, by now, it has gained considerable experimental support. According to QCD, hadrons are all made from quarks. There are various types of quark. These have different weak and electromagnetic interactions. The characteristic which distinguishes one quark from another is called *flavour*. The role of the bosons in the electroweak theory is played by the gluons, a family of eight massless bosons; the role of charge is replaced by a property of quarks and gluons called *colour*. At energies exceeding of the order of 200–300 MeV quarks are no longer bound into hadrons, and what appears is a quark–gluon plasma. The symmetry which the strong interactions respect is denoted SU(3).

The success of the unification of electromagnetic and weak interactions (Weinberg 1967; Salam 1968; Georgi and Glashow 1974) by the device of a restoration of a symmetry which is broken at low temperatures - i.e. $SU(2) \times U(1)$ - has encouraged many authors to attempt the unification of the strong interactions with the electroweak force. These theories are called GUTs (Grand Unified Theories); there exist many such theories and, as yet, no strong experimental evidence in their favour. In these theories other bosons, the superheavy bosons (with masses around 10^{15} GeV), are responsible for mediating the unified force; the Higgs boson is responsible for breaking the GUT symmetry. Amongst other things, such theories predict that protons should decay, with a mean lifetime of around 10^{32}–10^{33} years; various experiments are in progress to test this prediction, and it is possible that it will be verified or ruled out in the not too distant future. The simplest version of a GUT respects the SU(5) symmetry group which is spontaneously broken at an energy $E_{\mathrm{GUT}} \simeq 10^{15}$ GeV, so that $SU(5) \rightarrow SU(3) \times SU(2) \times U(1)$ (even though the SU(5) version seems to be rejected on the grounds of the mean lifetime of the proton, we refer to it here because it is the simplest model). For a certain choice of parameters the original SU(5) symmetry breaks instead to $SU(4) \times U(1)$ around 10^{14} GeV, which disappears around 10^{13} GeV giving the usual $SU(3) \times SU(2) \times U(1)$. The possible symmetry breaking occurs as a result of a first-order phase transition (which we shall discuss in the next section) and forms the basis of the first version of the inflationary universe model produced by Guth in 1981. It is, of course, possible that there is more than one phase transition between 10^{15} GeV and 100 GeV. At energies above E_{GUT}, in the simplest SU(5) model, the number of particle types corresponds to $g^*(T) \simeq 160$.

The fourth fundamental interaction is the *gravitational interaction*, which is described classically by *general relativity*. We have discussed some of the limitations of this theory in the previous chapter. The boson which mediates the gravitational force is usually called the graviton. It is interesting in the context of this chapter to ask whether we will ever arrive at a unification of all four interactions. Some attempts to construct a theory unifying gravity with the other forces involve the idea of *supersymmetry*; an example of such a theory is supergravity. This theory, amongst other things, unifies the fermions and bosons in a unique multiplet. More recently, great theoretical attention has been paid to the idea of *superstrings*, which we mentioned in the previous chapter. Whether these ideas will lead to significant progress towards a 'theory of everything' (TOE) remains an open question.

7.3 Physics of Phase Transitions

In certain many-particle systems one can find processes which can involve, schematically, the disappearance of some disordered phase, characterised by a certain symmetry, and the appearance of an ordered phase with a smaller degree of symmetry. In this type of order–disorder transition, called a *phase transition*, some macroscopic quantity, called the order parameter and denoted by Φ in this discussion, grows from its original value of zero in the disordered phase. The simplest physical examples of materials exhibiting these transitions are ferromagnetic substances and crystalline matter. In ferromagnets, for $T > T_c$ (the Curie temperature), the stable phase is disordered with net magnetisation $\mathcal{M} = 0$ (the quantity $\mathcal{M}$ in this case represents the order parameter); at $T < T_c$ a non-zero magnetisation appears in different domains (called the Weiss domains) and its direction in each domain breaks the rotational symmetry possessed by the disordered phase at $T > T_c$. In the crystalline phase of solids the order parameter is the deviation of the spatial distribution of ions from the homogeneous distribution they have at $T > T_f$, the melting point. At $T < T_f$ the ions are arranged on a regular lattice. One can also see an interesting example of a phase transition in the superconductivity properties of metals.

The lowering of the degree of symmetry of the system takes place even though the Hamiltonian which describes its evolution maintains the same degree of symmetry, even after the phase transition. For example, the macroscopic equations of the theory of ferromagnetism and the equations in solid-state physics do not pick out any particular spatial position or direction. The ordered states that emerge from such phase transitions have a degree of symmetry which is less than that governing the system. In fact, one can say that the solutions corresponding to the ordered state form a degenerate set of solutions (solutions with the same energy), which has the same degree of symmetry as the Hamiltonian. Returning to the above examples, the magnetisation $\mathcal{M}$ can in theory assume any direction. Likewise, the positioning of the ions in the crystalline lattice can be done in an infinite number of different ways. Taking into account all these possibilities we again obtain a homogeneous and isotropic state. Any small fluctuation, in the magnetic field of the domain for a ferromagnet or in the local electric field for a crystal, will pick out one preferred solution from this degenerate set and the system will end up in the state corresponding to that fluctuation. Repeating the phase transition with random fluctuations will produce randomly aligned final states. This is a little like the case of a free particle, described in Newtonian mechanics by $\dot{\boldsymbol{v}} = \mathbf{0}$, which has both translation and rotational symmetries. The solutions $\boldsymbol{r} = \boldsymbol{r}_0 + \boldsymbol{v}_0 t$, with $\boldsymbol{r}_0$ and $\boldsymbol{v}_0$ arbitrary, form a set which respects the symmetry of the original equation. But it is really just the initial conditions $\boldsymbol{r}_0$ and $\boldsymbol{v}_0$ which, at one particular time, select a solution from this set and this solution does not have the same degree of symmetry as that of the equations of motion.

A symmetry-breaking transition, during which the order parameter Φ grows significantly, can be caused by external influences of sufficient intensity: for example, an intense magnetic field can produce magnetisation of a ferromagnet even above T_c. Such phenomena are called *induced symmetry-breaking* processes, to distinguish them from *spontaneous symmetry breaking.* The spontaneous breaking of a symmetry comes from a gradual change of the parameters of the system itself. On this subject, it is convenient to consider the free energy of the system $F = U - TS$ (U is internal energy; T is temperature; S is entropy). Recall that the condition for existence of an equilibrium state of a system is that F must have a minimum. The free energy coincides with the internal energy only at $T = 0$. At higher temperatures, whatever the form of U, an increase in entropy (i.e. disorder) generally leads to a decrease in the free energy F, and is therefore favourable. For systems in which there is a phase transition, F is a function of the order parameter Φ. Given that Φ must respect the symmetry of the Hamiltonian of the system, it must be expressible in a manner which remains invariant with respect to transformations which leave the Hamiltonian itself unchanged. Under certain conditions F must have a minimum at $\Phi = 0$ (disordered state), while in others it must have a minimum with $\Phi \neq 0$ (ordered state).

Let us consider the simplest example. If the Hamiltonian has a reflection symmetry which is broken by the appearance of an order parameter Φ or, equivalently in this case, $-\Phi$, the free energy must be a function only of Φ^2 (in this example Φ is assumed to be a real, scalar variable). If Φ is not too large we can develop F in a power series

$$F(\Phi) \simeq F_0 + \alpha\Phi^2 + \beta\Phi^4, \tag{7.3.1}$$

where the coefficients α and β depend on the parameters of the system, such as its temperature. For $\alpha > 0$ and $\beta > 0$ we have a curve of the type marked '1' in Figure 7.1, while for $\alpha < 0$ and $\beta > 0$ we have a curve of type '2'.

Curve 1 corresponds to a disordered state; the system is in the minimum at $\Phi = 0$. Curve 2 has two minima at $\Phi_m = \pm(-\alpha/2\beta)^{1/2}$ and a maximum at $\Phi = 0$; in this case the disordered state is unstable, while the minima correspond to ordered states with the same probability: any small external perturbation, which renders one of the two minima Φ_m slightly deeper or nudges the system towards it, can make the system evolve towards this one rather than the other with $\Phi = -\Phi_m$. In this way one achieves a spontaneous symmetry breaking. If there is only one parameter describing the system, say the temperature, and the coefficient α is written as $\alpha = a(T - T_c)$, with $a > 0$, we have a situation represented by curve 2 for $T < T_c$. While T grows towards T_c the order parameter Φ decreases slowly and is zero at T_c. This type of transition, like its inverse, is called a *second-order phase transition* and it proceeds by a process known as *spinodal decomposition*: the order parameter appears or disappears gradually and the difference ΔF between $T > T_c$ and $T < T_c$ at $T \simeq T_c$ is infinitesimal.

There are also *first-order phase transitions*, in which at $T \simeq T_c$ the order parameter appears or disappears rapidly and the difference ΔF is finite. This difference

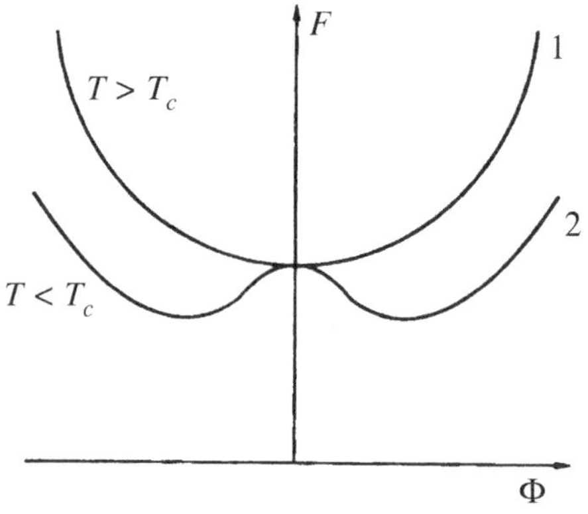

Figure 7.1 Free energy F of a system which undergoes a spontaneous symmetry breaking at a phase transition of second order in the order parameter Φ. The minimum of curve 1, corresponding to a temperature $T > T_c$, represents the equilibrium disordered state; the transition occurs at $T = T_c$; one of the two minima of curve 2, corresponding to the temperature $T < T_c$, represents the equilibrium ordered state which appears after the transition.

is called the *latent heat* of the phase transition. One would have this type of transition if, for example, in Equation (7.3.1) one added an extra term $\gamma(\Phi^2)^{3/2}$, with $\gamma < 0$, to the right-hand side. We now have the type of behaviour represented in Figure 7.2: in this case F acquires two new minima which become equal or less than $F_0 = F(0)$ for $T \leqslant T_c$.

In first-order phase transitions, when T changes from the situation represented by curve 1 of Figure 7.2 to that represented by curve 3, the phenomenon of *supercooling* can occur: the system remains in the disordered state represented by $\Phi = 0$ even when $T < T_c$ (state A); this represents a metastable equilibrium. As T decreases further, or the system is perturbed by either internal or external fluctuations, the system rapidly evolves into state B, which is energetically stable, liberating latent heat in the process. The system, still in the ordered state, is heated again up to a temperature of order T_c by the release of this latent heat, a phenomenon called *reheating*.

7.4 Cosmological Phase Transitions

The model of spontaneous symmetry breaking has been widely applied to the behaviour of particle interactions in the theories outlined in Section 7.2. Because phase transitions of this type appear generically in the early Universe according to standard particle physics models, the initial stages of the Big Bang are often described as the *era of phase transitions*. One important idea, which we shall refer to later, is that we can identify the order parameter Φ with the value of some scalar quantum field, most importantly the Higgs field at GUT scales, and the free energy F can then be related to the effective potential describing the interactions of that field, $V(\Phi)$. We shall elaborate on this in Sections 7.7 and 7.10.

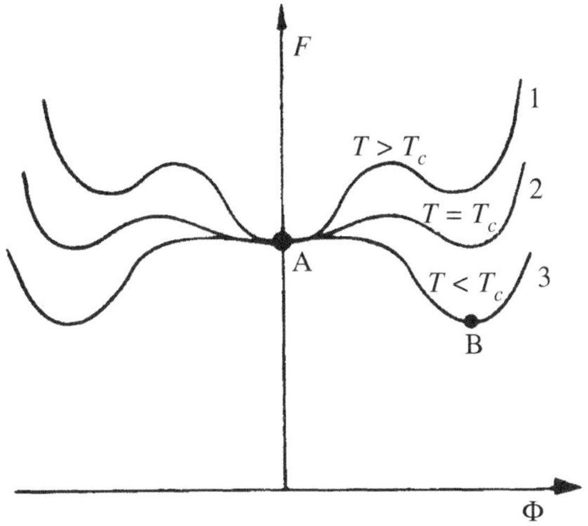

Figure 7.2 Free energy F of a system which undergoes a spontaneous symmetry breaking at a phase transition of first order in the order parameter Φ. The absolute minimum of curve 1, corresponding to a temperature $T > T_c$, represents the equilibrium disordered state; the transition does not happen at $T = T_c$ (curve 2), but at $T < T_c$, when the barrier between the central minimum and the two others becomes negligible (curve 3).

The period from $t_P \simeq 10^{-43}$ s, corresponding to a temperature $T_P \simeq 10^{19}$ GeV, to the moment at which quarks become confined in hadrons at $T \simeq$ 200–300 MeV, can be divided into various intervals according to the phase transitions which characterise them.

1. $T_P \simeq 10^{19}$ GeV $> T > T_{GUT} \simeq 10^{15}$ GeV. In this period quantum gravitational effects become negligible and the particles are held in thermal equilibrium for $T \leqslant 10^{16}$ GeV by means of interactions described by a GUT. Thanks to the fact that baryon number is not conserved in GUTs, any excess of baryons over antibaryons can be removed at high energies; at $T \simeq 10^{15}$ GeV the Universe is baryon-symmetric, i.e. quarks and antiquarks are equivalent. It is possibly also the case that viscosity effects at the GUT scale can lead to a reduction in the level of inhomogeneity of the Universe at this time. At temperatures $T_{GUT} \simeq 10^{15}$ GeV, corresponding to $t \simeq 10^{-37}$ s, we will take the simplest GUT symmetry of SU(5).

2. $T \simeq 10^{15}$ GeV. At $T \simeq 10^{15}$ GeV there is a spontaneous breaking of the SU(5) symmetry into SU(3) × SU(2) × U(1) or perhaps some other symmetry for some intervening period. As we shall see in detail later, the GUT phase transition at T_{GUT} results in the formation of *magnetic monopoles*: this is a problem of the standard model which is discussed in Section 7.6 and which may be solved by *inflation*, which is usually assumed to occur in this epoch. A GUT which unifies the electroweak interactions with the strong interactions, puts leptons and hadrons on the same footing and thus allows processes which do not conserve baryon number B (violation of baryon number conservation

is not allowed in either QCD or electroweak theory). It is thought therefore that processes could occur at T_{GUT}, which might create a baryon–antibaryon asymmetry which is observed now in the form of the very large ratio n_γ / n_b, as we explained in Section 5.5. In order to create an excess of baryons from a situation which is initially baryon-symmetric at $T > 10^{15}$ GeV, i.e. to realise a process of *baryosynthesis*, it is necessary to have

(a) processes which violate B conservation;

(b) violation of C or CP symmetry (C is charge conjugation; P is parity conjugation; violation of symmetry under these operations has been observed in electroweak interactions), otherwise, for any process which violates B-conservation, there would be another process with the same rate happening to the anti-baryons and thus cancelling the net effect;

(c) processes which do violate B-conservation must occur out of equilibrium because a theorem of statistical mechanics shows that an equilibrium distribution with B = 0 remains so regardless of whether B, C and CP are violated – this theorem shows that equilibrium distributions cannot be modified by collisions even if the invariance under time-reversal is violated.

It is interesting to note that the three conditions above, necessary for the creation of a baryon–antibaryon asymmetry, were given by Sakharov (1966). It seems that these conditions are valid at $T \simeq 10^{15}$ GeV, or slightly lower, depending on the particular version of GUT or other theory; it is even the case that baryosynthesis can occur at much lower energies, around the electroweak scale. Even though this problem is complicated and therefore rather controversial, with reasonable hypotheses one can arrive at a value of baryon–antibaryon asymmetry of order 10^{-8}–10^{-13}, which includes the observed value: the uncertainty here derives not only from the fact that one can obtain baryosynthesis in GUTs of various types, but also that in any individual GUT there are many free, or poorly determined, parameters. It is also worth noting that, if the Universe is initially lepton-symmetric, the reactions which violate B can also produce an excess of leptons over antileptons (equal in the case of SU(5) GUTs to that of the baryons over the antibaryons). This is simply because the GUTs unify quarks and leptons: this is one theoretical motivation for assumption, which we shall make in the next chapter: that the chemical potential for the leptons is very close to zero at the onset of nucleosynthesis. Notice finally that in a GUT the value of the baryon asymmetry actually produced depends only on microphysical parameters; this means that, even if the Universe is inhomogeneous, the value of the asymmetry should be the same in any region. Given that it is proportional to the entropy per baryon σ_{rad}, it turns out that any inhomogeneity produced must be of adiabatic type (i.e. leaving σ_{rad} unchanged relative to an unperturbed region). In some very special situations, which we shall not go into here, it is possible however to generate isothermal fluctuations. We shall discuss adiabatic and isothermal perturbations in much more detail in Chapter 12.

3. $T_{GUT} > T > T_{EW}$. When the temperature falls below 10^{15} GeV, the unification of the strong and electroweak interactions no longer holds. The superheavy bosons rapidly disappear through annihilation or decay processes. In the moment of symmetry breaking the order parameter Φ, whose appearance signals the phase transition proper, can assume a different 'sign' or 'direction' in adjoining spatial regions: it is possible in this way to create places where Φ changes rapidly with spatial position, as one moves between different regions, similar to the 'Bloch walls' which, in a ferromagnet, separate the different domains of magnetisation. These 'singular' regions where Φ is discontinuous have a structure which depends critically upon the symmetry which has been broken; we shall return to this in Section 7.6. The period we are discussing here lasts from $t_{GUT} \simeq 10^{-37}$ s to $t_{EW} \simeq 10^{-11}$ s: in logarithmic terms this is a very long time indeed. It is probable that phase transitions occur in this period which are not yet well understood. This corresponds to an energy range from 100-10^{15} GeV; within the framework of the SU(5) model discussed above there are no particles predicted to have masses in this range of energies, which is, consequently, called the 'grand desert'. Nevertheless, there remain many unresolved questions regarding this epoch. In any case, towards the end of this period one can safely say that, to a good approximation, the Universe is filled with an ideal gas of leptons and antileptons, the four vector bosons, quarks and antiquarks and gluons; in all this corresponds to $g^* \simeq 10^2$. At the end of this period the size of the cosmological horizon is around one centimetre and contains around 10^{19} particles.
4. $T_{EW} > T > T_{QH} \simeq$ 200–300 MeV. At $T \simeq 10^2$ GeV there will be a spontaneous breaking of the $SU(2) \times U(1)$ symmetry, through a phase transition which is probably of first order but very weakly so. All the leptons acquire masses (with the probable exception of the neutrinos) while the intermediate vector bosons give rise to the massive bosons W^+, W^- and Z_0 and photons. The massive bosons disappear rapidly through decay and annihilation processes when the temperature falls below around 90 GeV. For a temperature $T_{QH} \simeq$ 200–300 MeV, however, we have a final phase transition in the framework of QCD theory: the strong interactions do indeed become very strong and lead to the confinement of quarks into hadrons, the *quark–hadron phase transition*. There thus begins the (very short) hadron era, which we shall discuss in the next chapter. When the temperature reaches T_{QH}, the cosmological time is $t_{QH} \simeq 10^{-5}$ s and the cosmological horizon is around a kilometre in size.

7.5 Problems of the Standard Model

The standard model of the hot Big Bang is based on the following assumptions.

1. That the laws of physics which have been verified at the present time by laboratory experiments are also valid in the early Universe (this does not

include such theories as GUT, supersymmetry and the like which we refer to as 'new physics') and that gravity is described by the theory of general relativity without a cosmological constant.

2. That the Cosmological Principle holds.
3. That the appropriate 'initial conditions', which may in principle be predicted by a more general theory, are that the temperature at some early time $t_{\rm i}$ is such that $T_{\rm i} > 10^{12}$ K and the contents of the Universe are in thermal equilibrium, that there is (somehow) a baryon asymmetry consistent with the observed value of $\sigma_{\rm rad}$, that $\Omega(t_{\rm i})$ is very close to unity (see below), and, finally, that there is some spectrum of initial density fluctuations which give rise to structure formation at late times.

This standard cosmology has achieved four outstanding successes:

1. the predictions of light-element abundances produced during cosmological nucleosynthesis agree with observations, as we shall see in the next chapter;
2. the cosmic microwave background is naturally explained as a relic of the initial 'hot' thermal phase;
3. it accounts naturally for the expansion of the Universe; and
4. it provides a framework within which one can understand the formation of galaxies and other cosmic structures.

There remain, however, certain problems (or, at least, unexplained features) connected with the Big Bang cosmology:

1. the origin of the Universe or, in less elevated language, the evolution of the Universe before the Planck time;
2. the cosmological horizon, which we discuss below;
3. the question of why the Universe is close to being flat, again discussed below;
4. the baryosynthesis or, in other words, the origin of the baryon asymmetry;
5. the evolution of the Universe at energies greater than $T > 100$ GeV;
6. the origin of the primordial spectrum of density fluctuations, whatever it is;
7. the apparently 'excessive' degree of homogeneity and isotropy of the Universe; and
8. the nature of the ubiquitous dark matter.

Notice that there are, apparently, more 'problems' than 'solutions'!

The incorporation of 'new physics' into the Big Bang model holds out the possibility of resolving some of these outstanding issues, though this has so far only been achieved in a qualitative manner. The assumptions made in what one might call the 'revised standard model' would then be that

1. known physics and theories of particle physics ('new physics') are valid, as is general relativity with Λ not necessarily zero;

2. the Cosmological Principle is valid; and
3. the same initial conditions hold as in the standard model at $T_i \simeq 10^{19}$ GeV, except that the baryon asymmetry is accounted for (in principle) by the new physics we have accepted into the framework.

Successes of the 'revised standard model' are

1. all the advantages of the standard model;
2. a relatively clear understanding of the evolution of the Universe at $T > 10^{12}$ K;
3. the possible existence of non-baryonic particles as candidates for the dark matter;
4. the explanation of baryosynthesis (though, as yet, only qualitatively); and
5. a consolidation of the theory of structure formation by virtue of the existence of non-baryonic particles through (3).

This modernised version of the Big Bang therefore eliminates many of the problems of the standard model, particularly the fourth, fifth and eighth of the previous list, but leaves some and, indeed, adds some others. Two new problems which appear in this model are concerned with: (1) the possible production of magnetic monopoles and (2) the cosmological constant. We shall discuss these in Sections 7.6 and 7.7. We shall see later in this chapter that the theory of inflation can 'solve' the monopole, flatness and horizon problems.

7.6 The Monopole Problem

Any GUT in which electromagnetism, which has a U(1) gauge group, is contained within a gauge theory involving a spontaneous symmetry breaking of a higher symmetry, such as SU(5), provides a natural explanation for the quantisation of electrical charge and this implies the existence of *magnetic monopoles*. These monopoles are point-like defects in the Higgs field Φ which appears in GUTs. Defects are represented schematically in Figure 7.3, in which the arrows indicate the orientation of Φ in the internal symmetry space of the theory, while the location of the arrows represents a position in ordinary space. Monopoles are zero dimensional; higher-dimensional analogues are also possible and are called *strings* (one dimensional), *domain walls* (two dimensional) and *textures* (three dimensional).

In this discussion we shall use electrostatic units. Monopoles have a magnetic charge

$$g_n = n g_D, \tag{7.6.1}$$

which is a multiple of the Dirac charge g_D,

$$g_D = \frac{\hbar c}{2e} = 68.5e; \tag{7.6.2}$$

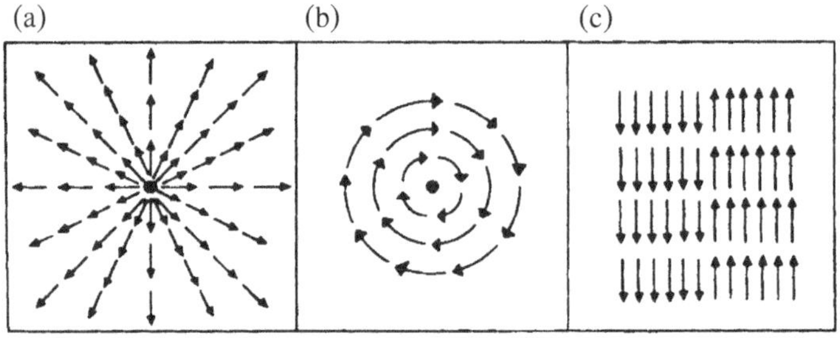

Figure 7.3 Schematic representation of topological defects in the Higgs field: a monopole (a); a string (b); a domain wall (c). The three-dimensional analogue of these defects is called a texture, but we cannot draw this in two dimensions! The arrows represent the orientation of the field Φ in an internal symmetry space, while their position indicates location in real space.

a mass

$$m_{\rm M} \simeq 4\pi \frac{\hbar c}{e^2} m_{\rm X} \simeq 10^3 m_{\rm X}, \tag{7.6.3}$$

where X is the boson that mediates the GUT interaction, called the Higgs boson, with mass

$$m_{\rm X} \simeq e(\hbar c)^{1/2} m_{\rm GUT} \simeq 10^{-1} m_{\rm GUT} \tag{7.6.4}$$

($m_{\rm GUT}$ is the energy corresponding to the spontaneous breaking of the GUT symmetry); the size of the monopoles is

$$r_{\rm M} \simeq \frac{\hbar}{m_{\rm X} c}. \tag{7.6.5}$$

For typical GUTs, such as SU(5), we have $m_{\rm GUT} \simeq 10^{14}$–$10^{15}$ GeV, so that $m_{\rm M} \simeq 10^{16}$ GeV ($\simeq 10^{-8}$ g) and $r_{\rm M} \simeq 10^{-28}$ cm.

The other types of topological defects in the Higgs field shown in Figure 7.3 are also predicted by certain GUTs. The type of defect appearing in a phase transition depends on the symmetry and how it is broken in a complicated fashion, which we shall not discuss here. From a cosmological perspective, domain walls, if they exist, represent a problem just as monopoles do and which we shall discuss a little later. Cosmic strings, however, again assuming they exist, may be a solution rather than a problem because they may be responsible for generating primordial fluctuations which give rise to galaxies and clusters of galaxies, though this is believed only by a minority of cosmologists; we shall discuss this option briefly in Section 13.9.

Now let us explain the *cosmological monopole problem*. In the course of its evolution the Universe suffers a spontaneous breaking of the GUT symmetry at $T_{\rm GUT}$, for example via SU(5) $\rightarrow$ SU(3)$\times$SU(2)$\times$U(1). As we discussed in Section 7.3 it therefore moves from a disordered phase to an ordered phase characterised by an order parameter $\Phi \neq 0$, which in this case is just the value of the Higgs field. During this transition monopoles will be formed. The number of monopoles can be estimated in the following manner: if ξ is the characteristic dimension of the domains which form during the breaking of the symmetry (ξ is also sometimes called the correlation length of the Higgs field), the maximum number density of monopoles $n_{\mathrm{M,max}}$ is of the order ξ^{-3}. In reality, not all the intersections between domains give rise to monopoles: one expects that this reduces the above estimate by a factor $p \simeq \frac{1}{10}$. Given that the points within any single domain are causally connected, we must have

$$\xi < r_{\mathrm{H}}(t) \simeq 2ct \simeq 0.6 g^*(T)^{-1/2} \frac{\hbar T_{\mathrm{P}} c}{k_{\mathrm{B}} T^2}, \tag{7.6.6}$$

where T_{P} is the Planck temperature. It turns out therefore that, at T_{GUT},

$$n_{\mathrm{M}} > p \left[\frac{g^*(T_{\mathrm{GUT}})^{1/2} T_{\mathrm{GUT}}}{0.6 T_{\mathrm{P}}} \right]^3 n_{\gamma}(T_{\mathrm{GUT}}), \tag{7.6.7}$$

which, for $T_{\mathrm{GUT}} \simeq 10^{15}$ GeV, gives

$$n_{\mathrm{M}} > 10^{-10} n_{\gamma}. \tag{7.6.8}$$

Any subsequent physical processes are expected to be very inefficient at reducing the ratio $n_{\mathrm{M}}/n_{\gamma}$. The present density of monopoles per unit volume is therefore expected to be

$$n_{0M} > 10^{-10} n_{0\gamma} \simeq n_{0\mathrm{b}}, \tag{7.6.9}$$

which is of order, or greater than, that of the baryons and which corresponds to a density parameter in monopoles of order

$$\Omega_{\mathrm{M}} > \frac{m_{\mathrm{M}}}{m_{\mathrm{p}}} \Omega_{\mathrm{b}} \simeq 10^{16}, \tag{7.6.10}$$

clearly absurdly large.

The problem of the domain walls, in cases where they are predicted by GUTs, is of the same character. The problem of cosmological monopole production, which to some extent negates the successes of cosmologies incorporating the 'new physics', was the essential stimulus which gave rise to the inflationary cosmology we shall discuss later in this chapter.

7.7 The Cosmological Constant Problem

As we saw in Chapter 1, the Einstein equations with $\Lambda \neq 0$, having

$$T_{ij}^{(\Lambda)} = -p_{\Lambda} g_{ij} + (p_{\Lambda} + \rho_{\Lambda} c^2) U_i U_j, \tag{7.7.1}$$

where

$$\rho_{\Lambda} = -\frac{p_{\Lambda}}{c^2} \equiv \frac{\Lambda c^2}{8\pi G}, \tag{7.7.2}$$

yield for the case of a homogeneous and isotropic universe the relations

$$\dot{a}^2 = \tfrac{8}{3}\pi G(\rho + \rho_\Lambda)a^2 - Kc^2, \tag{7.7.3 a}$$

$$\ddot{a} = -\tfrac{4}{3}\pi G\left(\rho + 3\frac{p}{c^2} - 2\rho_\Lambda\right)a. \tag{7.7.3 b}$$

From these equations at $t = t_0$, putting $p_0 \simeq 0$, we obtain

$$\frac{K}{a_0^2} = \frac{H_0^2}{c^2}(\Omega_0 + \Omega_\Lambda - 1), \tag{7.7.4 a}$$

$$q_0 = \tfrac{1}{2}\Omega_0 - \Omega_\Lambda, \tag{7.7.4 b}$$

where $\Omega_\Lambda \equiv \rho_\Lambda/\rho_{0c}$. The observational limits on Ω_0 and q_0 yield

$$|\rho_\Lambda| < 2\rho_{0c} \simeq 4 \times 10^{-29}\ \mathrm{g\ cm^{-3}} \simeq 10^{-46}\frac{m_n^4}{(\hbar/c)^3} \simeq 10^{-48}\ \mathrm{GeV}^4 \tag{7.7.5}$$

(m_n is the mass of a nucleon; in the last relation we have used 'natural' units in which $\hbar = c = 1$), corresponding to

$$|\Lambda| < 10^{-55}\ \mathrm{cm}^{-2}. \tag{7.7.6}$$

From Λ one can also construct a quantity which has the dimensions of a mass

$$m_\Lambda = \left[|\rho_\Lambda|\left(\frac{\hbar}{c}\right)^3\right]^{1/4} = \left(\frac{\hbar^3}{8\pi Gc}|\Lambda|\right)^{1/4} < 10^{-32}\ \mathrm{eV} \tag{7.7.7}$$

(to be compared with the upper limit on the mass of the photon: according to recent estimates this is $m_\gamma < 3 \times 10^{-27}$ eV). The problem of the cosmological constant lies in the fact that the quantities $|\Lambda|$, $|\rho_\Lambda|$ and $|m_\Lambda|$ are so amazingly and, apparently, 'unnaturally' small.

The modern interpretation of Λ is the following: ρ_Λ and p_Λ represent the density and pressure of the *vacuum*, which is understood to be like the ground state of a quantum system:

$$\rho_\Lambda \equiv \rho_v, \qquad p_\Lambda \equiv p_v = -\rho_v c^2 \tag{7.7.8}$$

(the equation of state $p_v = -\rho_v c^2$ comes from the Lorentz-invariance of the energy-momentum tensor of the vacuum). In modern theories of elementary particles with spontaneous symmetry breaking it turns out that

$$\rho_v \simeq V(\Phi, T), \tag{7.7.9}$$

where $V(\Phi, T)$ is the effective potential for the theory. This is the analogous quantity to the free energy F discussed above in the simple (non-quantum) thermodynamical case of Section 7.3; its variation with T determines the spontaneous breaking of the symmetry; Φ is the Higgs field, the expectation value of which is analogous to the order parameter in the thermodynamical case. An important consequence of Equation (7.7.9) is that the cosmological 'constant' depends on time through its dependence upon T. This fact is essentially the basis of the inflationary model we shall come to shortly.

Modern gauge theories predict that

$$\rho_{\mathrm{v}} \simeq \frac{m^4}{(\hbar/c)^3} + \text{const.}, \tag{7.7.10}$$

where m is the energy at which the transition occurs (10^{15} GeV for GUT transitions, 10^2 GeV for the electroweak transition, 10^{-1} GeV for the quark-hadron transition and (perhaps) 10^3 GeV for a supersymmetric transition). The constant in Equation (7.7.10) is arbitrary (although its value might be accounted for in supersymmetric theories). In the symmetry-breaking phase one has a decrease of ρ_{v} of order

$$\Delta\rho_{\mathrm{v}} \simeq \frac{m^4}{(\hbar/c)^3}, \tag{7.7.11}$$

corresponding to 10^{60} GeV^4 for the GUT, 10^{12} GeV^4 for supersymmetry, 10^8 GeV^4 for the electroweak transition, and 10^{-4} GeV^4 for QCD.

In light of these previous comments the cosmological constant problem can be posed in a clearer form:

$$\rho_{\mathrm{v}}(t_{\mathrm{P}}) = \rho_{\mathrm{v}}(t_0) + \sum_i \Delta\rho_{\mathrm{v}}(m_i) \simeq 10^{-48}\ \mathrm{GeV}^4 + 10^{60}\ \mathrm{GeV}^4 = \sum_i \Delta\rho_{\mathrm{v}}(m_i)(1+10^{-108}), \tag{7.7.12}$$

where $\rho_{\mathrm{v}}(t_{\mathrm{P}})$ and $\rho_{\mathrm{v}}(t_0)$ are the vacuum density at the Planck and present times, respectively, and m_i represents the energies of the various phase transitions which occur between t_{P} and t_0. Equation (7.7.12) can be phrased in two ways: $\rho_{\mathrm{v}}(t_{\mathrm{P}})$ must differ from $\sum_i \Delta\rho_{\mathrm{v}}(m_i)$ over the successive phase transitions by only one part in 10^{108}; or the sum $\sum_i \Delta\rho_{\mathrm{v}}(m_i)$ must, in some way, arrange itself so as to satisfy (7.7.12). Either way, there is definitely a problem of extreme 'fine-tuning' in terms of $\rho_{\mathrm{v}}(t_{\mathrm{P}})$ or $\sum_i \Delta\rho_{\mathrm{v}}(m_i)$.

At the moment, there exist only a few theoretical models which even attempt to resolve the problem of the cosmological constant. Indeed, many cosmologists regard this problem as the most serious one in all cosmology. This is strictly connected with the theory of particle physics and, in some way, to quantum gravity. Inflation, we shall see, does not solve this problem; indeed, one could say that inflation is founded upon it.

7.8 The Cosmological Horizon Problem

7.8.1 The problem

Recall that one of the fundamental assumptions of the Big Bang theory is the Cosmological Principle, which, as we explained in Chapter 6, is intimately connected with the existence of the initial singularity. As we saw in Chapter 2, all the Friedmann models with equation of state in the form $p = w\rho c^2$, with $w \geqslant 0$, possess a particle horizon. This result can also be extended to other equations of state with $p \geqslant 0$ and $\rho \geqslant 0$. If the expansion parameter tends to zero at early times like t^β (with $\beta > 0$), then the particle horizon at time t,

$$R_{\mathrm{H}}(t) = a(t) \int_0^t \frac{c\,\mathrm{d}t'}{a(t')}, \tag{7.8.1}$$

exists if $\beta < 1$. From Equation (6.1.1), with $a \propto t^\beta$, we obtain

$$\beta(\beta - 1) = -\tfrac{4}{3}\pi G\left(\rho + 3\frac{p}{c^2}\right)t^2 \propto \ddot{a}. \tag{7.8.2}$$

This demonstrates that the condition for the existence of the Big Bang singularity, $\ddot{a} < 0$, requires that $0 < \beta < 1$ and that there must therefore also be a particle horizon.

The existence of a cosmological horizon makes it difficult to accept the Cosmological Principle. This principle requires that there should be a correlation (a very strong correlation) of the physical conditions in regions which are outside each other's particle horizons and which, therefore, have never been able to communicate by causal processes. For example, the observed isotropy of the microwave background implies that this radiation was homogeneous and isotropic in regions on the last scattering surface (i.e. the spherical surface centred upon us which is at a distance corresponding to the look-back time to the era at which this radiation was last scattered by matter). As we shall see in Chapter 9, last scattering probably took place at an epoch, t_{ls}, corresponding to a redshift $z_{\mathrm{ls}} \simeq 1000$. At that epoch the last scattering surface had a radius

$$r_{\mathrm{ls}} \simeq \frac{c(t_0 - t_{\mathrm{ls}})}{(1 + z_{\mathrm{ls}})} \simeq \frac{ct_0}{z_{\mathrm{ls}}}, \tag{7.8.3}$$

because $z_{\mathrm{ls}} \gg 1$. The radius of the particle horizon at this epoch is given by Equation (2.7.3) with $w = 0$,

$$R_{\mathrm{H}}(z_{\mathrm{ls}}) \simeq 3ct_0 z_{\mathrm{ls}}^{-3/2} \simeq 3r_{\mathrm{ls}} z_{\mathrm{ls}}^{-1/2} \simeq 10^{-1} r_{\mathrm{ls}} \ll r_{\mathrm{ls}}; \tag{7.8.4}$$

at z_{ls} the microwave background was homogeneous and isotropic over a sphere with radius at least ten times larger than that of the particle horizon.

Various routes have been explored in attempts to find a resolution of this problem. Some homogeneous but anisotropic models do not have a particle horizon at all. One famous example is the mix-master model proposed by Misner (1968), which we mentioned in Chapters 1 and 3. Other possibilities are to invoke some kind of isotropisation process connected with the creation of particles at the Planck epoch, or a modification of Einstein's equations to remove the Big Bang singularity and its associated horizon.

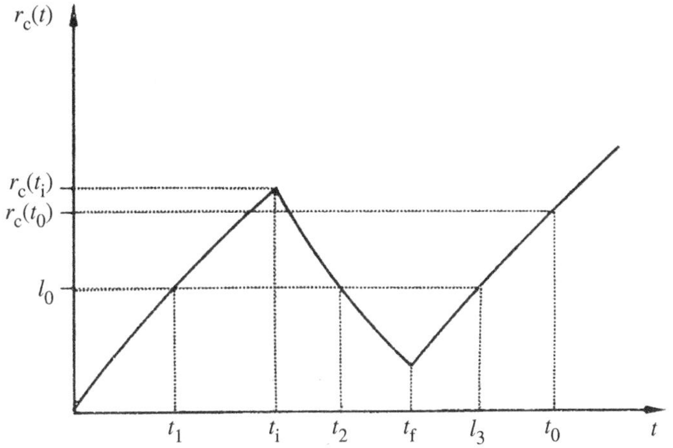

Figure 7.4 Evolution of the comoving cosmological horizon $r_c(t)$ in a universe characterised by a phase with an accelerated expansion (inflation) from t_i to t_f. The scale l_0 enters the horizon at t_1, leaves at t_2 and re-enters at t_3. In a model without inflation the horizon scale would never decrease so scales entering at t_0 could never have been in causal contact before. The horizon problem is resolved if $r_c(t_0) \leqslant r_c(t_i)$.

7.8.2 The inflationary solution

The inflationary universe model also resolves the cosmological horizon problem in an elegant fashion. We shall discuss inflation in detail in Sections 7.10 and 7.11, but this is a good place to introduce the basic idea. Recall that the horizon problem is essentially the fact that a region of proper size l can only become causally connected when the horizon $R_H = l$. In the usual Friedmann models at early times the horizon grows like t, while the proper size of a region of fixed comoving size scales as t^β with $\beta < 1$. In the context of inflation it is more illuminating to deal with the radius of the Hubble sphere (which determines causality properties at a particular epoch) rather than the particle horizon itself. As in Section 2.7 we shall refer to this as the *cosmological horizon* for the rest of this chapter; its proper size is $R_c = c/H = ca/\dot{a}$ and its comoving size is $r_c = R_c(a_0/a) = ca_0/\dot{a}$. The comoving scale l_0 enters the cosmological horizon at time $t_H(l_0) \neq 0$ because r_c grows with time. Processes occurring at the epoch t cannot connect the region of size l_0 causally until $t \geqslant t_H(l_0)$. In the 'standard' models, with $p/\rho c^2 = w =$ const. and $w > -\frac{1}{3}$, we have at early times

$$r_H = \frac{a_0}{a} R_H(t) = a_0 \int_0^t \frac{c\, dt'}{a(t')} \simeq \frac{3(1+w)}{(1+3w)} ct_0 \left(\frac{t}{t_0}\right)^{(1+3w)/3(1+w)} \simeq c\frac{a_0}{\dot{a}}, \quad (7.8.5)$$

so that

$$r_H \simeq r_c; \quad (7.8.6)$$

one therefore finds that $\dot{r}_H \propto -\ddot{a} \propto (1+3w) > 0$.

Imagine that there exists a period $t_i < t < t_f$ sometime during the expansion of the Universe, in which the comoving scale l_0, which has already been causally connected, somehow manages to escape from the horizon, in the sense that any physical processes occurring in this interval can no longer operate over the scale l_0. We stress that it is not possible to 'escape' in this way from a particle horizon (or event horizon), but the cosmological horizon is not a true horizon in the formal sense explained in Section 2.7. Such an escape occurs if

$$l_0 > r_c. \tag{7.8.7}$$

This inequality can only be valid if the comoving horizon $ca_0/\dot{a}$ decreases with time, which requires an accelerated expansion, $\ddot{a} > 0$. After t_f we suppose that the Universe resumes the usual decelerated expansion. The behaviour of r_c in such a model is shown graphically in Figure 7.4. The scale l_0 is not causally connected before t_1. It becomes connected in the interval $t_1 < t < t_2$; at t_2 it leaves the horizon; in the interval $t_2 < t < t_3$ its properties cannot be altered by (causal) physical processes; at t_3 it enters the horizon once more, in the sense that causal processes can affect the physical properties of regions on the scale l_0 after this time. An observer at time t_3, who was unaware of the existence of the period of accelerated expansion, would think the scale l_0 was coming inside the horizon for the first time and would be surprised if it were homogeneous. This observer would thus worry about the horizon problem. The problem is, however, non-existent if there is an accelerated expansion and if the maximum scale which is causally connected is greater than the present scale of the horizon, i.e.

$$r_c(t_0) \leqslant r_c(t_i). \tag{7.8.8}$$

To be more precise, unless we accept it as a coincidence that these two comoving scales should be similar, a solution is only really obtained if the inequality (7.8.8) is strong, i.e. $r_c(t_0) \ll r_c(t_i)$.

In any case the solution is furnished by a period $t_i < t < t_f$ of appropriate duration, in which the universe suffers an accelerated expansion: this is the definition of *inflation*. In such an interval we must therefore have $p < -\rho c^2/3$; in particular if $p = w\rho c^2$, with constant w, we must have $w < -\frac{1}{3}$. From the Friedmann equations in this case we recover, for $t_f > t > t_i$,

$$a \simeq a(t_i)\left[1 + \frac{1}{q}H(t_i)(t - t_i)\right]^q, \qquad q = \frac{2}{3(1 + w)} \tag{7.8.9}$$

(this solution is exact when the curvature parameter $K = 0$). For $H(t_i)t \gg 1$ one has

$$a \propto t^q \qquad (-\tfrac{1}{3} > w > -1), \tag{7.8.10a}$$

$$a \propto \exp(t/\tau) \qquad (w = -1), \tag{7.8.10b}$$

$$a \propto (t_a - t)^q \qquad (w < -1); \tag{7.8.10c}$$

the exponent q is greater than one in the first case and negative in the last case; $\tau = (a/\dot{a})_{t=t_i}$ and $t_a = t_i - [2/(3(1 + w))]H(t_i)^{-1} > t_i$. The types of expansion described by these equations are particular cases of an accelerated expansion. One can verify that the condition for inflation can be expressed as

$$\ddot{a} = a(H^2 + \dot{H}) > 0; \tag{7.8.11}$$

sometimes one uses the terms *sub-inflation* for models in which $\dot{H} < 0$, *standard inflation* or *exponential inflation* for $\dot{H} = 0$, and *super-inflation* for $\dot{H} > 0$. The three solutions (7.8.10) correspond to these three cases, respectively; the type of inflation expressed by (7.8.10 *a*) is also called *power-law inflation.*

The requirement that they solve the horizon problem imposes certain conditions on inflationary models. Consider a simple model in which the time between some initial time t_i and the present time t_0 is divided into three intervals: (t_i, t_f), (t_f, t_{eq}), (t_{eq}, t_0). Let the equation-of-state parameter in any of these intervals be w_{ij}, where i and j stand for any of the three pairs of starting and finishing times. Let us take, for example, $w_{ij} = w < -\frac{1}{3}$ for the first interval, $w_{ij} = \frac{1}{3}$ for the second, and $w_{ij} = 0$ for the last. If $\Omega_{ij} \simeq 1$ in any interval, then

$$\frac{H_i a_i}{H_j a_j} \simeq \left(\frac{a_i}{a_j}\right)^{-(1+3w_{ij})/2} \tag{7.8.12}$$

from Equation (2.1.12). The requirement that

$$r_c(t_i) = c\frac{a_0}{\dot{a}_i} \gg r_c(t_0) = \frac{c}{H_0} \tag{7.8.13}$$

implies that $H_i a_i \ll H_0 a_0$. This, in turn, means that

$$\frac{H_i a_i}{H_f a_f} \ll \frac{H_0 a_0}{H_f a_f} = \frac{H_0 a_0}{H_{eq} a_{eq}} \frac{H_{eq} a_{eq}}{H_f a_f}, \tag{7.8.14}$$

so that, from (7.8.12), one gets

$$\left(\frac{a_f}{a_i}\right)^{-(1+3w)} \gg \left(\frac{a_0}{a_{eq}}\right)\left(\frac{a_{eq}}{a_f}\right)^2, \tag{7.8.15}$$

which yields, after some further manipulation,

$$\left(\frac{a_f}{a_i}\right)^{-(1+3w)} \gg 10^{60} z_{eq}^{-1} \left(\frac{T_f}{T_P}\right)^2 \tag{7.8.16}$$

($T_P \simeq 10^{32}$ K is the usual Planck temperature). This result requires that the number of e-foldings, $\mathcal{N} \equiv \ln(a_f/a_i)$, should be

$$\mathcal{N} \gg 60\left[\frac{2.3 + \frac{1}{30}\ln(T_f/T_P) - \frac{1}{60}\ln z_{eq}}{|1+3w|}\right]. \tag{7.8.17}$$

In most inflationary models which have been proposed, $w \simeq -1$ and the ratio T_f/T_P is contained in the interval between 10^{-5} and 1, so that this indeed requires $\mathcal{N} \gg 60$.

7.9 The Cosmological Flatness Problem

7.9.1 The problem

In the Friedmann equation without the cosmological constant term

$$\left(\frac{\dot{a}}{a}\right)^2 = \frac{8\pi}{3}G\rho - \frac{Kc^2}{a^2}, \tag{7.9.1}$$

when the universe is radiation dominated so that $\rho \propto T^4$, there is no obvious characteristic scale other than the Planck time

$$t_{\mathrm{P}} \simeq \left(\frac{G\hbar}{c^5}\right)^{1/2} \simeq 10^{-43}\ \mathrm{s}. \tag{7.9.2}$$

From a theoretical point of view, in a closed universe, one is led to expect a time of maximum expansion t_{m} which is of order t_{P} followed by a subsequent rapid collapse. On the other hand, in an open universe, the curvature term Kc^2/a^2 is expected to dominate over the gravitational term $8\pi G\rho/3$ in a time $t^* \simeq t_{\mathrm{P}}$. In this second case, given that, as one can deduce from Equation (2.3.9), for $t > t^*$ we have

$$\frac{a(t)}{a(t_{\mathrm{P}})} \simeq \frac{t}{t_{\mathrm{P}}} \simeq \frac{T_{\mathrm{P}}}{T}, \tag{7.9.3}$$

we obtain

$$t_0 \simeq t_{\mathrm{P}}\frac{T_{\mathrm{P}}}{T_{0\mathrm{r}}} \simeq 10^{-11}\ \mathrm{s}. \tag{7.9.4}$$

The Universe has probably survived for a time of order 10^{10} years, corresponding to around $10^{60}t_{\mathrm{P}}$, meaning that at very early times the kinetic term $(\dot{a}/a)^2$ must have differed from the gravitational term $8\pi G\rho/3$ by a very small amount indeed. In other words, the density at a time $t \simeq t_{\mathrm{P}}$ must have been very close to the critical density.

As we shall see shortly, we have

$$\Omega(t_{\mathrm{P}}) \simeq 1 + (\Omega_0 - 1)10^{-60}. \tag{7.9.5}$$

The kinetic term at t_{P} must have differed from the gravitational term by about one part in 10^{60}. This is another 'fine-tuning' problem. Why are these two terms tuned in such a way as to allow the Universe to survive for 10^{10} years? On the other hand the kinetic and gravitational terms are now comparable because a very conservative estimate gives

$$10^{-2} < \Omega_0 < 2. \tag{7.9.6}$$

This problem is referred to as the *age problem* (how did the Universe survive so long?) or the (near) *flatness problem* (why is the density so close to the critical density?).

There is yet another way to present this problem. The Friedmann equation, divided by the square of the constant $Ta = T_{0r}a_0$, becomes

$$\left(\frac{H_0}{T_{0r}}\right)^2(\Omega_0 - 1) = \left(\frac{H}{T_r}\right)^2(\Omega - 1) = \text{const.} = \frac{Kc^2}{(a_0T_{0r})^2}; \tag{7.9.7}$$

this constant can be rendered dimensionless by multiplying by the quantity $(\hbar/k_B)^2$. We thus obtain

$$|\epsilon(T)| \equiv |K|\left(\frac{\hbar c}{ak_BT}\right)^2 = \left(\frac{\hbar H_0}{k_BT_{0r}}\right)^2|\Omega_0 - 1| \simeq |\Omega_0 - 1|10^{-58} < 10^{-57}; \tag{7.9.8}$$

the dimensionless constant we have introduced remains constant at a very small value throughout the evolution of the Universe. The flatness problem can be regarded as the problem of why $|\epsilon(T)|$ is so small. Perhaps one might think that the correct resolution is that $\epsilon(T) = 0$ exactly, so that $K = 0$. However, one should bear in mind that the Universe is not exactly described by a Robertson–Walker metric because it is not perfectly homogeneous and isotropic; it is therefore difficult to see how to construct a physical principle which requires that a parameter such as $\epsilon(T)$ should be exactly zero.

It is worth noting that $\epsilon(T)$ is related to the entropy S_r of the radiation of the Universe. Supposing that $K \neq 0$, the dimensionless entropy contained inside a sphere of radius $a(t)$ (the curvature radius) is

$$\sigma_U = \frac{S_r}{k_B} \simeq \left(\frac{k_BTa}{\hbar c}\right)^3 \simeq |\epsilon(T)|^{-3/2} = \left(\frac{k_BT_{0r}}{\hbar H_0}\right)^3|\Omega_0 - 1|^{-3/2} > 10^{86}. \tag{7.9.9}$$

Given that the entropy of the matter is negligible compared with that of the radiation and of the massless neutrinos (S_ν is of order S_r), the quantity σ_U can be defined as the dimensionless entropy of the Universe (a_0 is often called the 'radius of the universe'). This also represents the number of particles (in practice, photons and neutrinos) inside the curvature radius. What is the explanation for this enormous value of σ_U? This is, in fact, just another statement of the flatness problem. It is therefore clear that any model which explains the high value of σ_U also solves this problem. As we shall see, inflationary universe models do resolve this issue; indeed they generally predict that Ω_0 should be very close to unity, which may be difficult to reconcile with observations.

It is now an appropriate time to return in a little more detail to Equation (7.9.5). From the Friedmann equation

$$\dot{a}^2 - \tfrac{8}{3}\pi G\rho a^2 = -Kc^2, \tag{7.9.10}$$

one easily finds that during the evolution of the Universe we have

$$(\Omega^{-1} - 1)\rho(t)a(t)^2 = (\Omega_0^{-1} - 1)\rho_0a_0^2 = \text{const.} \tag{7.9.11}$$

The standard picture of the Universe (without inflation) is well described by a radiative model until z_{eq} and by a matter-dominated model from then until now. From Equation (7.9.11) and the usual formulae

$$\rho = \rho_{eq}\left(\frac{a_{eq}}{a}\right)^4 \quad (z > z_{eq}), \qquad \rho = \rho_0\left(\frac{a_0}{a}\right)^3 \quad (z < z_{eq}), \tag{7.9.12}$$

we can easily obtain the relationship between Ω, corresponding to a time $t \ll t_{eq}$ when the temperature is T, and Ω_0:

$$(\Omega^{-1} - 1) = (\Omega_0^{-1} - 1)(1 + z_{eq})^{-1}\left(\frac{T_{eq}}{T}\right)^2 = (\Omega_0^{-1} - 1)10^{-60}\left(\frac{T_P}{T}\right)^2. \tag{7.9.13}$$

If we accept that $|\Omega_0^{-1} - 1| \simeq 1$, this implies that Ω must have been extremely close to unity during primordial times. For example, at t_P we have $|\Omega_P^{-1} - 1| \simeq 10^{-60}$, as we have already stated in Equation (7.9.5).

7.9.2 The inflationary solution

Now we suppose that there is a period of accelerated expansion between t_i and t_f. Following the same philosophy as we did in Section 7.8, we divide the history of the Universe into the same three intervals (t_i, t_f), (t_f, t_{eq}) and (t_{eq}, t_0), where $\rho \propto a^{-3(1+w_{ij})}$, with $w_{ij} = w < -\frac{1}{3}$, $w_{ij} = \frac{1}{3}$ and $w_{ij} = 0$, respectively. We find, from Equation (7.9.11),

$$(\Omega_i^{-1} - 1)\rho_i a_i^2 = (\Omega_f^{-1} - 1)\rho_f a_f^2 = (\Omega_{eq}^{-1} - 1)\rho_{eq} a_{eq}^2 = (\Omega_0^{-1} - 1)\rho_0 a_0^2, \tag{7.9.14}$$

so that

$$\frac{\Omega_i^{-1} - 1}{\Omega_0^{-1} - 1} = \frac{\rho_0 a_0^2}{\rho_i a_i^2} = \frac{\rho_0 a_0^2}{\rho_{eq} a_{eq}^2}\frac{\rho_{eq} a_{eq}^2}{\rho_f a_f^2}\frac{\rho_f a_f^2}{\rho_i a_i^2}, \tag{7.9.15}$$

which gives, in a similar manner to Equation (7.8.15),

$$\left(\frac{a_f}{a_i}\right)^{-(1+3w)} = \left(\frac{\Omega_i^{-1} - 1}{\Omega_0^{-1} - 1}\right)\left(\frac{a_0}{a_{eq}}\right)\left(\frac{a_{eq}}{a_f}\right)^2. \tag{7.9.16}$$

After some further manipulation we find

$$\left(\frac{a_f}{a_i}\right)^{-(1+3w)} = \left(\frac{1 - \Omega_i^{-1}}{1 - \Omega_0^{-1}}\right)10^{60} z_{eq}^{-1}\left(\frac{T_f}{T_P}\right)^2. \tag{7.9.17}$$

One can assume that the flatness problem is resolved as long as the following inequality is valid:

$$\frac{1 - \Omega_i^{-1}}{1 - \Omega_0^{-1}} \geqslant 1, \tag{7.9.18}$$

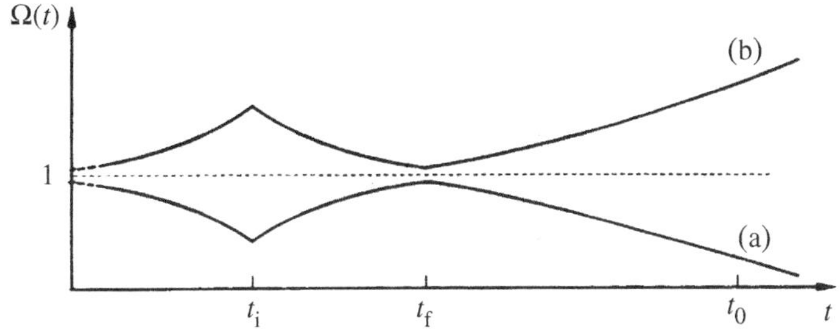

Figure 7.5 Evolution of $\Omega(t)$ for an open universe (a) and closed universe (b) characterised by three periods $(0, t_i)$, (t_i, t_f), (t_f, t_0). During the first and last of these periods $p/\rho c^2 = w > -\frac{1}{3}$ (decelerated expansion), while in the second $w < -\frac{1}{3}$ (accelerated expansion). If the inflationary period is sufficiently dramatic, the later divergence of the trajectories from $\Omega = 1$ is delayed until well beyond t_0.

in other words Ω_0 is no closer to unity now than Ω_i was. The condition (7.9.18), expressed in terms of the number of e-foldings $\mathcal{N}$, becomes

$$\mathcal{N} \geqslant 60\left[\frac{2.3 + \frac{1}{30}\ln(T_f/T_P) - \frac{1}{60}\ln z_{eq}}{|1 + 3w|}\right]. \tag{7.9.19}$$

For example, in the case where $w \simeq -1$ the solution of the horizon problem $\mathcal{N} \simeq p\mathcal{N}_{min} = p30[2.3 + \frac{1}{30}\ln(T_f/T_P)]$, with $p > 1$, implies a relationship between Ω_i and Ω_0

$$(1 - \Omega_0^{-1}) = \frac{(1 - \Omega_i^{-1})}{\exp[2(p-1)\mathcal{N}_{min}]}. \tag{7.9.20}$$

If $|1 - \Omega_i^{-1}| \simeq 1$, even if $p = 2$, one obtains

$$|1 - \Omega_0^{-1}| \simeq 10^{-[60+\ln(T_f/T_P)]} \ll 1. \tag{7.9.21}$$

In general, therefore, an adequate solution of the horizon problem ($p \gg 1$) would imply that Ω_0 would be very close to unity for a universe with $|1 - \Omega_i^{-1}| \simeq 1$. In other words, in this case inflation would automatically take care of the flatness problem as well.

This argument may explain why Ω is close to unity today, but it also poses a problem of its own. If $\Omega_0 \simeq 1$ to high accuracy, what is the bulk of the matter made from, and why do dynamical estimates of Ω_0 yield typical values of order 0.2? If it turns out that Ω_0 is actually of this order, then much of the motivation for inflationary models will have been lost. We should also point out that inflation does not predict an exactly smooth Universe; small-amplitude fluctuations appear in a manner described in Chapter 14. These fluctuations mean that, on the scale of our observable Universe, the density parameter would be uncertain by the amount of the density fluctuation on that scale. In most models the fractional fluctuation is of order 10^{-5}, so it does not make sense to claim that Ω_0 is predicted to be unity with any greater accuracy than this.

7.10 The Inflationary Universe

The previous sections have given some motivation for imagining that there might have been an epoch during the evolution of the Universe in which it underwent an accelerated expansion phase. This would resolve the flatness and horizon problems. It would also possibly resolve the problem of topological defects because, as long as inflation happens after (or during) the phase transition producing the defects, they will be diluted by the enormous increase of the scale factor. Beginning in 1982, various authors have also addressed another question in the framework of the inflationary universe which is directly relevant to the main subject of this book. The idea here is that quantum fluctuations on microscopic scales during the inflationary epoch can, again by virtue of the enormous expansion, lead to fluctuations on very large scales today. It is possible that this 'quantum noise' might therefore be the source of the primordial fluctuation spectrum we require to make models of structure-formation work. In fact, as we shall see in Section 14.6, one obtains a primordial spectrum which is slightly dependent upon the form of the inflationary model, but is usually close to the so-called *Harrison–Zel'dovich spectrum* which was proposed, for different reasons, by Harrison, Zel'dovich and also Peebles and Yu, around 1970.

Assuming that we accept that an epoch of inflation is in some sense desirable, how can we achieve such an epoch physically? The answer to this question lies in the field of high-energy particle physics, so from now until the end of this chapter we shall use the language of natural units with $c = \hbar = 1$.

The idea at the foundation of most models of inflation is that there was an epoch in the early stages of the evolution of the Universe in which the energy density of the vacuum state of a scalar field $\rho_V \simeq V(\phi)$ is the dominant contribution to the energy density. In this phase the expansion factor a grows in an accelerated fashion which is nearly exponential if $V \simeq$ const. This, in turn, means that a small causally connected region with an original dimension of order H^{-1} can grow to such a size that it exceeds the size of our present observable Universe, which has a dimension of order H_0^{-1}.

There exist many different versions of the inflationary universe. The first was formulated by Guth (1981), although many of his ideas had been presented previously by Starobinsky (1979). In Guth's model inflation was assumed to occur while the universe is trapped in a false vacuum with $\Phi = 0$ corresponding to the first-order phase transition which characterises the breaking of an SU(5) symmetry into $\mathrm{SU}(4) \times \mathrm{U}(1)$. This model was subsequently abandoned for reasons which we shall mention below.

The next generation of inflationary models shared the characteristics of a model called the *new inflationary universe*, which was suggested independently by Linde (1982a,b) and Albrecht and Steinhardt (1982). In models of this type, inflation occurs during a phase in which the region which grows to include our observable 'patch' evolves slowly from a 'false' vacuum with $\Phi = 0$ towards a 'true' vacuum with $\Phi = \Phi_0$. In fact, it was later seen that this kind of inflation could also be achieved in many different contexts, not necessarily requiring the existence of a phase transition or a spontaneous symmetry breaking. Anyway, from an explanatory point of view, this model appears to be the clearest. It is based on a certain choice of parameters for an SU(5) theory which, in the absence of any experimental constraints, appears a little arbitrary. This problem is common also to other inflationary models based on theories like supersymmetry, superstrings or supergravity which have not yet received any experimental confirmation or, indeed, are likely to in the foreseeable future. It is fair to say that the inflationary model has become a sort of 'paradigm' for resolving some of the difficulties with the standard model, but no particular version of it has received any strong physical support from particle physics theories.

Let us concentrate for a while on the physics of generic inflationary models involving symmetry breaking during a phase transition. In general, gauge theories of elementary particle interactions involve an order parameter Φ, determining the breaking of the symmetry, which is the expectation value of the scalar field which appears in the classical Lagrangian L_Φ

$$L_\Phi = \tfrac{1}{2}\dot{\Phi}^2 - V(\Phi; T). \tag{7.10.1}$$

As we mentioned in Section 6.1, the first term in Equation (7.10.1) is called the kinetic term and the second is the effective potential, which is a function of temperature. In Equation (7.10.1) for simplicity we have assumed that the expectation value of Φ is homogeneous and isotropic with respect to spatial position. As we have already explained in Section 6.1, the energy–momentum tensor of a scalar field can be characterised by an effective energy density ρ_Φ and by an effective pressure p_Φ given by

$$\rho_\Phi = \tfrac{1}{2}\dot{\Phi}^2 + V(\Phi; T), \tag{7.10.2 a}$$

$$p_\Phi = \tfrac{1}{2}\dot{\Phi}^2 - V(\Phi; T), \tag{7.10.2 b}$$

respectively. The potential $V(\Phi; T)$ plays the part of the free energy F of the system, which displays the breaking symmetry described in Section 7.3; in particular, Figure 7.2 is a useful reference for the following comments. This figure refers to a first-order phase transition, so what follows is relevant to the case of Guth's original 'old' inflation model. The potential has an absolute minimum at $\Phi = 0$ for $T \gg T_c$, this is what will correspond to the 'false' vacuum phase. As T nears T_c the potential develops another two minima at $\Phi = \pm\Phi_0$, which for $T \simeq T_c$ have a value of order $V(0; T_c)$: the three minima are degenerate. We shall now assume that the transition 'chooses' the minimum at Φ_0; at $T \ll T_c$ this minimum becomes absolute and represents the true vacuum after the transition; at these energies we can ignore the dependence of the potential upon temperature. We also assume, for reasons which will become clear later, that $V(\Phi_0; 0) = 0$. In this case the transition does not occur instantaneously at T_c because of the potential barrier between the false and true vacua; in other words, the system undergoes a supercooling while the system remains trapped in the false vacuum. Only at some later temperature $T_b < T_c$ can thermal fluctuations or quantum tunnelling effects shift the Φ field over the barrier and down into the true vacuum. Let us indicate by Φ_b the value assumed by the order parameter at this event. The dynamics of this process depends on the shape of the potential. If the potential is such that the transition is first order (as in Figure 7.2), the new phase appears as bubbles nucleating within the false vacuum background; these then grow and coalesce so as to fill space with the new phase when the transition is complete. If the transition is second order, one generates domains rather than bubbles, like the Weiss domains in a ferromagnet. One such region (bubble or domain) eventually ends up including our local patch of the Universe.

The energy-momentum tensor of the whole system, T_{ij}, also contains, in addition to terms due to the Φ field, terms corresponding to interacting particles, which can be interpreted as thermal excitations above the minimum of the potential, with an energy density ρ and pressure p; in this period we have $p = \rho/3$. The Friedmann equations therefore become

$$\left(\frac{\dot{a}}{a}\right)^2 = \tfrac{8}{3}\pi G(\rho_\Phi + \rho) - \frac{K}{a^2}, \qquad (7.10.3\,a)$$

$$\ddot{a} = -\tfrac{4}{3}\pi G[\rho_\Phi + \rho + 3(p_\Phi + p)]a = \tfrac{8}{3}\pi G[V(\Phi) - \dot{\Phi}^2 - \rho]a. \qquad (7.10.3\,b)$$

The evolution of Φ is obtained from the equation of motion for a scalar field:

$$\frac{\mathrm{d}}{\mathrm{d}t}\frac{\partial(L_\Phi a^3)}{\partial\dot{\Phi}} - \frac{\partial(L_\Phi a^3)}{\partial\Phi} = 0, \qquad (7.10.4)$$

which gives

$$\ddot{\Phi} + 3\frac{\dot{a}}{a}\dot{\Phi} + \frac{\partial V(\Phi)}{\partial\Phi} = 0. \qquad (7.10.5)$$

This equation is similar to that describing a ball moving under the action of the force $-\partial V/\partial\Phi$ against a source of friction described by the viscosity term proportional to $3\dot{a}/a$; in the usual language, one talks of the Φ field 'rolling down' the potential towards the minimum at Φ_0. Let us consider potentials which have a large interval (Φ_i, Φ_f) with $\Phi_b < \Phi_i \leqslant \Phi \leqslant \Phi_f < \Phi_0$ in which $V(\Phi; T)$ remains roughly constant; this property ensures a very slow evolution of Φ towards Φ_0, usually called the *slow-rolling phase* because, in this interval, the kinetic term $\dot{\Phi}^2/2$ is negligible compared with the potential $V(\Phi; T)$ in Equation (7.10.3 *b*) and the $\ddot{\Phi}$ term is negligible in Equation (7.10.5). One could say that the motion of the field is in this case dominated by friction, so that the motion of the field resembles the behaviour of particles during sedimentation.

In order to have inflation one must assume that, at some time, the Universe contains some rapidly expanding regions in thermal equilibrium at a temperature $T > T_c$ which can eventually cool below T_c before any gravitational recollapse can occur. Let us assume that such a region, initially trapped in the false vacuum phase, is sufficiently homogeneous and isotropic to be described by a Robertson–Walker metric. In this case the evolution of the patch is described by

Equation (7.10.3 *a*). The expansion rapidly causes ρ and K/a^2 to become negligible with respect to ρ_Φ, which is varying slowly. One can therefore assume that Equation (7.10.3 *a*) is then

$$\left(\frac{\dot{a}}{a}\right)^2 \simeq \tfrac{8}{3}\pi G \rho_\Phi. \tag{7.10.6}$$

In the approximation $\dot{\Phi}^2 \ll V(\Phi; T) \simeq$ const., which is valid during the slow-rolling phase, this equation has the de Sitter universe solution

$$a \propto \exp\left(\frac{t}{\tau}\right), \tag{7.10.7}$$

with

$$\tau \simeq \left[\frac{3}{8\pi G V(\Phi; T_\mathrm{b})}\right]^{1/2}, \tag{7.10.8}$$

which is of order 10^{-34} s in typical models. Let us now fix our attention upon one such region, which has dimensions of order $1/H(t_\mathrm{b})$ at the start of the slow-rolling phase and is therefore causally connected. This region expands by an enormous factor in a very short time τ; any inhomogeneity and anisotropy present at the initial time will be smoothed out so that the region loses all memory of its initial structure. This effect is, in fact, a general property of inflationary universes and it is described by the so-called *cosmic no-hair theorem*. The number of e-foldings of the inflationary expansion during the interval $(t_\mathrm{i}, t_\mathrm{f})$ depends on the potential:

$$\mathcal{N} = \ln\left[\frac{a(t_\mathrm{f})}{a(t_\mathrm{i})}\right] \simeq -8\pi G \int_{\Phi_\mathrm{i}}^{\Phi_\mathrm{f}} \left(\frac{\mathrm{d}\ln V(\Phi; T)}{\mathrm{d}\Phi}\right)^{-1} \mathrm{d}\Phi; \tag{7.10.9}$$

if this number is sufficiently large, the horizon and flatness problems can be solved. The initial region is expanded by such a large factor that it encompasses our present observable Universe.

Because of the large expansion, the patch we have been following also becomes practically devoid of particles. This also solves the monopole problem (and also the problem of domain walls, if they are predicted) because any defects formed during the transition will be drastically diluted as the Universe expands so that their present density will be negligible. After the slow-rolling phase the field Φ falls rapidly into the minimum at Φ_0 and there undergoes oscillations: while this happens there is a rapid liberation of energy which was trapped in the term $V \simeq V(\Phi_\mathrm{f}; T_\mathrm{f})$, i.e. the 'latent heat' of the transition. The oscillations are damped by the creation of particles coupled to the Φ field and the liberation of the latent heat thus raises the temperature to some value $T_\mathrm{rh} \leqslant T_\mathrm{c}$: this phenomenon is called *reheating*, and T_rh is the reheating temperature. The region thus acquires virtually all the energy and entropy that originally resided in the quantum vacuum by particle creation.

Once the temperature has reached T_rh, the evolution of the patch again takes the character of the usual radiative Friedmann models without a cosmological constant; this latter condition is, however, only guaranteed if $V(\Phi_0; 0) = 0$ because

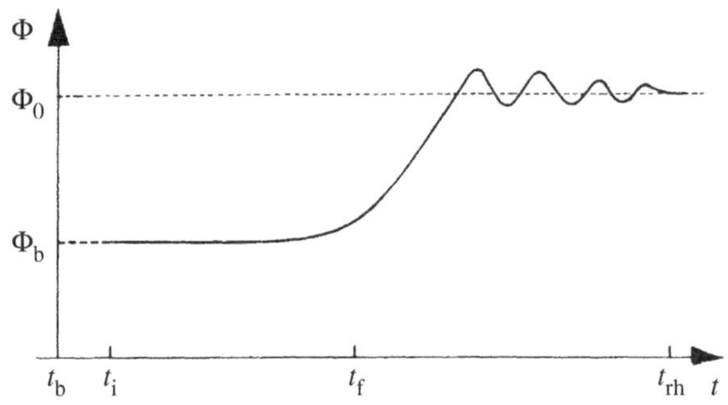

Figure 7.6 Evolution of Φ inside a 'patch' of the Universe. In the beginning we have the slow-rolling phase between t_i and t_f, followed by the rapid fall into the minimum at Φ_0, representing the true vacuum, and subsequent rapid oscillations which are eventually smeared out by particle creation leading to reheating of the Universe.

any zero-point energy in the vacuum would play the role of an effective cosmological constant. We shall return to this question in the next section.

It is important that the inflationary model should predict a reheating temperature sufficiently high that GUT processes which violate conservation of baryon number can take place so as to allow the creation of a baryon asymmetry.

As far as its global properties are concerned, our Universe is reborn into a new life after reheating: it is now highly homogeneous, and has negligible curvature. This latter prediction may be a problem for, as we have seen, there is little strong evidence that Ω_0 is very close to unity.

Another general property of inflationary models, which we have not described here, is that fluctuations in the quantum field driving inflation can, in principle, generate a primordial spectrum of density fluctuations capable of seeding the formation of galaxies and clusters. We shall postpone a discussion of this possibility until Section 14.6.

7.11 Types of Inflation

We have already explained that there are many versions of the inflationary model which are based on slightly different assumptions about the nature of the scalar field and the form of the phase transition. Let us mention some of them here.

7.11.1 Old inflation

The first inflationary model, suggested by Guth (1981), is usually now called *old inflation*. This model is based on a scalar field theory which undergoes a first-order phase transition. The problem is that, being a first-order transition, it occurs by a process of bubble nucleation. It turns out, however, that these bubbles would be too small to be identified with our observable Universe and would be carried apart by the expanding phase too quickly for them to coalesce and produce a large bubble which one could identify in this way. The end state of this model would therefore be a highly chaotic universe, quite the opposite of what is intended. This model was therefore abandoned soon after it was suggested.

7.11.2 New inflation

The successor to old inflation was *new inflation* (Linde 1982a,b; Albrecht and Steinhardt 1982). This is again a theory based on a scalar field, but this time the potential is qualitatively similar to Figure 7.1, rather than 7.2. The field is originally in the false vacuum state at $\Phi = 0$, but as the temperature lowers it begins to roll down into one of the two degenerate minima. There is no potential barrier, so the phase transition is second order. The process of spinodal decomposition which accompanies a second-order phase transition usually leaves one with larger coherent domains and one therefore ends up with relatively large space-filling domains.

The problem with new inflation is that it suffers from severe fine-tuning problems. One such problem is that the potential must be very flat near the origin to produce enough inflation and to avoid excessive fluctuations due to the quantum field. Another is that the field Φ is assumed to be in thermal equilibrium with the other matter fields before the onset of inflation; this requires that Φ be coupled fairly strongly to the other fields. But the coupling constant would induce corrections to the potential which would violate the previous constraint. It seems unlikely therefore that one can achieve thermal equilibrium in a self-consistent way before inflation starts under the conditions necessary for inflation to happen.

7.11.3 Chaotic inflation

One of the most popular inflationary models is *chaotic inflation*, due to Linde (1983). Again, this is a theory based on a scalar field, but it does not require any phase transitions. The basis of this model is that, whatever the detailed shape of the effective potential, a patch of the Universe in which Φ is large, uniform and static will automatically lead to inflation. For example, consider the simple quadratic potential

$$V(\Phi) = \tfrac{1}{2}m^2\Phi^2, \tag{7.11.1}$$

where m is an arbitrary parameter describing the mass of the scalar field. Assume that, at $t = t_i$, the field $\Phi = \Phi_i$ is uniform over a scale $\sim H^{-1}(t_i)$ and that

$$\dot{\Phi}_i^2 \ll V(\Phi_i). \tag{7.11.2}$$

The equation of motion of the scalar field then simply becomes

$$\ddot{\Phi} + 3H\dot{\Phi} = -m^2\Phi, \tag{7.11.3}$$

which, with the slow-rolling approximation, is just

$$3H\dot{\Phi} \simeq -m^2\Phi. \tag{7.11.4}$$

Since $H \propto V^{1/2} \propto \Phi$, this equation is easy to solve and it turns out that, in order to get sufficient inflation to solve the flatness and horizon problems, one needs $\Phi > 3m_P$ in the patch.

In chaotic inflation one assumes that at some initial time, perhaps just after the Planck time, the Φ field varied from place to place in an arbitrary manner. If any region satisfies the above conditions it will inflate and eventually encompass our observable Universe. While the end result of chaotic inflation is locally flat and homogeneous in our observable 'patch', on scales larger than the horizon the Universe is highly curved and inhomogeneous. Chaotic inflation is therefore very different from both old and new inflationary models. This is reinforced by the fact that no mention of GUT or supersymmetry theories appears in this analysis. The field Φ which describes chaotic inflation at the Planck time is completely decoupled from all other physics.

7.11.4 Stochastic inflation

The natural extension of Linde's chaotic inflationary model is called *stochastic inflation* or, sometimes, *eternal inflation* (Linde *et al.* 1994). The basic idea is the same as chaotic inflation in that the Universe is globally extremely inhomogeneous. The stochastic inflation model, however, takes into account quantum fluctuations during the evolution of Φ. One finds in this case that the Universe at any time will contain regions which are just entering into an inflationary phase. One can picture the Universe as a continuous 'branching' process in which new 'miniuniverses' expand to produce locally smooth Hubble patches within a highly chaotic background Universe. This picture is like a Big Bang on the scale of each miniuniverse, but globally is reminiscent of the steady-state universe. The continual birth and rebirth of these miniuniverses is often called, rather poetically, the 'Phoenix Universe' model.

7.11.5 Open inflation

In the mid-1990s there was a growing realisation among cosmologists that evidence for a critical matter density was not forthcoming (e.g. Coles and Ellis 1994). This even reached inflation theorists, who defied the original motivation for inflation and came up with versions of inflation that would homogeneous but curved universes. Usually inflation stretches the curvature as well as smoothing lumpiness, so this seems at first sight a very difficult task for inflation.

Open inflation models square the circle by invoking a kind of quantum tunnelling from a metastable false vacuum state immediately followed by a second phase of inflation, an idea originally due to Gott (1982). The tunnelling creates a bubble inside which the space-time resembles an open universe.

Although it is possible to engineer an inflationary model that produces $\Omega_0 \simeq 0.2$ at the present epoch, it certainly seems to require more complexity than models that produce flat spatial sections. Recent evidence from microwave background observations that the Universe seems to be flat even if it does not have a critical density have reduced interest in these open inflation models too; see Chapter 18.

7.11.6 Other models

At this point it is appropriate to point out that there are very many inflationary models about. Indeed, inflation is in some sense a generic prediction of most theories of the early Universe. We have no space to describe all of these models, but we can briefly mention some of the most important ones.

Firstly, one can obtain inflation by modifying the classical Lagrangian for gravity itself, as mentioned in Chapter 6. If one adds a term proportional to R^2 to the usual Lagrangian, then the equations of motion that result are equivalent to ordinary general relativity in the presence of a scalar field with some particular action. This 'effective' scalar field can drive inflation in the same way as a real field can.

An alternative way to modify gravity might be to adopt the Brans-Dicke (scalar-tensor) theory of gravity described in Section 3.4. The crucial point here is that an effective equation of state of the form $p = -\rho c^2$ in this theory produces a power-law, rather than exponential, inflationary epoch. This even allows 'old inflation' to succeed: the bubbles which nucleate the new phase can be made to merge and fill space if inflation proceeds as a power law in time rather than an exponential (Lucchin and Matarrese 1985). Theories based on Brans-Dicke modified gravity are usually called *extended inflation.*

Another possibility relies on the fact that many unified theories, such as supergravity and superstrings, are only defined in space-times of considerably higher dimensionality than those we are used to. The extra dimensions involved in these theories must somehow have been compactified to a scale of order the Planck length so that we cannot perceive them now. The contraction of extra spatial dimensions can lead to an expansion of the three spatial dimensions which must survive, thus leading to inflation. This is the idea behind so-called *Kaluza-Klein theories.*

There are many other possibilities: models with more than one scalar field, with modified gravity and a scalar field, models based on more complicated potentials, on supersymmetric GUTs, supergravity and so on. Inflation has led to an almost exponential increase in the number of inflationary models since 1981!

7.12 Successes and Problems of Inflation

As we have explained, the inflationary model provides a conceptual explanation of the horizon problem and the flatness problem. It may also rescue grand unified theories which predict a large present-day abundance of monopoles or other topological defects.

We have seen how inflationary models have evolved to avoid problems with earlier versions. Some models are intrinsically flawed (e.g. old inflation) but can be salvaged in some modified form (extended inflation). The density and gravitational wave fluctuations they produce may also be too high for some parameter choices, as we discuss in Chapter 14. For example, the requirement that density fluctuations be acceptably small places a strong constraint on m in Equation (7.11.1) corresponding to the chaotic inflation model. This, however, requires a fine-tuning of the scalar field mass m which does not seem to have any strong physical motivation. Such fine-tunings are worrying but not fatal flaws in these models.

There are, however, much more serious problems associated with these scenarios. Perhaps the most important is one we have mentioned before and which is intimately connected with one of the successes. Most inflationary models predict that spatial sections at the present epoch should be almost flat. In the absence of a cosmological constant this means that $\Omega_0 \simeq 1$. However, evidence from galaxy-clustering studies suggests this is not the case: the apparent density of matter is less than the critical density. It is possible to produce a low-density universe after inflation, but it requires very particular models. On the other hand, one could reconcile a low-density universe with apparently more natural inflationary models by appealing to a relic cosmological constant: the requirement that spatial sections should be (almost) flat simply translates into $\Omega_0 + \Omega_{0\Lambda} \simeq 1$. This seems to be that a potentially successful model of structure formation, as well as allowing accounting for the behaviour of high-redshift supernovae (Chapter 4) and cosmic microwave background fluctuations (Chapter 18).

One also worries about the status of inflation as a physical theory. To what extent is inflation predictive? Is it testable? One might argue that inflation does predict that $\Omega_0 \simeq 1$. This may be true, but one can have Ω_0 close to unity without inflation if some process connected with quantum gravity can arrange it. Likewise one can have $\Omega_0 < 1$ either with inflation or without it. Inflationary models also produce density fluctuations and gravitational waves. If these are observed to have the correct properties, they may eventually constitute a test of inflation, but this is not the case at present. All we can say is the COBE fluctuations in the microwave background do indeed seem to be consistent with the usual inflationary models. At the moment, therefore, inflation has a status somewhere between a theory and a paradigm, but we are still a long way from being able to use these ideas to test GUT scale physics and beyond in any definite way.

7.13 The Anthropic Cosmological Principle

We began this book with a discussion of the importance of the *Cosmological Principle*, which, as we have seen in the first two chapters, has an important role to play in the construction of the Friedmann models. This principle, in light of the cosmological horizon problem, has more recently led to the idea of the inflationary universe we have explored in this chapter. The Cosmological Principle is a development of the *Copernican Principle*, asserting that, on a large scale, all spatial positions in the Universe are equivalent. At this point in the book it is worth mentioning an alternative Cosmological Principle – *the Anthropic Cosmological Principle* – which seeks to explore the connection between the physical structure of the Universe and the development of intelligent life within it. There are, in fact, many versions of the Anthropic Principle. The *Weak Anthropic Principle* merely cautions that the fact of our own existence implies that we do occupy some sort of special place in the Universe. For example, as noted by Dicke (1961), human life requires the existence of heavy elements such as Carbon and Oxygen which must be synthesised by stars. We could not possibly have evolved to observe the Universe in a time less than or of order the main sequence lifetime of a star, i.e. around 10^{10} years in the Big Bang picture. This observation is itself sufficient to explain the large-number coincidences described in Chapter 3 which puzzled Dirac so much. In fact, the Weak Anthropic Principle is not a 'principle' in the same sense as the Cosmological Principle: it is merely a reminder that one should be aware of all selection effects when interpreting cosmological data.

It is important to stress that the Weak Anthropic Principle is not a tautology, but has real cognitive value. We mentioned in Chapter 3 that, in the steady-state model, there is no reason why the age of astronomical objects should be related to the expansion timescale H_0^{-1}. In fact, although both these timescales are uncertain, we know that they are equal to within an order of magnitude. In the Big Bang model this is naturally explained in terms of the requirement that life should have evolved by the present epoch. The Weak Anthropic Principle therefore supplies a good argument whereby one should favour the Big Bang over the steady state: the latter has an unresolved 'coincidence' that the former explains quite naturally.

An entirely different status is held by the *Strong Anthropic Principle* and its variants. This version asserts a teleological argument (i.e. an argument based on notions of 'purpose' or 'design') to account for the fact that the Universe seems to have some properties which are finely tuned to allow the development of life. Slight variations in the 'pure' numbers of atomic physics, such as the fine-structure constant, would lead to a world in which chemistry, and presumably life, as we know it, could not have developed. These coincidences seem to some physicists to be so striking that only a design argument can explain them. One can, however, construct models of the Universe in which a weak explanation will suffice. For example, suppose that the Universe is constructed as a set of causally disjoint 'domains' and, within each such domain, the various symmetries of particle physics have been broken in different ways. A concrete implementation of this idea may be realised using Linde's eternal chaotic inflation model which we discussed earlier. Physics in some of these domains would be similar to our Universe; in particular, the physical parameters would be such as to allow the development of life. In other domains, perhaps in the vast majority of them, the laws of physics would be so different that life could never evolve in them. The Weak Anthropic Principle instructs us to remember that we must inhabit one of the former domains, rather than one of the latter ones. This idea is, of course, speculative but it does have the virtue of avoiding an explicitly teleological language.

The status of the Strong Anthropic Principle is rightly controversial and we shall not explore it further in this book. It is interesting to note, however, that after centuries of adherence to the Copernican Principle and its developments, cosmology is now seeing the return of a form of Ptolemaic reasoning (the Strong Anthropic Principle), in which man is again placed firmly at the centre of the Universe.

Bibliographic Notes on Chapter 7

More detailed treatments of elementary particle physics can be found in Chaichian and Nelipe (1984); Collins *et al.* (1989); Dominguez-Tenreiro and Quiros (1987); Hughes (1985); Kolb and Turner (1990) and Roos (1994). A more technical treatment of particle cosmology can be found in Barrow (1983). Weinberg (1988) gives an authoritative review of the cosmological constant problem. A nice introductory account of inflation can be found in Narlikar and Padmanabhan (1991) of Linde (1990); a more technical review is Linde (1984). The definitive treatment of the anthropic principles is Barrow and Tipler (1986).

Problems

The following problems all concern a simplified model of the history of a flat universe involving a period of inflation. The history is split into four periods: (a) $0 < t < t_3$ radiation only; (b) $t_3 < t < t_2$ vacuum energy dominates, with an effective cosmological constant $\Lambda = \frac{3}{4}t_3^2$; (c) $t_2 < t < t_1$ a period of radiation domination; and (d) $t_1 < t < t_0$ matter domination.

1. Show that in epoch (c) $\rho(t) = \rho_{\rm r}(t) = \frac{3}{32}\pi G t^2$, and in (d) $\rho(t) = \rho_{\rm m}(t) = \frac{1}{6}\pi G t^2$.

2. Give simple analytical formulae for $a(t)$ which are approximately true in these four phases.

3. Show that, during the inflationary phase (b) the universe expands by a factor

$$\frac{a(t_2)}{a(t_3)} = \exp\left(\frac{t_2 - t_3}{2t_3}\right).$$

4. Derive an expression for Λ in terms of t_2, t_3 and $\rho(t_2)$.

5. Show that

$$\frac{\rho_{\rm r}(t_0)}{\rho_{\rm m}(t_0)} = \frac{9}{16}\left(\frac{t_1}{t_0}\right)^{2/3}.$$

6. If $t_3 = 10^{-35}$ s, $t_2 = 10^{-32}$ s, $t_1 = 10^4$ years and $t_0 = 10^{10}$ years, give a sketch of $\log a$ against $\log t$ marking any important epochs.

8

The Lepton Era

8.1 The Quark–Hadron Transition

At very high temperatures, the matter in the Universe exists in the form of a quark–gluon plasma. When the temperature falls to around $T_{QH} \simeq 200\text{–}300$ MeV the quarks are no longer free, but become confined in composite particles called hadrons. These particles are generally short lived (with the exception of the proton and neutron), so there is only a brief period in which the hadrons flourish. This period is often called the hadron era, but that is a somewhat misleading term because the hadrons even in this era do not dominate the energy density of the Universe. At the energy corresponding to a temperature T_{QH}, the Universe - which was composed of photons, gluons, lepton–antilepton pairs and quark–antiquark pairs before - undergoes a (probably first-order) phase transition through which the quark–antiquark pairs join together to form the hadrons, including pions and nucleons. In this period pion–pion interactions are very important and, consequently, the equation of state of the hadron fluid becomes very complicated: one can certainly not apply the ideal gas approximation (Section 7.1) to hadrons in this era. The end of this era occurs when $T \simeq 130$ MeV at which point the pions annihilate.

At a temperature just a little greater than 100 MeV the Universe comprises three types of pion (π^+, π^-, π_0); small numbers of protons, antiprotons, neutrons and antineutrons (these particles are no longer relativistic at this temperature); charged leptons (muons, antimuons, electrons, positrons - the tau leptons will have annihilated at this stage) and their respective neutrinos (ν_μ, $\bar{\nu}_\mu$, ν_e, $\bar{\nu}_e$, ν_τ, $\bar{\nu}_\tau$); and photons. At a temperature of $T \simeq 130$ MeV the π^+-π^- pairs rapidly annihilate and the neutral pions π^0 decay into photons. This is the last act of the brief era of the hadrons. After this, there remain only leptons, antileptons, photons and the small excess of baryons (protons and neutrons) that we discussed in relation to the radiation entropy per baryon in Chapter 5; this, as we have explained, is probably due to processes which violated baryon number conservation while the temperature was around $T \simeq 10^{15}$ GeV. These baryons have a number density n given by the Boltzmann distribution:

$$n_{\mathrm{p(n)}} \simeq 2\left(\frac{m_{\mathrm{p(n)}} k_{\mathrm{B}} T}{2\pi\hbar^2}\right)^{3/2} \exp\left(-\frac{m_{p(n)}c^2}{k_{\mathrm{B}}T}\right), \tag{8.1.1}$$

where the suffixes 'n' and 'p' denote neutrons and protons, respectively. In Equation (8.1.1) we have neglected the chemical potential of the protons and neutrons $\mu_{\mathrm{p(n)}}$; we shall return to this matter in Section 8.2. From Equation (8.1.1) one finds that the ratio between the numbers of protons and neutrons is

$$\frac{n_{\mathrm{n}}}{n_{\mathrm{p}}} \simeq \left(\frac{m_{\mathrm{n}}}{m_{\mathrm{p}}}\right)^{3/2} \exp\left(-\frac{Q}{k_{\mathrm{B}}T}\right) \simeq \exp\left(-\frac{Q}{k_{\mathrm{B}}T}\right), \tag{8.1.2}$$

where

$$Q = (m_{\mathrm{n}} - m_{\mathrm{p}})c^2 \simeq 1.3 \text{ MeV} \tag{8.1.3}$$

is the difference in rest-mass energy between 'n' and 'p', corresponding to a temperature $T_{\mathrm{pn}} \equiv Q/k_{\mathrm{B}} \simeq 1.5 \times 10^{10}$ K. For $T \gg T_{\mathrm{pn}}$, the number of protons is virtually identical to the number of neutrons.

8.2 Chemical Potentials

Throughout this chapter we shall need to keep track of the effective number of particle species which are relativistic at temperature T. This is done through the quantity $g^*(T)$, the number of degrees of freedom as a function of temperature. We need to consider thermodynamic aspects of the particle interactions in order to make progress. In particular we need to consider the chemical potentials μ relevant to the different particle species. Recall that the chemical potential, roughly speaking, defines the way in which the internal energy of a system changes as the number of particles is changed.

In the case of an ideal gas the chemical potential μ_i for the ith particle type (which we assume to have statistical weight g_i) affects the equilibrium number density n_i according to

$$n_i = \frac{g_i}{2\pi^2\hbar^3} \int_0^\infty \left[\exp\left(\frac{pc - \mu_i}{k_{\mathrm{B}}T}\right) \pm 1\right]^{-1} p^2 \, \mathrm{d}p, \tag{8.2.1}$$

where the '+' sign applies to fermions, and the '−' sign to bosons. The existence of a non-zero chemical potential signifies the existence of *degeneracy*. It is a basic tenet of the theory of statistical mechanics that one conserves the chemical potentials of ingoing and outcoming particles during a reaction when the reaction is in equilibrium; also, the chemical potential of photons is zero.

In what follows we shall assume that the appropriate chemical potentials describing the thermodynamics of the particle interactions are zero. It is necessary to make some remarks to justify this assumption. As we shall see, the reason for this is basically founded upon the conservation of electric charge Q, baryon number B and lepton numbers L_e and L_μ (the former for the electron, the latter for the muon). For simplicity we shall omit other lepton families, although there is one more lepton called the tau particle. As we have already stated, B and L are conserved in any reaction after the GUT phase transition at T_{GUT}.

Let us now consider the hadron era ($T \simeq 10^2$ GeV). We take the contents of the Universe to be hadrons (nucleons and pions), leptons and photons. These particles interact via *electromagnetic interactions* such as

$$\mathrm{p} + \bar{\mathrm{p}} \rightleftharpoons \mathrm{n} + \bar{\mathrm{n}} \rightleftharpoons \pi^+ + \pi^- \rightleftharpoons \mu^+ + \mu^- \rightleftharpoons \mathrm{e}^+ + \mathrm{e}^- \rightleftharpoons \pi_0 \rightleftharpoons 2\gamma, \tag{8.2.2}$$

weak interactions, such as

$$\mathrm{e}^- + \mu^+ \rightleftharpoons \nu_e + \nu_\mu, \qquad \mathrm{e}^- + p \rightleftharpoons \nu_e + \mathrm{n}, \qquad \mu^- + \mathrm{p} \rightleftharpoons \bar{\nu}_\mu + \mathrm{n}, \qquad \dots, \tag{8.2.3 a}$$

$$\mathrm{e}^+ + \mathrm{e}^- \rightleftharpoons \nu_e + \bar{\nu}_e, \qquad \mathrm{e}^\pm + \nu_e \rightleftharpoons \mathrm{e}^\pm + \nu_e, \qquad \dots, \tag{8.2.3 b}$$

and the hadrons undergo *strong interactions* with each other. The relevant cross-section for the electromagnetic interactions is the Thomson cross-section, whose value in electrostatic units is given by

$$\sigma_T = \frac{8\pi}{3}\left(\frac{e^2}{mc^2}\right)^2 \simeq 6.65 \times 10^{-25}\left(\frac{m_e}{m}\right)^2 \text{ cm}^2, \tag{8.2.4}$$

where m is the mass of a generic particle. The weak interactions have a cross-section

$$\sigma_{wk} \simeq g_{wk}^2\left[\frac{k_B T}{(\hbar c)^2}\right]^2, \tag{8.2.5}$$

in which (g_{wk} is the weak interaction coupling constant which takes a value $g_{wk} \simeq 1.4 \times 10^{49}$ erg cm^3). The electromagnetic and weak interactions guarantee that in this period there is thermal equilibrium between these particles, because $\tau_H \gg \tau_{coll}$. Later on, we shall verify this condition for the neutrinos.

From (8.2.2) and Equations (8.2.3 *a*) and (8.2.3 *b*) it is clear that the chemical potentials of particles and antiparticles must be equal in magnitude and opposite in sign, and that the chemical potential for π_0 must be zero. The other thing to take into account when determining μ_i is the set of conserved quantities we mentioned above: electric charge Q, baryon number B and lepton numbers L_e and L_μ. Recall that p and n ($\bar{\mathrm{p}}$ and $\bar{\mathrm{n}}$) have $B = 1$ (-1); e^- and ν_e (e^+ and $\bar{\nu}_e$) have $L_e = 1$ (-1); μ^+ and ν_μ (μ^- and $\bar{\nu}_\mu$) have $L_\mu = 1$ (-1); also $B \neq 0$ implies $L_e = L_\mu = 0$ and so on. In particular, we assume that the chemical potentials of all the particle species are zero. For simplicity, let us neglect the pions and their corresponding strong interactions; more detailed treatments show that this is a good approximation.

The conservation of Q requires

$$n_Q = (n_p + n_{e^+} + n_{\mu^+}) - (n_{\bar{p}} + n_{e^-} + n_{\mu^-}) = 0, \tag{8.2.6}$$

so that the Universe is electrically neutral. Introducing the function

$$f(x) = \int_0^\infty \{[\exp(y - x) + 1]^{-1} - [\exp(y + x) + 1]^{-1}\} y^2 \, dy, \tag{8.2.7}$$

which is symmetrical about the origin and in which the dimensionless quantities $x_i = \mu_i / k_B T$ are called the degeneracy parameters, Equation (8.2.6) becomes

$$f(x_p) + f(x_{e^+}) + f(x_{\mu^+}) = 0. \tag{8.2.8}$$

The conservation of B, valid from the epoch we are considering until the present time, yields

$$n_B a^3 = \pi^{-2} \left(\frac{k_B T}{\hbar c} \right)^3 [f(x_p) + f(x_n)] a^3 = n_{0B} a_0^3. \tag{8.2.9}$$

Introducing the radiation entropy per baryon $\sigma_{\rm rad}$ we discussed in Chapter 5, this becomes

$$\sigma_{\rm rad}^{-1} n_{0\gamma} a_0^3 \simeq \sigma_{\rm rad}^{-1} \left(\frac{k_B T_{0r} a_0}{\hbar c} \right)^3 \simeq \sigma_{\rm rad}^{-1} \left(\frac{k_B T a}{\hbar c} \right)^3, \tag{8.2.10}$$

because the high value of $\sigma_{\rm rad}$ means that $T_{0r} a_0 \simeq Ta$. This relation is therefore equivalent to

$$f(x_p) + f(x_n) \simeq \sigma_{0r}^{-1} \simeq 0. \tag{8.2.11}$$

As far as L_e and L_μ are concerned, we shall assume that the density of the appropriate lepton numbers are very small, as is the baryon number density. We shall justify this approximation for the leptons only partially, and in an *a posteriori* manner, when we look at nucleosynthesis. The assumption is nevertheless quite strongly motivated in the framework of GUT theories in which one might expect the lepton and baryon asymmetries to be similar. In analogy with Equation (8.2.11) we therefore have

$$f(x_{e^+}) + \tfrac{1}{2} f(x_{\bar{\nu}_e}) \simeq 0, \tag{8.2.12 a}$$

$$f(x_\mu) + \tfrac{1}{2} f(x_{\nu_\mu}) \simeq 0, \tag{8.2.12 b}$$

where the factor $\frac{1}{2}$ comes from the relation $g_\mu = g_e = 2g_\nu = 2$. From Equation (8.2.3) and from the relation $\mu_i = -\mu_{\bar{i}}$ we have

$$x_n = x_p - x_{e^+} + x_{\bar{\nu}_e}, \tag{8.2.13 a}$$

$$x_{\mu^+} = x_{e^+} - x_{\bar{\nu}_e} + x_{\nu_\mu}, \tag{8.2.13 b}$$

which, with Equations (8.2.9)–(8.2.12), furnishes a set of six equations for the six unknowns x_p, x_n, x_{e^+}, x_{μ^+}, $x_{\bar{\nu}_e}$, x_{ν_μ}. If this system has a solution x_i^* (i = p, n, e^+, μ^+, $\bar{\nu}_e$, ν_μ), then it also admits the symmetric solution $-x_i^*$. To have physical significance, however, the solution must be unique; this means that $x_i^* = 0$. The six chemical potentials we have mentioned and, therefore, the others related to them by symmetry, are all zero.

Before ending this discussion it is appropriate to underline again the fact that the hypothesis that we can neglect the lepton number density with respect to n_γ is only partially justified by the observations of cosmic abundances which the standard nucleosynthesis model predicts and which we discuss later in this chapter. The greatest justification for this hypothesis is actually the enormous simplification one achieves by using it, as well as a theoretical predisposition towards vanishing L_e and L_μ (as with B) on grounds of symmetry, particularly in the framework of GUTs. One can, however, obtain a firm upper limit on the chemical potential of the cosmic neutrino background from the condition that the global value of Ω_0 cannot be greater than a few. Assuming that there are only three neutrino flavours, and that neutrinos are massless, one can derive the following constraint:

$$\frac{\sum_{i=1}^{3} \mu_{\nu_i,0}^4}{8\pi^2(\hbar c)^3} \simeq \rho_{0\nu} c^2 < 2\rho_{0c} c^2. \tag{8.2.14}$$

This limit corresponds to a present value of the degeneracy parameter which is much greater than we suggested above: if the $\mu_{\nu_i,0}$ are all equal, and if $T_{0\nu_i} \simeq 2$ K (as we will find later), this limit corresponds to a degeneracy parameter of the order of 40.

8.3 The Lepton Era

The lepton era lasts from the time the pions either annihilate or decay into photons, i.e. from $T_\pi \simeq 130$ MeV $\simeq 10^{12}$ K, to the time in which the $e^+ - e^-$ pairs annihilate at a temperature $T_e \simeq 5 \times 10^9$ K $\simeq 0.5$ MeV. At the beginning of the lepton era the Universe comprises photons, a small number of baryons and the leptons e^-, e^+, μ^+, μ^- (and probably τ^+ and τ^-), with their respective neutrinos. If the τ particles are much more massive than muons, then they will already have annihilated by this epoch, but the corresponding neutrinos will remain. Neglecting the (non-relativistic) baryon component, the number of degrees of freedom at the start of the lepton era is $g^*(T < T_\pi) = 4 \times 2 \times \frac{7}{8} + N_\nu \times 2 \times \frac{7}{8} + 2 \simeq 14.25$ (if the number of neutrino types is $N_\nu = 3$), corresponding to a cosmological time $t_\pi \simeq 10^{-5}$ s. We will study the Universe during the lepton era under the hypothesis which we have just discussed in the previous section, namely that all the relevant chemical potentials are zero.

At the start of the lepton era, all the constituent particles mentioned above are still in thermal equilibrium because the relevant collision time τ_{coll} is much smaller than τ_{H}, the Hubble time. For example, at $T \simeq 10^{11}$ K ($t \simeq 10^{-4}$ s) the collision time between photons and electrons is $\tau_{\mathrm{coll}} \simeq (\sigma_{\mathrm{T}} n_e c)^{-1} \simeq 10^{-21}$ s. The same can be said for the neutrinos for $T > 10^{10}$ K, which is the temperature at which they decouple from the rest of the Universe as we shall show.

Other important facts during the lepton era are the annihilation of muons at $T_\mu < 10^{12}$ K, which happens early on, the annihilation of the electron–positron pairs, which happens at the end, and cosmological nucleosynthesis, which begins at around $T \simeq 10^9$ K, at the beginning of the radiative era. Because the conditions for nucleosynthesis are prepared during the lepton era, we shall cover nucleosynthesis in this chapter, rather than in the next.

During the evolution of the Universe we assume that entropy is conserved for components still in thermal equilibrium. This hypothesis is justified by the slow rate of the relevant processes: one has to deal with phenomena which are essentially reversible adiabatic processes. The relativistic components contribute virtually all of the entropy in a generic volume V, so that

$$S = \frac{(\rho c^2 + p)V}{T} = \frac{4}{3}\frac{\rho c^2 V}{T} = \tfrac{4}{3} g^*(T) \tfrac{1}{2}\sigma T^3 V. \tag{8.3.1}$$

If pair annihilation occurs at a temperature T, for example the electron–positron annihilation at T_e, then let us indicate with the symbols $(-)$ and $(+)$ appropriate quantities before and after T. From conservation of entropy we obtain

$$S_{(-)} = \tfrac{2}{3} g^*_{(-)} \sigma T^3_{(-)} V = S_{(+)} = \tfrac{2}{3} g^*_{(+)} \sigma T^3_{(+)} V. \tag{8.3.2}$$

Because of the removal of the pairs we have $g^*_{(+)} < g^*_{(-)}$ and, therefore,

$$T_{(+)} = \left(\frac{g^*_{(-)}}{g^*_{(+)}}\right)^{1/3} T_{(-)} > T_{(-)}: \tag{8.3.3}$$

the annihilation of the pairs produces an increase in the temperature of the components which remain in thermal equilibrium. For this reason the relation $T \propto a^{-1}$ is not exact: the correct relation is of the form

$$T = T_P \frac{a(t_P)}{a(t)} \left[\frac{g^*(T_P)}{g^*(T)}\right]^{1/3}, \tag{8.3.4}$$

where T_P is the Planck temperature and t_P the Planck time. However, the error in using the simpler formula is small because $g^*(T)$ never changes by more than an order of magnitude, while T changes by more than 30 orders of magnitude. For this reason Equation (8.3.4) reduces in practice to $T \propto a^{-1}$.

8.4 Neutrino Decoupling

Before the annihilation of μ^+-μ^- pairs at $T \simeq 10^{12}$ K, the Universe is composed mainly of e^-, e^+, μ^-, μ^+, ν_e, $\bar{\nu}_e$, ν_μ, $\bar{\nu}_\mu$, ν_τ, $\bar{\nu}_\tau$ and γ. The neutrinos are still in thermal equilibrium through scattering reactions of the form

$$\nu_e + \mu^- \rightleftharpoons \bar{\nu}_\mu + e^-, \qquad \bar{\nu}_\mu + \mu^+ \rightleftharpoons \nu_e + e^+, \qquad \dots. \tag{8.4.1}$$

For this reason the relevant cross-section is σ_{wk} mentioned above. When the rate of these interactions falls below the expansion rate they can no longer maintain equilibrium and the neutrinos become decoupled. The condition for neutrino decoupling to occur is therefore

$$\tau_H = \frac{a}{\dot{a}} \simeq 2t \simeq 2\left(\frac{3}{32\pi G\rho}\right)^{1/2} < \tau_{coll} \simeq \frac{1}{n_l \sigma_{wk} c}, \tag{8.4.2}$$

where n_l is the number density of a generic lepton, given by

$$n_l \simeq 0.1\langle g_l\rangle\left(\frac{k_B T}{\hbar c}\right)^3 \simeq 0.2\left(\frac{k_B T}{\hbar c}\right)^3 \tag{8.4.3}$$

($\langle g_l\rangle$ is the mean statistical weight of the leptons), while ρ is given by

$$\rho \simeq g^*(T)\frac{\sigma T^4}{2} \simeq \frac{5\pi^2}{12}\left(\frac{k_B T}{\hbar c}\right)^3 k_B T \simeq 4\frac{(k_B T)^4}{(\hbar c)^3}. \tag{8.4.4}$$

The condition (8.4.2) therefore becomes

$$\frac{\tau_H}{\tau_{coll}} \simeq 5\times 10^{-2} G^{-1/2}(\hbar c)^{-11/2} c g_{wk}^2 (k_B T)^3 \simeq \left(\frac{T}{3\times 10^{10}\ \mathrm{K}}\right)^3 < 1: \tag{8.4.5}$$

neutrino decoupling is then at $T_{d\nu} \simeq 3\times 10^{10}$ K. It is noteworthy that in any case the decoupling of the neutrinos happens after the annihilation of the μ^+-μ^- pairs and before the annihilation of the e^+-e^- pairs: this is important for calculating the properties of the cosmic neutrino background, as we show in the next section.

8.5 The Cosmic Neutrino Background

At the time of their decoupling, the temperature of the neutrinos coincides with the temperature T of the other constituents of the Universe which are still in thermal equilibrium: e^+, e^- and γ. The neutrino 'gas' then expands adiabatically because no other component is in thermal contact with it: for such a gas one can assume an equation of state appropriate for radiative matter and one therefore finds the relation

$$T_\nu = T_{d\nu}\frac{a(t_{d\nu})}{a}. \tag{8.5.1}$$

Until the moment of e^+-e^- annihilation, the 'gas' composed of e^-, e^+ and γ also follows a law identical to Equation (8.5.1). The temperature T suffers an increase at the moment of pair annihilation, as was explained in Section 8.3. Applying Equation (8.3.3) one finds that at $T_e \simeq 5\times 10^9$ K the temperature T (which now is just T_r) becomes

$$T_r = T = \left(\tfrac{11}{4}\right)^{1/3} T_{(-)} \simeq 1.4 T_{(-)} = 1.4 T_\nu, \tag{8.5.2}$$

because for $T > T_e$ one has $g^*_{(-)} = \frac{11}{2}$, while for $T < T_e$ we have $g^*_{(+)} = 2$ (just photons). After pair annihilation the photon gas expands adiabatically and, for high values of σ_{0r}, we get

$$T = T_r \simeq T_{(+)}\frac{a(T_e)}{a}. \tag{8.5.3}$$

One thus finds that the temperature of the radiation background remains a factor of $(11/4)^{1/3}$ higher than the temperature of the neutrino background. One therefore finds

$$T_{0\nu} = \left(\tfrac{4}{11}\right)^{1/3} T_{0r} \simeq 1.9\ \mathrm{K}, \tag{8.5.4}$$

corresponding to a number density

$$n_{0\nu} = N_\nu \times 2 \times g_\nu \times \frac{3}{4}\frac{\zeta(3)}{\pi^2}\left(\frac{k_B T_{0\nu}}{\hbar c}\right)^3 \simeq N_\nu 108\ \text{cm}^{-3} \tag{8.5.5}$$

and to a density

$$\rho_{0\nu} = N_\nu \times 2 \times g_\nu \times \frac{7}{8}\frac{\sigma T_{0\nu}^4}{2c^2} \simeq N_\nu \times 10^{-34}\ \text{g cm}^{-3}, \tag{8.5.6}$$

to be compared with the analogous quantities for photons

$$n_{0\gamma} \simeq 420\ \text{cm}^{-3} \simeq 3.7 N_\nu^{-1} n_{0\nu}, \tag{8.5.7}$$

$$\rho_{0\gamma} \simeq 4.8 \times 10^{-34}\ \text{g cm}^{-3} \simeq 4.8 N_\nu^{-1} \rho_{0\nu}. \tag{8.5.8}$$

As we have explained, the number of neutrino species is probably $N_\nu = 3$; considerations based on cosmological nucleosynthesis have for some time ruled out the possibility that $N_\nu > 4$–5. In the case $N_\nu = 3$, where we have ν_e, ν_μ and ν_τ along with their respective antineutrinos, we get $n_{0\gamma} \simeq n_{0\nu}$ and $\rho_{0\gamma} \simeq \rho_{0\nu}$. We stress again that all these results are obtained under the assumption that the neutrinos are not degenerate and that they are massless.

Let us now discuss what happens to the cosmic neutrino background if the neutrinos have a mean mass of order 10 eV, parametrised by $\langle m_\nu \rangle = \sum_{i=1}^{N_\nu} m_{\nu_i}/N_\nu$. After decoupling, the number of neutrinos in a comoving volume does not change so that Equation (8.5.5) is still valid; this is due to the fact that for $T \simeq T_{d\nu}$ the neutrinos are still ultrarelativistic, so that the above considerations are still valid. We therefore obtain

$$\rho_{0\nu} = \langle m_\nu \rangle n_{0\nu} \simeq 1.92 \times N_\nu \times \frac{\langle m_\nu \rangle}{10\ \text{eV}} \times 10^{-30}\ \text{g cm}^{-3}, \tag{8.5.9}$$

corresponding to a density parameter

$$\Omega_{0\nu} \simeq 0.1 \times N_\nu \frac{\langle m_\nu \rangle}{10\ \text{eV}} \times h^{-2} \simeq 1; \tag{8.5.10}$$

the Universe would be dominated by neutrinos.

In the case of massive neutrinos, the quantity $T_{0\nu}$ is not so much a physical temperature, but more a kind of 'counter' for the number of particles; we shall come back to this shortly. The distribution function for neutrinos (number of particles per unit volume in a unit range of momentum) f_ν before the time $t_{d\nu}$ (which we suppose, for simplicity, is the same for all types) is the relativistic one because $T_{d\nu} \gg m_\nu c^2/3k_B = T_{n\nu} \simeq 1.3 \times 10^5 (m_\nu/10\ \text{eV})$ K (the epoch in which $T \simeq T_{n\nu}$ indicates the passage from the era when the neutrinos are relativistic to the era when they are no longer relativistic; in the above approximation this happens a little before equivalence). We therefore obtain

$$f_\nu \propto \left[\exp\left(\frac{p_\nu c}{k_B T_\nu}\right) + 1\right]^{-1}, \tag{8.5.11}$$

where p_ν is the neutrino momentum. After decoupling, because the neutrinos undergo a free expansion, one has $p_\nu \propto a^{-1}$ and the neutrino distribution is still described by Equation (8.5.11) if one uses the counter

$$T_\nu = T_\nu(t_{d\nu})\frac{a(t_{d\nu})}{a(t)} = (\tfrac{4}{11})^{1/3}T. \tag{8.5.12}$$

Notice that the 'temperature' varies as a^{-1} for the neutrinos, just as it does for radiation. As we mentioned above, this is not really a true physical temperature because the neutrinos are no longer relativistic at low redshifts, though their 'temperature' still varies in the same way as radiation. On the other hand, initially cold (non-relativistic) particles would have $T \propto a^{-2}$ in this regime due to the adiabatic expansion.

The energy density of neutrinos for $T < T_{d\nu}$ is given by

$$\rho_\nu \simeq N_\nu \times 2 \times g_\nu \times \frac{7}{8}\frac{\sigma}{2}\frac{T_\nu^4}{c^2} \simeq \frac{N_\nu \rho_\gamma}{4.4} \propto (1+z)^4, \tag{8.5.13}$$

for $T_\nu = (\frac{4}{11})^{1/3}T_{\mathrm{r}} \gg T_{\mathrm{n}\nu} \simeq T_{\mathrm{eq}}$, while it is evident that

$$\rho_\nu \simeq \rho_{0\nu}\left(\frac{T_\nu}{T_{0\nu}}\right)^3 \propto (1+z)^3, \tag{8.5.14}$$

for $T_\nu \ll T_{\mathrm{n}\nu}$.

Recent experimental measurements, such as those from SuperKamiokande (Fukuda *et al.* 1999) suggest that at least one of the neutrino flavours must have a non-zero mass. The physics behind these measurements stems from the realisation that the energy (or mass) eigenstates of the neutrinos might not coincide with the states of pure lepton number; a similar phenomenon called Cabibbo mixing occurs with quarks. To illustrate, let us consider only the electron neutrino ν_{e} and the muon version ν_μ. These are the lepton states with $L_{\mathrm{e}} = 1$ and $L_\mu = 1$, respectively. In general one might imagine that these are combinations of the mass eigenstates which we can call ν_1 and ν_2:

$$\begin{pmatrix} \nu_{\mathrm{e}} \\ \nu_\mu \end{pmatrix} = \begin{pmatrix} \cos\theta & \sin\theta \\ -\sin\theta & \cos\theta \end{pmatrix} \begin{pmatrix} \nu_1 \\ \nu_2 \end{pmatrix}, \tag{8.5.15}$$

where θ is a mixing angle. That means that a state of pure electron neutrino is a superposition of the ν_1 and ν_2 states:

$$|\nu_{\mathrm{e}}\rangle = \cos\theta|\nu_1\rangle + \sin\theta|\nu_2\rangle. \tag{8.5.16}$$

If the eigenvalues of the two energy eigenstates are E_1 and E_2, respectively, then the state will evolve according to

$$|\nu_{\mathrm{e}}(t)\rangle = \cos\theta|\nu_1\rangle\exp(-\mathrm{i}E_1 t/\hbar) + \sin\theta|\nu_2\rangle\exp(-\mathrm{i}E_2 t/\hbar). \tag{8.5.17}$$

It then follows that the probability of finding a pure electron neutrino state a time t after it is set up is

$$P(t) = 1 - \sin^2(2\theta)\sin^2[\tfrac{1}{2}(E_1 - E_2)t/\hbar], \tag{8.5.18}$$

hence the term neutrino oscillation: the particle precesses between electron-neutrino and mu-neutrino states. If both states have the same momentum, then the energy difference is just

$$E_2 - E_1 = \frac{(m_2^2 - m_1^2)c^4}{2E} = \frac{\Delta m^2 c^4}{2E}, \tag{8.5.19}$$

where $E = (E_1 + E_2)/2$. This then leads to a neat alternative form to (8.5.18),

$$P(t) = 1 - \sin^2(2\theta)\sin^2\left(\frac{\pi R}{L}\right), \tag{8.5.20}$$

for a beam of electron neutrinos travelling a distance R. The quantity L is the oscillation length

$$L = \frac{4\pi E\hbar}{\Delta m^2 c^3}, \tag{8.5.21}$$

which gives the typical scale of the oscillations. Note that oscillations do not occur if the two neutrinos have equal mass. The mixing length (8.5.21) is typically very large, so the best experiments involve solar neutrinos (produced by nuclear reactions in the Sun's core) or atmospheric neutrinos (produced by cosmic ray collisions in the atmosphere). Recent results agree on a positive detection, but there is some uncertainty in the neutrino masses that can be involved and also whether all three neutrino species (including the tau) can be massive. It seems unlikely, however, that the neutrinos have masses around 10 eV, which is the mass they would have to have in order to contribute significantly to the critical density.

8.6 Cosmological Nucleosynthesis

8.6.1 General considerations

We begin our treatment of cosmological nucleosynthesis in the framework of the Big Bang model with some definitions and orders of magnitude. We define the abundance by mass of a certain type of nucleus to be the ratio of the mass contained in such nuclei to the total mass of baryonic matter contained in a suitably large volume. The abundance of ^{4}He, usually indicated with the symbol Y, has a value $Y \simeq 0.25$, obtained from various observations (stellar spectra, cosmic rays, globular clusters, solar prominences, etc.) or about 6% of all nuclei. The abundance of ^{3}He corresponds to about $10^{-3}Y$, while that of deuterium D (^{2}H or, later on, d), is of order $2 \times 10^{-2}Y$.

In the standard cosmological model the nucleosynthesis of the light elements (that is, elements with nuclei no more massive than ^{7}Li) begins at the start of the radiative era. Nucleosynthesis of the elements of course occurs in stellar interiors, during the course of stellar evolution. Stellar processes, however, generally involve destruction of ^{2}H more quickly than it is produced, because of the very large cross-section for photodissociation reactions of the form

$$^2\mathrm{H} + \gamma \rightleftharpoons \mathrm{p} + \mathrm{n}. \tag{8.6.1}$$

Nuclei heavier than ^{7}Li are essentially only made in stars. In fact there are no stable nuclei with atomic weight 5 or 8 so it is difficult to construct elements heavier than helium by means of $\mathrm{p} + \alpha$ and $\alpha + \alpha$ collisions (α represents a ^{4}He nucleus). In stars, however, $\alpha + \alpha$ collisions do produce small quantities of unstable ^{8}Be, from which one can make ^{12}C by $^8\mathrm{Be} + \alpha$ collisions; a chain of synthesis reactions can therefore develop leading to heavier elements. In the cosmological context, at the temperature of 10^9 K characteristic of the onset of nucleosynthesis, the density of the Universe is too low to permit the synthesis of significant amounts of ^{12}C from $^8\mathrm{Be} + \alpha$ collisions. It turns out therefore that the elements heavier than ^{4}He are made mostly in stellar interiors. On the other hand, the percentage of helium observed is too high to be explained by the usual predictions of stellar evolution. For example, if our Galaxy maintained a constant luminosity for the order of 10^{10} years, the total energy radiated would correspond to the fusion of 1% of the original nucleons, in contrast to the 6% which is observed.

It is interesting to note that the difficulty in explaining the nucleosynthesis of helium by stellar processes alone was recognised by Gamow (1946) and by Alpher *et al.* (1948), who themselves proposed a model of cosmological nucleosynthesis. Difficulties with this model, in particular an excessive production of helium, persuaded Alpher and Herman (1948) to consider the idea that there might have been a significant radiation background at the epoch of nucleosynthesis; they estimated that this background should have a present temperature of around 5 K, not far from the value it is now known to have ($T_{0\mathrm{r}} \simeq 2.73$ K), although some 15 years were to pass before this background was discovered. For this reason one can safely say that the satisfactory calculations of primordial element abundances which emerge from the theory represent, along with the existence of the cosmic microwave background, one of the central pillars upon which the Big Bang model is based.

8.6.2 The standard nucleosynthesis model

The hypotheses usually made to explain the cosmological origin of the light elements are as follows.

1. The Universe has passed through a hot phase with $T \geqslant 10^{12}$ K, during which its components were in thermal equilibrium.
2. General Relativity and known laws of particle physics apply at this time.
3. The Universe is homogeneous and isotropic at the time of nucleosynthesis.
4. The number of neutrino types is not high (in fact we shall assume $N_\nu \simeq 3$).

5. The neutrinos have a negligible degeneracy parameter.
6. The Universe is not composed in such a way that some regions contain matter and others antimatter.
7. There is no appreciable magnetic field at the epoch of nucleosynthesis.
8. The density of any exotic particles (photinos, gravitinos, etc.) at T_e is negligible compared with the density of the photons.

As we shall see, these hypotheses agree pretty well with such facts as we know. The hypothesis (3) is made because at the moment of nucleosynthesis, $T^* \simeq 10^9$ K ($t^* \simeq 300$ s), the mass of baryons contained within the horizon is very small, i.e. $\sim 10^3 M_\odot$, while the light-element abundances one measures seem to be the same over scales of order tens of Mpc; the hypotheses (4) and (8) are necessary because an increase in the density of the Universe at the epoch of nucleosynthesis would lead, as we shall see, to an excessive production of helium; the hypothesis (6) is made because the gamma rays which would be produced at the edges where such regions touch would result in extensive photodissociation of the ^{2}H, and therefore a decrease in the production of ^{4}He.

Later on, we shall discuss briefly some of the consequences on the nucleosynthesis process of relaxing or changing some of these assumptions.

8.6.3 The neutron–proton ratio

In Section 8.1 we stated that the ratio between the number densities of neutrons and protons is given by the relation

$$\frac{n_\mathrm{n}}{n_\mathrm{p}} \simeq \exp\left(-\frac{Q}{k_\mathrm{B}T}\right) = \exp\left(-\frac{1.5\times 10^{10}\ \mathrm{K}}{T}\right) \tag{8.6.2}$$

as long as the protons and neutrons are in thermal equilibrium. This equilibrium is maintained by the weak interactions

$$\mathrm{n} + \nu_\mathrm{e} \rightleftharpoons \mathrm{p} + \mathrm{e}^-, \qquad \mathrm{n} + \mathrm{e}^+ \rightleftharpoons \mathrm{p} + \bar{\nu}_\mathrm{e}, \tag{8.6.3}$$

which occur on a characteristic timescale τ_coll of order that given by Equation (8.4.2); this timescale is much smaller than τ_H for $T \gtrsim T_{\mathrm{d}\nu} \simeq 10^{10}$ K, i.e. until the time when the neutrinos decouple. At $t_{\mathrm{d}\nu}$ the ratio

$$X_\mathrm{n} = \frac{n}{n+p} \simeq \frac{n}{n_\mathrm{tot}} \tag{8.6.4}$$

turns out to be, from Equation (8.6.2),

$$X_\mathrm{n}(t_{\mathrm{d}\nu}) \simeq [1+\exp(1.5)]^{-1} \simeq 0.17 = X_\mathrm{n}(0). \tag{8.6.5}$$

More accurate calculations (taking into account the only partial efficiency of the above reactions) lead one to the conclusion that the ratio X_n remains equal to the equilibrium value until $T_n \simeq 1.3 \times 10^9$ K ($t_n \simeq 20$ s), after which the neutrons can only transform into protons via the β-decay, $n \to p + e^- + \bar{\nu}_e$, which has a mean lifetime τ_n of order 900 s. After t_n the ratio X_n thus varies according to the law of radioactive decay:

$$X_n(t) \equiv X_n(0) \exp\left(-\frac{t - t_n}{\tau_n}\right) \simeq X_n(0), \tag{8.6.6}$$

for $t - t_n \simeq t < \tau_n$; the value of X_n remains frozen at the value $X_n(0) \simeq 0.17$ for the entire period we are interested in. As we shall see, nucleosynthesis effectively begins at $t^* \simeq 10^2$ s.

When the temperature is of order T_n, the relevant components of the Universe are photons, protons and neutrons in the ratio $n/p \simeq \exp(-1.5) \simeq 0.2$, corresponding to the value $X_n(0)$, and small amounts of heavier particles (besides the neutrinos which have already decoupled). The electrons and positrons annihilate at $T_e \simeq 5 \times 10^9$ K; the annihilation process is not very important for nucleosynthesis, it merely acts as a marker of the end of the lepton era and the beginning of the radiative era.

8.6.4 Nucleosynthesis of Helium

To build nuclei with atomic weight $A \geqslant 3$ one needs to have a certain amount of deuterium. The amount created is governed by the equation

$$n + p \rightleftharpoons d + \gamma; \tag{8.6.7}$$

one can easily verify that this reaction has a characteristic timescale $\tau_{coll} \ll \tau_H$ in the period under consideration. The particles n, p, d and γ therefore have a number density given by the statistical equilibrium relations under the Boltzmann approximation:

$$n_i \simeq g_i \left(\frac{m_i k_B T}{2\pi\hbar^2}\right)^{3/2} \exp\left(\frac{\mu_i - m_i c^2}{k_B T}\right), \tag{8.6.8}$$

with $i = $ n, p, d and $g_n = g_p = 2g_d/3 = 2$. For the chemical potentials we take the relationship already mentioned in Section 8.2, giving

$$\mu_n + \mu_p = \mu_d. \tag{8.6.9}$$

It is perhaps a good time to stress that the chemical potentials of these particles are negligible when $n_n \simeq n_{\bar{n}}$ and $n_p \simeq n_{\bar{p}}$, but this is certainly not the case at the present epoch, because the thermal conditions are now very different.

It is useful to introduce, alongside X_n, another quantity $X_p = p/n_{tot} \simeq 1 - X_n$ and $X_d = d/n_{tot}$. From Equations (8.6.7) and (8.6.8), one can derive the equilibrium relations between n, p and d:

$$X_d \simeq \frac{3}{n_{tot}} \left(\frac{m_d k_B T}{2\pi\hbar^2}\right)^{3/2} \exp\left[\frac{\mu_n + \mu_p - (m_n + m_p)c^2 + B_d}{k_B T}\right], \tag{8.6.10}$$

which can be expressed as

$$X_{\rm d} \simeq n_{\rm tot}\left(\frac{m_{\rm d}}{m_{\rm n}m_{\rm p}}\right)^{3/2}\frac{3}{4}\left(\frac{k_{\rm B}T}{2\pi\hbar^2}\right)^{-3/2} X_{\rm n}X_{\rm p}\exp\left(\frac{B_{\rm d}}{k_{\rm B}T}\right), \tag{8.6.11}$$

and consequently as

$$X_{\rm d} \simeq X_{\rm n}X_{\rm p}\exp\left[-29.33+\frac{25.82}{T_9}-\tfrac{3}{2}\ln T_9+\ln(\Omega_{0\rm b}h^2)\right], \tag{8.6.12}$$

where $T_9 = (T/10^9\ \mathrm{K})$, $\Omega_{0\mathrm{b}}$ is the present density parameter in baryonic material. In (8.6.12), $B_{\rm d}$ is the binding energy of deuterium:

$$B_{\rm d} = (m_{\rm n}+m_{\rm p}-m_{\rm d})c^2 \simeq 2.225\ \mathrm{MeV} \simeq 2.5\times10^{10}\ \mathrm{K}. \tag{8.6.13}$$

The function $X_{\rm d}$ depends only weakly on Ωh^2.

For $T_9 \geqslant 10$ the value of $X_{\rm d}$ is negligible: all the nucleons are still free because the high energy of the ambient photons favours the photodissociation reaction. The fact that nucleosynthesis cannot proceed until $X_{\rm d}$ grows sufficiently large is usually called the *deuterium bottleneck* and is an important influence on the eventual helium abundance. The value of $X_{\rm d}$ is no longer negligible when $T_9 \simeq 1$. At $T_9^* \simeq 0.9$ for $\Omega = 1$ ($t^* \simeq 300$ s) or at $T_9^* \simeq 0.8$ for $\Omega \simeq 0.02$ ($t^* \simeq 200$ s) $X_{\rm d} \simeq X_{\rm n}X_{\rm p}$. For $T < T_9^*$ the value of $X_{\rm d}$ becomes significant. At lower temperatures all the neutrons might be expected to be captured to form deuterium. This deuterium does not appear, however, because reactions of the form

$$\mathrm{d}+\mathrm{d} \to {}^3\mathrm{He}+\mathrm{n}, \qquad {}^3\mathrm{He}+\mathrm{d} \to {}^4\mathrm{He}+\mathrm{p}, \tag{8.6.14}$$

which have a large cross-section and are therefore very rapid, mop up any free neutrons into ^{4}He. Thus, the abundance of helium that forms is

$$Y \simeq Y(T^*) = 2X_{\rm n}(T^*) = 2X_{\rm n}(T_{\rm n})\exp\left(-\frac{t^*-t_{\rm n}}{\tau_{\rm n}}\right) \simeq 0.25, \tag{8.6.15}$$

in reasonable accord with that given by observations. In Equation (8.6.13), the factor 2 takes account of the fact that, after helium synthesis, there are practically only free protons and helium nuclei, so that

$$Y = \frac{m_{\rm He}}{m_{\rm tot}} = 4\frac{n_{\rm He}}{n_{\rm tot}} \simeq 4\times\frac{1}{2}\frac{n_{\rm n}}{n_{\rm tot}} = 2X_{\rm n}. \tag{8.6.16}$$

The value of Y obtained is roughly independent of Ω. This is essentially due to two reasons:

1. the value of $X_{\rm n}$ before nucleosynthesis does not depend on Ω because it is determined by weak interactions between nucleons and leptons and not by strong interactions between nucleons; and
2. the start of nucleosynthesis is determined by the temperature rather than the density of the nucleons.

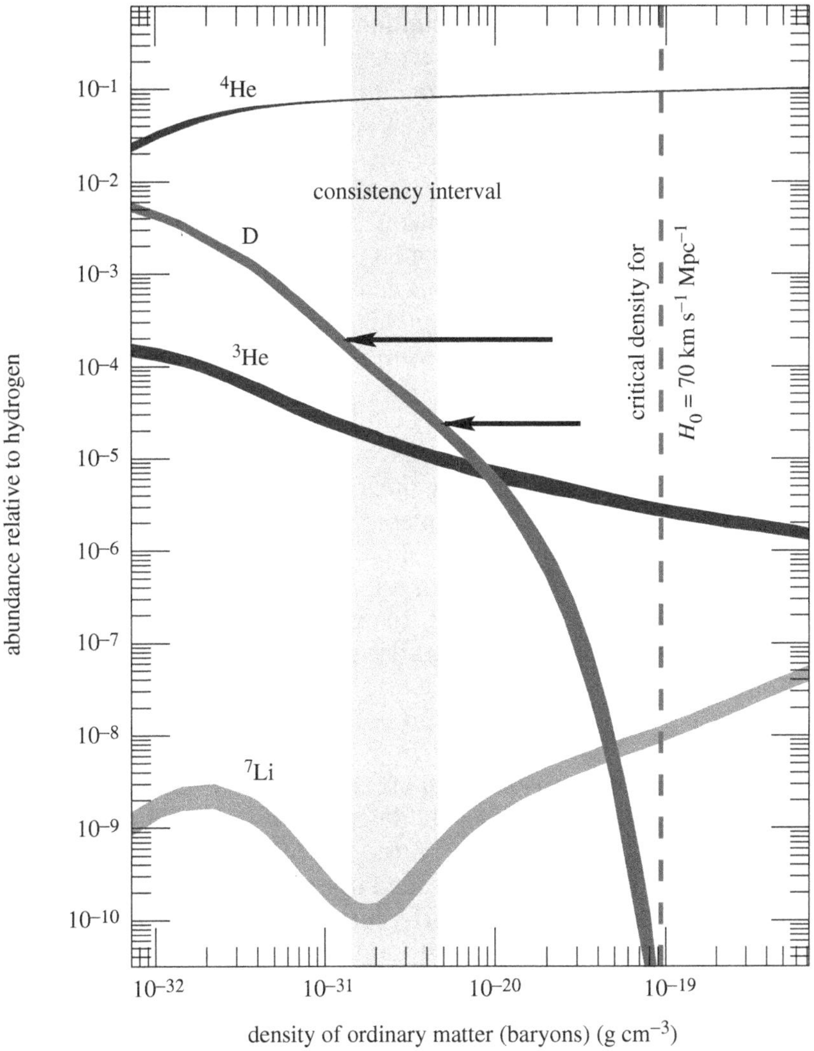

Figure 8.1 Light-element abundance determined by numerical calculations as functions of the matter density, as explained in the text. The arrows mark the possible deuterium abundance. From Schramm and Turner (1996). Picture courtesy of Mike Turner.

8.6.5 Other elements

As far as the abundances of other light elements are concerned one needs to perform a detailed numerical integration of all the rate equations describing the reaction network involved in building up heavier nuclei than ^{4}He. We have no space to discuss the details of these calculations here, but the main results are illustrated in Figure 8.1.

The figure shows the computed abundance of ^{4}He (usually denoted by Y_P), depending on the number of neutrino types. Note that some helium is certainly made in stars so that a correction must be made to the observed abundance Y

in order to estimate the primordial abundance which is Y_P. The error bar on the central line indicates the effect of an error of ± 0.2 min in the neutron half-life. The other curves show the relative abundances (compared with ^{1}H) of deuterium D, ^{3}He, ^{3}He + D and ^{7}Li. The abundances are all shown as a function of η, the baryon-to-photon ratio which is related to Ω_b by $\Omega_b \simeq 0.004h^{-2}\eta/10^{-10}$.

The abundances of deuterium and ^{3}He are about three orders of magnitude below ^{4}He, while ^{7}Li is nine orders of magnitude smaller than this; all other nuclei are less abundant than this. The basic effect one can see is that, since the abundance of ^{4}He increases slowly with η (because nucleosynthesis starts slightly earlier and burning into ^{4}He is more complete), the abundances of the 'incomplete' products D and ^{3}He decrease in compensation. The abundance of ^{7}Li is more complicated because of the two possible formation mechanisms: direct formation via fusion of ^{4}He and ^{3}H dominates at low η, while electron capture by ^{7}Be dominates at high η. In between, the 'dip' is caused by the destruction reaction involving proton capture and decay into two ^{4}He nuclei.

So how do these computations compare with observations? At the outset we should stress that relevant observational data in this field are difficult to obtain. The situation with regard to ^{4}He is perhaps the clearest but, although the expected abundance is large, the dependence of this abundance on cosmological parameters is not strong. Precise measurements are therefore required to test the theory. For the other elements shown in Figure 8.1, the parameter dependence is strong and is dominated by the dependence on η, but the expected abundances, as we have shown, are tiny. Moreover, any material we can observe has been at least partly processed through stars. Burning of H into ^{4}He is the main source of energy for stars. Deuterium can be very easily destroyed in stars (but cannot be made there). The other isotopes ^{3}He and ^{7}Li can be both created and destroyed in stars. The processing of material by stars is called *astration* and it means that uncertain corrections have to be introduced to derive 'primordial' abundances from present-day observations. One should also mention that *fractionation* (either physical or chemical in origin) may mean that the abundances in one part of an astronomical object may not be typical of the object as a whole; such effects are known to be important, for example, in determining the abundance of deuterium in the Earth's oceans.

Despite these difficulties, there is a considerable industry involved in comparing observed abundances with these theoretical predictions. Relevant data can be obtained from stellar atmospheres, interstellar emission and absorption lines (and intergalactic ones), planetary atmospheres, meteorites and from terrestrial measurements. Abundances of elements other than ^{4}He determined by these different methods differ by a factor of five or more, presumably because of astration and/or fractionation.

8.6.6 Observations: Helium 4

It is relatively well established that the abundance of ^{4}He is everywhere close to 25% and this in itself is good evidence that the basic model is correct. To get the primordial helium abundance more accurately than this rough figure, it is necessary to correct for the processing of hydrogen into helium in stars. This is generally done by taking account of the fact that stars with higher metallicity have a slightly higher ^{4}He abundance, and extrapolating to zero metallicity; metals are assumed to be a byproduct of the fusion of hydrogen into helium. One therefore generally requires an index of metallicity in the form of either O/H or N/H determinations. Good data on these abundances have been obtained for around 50 extragalactic HII regions (Pagel *et al.* 1992; Skillman *et al.* 1993; Izotov *et al.* 1994). Olive and Steigman (1995) and Olive and Scully (1995), for example, have found on the basis of these data that there is evidence for a linear correlation of Y with O/H and N/H; the intercept of this relation yields

$$Y_{\rm p} = 0.234 \pm 0.003 \pm 0.005. \tag{8.6.17}$$

The first error is purely statistical and the second is an estimate of the systematic uncertainty in the abundance determinations.

8.6.7 Observations: Deuterium

The abundance of deuterium has been the subject of intense investigation in recent months. Prior to this period, deuterium abundance information was based on interstellar medium (ISM) observations and Solar System data. From the ISM, one gets

$$\mathrm{D/H} \simeq 1.60 \times 10^{-5} \tag{8.6.18}$$

with an uncertainty of about 10% (Linsky *et al.* 1993, 1995). This value may or may not be close to universal, as it is possible that the abundances in the ISM are inhomogeneous. Solar System investigations based on properties of meteoritic rock involve a more circuitous route through ^{3}He (which one assumes was efficiently burned into D during the pre-main-sequence phase of the Sun). This argument leads to a value of

$$(\mathrm{D/H})_\odot \simeq 2.6 \times 10^{-5} \tag{8.6.19}$$

with an uncertainty of nearly 100% (Scully *et al.* 1996).

More recently, the rough consensus between these two estimates was shaken by claims of detections of deuterium absorption in the spectra of high-redshift quasars. The occurrence of gas at high redshift and in systems of low metallicity suggests that one might well expect to see a light-element abundance close to the primordial value. The first such observations yielded much higher values than (8.6.18) and (8.6.19) by about a factor of 10 (Carswell *et al.* 1994; Songaila *et al.* 1994), i.e.

$$(\mathrm{D/H}) \simeq 2 \times 10^{-4}; \tag{8.6.20}$$

other measurements seemed to confirm these high values (Rugers and Hogan 1996a,b; Carswell *et al.* 1996; Wampler *et al.* 1996).

On the other hand, significantly lower deuterium abundances have been found by other workers in similar systems (Tytler *et al.* 1996; Burles and Tytler 1996). This raises the suspicion that the high inferred deuterium abundances may be a mistake, perhaps from a misidentified absorption feature (e.g. Steigman 1994). On the other hand, one does expect deuterium to be destroyed by astration and, on these grounds, one is tempted to identify the higher values of D/H with the primordial value.

Over the last few years, evidence has gathered that the low deuterium abundance is more secure and that previous high values may have been due to observational problems. The recent published estimate by Burles and Tytler gives

$$(\mathrm{D/H}) \simeq (3.3 \pm 0.6) \times 10^{-5}, \tag{8.6.21}$$

although this may not be the end of the story.

8.6.8 Helium 3

There are various ways in which the primordial ^{3}He abundance can be estimated. For a start, the Solar System deuterium estimate entails an estimate of the ^{3}He abundance which generally comes out around 1.5×10^{-5}. ISM observations and galactic HII regions yield values with a wide dispersion:

$$(^3\mathrm{He/H}) \simeq 2.5 \times 10^{-5}; \tag{8.6.22}$$

the spread is around a factor of 2 either side of this value.

The primordial ^{3}He, however, is modified by the competition between stellar production and destruction processes, and a detailed evolution model is required to relate the observed abundances, themselves highly uncertain, with their inferred primordial values. As we mentioned above, one may be helped in this task by using the combined abundance of D and ^{3}He (e.g. Steigman and Tosi 1995). The simplest way to use these data employs the argument that when deuterium is processed into stars it is basically turned into ^{3}He, which can be processed further, but which burns at a higher temperature. Stars of different masses therefore differ in their net conversion between these two species. But since *all* stars do destroy deuterium to some extent and at least *some* ^{3}He survives stellar processing, the primordial combination of D + ^{3}He might well be expected to be bounded above by the observed value. Attempts to go further introduce further model-dependent parameters and corresponding uncertainties into the analysis. For reference, a rough figure for the combined abundance is

$$(\mathrm{D} + {}^3\mathrm{He})/\mathrm{H} \simeq 4.1 \times 10^{-5}, \tag{8.6.23}$$

with an uncertainty of about 50%.

8.6.9 Lithium 7

In old hot stars (Population II), the lithium abundance is found to be nearly uniform (Molaro *et al.* 1995; Spite *et al.* 1996). Indeed there appears to be little variation from star to star in a sample of 100 halo stars, over and above that expected from the statistical errors in the abundance determinations. The problem with the interpretation of such data, however, is in the fact that astrophysical processes can both create and destroy lithium. Up to about half the primordial ^{7}Li abundance may have been destroyed in stellar processes, while it is estimated that up to 30% of the observed abundance might have been produced by cosmic ray collisions. The resulting best guess for the primordial abundance is

$$\mathrm{Li/H} \simeq 1.6 \times 10^{-10}, \tag{8.6.24}$$

but the uncertainty, dominated by unknown parameters of the model used to process the primordial abundance, is at least 50% and is itself highly uncertain (Walker *et al.* 1993; Olive and Schramm 1992; Steigman *et al.* 1993).

8.6.10 Observations versus theory

We have tried to be realistic about the uncertainties in both the observations and the extrapolation of those observations back to the primordial abundances. Going into the detailed models of galactic chemical evolution that are required to handle D, ^{3}He and ^{7}Li opens up a rather large can of model-dependent worms, so we shall simply sketch out the general consensus about what these results mean for η and Ω_b.

The estimates of the primordial values of the relative abundances of deuterium (D), ^{3}He, ^{4}He and ^{7}Li all appear to be in accord with nucleosynthesis predictions, but only if the density parameter in baryonic material is

$$\Omega_{0\mathrm{b}} h^2 \simeq 0.02 \tag{8.6.25}$$

(e.g. Walker *et al.* 1991; Smith *et al.* 1993). This roughly corresponds to $3 \leqslant \eta_{10} \leqslant 4$. A baryon density higher than this would produce too much ^{7}Li, while a lower value would produce too much deuterium and ^{3}He. Copi *et al.* (1995a,b) suggest a somewhat wider range of allowed systematic errors, leading to $2 \leqslant \eta_{10} \leqslant 6.5$, which translates into

$$0.005 < \Omega_\mathrm{b} h^2 < 0.026. \tag{8.6.26}$$

The dependence of ^{4}He is so weak that it can really only be used as a consistency check on the scheme.

This strong constraint on Ω_b is the main argument for the existence of *non-baryonic dark matter*, which we discuss in more detail in the second half of this book.

8.7 Non-standard Nucleosynthesis

We have seen that standard nucleosynthesis seems to account reasonably well for the observed light-element abundances and also places strong constraints on the allowed range of the density parameter. To what extent do these results rule out alternative models for nucleosynthesis, and what constraints can we place on models which violate the conditions (1)–(8) of the previous section? We shall make some comments on this question by describing some attempts that have been made to vary the conditions pertaining to the standard model.

First, one could change the expansion rate τ_H at the start of nucleosynthesis. A decrease of τ_H (i.e. a faster expansion rate) can be obtained if the Universe contains other types of particles in equilibrium at the epoch under consideration. These could include new types of neutrino, or supersymmetric particles like photinos and gravitinos: in general, $\tau_H \simeq t \propto (g^* T^4)^{-1/2}$. A small reduction of τ_H reduces the time available for the neutrons to decay into protons, so that the value of X_n tends to move towards its primordial value of $X_n \simeq 0.5$; the reduction of τ_H does not, however, influence the time of onset of nucleosynthesis to any great extent so that this still occurs at $T \simeq 10^9$ K. The net result is an increase in the amount of helium produced. As we have mentioned above, these results have for a long time led cosmologists to rule out the possibility than N_ν might be larger than 4 or 5. Now we know that $N_\nu = 3$ from particle experiments; nucleosynthesis still rules out the existence of any other relativistic particle species at the appropriate epoch. A large reduction in τ_H, however, tends to reduce the abundance of helium: the reactions (8.6.12) have too little time to produce significant helium because the density of the Universe falls rapidly. A decrease in the expansion rate allows a larger number of neutrons to decay into protons so that the ratio $X_n(T^*)$ becomes smaller. Since basically all the neutrons end up in helium, the production of this element is decreased.

Another modification one can consider concerns the hypothesis that the neutrinos are not degenerate. If the chemical potential of ν_e is such that

$$40 > \left| \frac{\mu_{\nu_e}}{k_B T} \right| = |x_{\nu_e}| \geqslant 1 \tag{8.7.1}$$

(the upper limit was derived in Section 8.2), the obvious relation

$$\mu_p - \mu_n = \mu_{\nu_e} - \mu_{e^-} \simeq \mu_{\nu_e} = x_{\nu_e} k_B T \tag{8.7.2}$$

(because at $T \simeq 10^{10}$ K we have $\mu_{e^+} \simeq \mu_{e^-} \simeq 0$ through the requirement of electrical neutrality) leads one to the conclusion that

$$X_n(T) \simeq \left[1 + \exp\left(x_{\nu_e} + \frac{Q}{k_B T} \right) \right]^{-1}, \tag{8.7.3}$$

for $T \geqslant T_{d\nu} \simeq 10^{10}$ K. For $x_{\nu_e} \gg 0$ (degeneracy of the ν_e), the value of $X_n(T_{d\nu})$ is much less than 0.5, so that one makes hardly any helium. If $x_{\nu_e} \ll 0$ (degeneracy of

$\bar{\nu}_e$), the high number of neutrons, because $X_n(T_{d\nu}) < 1$, and the consequent low number of protons prevents the formation of deuterium and therefore helium. Deuterium would be formed much later when the expansion of the Universe had diluted the $\bar{\nu}_e$ and some neutrons could have decayed into protons. But at this point the density would be too low to permit significant nucleosynthesis, unless $\Omega \geqslant 1$. In the case when $x_{\nu_e} \simeq -1$, one can have $X_n \simeq 0.5$ at the moment of nucleosynthesis, so that all the neutrons end up in helium. This would mean that essentially all the baryonic matter in the Universe would be in the form of helium. In the case where the neutrinos or antineutrinos are degenerate there is another complication in the theory of nucleosynthesis: the total density of neutrinos and antineutrinos would be greater than one would think if there were such a degeneracy. For example, if $|x_{\nu_e}| \ll 1$, we have

$$\rho(\nu_e) + \rho(\bar{\nu}_e) \simeq \frac{\sigma T_\nu^4}{c^2}\left(\frac{7}{8} + \frac{15}{4\pi^2}x_{\nu_e}^2 + \frac{15}{8\pi^4}x_{\nu_e}^4\right). \tag{8.7.4}$$

This fact gives rise to a decrease in the characteristic time for the expansion τ_H, with the corresponding consequences for nucleosynthesis. One can therefore conclude that the problems connected with a significant neutrino degeneracy are large, and one might be tempted to reject them on the grounds that models invoking such a degeneracy are also much more complicated than the standard model.

Even graver difficulties face the idea of nucleosynthesis in a cold universe, i.e. a model in which the background radiation is not all of cosmological origin and in models where the universal expansion is not isotropic.

We should also mention that it has been suggested, and still is suggested by (the few remaining) advocates of the steady-state theory, that a radically alternative but possibly attractive model of nucleosynthesis might be one in which the light elements were formed in an initial highly luminous phase of galaxy formation or, perhaps, in primordial 'stars' of very high mass, the so-called Population III objects. The constraints on these models from observations of the infrared background are, however, severe.

Probably the best argument for non-standard nucleosynthesis is the suggestion that the standard model itself may be flawed. If the quark–hadron phase transition is a first-order transition, then, as the Universe cools, one would produce bubbles of the hadron phase inside the quark plasma. The transition proceeds only after the nucleation of these bubbles, and results in a very inhomogeneous distribution of hadrons with an almost uniform radiation background. In this situation, both protons and neutrons are strongly coupled to the radiation because of the efficiency of 'charged-current' interactions. These reactions, however, freeze out at $T \simeq 1$ MeV so that the neutrons can then diffuse while the protons remain locked to the radiation field. The result of all this is that the n/p ratio, which is one of the fundamental determinants of the ^{4}He abundance, could vary substantially from place to place. In regions of relatively high proton density, every neutron will end up in a ^{4}He nucleus. In neutron-rich regions, however, the neutrons have to undergo β-decay before they can begin to fuse. The net result is less ^{4}He and more D than in the standard model, for the same value of η. The observed limits on cosmological abundances do not therefore imply such a strong upper limit on Ω_b. It has even been suggested that such a mechanism may allow a critical density of baryons, $\Omega_b = 1$, to be compatible with observed elemental abundances. This idea is certainly interesting, but to find out whether it is correct one needs to perform a detailed numerical solution of the neutron transport and nucleosynthesis reactions, allowing for a strong spatial variation. In recent years, attempts have been made to perform such calculations but they have not been able to show convincingly that the standard model needs to be modified and the limits (8.6.25) weakened.

In conclusion we would like to suggest that, even if the standard model of nucleosynthesis is in accord with observations (which is quite remarkable, given the simplicity of the model), the constraints particularly on Ω_b emerging from these calculations are so fundamental to so many things that one should always keep an open mind about alternative, non-standard models which, as far as we are aware, are not completely excluded by observations.

Bibliographic Notes on Chapter 8

Bernstein (1988) is a detailed monograph on relativistic statistical mechanics, which is also well covered by Kolb and Turner (1990). The physics of the quark-hadron transition is discussed by Applegate and Hogan (1985) and Bonometto and Pantano (1993).

For more extensive discussions of both theoretical and observational aspects of cosmological nucleosynthesis, see the technical review articles of Schramm and Wagoner (1979), Merchant Boesgaard and Steigman (1985), Bernstein *et al.* (1988), Walker *et al.* (1991) and Smith *et al.* (1993) and the book by Börner (1988). An important paper in the historical development of this field is Hoyle and Tayler (1964).

Problems

1. Cross-sections for weak interactions at an energy E increase with E as E^2. Show that the rate of weak interactions in the early Universe depends on the temperature T as $\sigma_{wk} \propto T^5$. Using an appropriate model, estimate the temperature at which weak interactions freeze out in the Big Bang.

2. Let t_1 be the epoch when electron-positron annihilation is completed and t_2 be the epoch when helium fusion begins. You may assume that these two events take place at temperatures of 5×10^9 and 10^9 K, respectively. Assuming a simplified model in which $\Lambda = k = 0$ and which is radiation dominated before $t_{eq} = 3 \times 10^5$ years and matter dominated from t_{eq} until the present time (which you can take to be 10^{10} years), use the present temperature of the cosmic microwave background, 2.7 K, to infer values of t_1 and t_2.

3. If the abundance of neutrons, X_n, declines by beta decay in the interval between t_1 and t_2 (given in Question 2) according to

$$X_n = 0.16 \exp\left(-\frac{\Delta t}{1013\ \mathrm{s}}\right),$$

derive an estimate of X_n at the time helium fusion begins.

9

The Plasma Era

9.1 The Radiative Era

The radiative era begins at the moment of the annihilation of electron–positron pairs (e^+–e^-). This occurs, as we have explained, at a temperature $T_e \simeq 5 \times 10^9$ K, corresponding to a time $t_e \simeq 10$ s. After this event, the contents of the Universe are photons and neutrinos (which have already decoupled from the background and which in this chapter we shall assume to be massless) and matter (which we take to be essentially protons, electrons and helium nuclei after nucleosynthesis; the possible existence of non-baryonic dark matter is not relevant to the following considerations and we shall therefore use Ω_0 to mean Ω_{0b} throughout this chapter).

The density of photons and neutrinos (the relativistic particles) is

$$\rho_{\gamma,\nu} = \rho_{0r}\left(\frac{T}{T_{0r}}\right)^4 + \rho_{0\nu}\left(\frac{T_\nu}{T_{0\nu}}\right)^4 \simeq \rho_{0r}(1 + 0.227N_\nu)\left(\frac{T}{T_{0r}}\right)^4 = \rho_{0r}K_0(1+z)^4 \quad (9.1.1)$$

(as we have explained, $K_0 \simeq 1.68$ if $N_\nu = 3$). The density of matter is

$$\rho_m = \rho_{0c}\Omega_{0m}(1+z)^3 \simeq \rho_{0c}\Omega_0(1+z)^3. \quad (9.1.2)$$

The end of the radiative era occurs when the density of matter coincides with that of the relativistic particles, corresponding to a redshift

$$1 + z_{eq} = \frac{\rho_{0c}\Omega_0}{K_0\rho_{0r}} \simeq \frac{4.3}{K_0} \times 10^4 \Omega_0 h^2 \quad (9.1.3)$$

and a temperature

$$T_{eq} = T_{0r}(1 + z_{eq}) \simeq \frac{10^5 \Omega_0 h^2}{K_0} \text{ K}. \quad (9.1.4)$$

At high temperatures both the hydrogen and helium are fully ionised, and exist in the form of ions (H^+, He^{++}). Gradually, as the temperature cools, the number of He^+ ions and neutral H and He atoms grows according to the equilibrium reactions

$$H^+ + e^- \rightleftharpoons H + \gamma, \qquad He^{++} + e^- \rightleftharpoons He^+ + \gamma, \qquad He^+ + e^- \rightleftharpoons He + \gamma, \tag{9.1.5}$$

in which the density of the individual components is governed by the *Saha equation* which we saw in a different context in Section 8.6. We shall study in Section 9.3 in particular the equilibrium with regard to hydrogen recombination. It has been calculated that at $T \simeq 10^4$ K the helium content is 50% in the form He^{++} and 50% He^+, while the hydrogen is 100% H^+; at $T \simeq 7 \times 10^3$ K one has 50% He^+ and 50% He but still 100% H^+; at $T \simeq 4 \times 10^3$, corresponding to $z \simeq 1500$, one has 100% He, 50% H^+ and 50% H. One usually takes the epoch of recombination to be that corresponding to a temperature of around $T_{\rm rec} \simeq 4000$ K when 50% of the matter is in the form of neutral atoms to a good approximation. Usually, in fact, one ignores the existence of helium during the period in which $T > T_{\rm rec}$; this period is usually called the *plasma epoch*.

9.2 The Plasma Epoch

The plasma we consider is composed of protons, electrons and photons at a temperature $T > T_{\rm rec}$. In this situation the plasma is an example of a 'good plasma', in the sense that the energy contributed by Coulomb interactions between the particles is much less than their thermal energy. This criterion is expressed by the inequality

$$\lambda_{\rm D} \gg \lambda, \tag{9.2.1}$$

where $\lambda_{\rm D}$ is the *Debye radius*

$$\lambda_{\rm D} = \left(\frac{k_{\rm B}T}{4\pi n_{\rm e} e^2}\right)^{1/2}, \tag{9.2.2}$$

in which $n_{\rm e}$ is the number-density of ions from which one can obtain the mean separation

$$\lambda \simeq n_{\rm e}^{-1/3} \simeq \left(\frac{m_{\rm p}}{\rho_{0{\rm c}}\Omega_0}\right)^{1/3}\left(\frac{T_{0{\rm r}}}{T}\right). \tag{9.2.3}$$

In these equations, and throughout this section, e is expressed in electrostatic units. In the cosmological case we find that

$$\frac{\lambda_{\rm D}}{\lambda} \simeq 10^2(\Omega_0 h^2)^{-1/6}. \tag{9.2.4}$$

An equivalent way to express (9.2.1) is to assert that the number of ions $N_{\rm D}$ inside a sphere of radius $\lambda_{\rm D}$ is large ('screening' effects are negligible). One can show that

$$N_{\rm D} = \tfrac{4}{3}\pi n_{\rm e}\lambda_{\rm D}^3 \simeq 1.8 \times 10^6(\Omega_0 h^2)^{-1/2}. \tag{9.2.5}$$

The Coulomb interaction between an electron and a proton is felt only while the electron traverses the Debye sphere of radius λ_D around an ion. The typical time taken to cross the Debye sphere is

$$\tau_e = \omega_e^{-1} = \left(\frac{m_e}{4\pi n_e e^2}\right)^{1/2} \simeq 2.2 \times 10^8 T^{-3/2} \text{ s}, \tag{9.2.6}$$

where ω_e is the *plasma frequency*. The time τ_e can be compared with the characteristic time for an electron to lose its momentum by electron–photon scattering

$$\tau'_{e\gamma} = \frac{3m_e}{4\sigma_T \rho_r c} = 4.4 \times 10^{21} T^{-4} \text{ s}; \tag{9.2.7}$$

the result is that $\tau_e \ll \tau'_{e\gamma}$ for $z \ll 2 \times 10^7 (\Omega_0 h^2)^{1/5}$, which is true for virtually the entire period in which we are interested here. The fact that $\tau_e \ll \tau'_{e\gamma}$ means that collective plasma effects are insignificant in this case, i.e. there is a very small probability of an electron–photon collision during the time of an electron–proton collision. On the other hand, for $z \gg 2 \times 10^7 (\Omega_0 h^2)^{1/5}$ electrons and photons are effectively 'glued' together ($\tau'_{e\gamma} \gg \tau_e$ in this period). One must therefore assign the electron an 'effective mass' $m_e^* = m_e + (\rho_r + p_r/c^2)/n_e \simeq \frac{4}{3}\rho_r/n_e \gg m_e$ when describing an electron–proton collision. Returning to the case where $z \ll 2 \times 10^7 (\Omega_0 h^2)^{1/5}$, the electrons and protons are strongly coupled and effectively stuck together; the characteristic time for electron–photon scattering is

$$\tau_{e\gamma} = \frac{3}{4}\frac{m_e + m_p}{\sigma_T \rho_r c} \simeq \frac{3}{4}\frac{m_p}{\sigma_T \rho_r c} \simeq 9 \times 10^{24} T^{-4} \text{ s}, \tag{9.2.8}$$

which we refer to in Section 12.8. One should mention here that the factor $\frac{3}{4}$ in Equations (9.2.7) and (9.2.8) comes from the fact that, as well as the inertia $\rho_r c^2$ of the radiation, one must also include the pressure $p_r = \rho_r c^2/3$. Another timescale of interest is the timescale for photon–electron scattering; this is of order

$$\tau_{\gamma e} = \frac{1}{n_e \sigma_T c} = \frac{m_p}{\rho_m \sigma_T c} = \frac{4}{3}\tau_{e\gamma}\frac{\rho_r}{\rho_m} \simeq 10^{20} (\Omega_0 h^2)^{-1} T^{-3} \text{ s}. \tag{9.2.9}$$

The relaxation time for thermal equilibrium between the protons and electrons to be reached is

$$\tau_{ep} \simeq 10^6 (\Omega_0 h^2)^{-1} T^{-3/2} \text{ s}, \tag{9.2.10}$$

which is much smaller than the characteristic time for the expansion of the Universe during this period. One can therefore assume that protons and electrons have the same temperature. In the cosmological plasma, Compton scattering is the dominant form of interaction. In the absence of sources of heat, this scattering maintains the plasma in thermal equilibrium with the radiation. This is the basic reason why we expect to see a thermal black-body radiation spectrum. As we shall discuss in Section 9.5, energy injected into the plasma at a redshift $z > z_t \simeq 10^7$–10^8 will be completely thermalised on a very short timescale. One cannot therefore obtain information about energy sources at $z > z_t$ from the observed spectrum of the radiation. On the other hand, energy injected after z_t may not be thermalised, and one might expect to see some signal of this injection in the spectrum of relic radiation.

9.3 Hydrogen Recombination

During the final stages of the plasma epoch, the particles p, e^-, H and γ (ignoring the helium for simplicity) are coupled together via the reactions (9.1.5). Supposing that these reactions hold the particles in thermal equilibrium, we can study the process of hydrogen recombination, which marks the end of the plasma era and the beginning of the era of neutral matter. Let us concentrate on the *ionisation fraction*

$$x = \frac{n_e}{n_p + n_H} \simeq \frac{n_e}{n_{tot}}. \tag{9.3.1}$$

Neutral hydrogen has a binding energy $B_H \simeq 13.6$ eV (corresponding to a temperature $T_H \simeq 1.6 \times 10^5$ K). At a temperatures of the order of $T \simeq 10^4$ K all the particles involved are non-relativistic, and one can therefore apply simple Boltzmann statistics to the plasma. We therefore obtain the number-density of the ith particle species in the form

$$n_i \simeq g_i \left(\frac{m_i k_B T}{2\pi\hbar^2}\right)^{3/2} \exp\left(\frac{\mu_i - m_i c^2}{k_B T}\right) \tag{9.3.2}$$

(cf. Section 8.6). The relevant chemical potentials are related by

$$\mu_p + \mu_{e^-} = \mu_H : \tag{9.3.3}$$

the photons are in equilibrium and therefore have zero chemical potential. The statistical weights of the particles we are considering are $g_p = g_{e^-} = \frac{1}{2} g_H = 2$. The masses of the proton, the electron and the neutral hydrogen atoms are related by

$$m_H c^2 = (m_p + m_e)c^2 - B_H. \tag{9.3.4}$$

From the preceding equations, noting that global charge neutrality requires $n_e = n_p$, we obtain the relation

$$\frac{n_e n_p}{n_H n_{tot}} = \frac{n_e^2}{(n_{tot} - n_e) n_{tot}} = \frac{x^2}{1-x} = \frac{1}{n_{tot}} \left(\frac{m_e k_B T}{2\pi\hbar^2}\right)^{3/2} \exp\left(-\frac{B_H}{k_B T}\right), \tag{9.3.5}$$

which is called the *Saha formula* corresponding to the hydrogen recombination reaction. In Table 9.1 we give some examples of the behaviour of the hydrogen ionisation fraction x as a function of redshift z and temperature $T = T_{0r}(1+z)$ for various values of the density parameter in the form $\Omega_0 h^2$. As one can see from Table 9.1, the process of hydrogen recombination does not begin at T_H because

Table 9.1 Ionisation fractions as function of z (or T) and $\Omega_0 h^2$.

z	2000	1800	1600	1400	1200	1000
T (K)	5400	4860	3780	3240	2970	2700
$\Omega_0 h^2$						
10	0.995	0.914	0.358	0.004	0.001	1×10^{-5}
1	0.999	0.990	0.732	0.108	0.004	4×10^{-5}
0.1	1.0	1.0	0.954	0.303	0.012	1×10^{-4}
0.01	1.0	1.0	0.995	0.664	0.039	3×10^{-4}

of the relatively large numerical factor appearing in front of the exponential in Equation (9.3.5). The redshift at which the ionisation fraction falls to 0.5 does not vary much with the parameter $\Omega_0 h^2$ and is always contained in the interval 1400–1600. It is a good approximation therefore to assume a redshift $z_{\rm rec} \simeq 1500$ as characteristic of the recombination epoch.

The Saha formula is valid as long as thermal equilibrium holds. In an approximate way, one can say that this condition is true as long as the characteristic timescale for recombination $\tau_{\rm rec} \simeq x/\dot{x}$ is much smaller than the timescale for the expansion of the Universe, $\tau_{\rm H}$. This latter condition is true for $z > 2000(\Omega_0 h^2)^{-1}$, only when the ionisation fraction is still of order unity. It is possible therefore that physical processes acting out of thermal equilibrium could have significantly modified the cosmological ionisation history. For this reason, many authors have investigated non-equilibrium thermodynamical processes during the plasma epoch. These studies are much more complex than the quasi-equilibrium treatment we have described here, and to make any progress requires certain approximations. There is nevertheless a consensus that the value of x during recombination ($z \simeq 1000$) is probably a factor of order 100 greater than that predicted by the Saha Equation (9.3.5). In fact, in the interval $900 < z < 1500$, the following approximate expression for $x(z)$, due to Sunyaev and Zel'dovich, holds:

$$x(z) \simeq 5.9 \times 10^6 (\Omega_0 h^2)^{-1/2} (1+z)^{-1} \exp\left(-\frac{B_{\rm H}}{k_{\rm B} T_{0\rm r} z}\right). \tag{9.3.6}$$

All calculations predict that the ionisation fraction tends to a value in the range 10^{-4}–10^{-5} for $z \to 0$. As we shall see in Chapter 19, the ionisation fraction of intergalactic matter at $t = t_0$ is actually much higher than this, probably due to the injection of energy by early structure formation after $z_{\rm rec}$.

9.4 The Matter Era

The matter era begins at $z_{\rm eq}$. As we have already explained, assuming a value of $z_{\rm rec} \simeq 1500$, one concludes that $z_{\rm eq} > z_{\rm rec}$ for $\Omega_0 h^2 \geqslant 0.04$. During the matter era the relations (9.1.1) and (9.1.2) are still valid for the radiation and matter densities, respectively, and the radiation temperature is given by $T_{\rm r} = T_{0\rm r}(1+z)$. As far as the matter temperature is concerned, this remains approximately equal to the radiation temperature until $z \simeq 300$, thanks to the residual ionisation which allows an exchange of energy between matter and radiation via Compton diffusion. The characteristic timescale differs by a factor $1/x$ from that given by Equation (9.2.9) due to the partial ionisation. The timescale $\tau_{e\gamma}$ can be compared with the characteristic time for the expansion of the Universe which, for $z_{eq} \gg z \gg \Omega_0^{-1}$, is given by

$$\tau_H = \tfrac{3}{2} t_{0c} (\Omega_0 h^2)^{-1/2} (1+z)^{-3/2} \simeq 3.15 \times 10^{17} (\Omega_0 h^2)^{-1/2} (1+z)^{-3/2}\ \text{s} \quad (9.4.1)$$

(cf. Equation (5.6.11)). One finds that $\tau_H < \tau_{e\gamma}$ for $z < 10^2 (\Omega_0 h^2)^5$. After this redshift the thermal interaction between matter and radiation becomes insignificant, so that the matter component cools adiabatically with a law $T_m \propto (1+z)^2$. The epoch $z_d \simeq 300$ is the order of magnitude of the epoch of decoupling.

After decoupling, any primordial fluctuations in the matter component that survive the radiative era can grow and eventually give rise to cosmic structures: stars, galaxies and clusters of galaxies. The part of the gas that does not end up in such structures may be reheated and partly reionised by star and galaxy formation. This partial *reionisation* is called reheating, but should not be confused with the process of reheating which happens at the end of inflation.

An important consideration in the post-recombination epoch is the issue of the *optical depth* τ of the Universe due to Compton scattering. This is a dimensionless quantity such that $\exp(-\tau)$ (often called the *visibility*) describes the attenuation of the photon flux as it traverses a certain length. The probability dP that a photon has suffered a scattering event from an electron while travelling a distance $c\,dt$ is given by

$$dP = -\frac{dN_\gamma}{N_\gamma} = -\frac{dI}{I} = \frac{dt}{\tau_{\gamma e}} = n_e \sigma_T c\, dt = -\frac{x \rho_m}{m_p} \sigma_T c \frac{dt}{dz}\, dz = -d\tau, \quad (9.4.2)$$

where N_γ is the photon flux, so that

$$I(t_0, z) = I(t) \exp\left(-\int_0^z \frac{x\rho_m}{m_p} \sigma_T c \frac{dt}{dz}\, dz\right) = I(t) \exp[-\tau(z)]; \quad (9.4.3)$$

$I(t_0, z)$ is the intensity of the background radiation reaching the observer at time t_0 with a redshift z if it is incident on a region at a redshift z with intensity $I[t(z)]$; $\tau(z)$ is called the optical depth of such a region. The probability that a photon, which arrives at the observer at the present epoch, suffered its last scattering event between z and $z - dz$ is

$$-\frac{d}{dz}\{1 - \exp[-\tau(z)]\}\, dz = \exp[-\tau(z)]\, d\tau = g(z)\, dz. \quad (9.4.4)$$

The quantity $g(z)$ is called the *differential visibility* or *effective width* of the surface of last scattering; with a behaviour of the ionisation fraction given by (9.3.6) for $z > 900$ and a residual value $x(z) \simeq 10^{-4}$–10^{-5} for $z < 900$, one finds that $g(z)$

is well approximated by a Gaussian with peak at $z_{ls} \simeq 1100$ and width $\Delta z \simeq 400$, which corresponds to a (comoving) length scale of around $40h^{-1}$ Mpc or to an angular scale subtended on the last scattering surface of $10\Omega_0^{1/2}$ arcmin. (Incidentally, at z_{rec} the horizon is of order $200h^{-1}$ Mpc, which corresponds to an angular scale of around 2°.) The value of z_{ls} is not very sensitive to variations in $\Omega_0 h^2$. The integral of $g(z)$ over the range $0 \leqslant z \leqslant \infty$ is clearly unity. At redshift z_{ls} we also have $\tau(z) \simeq 1$. One usually takes the 'surface' of last scattering to be defined by the distance from the observer from which photons arrive with a redshift z_{ls}, due to the expansion of the Universe.

If there is a reionisation of the intergalactic gas, in the manner we have described above, at $z_{reh} < z_{rec}$, we can put $x = 1$ in the interval $0 \leqslant z \leqslant z_{reh}$ and obtain, from Equations (2.4.16) and (9.4.2),

$$\tau(z) = \frac{\rho_{0c}\Omega_0\sigma_T c}{m_p H_0}\int_0^z \frac{(1+z)}{(1+\Omega_0 z)^{1/2}}\,dz. \tag{9.4.5}$$

If $\Omega_0 z \gg 1$, we get the approximate result

$$\tau(z) \simeq 10^{-2}(\Omega_0 h^2)^{1/2} z^{3/2}; \tag{9.4.6}$$

in this case $\tau(z)$ is unity at $z_{ls} \simeq 20(\Omega_0 h^2)^{1/3}$, which is reasonably exact for acceptable values of $\Omega_0 h^2$. In conclusion, we can see that, if $z_{reh} > 20(\Omega_0 h^2)^{1/3}$, then the redshift of last scattering is given by $z_{ls} \simeq 20(\Omega_0 h^2)^{-1/3}$; if, however, $z_{reh} < 2$, the redshift of last scattering is of order 10^3 and we have a 'standard' ionisation history. In either case the study of the isotropy of the radiation background can give information on the state of the Universe only as far as regions at distances corresponding to z_{ls}.

9.5 Evolution of the CMB Spectrum

Assuming that radiation is held in thermal equilibrium at some temperature T_i, the intensity of the radiation (defined as power received per unit frequency per unit area per steradian) is given by a *black-body spectrum*:

$$I(t_i, \nu) = \frac{4\pi\hbar\nu^3}{c}\left[\exp\left(\frac{h\nu}{k_B T_i}\right) - 1\right]^{-1}. \tag{9.5.1}$$

One can easily show that in the course of an adiabatic expansion of the Universe, after all processes creating or absorbing photons have become insignificant, the form of the spectrum $I(t, \nu)$ remains the same with the replacement of T_i by

$$T = T_i\frac{a(t_i)}{a(t)}. \tag{9.5.2}$$

This can be understood because the number of photons per unit frequency in volume $V \propto a(t)^3$ is given by

$$N_\nu = \left[\exp\left(\frac{h\nu}{k_B T}\right) - 1\right]^{-1}; \tag{9.5.3}$$

the expansion creates a variation of $\nu \propto a(t)^{-1}$ and, because N_ν must be conserved, T must also vary as $a(t)^{-1}$. In fact, one can use a similar argument to show that a thermal Maxwell–Boltzmann distribution of particle velocities also remains constant during the expansion of the Universe but the effective temperature varies as $T \propto a(t)^{-2}$.

The FIRAS instrument on the COBE satellite (Mather *et al.* 1994) obtained the results shown in Figure 9.1, together with results in different wavelength regions from other experiments. The fit to the black-body spectrum is extremely good, providing clear evidence that this radiation is indeed relic thermal radiation from a primordial fireball.

In fact, the quality of the fit of the observed CMB spectrum to a black-body curve does more than confirm the Big Bang picture. It places important constraints on processes which might be expected to occur within the Big Bang model itself and which would lead to slight distortions in the black-body shape. For example, even in the idealised equilibrium model of hydrogen recombination, the physical nature of this process is expected to produce distortions of the spectrum. Recombination occurs when $T_r \simeq 4000$ K. Although the number-density of photons is some 10^9 times greater than the number-density of baryons at this time, the density of photons with $h\nu > 13.6$ eV is less than the number-density of baryons. Since the optical depth for absorption of Lyman series photons is very high, recombination occurs mainly through two-photon decay, which is relatively slow. (This is one of the reasons why the ionisation fraction is somewhat higher than the Saha equation predicts.) Although each recombination therefore produces several photons, since the number-density of baryons is so much smaller than that of the photons, these recombination photons cannot change the spectral shape very much near its peak. They can, however, lead to strong distortions in the far Wien ($h\nu \gg k_B T$) and far Rayleigh–Jeans ($h\nu \ll k_B T$) parts of the spectrum. Unfortunately, the spectrum is quite weak in this region and galactic dust makes it difficult to make observations to test these ideas.

A more significant distortion mechanism is associated with the injection of some form of energy into the plasma at some time. As we have explained, the relaxation time for non-thermal energy injection to be thermalised is usually very short. Nevertheless, certain types of energy release cannot be thermalised and could therefore lead to observable distortions.

After energy injection, the first thing that happens is that the electrons adjust their temperature to whatever the non-equilibrium spectrum is. This happens on a timescale determined by the number-density of electrons, which is much smaller than the number-density of photons. Next, the radiation spectrum is adjusted by multiple scattering processes which conserve the total number of photons. As a result, the total number of photons does not match the effective temperature of the spectrum; one finds instead a form

$$I(t_i, \nu) = \frac{4\pi\hbar\nu^3}{c}\left[\exp\left(\frac{h\nu}{k_B T_i} + \mu\right) - 1\right]^{-1}, \tag{9.5.4}$$

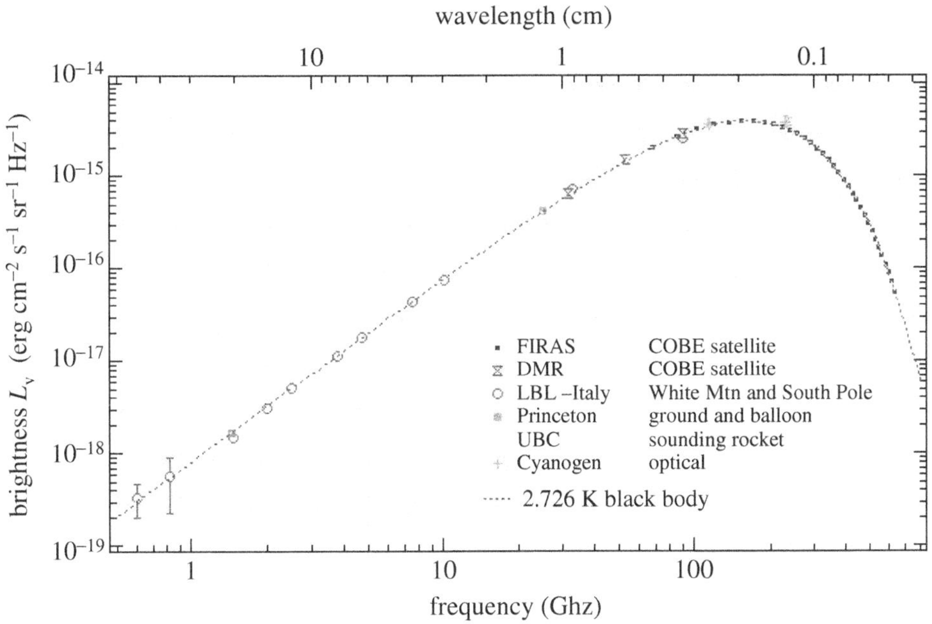

Figure 9.1 The spectrum of the cosmic microwave background as measured by the FIRAS instrument on the COBE satellite along with other experimental results. The best-fitting black-body spectrum has $T = 2.726 \pm 0.010$ K (95% confidence). Picture courtesy of George Smoot.

with a chemical potential $\mu \geqslant 0$; for convenience we shall take μ to be measured in units of $k_B T$ throughout the rest of this section. For $\mu \ll 1$ the difference between the spectrum (9.5.4) and the pure black-body (9.5.1) is largest for $h\nu \ll \mu k_B T$, i.e. in the Rayleigh–Jeans part of the spectrum. The final step in this process is the establishment of a full thermodynamical equilibrium at some new temperature T' compared with the original T; no trace of the injected energy remains at this stage.

Clearly, only the middle stage of this process which produces the μ-distorted spectrum (9.5.4) yields important information in this case. Accurate calculations of the relevant timescales show that energy injected at $z > 10^4$ (the limit is approximate) cannot be fully thermalised and would therefore be expected to produce a spectrum of the form (9.5.4). On the other hand, for energy injected at $z > 10^7$ the double Compton effect (radiation of an additional soft photon during Compton scattering) becomes important and this thermalises things very quickly. Observational constraints on μ therefore place an upper limit on any energy injection in the redshift window $10^7 > z > 10^4$; the current upper limit from COBE is $\mu < 3.3 \times 10^{-4}$. Possible sources of energy release in this window might be primordial back hole evaporation, decay of unstable particles, turbulence, superconducting cosmic strings or, less exotically, the damping of density fluctuations by photon diffusion, as described in Section 12.7.

Physical processes operating at $z < z_{\rm rec}$ can also distort the CMB spectrum, but here the distortion takes a slightly different form. If there exists a period of reionisation of the Universe, as indeed seems to be the case (see Section 19.3), Compton scattering of CMB photons by ionised material can distort the shape of the spectrum in a way that depends upon when the secondary heating occurred and how it affected the intergalactic gas. In many circumstances only one parameter is needed to describe the distortion, because the electron temperature $T_{\rm e}$ is greater than the radiation temperature $T_{\rm r}$. The relevant parameter is the y-parameter

$$y = \int_{t_{\min}}^{t_{\max}} \frac{k(T_{\rm e} - T_{\rm r})}{m_{\rm e}c^2} \sigma_{\rm T} n_{\rm e}(z) c \, {\rm d}t, \tag{9.5.5}$$

where the integral is taken over the time the photon takes to traverse the ionised medium. This is usually called the Sunyaev–Zel'dovich effect (Sunyaev and Zel'dovich 1970).

When CMB photons scatter through material which has been heated in this way the shape of the spectrum is distorted in both Rayleigh–Jeans and Wien regions. If $y < 0.25$ the shape of the Rayleigh–Jeans part of the spectrum does not change, but the effective temperature changes according to $T = T_{\rm r} \exp(-2y)$. At high frequencies the intensity actually increases. This can be understood in terms of low-frequency CMB photons being boosted in energy by Compton scattering and transferred to high-frequency parts of the spectrum. Strong constraints on the allowed y-distortions are also placed by the COBE satellite: $y \leqslant 3 \times 10^{-5}$. In Chapter 19 we explain how these observations can constrain theories of structure formation.

Bibliographic Notes on Chapter 9

A classic reference for the behaviour of the ionisation of the expanding Universe is Wyse and Jones (1985); Kaiser and Silk (1987) also contains an accessible discussion of optical depths and reionisation. Much of the other material is covered by standard texts; see in particular Peebles (1971, 1993).

Problems

1. Use the Saha formula (9.3.5) to compute the ionisation fraction of a pure hydrogen plasma at $T = 3000$ K if $\Omega_{0\rm b}h^2 = 0.01$.

2. Derive Equation (9.4.5), i.e. show that

$$\tau(z) = \frac{\rho_{0\rm c}\Omega_0\sigma_{\rm T}c}{m_{\rm p}H_0} \int_0^z \frac{(1+z)}{(1+\Omega_0 z)^{1/2}} \, {\rm d}z.$$

3. Using Equation (9.4.5), show that

$$\tau(z) \simeq A\frac{h}{\Omega_0}[(1+\Omega_0 z)^{1/2}(3\Omega_0 + \Omega_0 z - 2) - (3\Omega_0 - 2)],$$

and derive an expression for the constant A in terms of physical constants and cosmological parameters.

4. Low-energy photons from the cosmic microwave background pass through a cloud of hot plasma (at a temperature of order 10^8 K) before arriving at the observer. Show that the observer sees a fractional reduction in the temperature T of the microwave background in the direction of the cloud given by

$$\frac{\Delta T}{T} \simeq -2 \int \frac{\sigma_{\mathrm{T}} P_{\mathrm{e}}}{m_{\mathrm{e}} c} \, \mathrm{d}t.$$

PART 3

Theory of Structure Formation

10

Introduction to Jeans Theory

10.1 Gravitational Instability

In an attempt to understand the formation of stars and planets, Jeans (1902) demonstrated the existence of an important instability in evolving clouds of gas. This instability, now known as the *gravitational Jeans instability*, gravitational instability, or simply Jeans instability, is now the cornerstone of the standard model for the origin of galaxies and large-scale structure.

Jeans demonstrated that, starting from a homogeneous and isotropic 'mean' fluid, small fluctuations in the density, $\delta\rho$, and velocity, δv, could evolve with time. His calculations were done in the context of a static background fluid; the expansion of the Universe was not known at the time he was working and, in any case, is not relevant for the formation of stars and planets. In particular, he showed that density fluctuations can grow in time if the stabilising effect of pressure is much smaller than the tendency of the self-gravity of a density fluctuation to induce collapse. It is not surprising that such an effect should exist: gravity is an attractive force so, as long as pressure forces are negligible, an overdense region is expected to accrete material from its surroundings, thus becoming even more dense. The denser it becomes the more it will accrete, resulting in an instability which can ultimately cause the collapse of a fluctuation to a gravitationally bound object. The simple criterion needed to decide whether a fluctuation will grow with time is that the typical lengthscale of a fluctuation should be greater than the *Jeans length*, λ_J, for the fluid. Before we calculate the Jeans length in mathematical detail, we first give a simple order-of-magnitude argument to demonstrate its physical significance.

Imagine that, at a given instant, there is a spherical inhomogeneity of radius λ containing a small positive density fluctuation $\delta\rho > 0$ of mass M, sitting in a background fluid of mean density ρ. The fluctuation will grow (in the sense that $\delta\rho/\rho$ will increase) if the self-gravitational force per unit mass, F_{g}, exceeds the opposing force per unit mass arising from pressure, F_{p}:

$$F_{\mathrm{g}} \simeq \frac{GM}{\lambda^2} \simeq \frac{G\rho\lambda^3}{\lambda^2} > F_{\mathrm{p}} \simeq \frac{p\lambda^2}{\rho\lambda^3} \simeq \frac{v_{\mathrm{s}}^2}{\lambda}, \tag{10.1.1}$$

where v_{s} is the sound speed; this relation implies that growth occurs if $\lambda > v_{\mathrm{s}}(G\rho)^{-1/2}$. This establishes the existence of the Jeans length $\lambda_{\mathrm{J}} \simeq v_{\mathrm{s}}(G\rho)^{-1/2}$. Essentially the same result can be obtained by requiring that the gravitational self-energy per unit mass of the sphere, U, be greater than the kinetic energy of the thermal motion of the gas, again per unit mass, E_{T},

$$U \simeq \frac{G\rho\lambda^3}{\lambda} > E_{\mathrm{T}} \simeq v_{\mathrm{s}}^2, \tag{10.1.2}$$

or by requiring the gravitational free-fall time, τ_{ff}, to be less than the hydrodynamical time, τ_{h},

$$\tau_{\mathrm{ff}} \simeq \frac{1}{(G\rho)^{1/2}} < \tau_{\mathrm{h}} \simeq \frac{\lambda}{v_{\mathrm{s}}}. \tag{10.1.3}$$

When the conditions (10.1.2), (10.1.3) are not satisfied, the pressure forces inside the perturbation are greater than the self-gravity, and the perturbation then propagates like an acoustic wave with wavelength λ at velocity v_{s}.

In fact, as we shall see in Section 10.3, similar reasoning also turns out to hold for a collisionless fluid, as long as we replace v_{s}, the adiabatic sound speed, with v_*, which is of order the mean square velocity of the collisionless particles making up the fluid. In this case, for $\lambda > \lambda_{\mathrm{J}}$, the self-gravity overcomes the tendency of particles to stream at the velocity v_*, whereas if $\lambda < \lambda_{\mathrm{J}}$ the velocity dispersion of the particles is too large for them to be held by the self-gravity, and they undergo *free streaming*; in this case the fluid fluctuations do not behave like acoustic waves, but are smeared out and dissipated by this process. Before looking at collisionless fluids, however, let us investigate the collisional case more quantitatively.

10.2 Jeans Theory for Collisional Fluids

To investigate the Jeans instability and to find the Jeans length λ_{J} more accurately we need to look at the dynamics of a self-gravitating fluid. We shall begin by looking at the case Jeans himself studied, i.e. a collisional gas in a static background.

The equations of motion of such a fluid, in the Newtonian approximation, are

$$\frac{\partial \rho}{\partial t} + \nabla \cdot \rho \boldsymbol{v} = 0, \tag{10.2.1 a}$$

$$\frac{\partial \boldsymbol{v}}{\partial t} + (\boldsymbol{v} \cdot \nabla)\boldsymbol{v} + \frac{1}{\rho}\nabla p + \nabla \varphi = 0, \tag{10.2.1 b}$$

$$\nabla^2 \varphi - 4\pi G \rho = 0. \tag{10.2.1 c}$$

These are the *continuity equation*, the *Euler equation* and the *Poisson equation*, respectively. Throughout this chapter and the next we shall neglect any dissipative terms arising from viscosity or thermal conductivity. For this reason we must add another equation to the ones above, describing the conservation of entropy per unit mass s:

$$\frac{\partial s}{\partial t} + \boldsymbol{v} \cdot \nabla s = 0. \tag{10.2.1 d}$$

The system of Equations (10.2.1) admits the static solution with $\rho = \rho_0$, $\boldsymbol{v} = \mathbf{0}$, $s = s_0$, $p = p_0$ and $\nabla \varphi = 0$. Unfortunately, however, according to the system of Equations (10.2.1), if $\rho_0 \neq 0$, then the gravitational potential φ must vary spatially; in other words, a homogeneous distribution of ρ cannot be stationary, and must be globally either expanding or contracting. There is therefore nothing necessarily relativistic about the expansion of the Universe: the incompatibility of a static universe with the Cosmological Principle is also apparent in Newtonian gravity. This same effect is also the reason why the Einstein static universe is unstable. As we shall see, however, when we consider the case of an expanding universe, the results of Jeans remain qualitatively unchanged. We shall therefore proceed with Jeans' treatment, even though it does have this problem. It turns out to be an incorrect theory, which nevertheless can be 'reinterpreted' to give correct results. Its great advantage is that Newtonian gravity is more familiar to most students than general relativity.

Now let us look for a solution to (10.2.1) that represents a small perturbation of the (erroneous) static solution: $\rho = \rho_0 + \delta\rho$, $\boldsymbol{v} = \delta\boldsymbol{v}$, $p = p_0 + \delta p$, $s = s_0 + \delta s$, $\varphi = \varphi_0 + \delta\varphi$. Introducing these small quantities into the Equations (10.2.1) and neglecting terms of higher order in small quantities, we find

$$\frac{\partial \delta\rho}{\partial t} + \rho_0 \nabla \cdot \delta\boldsymbol{v} = 0, \tag{10.2.2 a}$$

$$\frac{\partial \delta\boldsymbol{v}}{\partial t} + \frac{1}{\rho_0}\left(\frac{\partial p}{\partial \rho}\right)_s \nabla\delta\rho + \frac{1}{\rho_0}\left(\frac{\partial p}{\partial s}\right)_\rho \nabla\delta s + \nabla\delta\varphi = 0, \tag{10.2.2 b}$$

$$\nabla^2 \delta\varphi - 4\pi G \delta\rho = 0, \tag{10.2.2 c}$$

$$\frac{\partial \delta s}{\partial t} = 0. \tag{10.2.2 d}$$

We now have to study all the solutions to this perturbed system of equations. Indeed, as we shall see, there are five solutions: two of *adiabatic* type, one of *entropic* type, and two *vortical* modes. To solve the Equations (10.2.2) we look for solutions in the form of plane waves

$$\delta u_i = \delta_\mathrm{i} \exp(\mathrm{i}\boldsymbol{k} \cdot \boldsymbol{r}), \tag{10.2.3}$$

where, $i = 1, 2, 3, 4$, and the perturbations δu_i stand for $\delta\rho$, $\delta\boldsymbol{v}$, $\delta\varphi$ and δs, respectively; the δ_i are functions only of time. Given that the unperturbed solutions do not depend upon position, one can search for solutions of the form

$$\delta_\mathrm{i}(t) = \delta_{0i} \exp(\mathrm{i}\omega t); \tag{10.2.4}$$

let us refer to the amplitudes δ_{0i} as D, $\boldsymbol{V}$, Φ and Σ. In the previous equations $\boldsymbol{r}$ is a position vector, $\boldsymbol{k}$ is a (real) wavevector, and ω is a frequency which is in general complex. Substituting from (10.2.3) and (10.2.4) into (10.2.2) and putting $v_\mathrm{s}^2 = (\partial p/\partial\rho)_S$ (v_s is the sound speed, as we mentioned above), and $\delta_0 = D/\rho_0$ we obtain

$$\omega\delta_0 + \boldsymbol{k} \cdot \boldsymbol{V} = 0, \tag{10.2.5 a}$$

$$\omega\boldsymbol{V} + \boldsymbol{k}v_\mathrm{s}^2\delta_0 + \frac{\boldsymbol{k}}{\rho_0}\left(\frac{\partial p}{\partial s}\right)_\rho \Sigma + \boldsymbol{k}\Phi = 0, \tag{10.2.5 b}$$

$$k^2\Phi + 4\pi G\rho_0\delta_0 = 0, \tag{10.2.5 c}$$

$$\omega\Sigma = 0. \tag{10.2.5 d}$$

Let us briefly consider at the start those solutions with $\omega = 0$, i.e. those that do not depend upon time. One such solution corresponds to $\Sigma = \Sigma^* \neq 0 = \mathrm{const}$. In the absence of viscosity and thermal conduction the perturbation to s is conserved in time; this is called the entropic solution. Another two solutions with $\omega = 0$ are obtained by putting $\Sigma = 0$ and $\boldsymbol{k} \cdot \boldsymbol{V} = 0$: these therefore have $\boldsymbol{k}$ perpendicular to $\boldsymbol{V}$ and represent vortical modes in which $\nabla \times \boldsymbol{v} \neq 0$, which does not imply any perturbations to the density, as is evident from (10.2.5 *b*) and (10.2.5 *c*).

The time-dependent solutions of (10.2.5), i.e. those with $\omega \neq 0$, are more interesting. In this case (10.2.5 *d*) implies that $\Sigma = 0$: the perturbations are adiabatic. From (10.2.5 *a*) one has that $\boldsymbol{k} \cdot \boldsymbol{V} \neq 0$. In this case, we can resolve into components parallel and perpendicular to $\boldsymbol{V}$. We mentioned above the consequence of having $\boldsymbol{k}$ perpendicular to $\boldsymbol{V}$, so now let us concentrate upon the parallel component. Perturbations with $\boldsymbol{k}$ and $\boldsymbol{V}$ parallel are longitudinal in character. Equations (10.2.5) now become

$$\omega\delta_0 + kV = 0, \tag{10.2.6 a}$$

$$\omega V + kv_\mathrm{s}^2\delta_0 + k\Phi = 0, \tag{10.2.6 b}$$

$$k^2\Phi + 4\pi G\rho_0\delta_0 = 0. \tag{10.2.6 c}$$

This system admits a non-zero solution for δ_0, V and Φ if and only if its determinant vanishes. This means that ω and k must satisfy the *dispersion relation*:

$$\omega^2 - v_\mathrm{s}^2k^2 + 4\pi G\rho_0 = 0. \tag{10.2.7}$$

The solutions are of two types, according to whether the wavelength $\lambda = 2\pi/k$ is greater than or less than

$$\lambda_J = v_s\left(\frac{\pi}{G\rho_0}\right)^{1/2}, \tag{10.2.8}$$

which is called the *Jeans length*. Notice the same dependence upon G, ρ_0 and v_s as the simple qualitative description given in Section 10.1.

In the case $\lambda < \lambda_J$ the angular frequency ω obtained from (10.2.7) is real:

$$\omega = \pm v_s k\left[1 - \left(\frac{\lambda}{\lambda_J}\right)^2\right]^{1/2}. \tag{10.2.9}$$

From Equations (10.2.3), (10.2.4) and (10.2.6) one obtains easily that

$$\frac{\delta\rho}{\rho_0} = \delta_0 \exp[\mathrm{i}(\boldsymbol{k}\cdot\boldsymbol{r} \pm |\omega|t)], \tag{10.2.10 a}$$

$$\delta\boldsymbol{v} = \mp\frac{\boldsymbol{k}}{k}v_s\delta_0\left[1 - \left(\frac{\lambda}{\lambda_J}\right)^2\right]^{1/2} \exp[\mathrm{i}(\boldsymbol{k}\cdot\boldsymbol{r} \pm |\omega|t)], \tag{10.2.10 b}$$

$$\delta\varphi = -\delta_0 v_s^2\left(\frac{\lambda}{\lambda_J}\right)^2 \exp[\mathrm{i}(\boldsymbol{k}\cdot\boldsymbol{r} \pm |\omega|t)], \tag{10.2.10 c}$$

which represent two sound waves in directions $\pm\boldsymbol{k}$, with a dispersion given by (10.2.9). The phase velocity tends to zero for $\lambda \to \lambda_J$.

When $\lambda > \lambda_J$ the frequency is imaginary:

$$\omega = \pm\mathrm{i}(4\pi G\rho_0)^{1/2}\left[1 - \left(\frac{\lambda_J}{\lambda}\right)^2\right]^{1/2}. \tag{10.2.11}$$

In this case we have

$$\frac{\delta\rho}{\rho_0} = \delta_0 \exp(\mathrm{i}\boldsymbol{k}\cdot\boldsymbol{r})\exp(\pm|\omega|t), \tag{10.2.12 a}$$

$$\delta\boldsymbol{v} = \mp\mathrm{i}\frac{\boldsymbol{k}\delta_0}{k^2}(4\pi G\rho_0)^{1/2}\left[1 - \left(\frac{\lambda_J}{\lambda}\right)^2\right]^{1/2} \exp(\mathrm{i}\boldsymbol{k}\cdot\boldsymbol{r} \pm |\omega|t), \tag{10.2.12 b}$$

$$\delta\varphi = -\delta_0 v_s^2\left(\frac{\lambda}{\lambda_J}\right)^2 \exp(\mathrm{i}\boldsymbol{k}\cdot\boldsymbol{r} \pm |\omega|t), \tag{10.2.12 c}$$

which represents a non-propagating solution (stationary wave) of either increasing or decreasing amplitude. The characteristic timescale for the evolution of this amplitude is

$$\tau \equiv |\omega|^{-1} = (4\pi G\rho_0)^{-1/2}\left[1 - \left(\frac{\lambda_J}{\lambda}\right)^2\right]^{-1/2}. \tag{10.2.13}$$

It is only this type of solution that exhibits the phenomenon we referred to above as the gravitational or Jeans instability. For scales $\lambda \gg \lambda_J$ the characteristic time τ coincides with the free-fall collapse time, $\tau_{\mathrm{ff}} \simeq (G\rho_0)^{-1/2}$, but for $\lambda \to \lambda_J$ this characteristic timescale diverges.

10.3 Jeans Instability in Collisionless Fluids

Let us now extend our analysis of the gravitational Jeans instability to a gas of collisionless particles. In a sense, the absence of collisions implies there is no pressure, so there would appear to be no analogy with the Jeans length in this case. However, collisionless particles do have velocities and these velocities are not necessarily represented as a single unique $\boldsymbol{v}$ at each position $\boldsymbol{x}$ as we assumed in Section 10.2 for an idealised fluid. Instead there is a distribution of random velocities at each point; in what follows we assume this distribution is isotropic. It is possible for a collisionless system to be well described by a fluid with zero pressure. That occurs when the fluid is extremely cold so that the resulting flow is nearly *laminar*, i.e. so that the particles always travel in nearly parallel trajectories that do not cross. In such a case it is a good approximation to suppose there is a unique velocity at every point. We shall return to this when we discuss cold dark matter. For simplicity we also assume all particles have the same mass m.

In the collisionless case, the Equations (10.2.1 a) and (10.2.1 b) should be replaced by the *Liouville equation*

$$\frac{\partial f}{\partial t} + \nabla \cdot f\boldsymbol{v} + \nabla_{\mathrm{v}} \cdot f\dot{\boldsymbol{v}} = 0, \tag{10.3.1}$$

where $\nabla_{\mathrm{v}} \equiv (\partial/\partial \boldsymbol{v})$ by analogy with $\nabla \equiv (\partial/\partial \boldsymbol{r})$. The function $f(\boldsymbol{r}, \boldsymbol{v}; t)$ is the phase-space distribution function for the particles; the phase space is six dimensional, and f also depends explicitly on time. The function f therefore represents the number-density of particles in a volume $\mathrm{d}\boldsymbol{r}$ at position $\boldsymbol{r}$ and with velocity in the volume $\mathrm{d}\boldsymbol{v}$ at $\boldsymbol{v}$; the actual number of particles in each of these volumes is given by $f(\boldsymbol{r}, \boldsymbol{v}; t)\, \mathrm{d}\boldsymbol{r}\, \mathrm{d}\boldsymbol{v}$. In our case, of a homogeneous and isotropic time-stationary background distribution, it can be shown that the distribution function is only a function of v^2.

We stress that the systems (10.2.1 a)–(10.2.1 c) and Equation (10.3.1) are both approximations to a full statistical mechanical treatment using a Boltzmann equation with a collisional term on the right-hand side of (10.3.1).

Equation (10.2.1 c) does not change in the collisionless situation, so we must bear in mind the comments we made above about the existence of stationary solutions. Nevertheless, let us consider Equation (10.2.2 c):

$$\nabla^2 \delta\varphi - 4\pi G \delta\rho = 0, \tag{10.3.2}$$

where we now have

$$\delta\rho = m \int \delta f \, \mathrm{d}\boldsymbol{v}; \tag{10.3.3}$$

δf is the perturbation of the distribution function and $\delta\varphi$ is the perturbation of the gravitational potential, related to the gravitational acceleration $\boldsymbol{g} = \dot{\boldsymbol{v}}$ by

$$\delta\boldsymbol{g} = -\nabla\delta\varphi. \tag{10.3.4}$$

Taking account of this last expression, Equation (10.3.1) becomes

$$\frac{\partial}{\partial t}\delta f + \boldsymbol{v} \cdot \nabla \delta f - \nabla \delta \varphi \cdot \nabla_{\mathrm{v}} f = 0. \tag{10.3.5}$$

By analogy with what we have done in the previous paragraph, we look for a solution to Equations (10.3.2) and (10.3.5) with δf, $\delta \varphi$ and $\delta \rho$ in the form of a plane wave. Without loss of generality, we can take the wavevector $\boldsymbol{k}$ to be in the x-direction. Applying the operator ∇ to (10.3.5) and using the fact that the operators ∇ and ∇_{v} commute, we obtain from (10.3.2) that

$$\delta f = 4\pi G \frac{\mathrm{d}f}{\mathrm{d}v^2} \frac{v_x}{k(\omega - k v_x)} \delta \rho. \tag{10.3.6}$$

This equation, after substitution in Equation (10.3.3), becomes the dispersion relation

$$k - 4\pi G m \int \frac{v_x}{\omega - k v_x} \frac{\mathrm{d}f}{\mathrm{d}v^2}\, \mathrm{d}\boldsymbol{v} = 0. \tag{10.3.7}$$

To find the solution appropriate to $k \to 0$ (long wavelengths) we can develop the dispersion relation as a power series in $k v_x / \omega$; keeping only the first two terms in such a series yields

$$\omega^2 \simeq \frac{4\pi G m \omega}{k} \int v_x \frac{\mathrm{d}f}{\mathrm{d}v^2}\, \mathrm{d}\boldsymbol{v} - 4\pi G m \int v_x^2 \frac{\mathrm{d}f}{\mathrm{d}v^2}\, \mathrm{d}\boldsymbol{v}. \tag{10.3.8}$$

The first term vanishes for reasons of symmetry, but the second can be evaluated by integration by parts (note that $f(v^2)$ tends to zero as $v \to \infty$): one has

$$\omega^2 \simeq -4\pi G \rho, \tag{10.3.9}$$

where ρ is obtained from a relation analogous to (10.3.3). This result shows that there is indeed a gravitational instability in this case, with characteristic timescale

$$\tau \simeq (4\pi G \rho)^{-1/2}, \tag{10.3.10}$$

identical to the previous expression (10.2.13) for $\lambda \gg \lambda_{\mathrm{J}}$.

The Jeans length λ_{J} can be obtained from (10.3.7) by putting $\omega = 0$, by analogy with what we have seen above; by similar reasoning to that which led to (10.3.10) we find

$$\lambda_{\mathrm{J}} = v_* \left(\frac{\pi}{G\rho}\right)^{1/2}, \tag{10.3.11}$$

where

$$v_*^{-2} = \frac{\int v^{-2} f\, \mathrm{d}^3\boldsymbol{v}}{\int f\, \mathrm{d}^3\boldsymbol{v}} \equiv \langle v^{-2} \rangle. \tag{10.3.12}$$

The velocity v_* replaces the velocity of sound v_s in (10.2.8). In the particular case of a Maxwellian distribution

$$f(v) = \frac{\rho}{(2\pi\sigma^2)^{3/2}} \exp\left(\frac{-v^2}{2\sigma^2}\right), \tag{10.3.13}$$

we have $v_* = \sigma$.

The analysis of the evolution of perturbations for $\lambda < \lambda_J$ is complicated and we shall not go into it further in this chapter. In fact, in this case, there is a rapid dissipation of fluctuations of wavelength λ in a time of order $\tau \simeq \lambda / v_*$ because of the diffusion of particles, a phenomenon known as 'free streaming', similar to the phenomenon known in collisionless plasma theory as 'Landau damping' or 'phase mixing'.

10.4 History of Jeans Theory in Cosmology

In the subsequent chapters we shall discuss how gravitational instability might take place in a cosmological context and how this theory furnishes a more-or-less complete picture of cosmic structure formation. We shall find a number of complications of the simple picture described by Jeans. For example, we shall have to take explicit account of the expansion of the Universe. We may also need to take into account how general relativity might alter the simple Newtonian analysis outlined above. We also need to understand how the relativistic and non-relativistic components of the fluid influence the growth of fluctuations, and what is the effect of dark matter in the form of weakly interacting particles. Before going on to cover this new ground in a mathematically complete way, it is instructive to give a brief historical outline of the application of Jeans theory in cosmology. This is an introductory survey only, and we shall give the arguments in greater technical detail in Chapters 12 and 13.

The first to tackle the problem of gravitational instability within the framework of general relativity was Lifshitz (1946). He studied the evolution of small fluctuations in the density of a Friedmann model. Curiously, it was not later that the evolution of perturbations in a Friedmann model with $p \ll \rho c^2$ was investigated in Newtonian theory by Bonnor (1957). In some ways the relativistic cosmological theory is more simple that the Newtonian analogue, which requires considerable mathematical subtlety.

These foundational studies were made at a time when the existence of the cosmic microwave background was not known. There was no generally accepted cosmological model within which to frame the problem of structure formation, and there was no way to test the gravitational instability hypothesis for the origin of structure. Nevertheless, it was clear at this time that if the Universe was evolving with time (as the Hubble expansion indicated), then it was possible, in principle, that structure may have evolved by some mechanism similar to the Jeans process. The discovery of the microwave background in the 1960s at last gave theorists a favoured model in which to study this problem: the hot Big Bang. The existence of the microwave background at the present time implied that there was a period in which the Universe comprised a plasma of matter and radiation in thermal equilibrium. Under these physical conditions, there are a number of processes, due to viscosity and thermal conduction in the radiative plasma, which could influence the evolution of a perturbation with wavelength less than λ_J. The pioneering works by Silk (1967, 1968), as well as Doroshkevich *et al.* (1967), Peebles and Yu (1970), Weinberg (1971), Chibisov (1972) and Field (1971), amongst many others, represented the first attempts to derive a theory of galaxy and structure formation within the framework of modern cosmology. At this time there was in fact a rival theory in which it was proposed that galaxies were formed as a result of primordial cosmic turbulence, i.e. large-scale vortical motions rather than longitudinal adiabatic perturbations. This theory, however, rapidly fell from fashion when it was realised that it should lead to large fluctuations in the temperature of the microwave background on the sky. In fact, this point about the microwave background was then and is now important in all theories of galaxy formation. If structure grows by gravitational instability, it is in principle possible to reconcile the present highly inhomogeneous Universe with a past Universe which was much smoother. The microwave background seemed to be at the same temperature in all directions to within about one part in a thousand in this period, indicating a comparable lack of inhomogeneity in the early Universe. If gravitational instability were the correct explanation for the origin of structure, however, there should be some fluctuations in the microwave background temperature. This initiated a search, which has only recently been successful, for fluctuations in the cosmic microwave background on the sky. But more of that later.

10.5 The Effect of Expansion: an Approximate Analysis

The original Jeans theory of gravitational instability, formulated in a static Universe, cannot be applied to an expanding cosmological model. We also have to contend with some features in the cosmological case which do not appear in the original analysis. For example, what happens to the Jeans instability if the Universe is radiation dominated? In this chapter our goal is to translate the usual language of gravitational instability into the context of the Friedmann models. We can then go on, in the next two chapters, to examine the physics of expanding universe models in more detail.

It is useful perhaps to outline the basic results we obtain later with an approximate argument that explains the basic physics. We assume for the moment that the Universe is dominated by pressureless material. The difficulty with the expanding Universe is that the density of matter varies with time according to the approximate relation

$$\rho \simeq \frac{1}{Gt^2}. \tag{10.5.1}$$

The characteristic time for this decrease in density is therefore

$$\tau = \frac{\rho}{\dot{\rho}} \simeq t \simeq \frac{1}{(G\rho)^{1/2}}, \tag{10.5.2}$$

which is the same order of magnitude as the characteristic time for the growth of long-wavelength density perturbations in the Jeans instability analysis, Equation (10.2.13). Qualitatively, we expect that any fluctuation on a scale less than λ_J would oscillate like an acoustic wave as before. A fluctuation with wavelength $\lambda > \lambda_J$ would be unstable but would grow at a reduced rate compared with the exponential form of the previous result. Let us suppose that there is in fact a small perturbation $\delta\rho > 0$ with wavelength $\lambda > \lambda_J$; the growth of the fluctuation must be slower than in the static case because the fluctuation must attract material from around itself which is moving away according to the general expansion of the Universe. In fact, we shall find later in this chapter that there are two modes of perturbation, one growing and one decaying, where $\delta = \delta\rho/\rho$ varies according to

$$\delta_+ \propto t^{2/3}, \qquad \delta_- \propto t^{-1}, \tag{10.5.3}$$

in a matter-dominated Einstein–de Sitter universe, and

$$\delta_+ \propto t, \qquad \delta_- \propto t^{-1}, \tag{10.5.4}$$

if the universe is flat and radiation dominated. We shall derive these results in more detail later on, but one can get a good physical understanding of how Equation (10.5.3) arises by using a simple semi-quantitative approximation. From Equation (10.2.12 *a*) we find formally that, for $\lambda \gg \lambda_J$, we have

$$\dot{\delta} = \pm|\omega|\delta = \pm(4\pi G\rho)^{1/2}\delta, \tag{10.5.5}$$

where we have now put ρ in place of ρ_0. The density ρ varies in a flat matter-dominated universe according to the relation

$$\rho = \frac{1}{6\pi G t^2}. \tag{10.5.6}$$

Substituting (10.5.6) into (10.5.5) and integrating yields

$$\delta_\pm = At^{\pm\sqrt{2/3}}, \tag{10.5.7}$$

where the 'constant' A can be interpreted as the amplitude of a wave of imaginary period, in the manner of Equation (10.2.12). In reality the amplitude of oscillation of a system varies if its parameters are variable in time. If these parameters vary slowly in time, one can apply the theory of *adiabatic invariants*. The critical assumption of this theory is that, in whatever oscillating system is being studied, physical parameters determining the period of oscillation (such as the length of a simple pendulum) vary on a timescale τ which is much longer than P, the period of the oscillations themselves. In a simple pendulum under these conditions, the energy E and the frequency of oscillations ν will vary in such a way that the ratio E/ν remains fixed; E/ν is thus called an adiabatic invariant. Applying this theory to the expanding Universe, we find that physical quantities determining the nature of oscillations vary on a timescale $\tau \simeq a/\dot{a} \simeq t$, so that one can hope to apply the theory of adiabatic invariants for length scales $\lambda = v_{\rm s}P < v_{\rm s}t \simeq \lambda_{\rm J}$ (for $\lambda > \lambda_{\rm J}$ there is an instability, which can be thought of as an oscillation with an imaginary period; in such a case we cannot apply the theory, because $|P| > t$).

The acoustic energy carried in a volume V by a sinusoidal wave is just

$$E = \left(\tfrac{1}{2}\rho\delta v^2 + \frac{v_{\rm s}^2}{2\rho}\delta\rho^2\right)V = \frac{v_{\rm s}^2\delta\rho^2}{\rho}V, \tag{10.5.8}$$

where δv and $\delta\rho$ are the amplitude and the velocity of a density wave, respectively. The last part of Equation (10.5.8) is implicit in Equation (10.2.10 *b*) of the previous chapter, for $\lambda \ll \lambda_{\rm J}$. The adiabatic invariant is then just

$$\frac{E}{\nu} \simeq E\frac{\lambda}{v_{\rm s}} = \text{const.} \tag{10.5.9}$$

If the Universe is sufficiently dense, there exists an interval between matter–radiation equivalence and recombination in which $\rho \simeq \rho_{\rm m}$ and $p \simeq p_{\rm r} \propto \rho_{\rm r} \propto \rho_{\rm m}^{4/3}$; here the acoustic waves we have been considering have a sound speed

$$v_{\rm s} \simeq \left(\frac{p_{\rm r}}{\rho_{\rm m}}\right)^{1/2} \propto \rho_{\rm m}^{1/6} \propto a^{-1/2}. \tag{10.5.10}$$

In this case Equations (10.5.8) and (10.5.9) give

$$\delta \propto t^{-1/6}, \tag{10.5.11}$$

which, if interpreted as being the correct growth law also for the amplitude of waves with $\lambda \gg \lambda_{\rm J}$, suggests that the quantity A in (10.5.7) should vary as $t^{-1/6}$ during the period between equivalence and recombination. If we assume that this law can be extrapolated also to late times (after recombination), one can obtain the following expressions for the growing and decreasing modes, respectively:

$$\delta_+ \propto t^{-1/6+\sqrt{2/3}} \simeq t^{0.65}, \tag{10.5.12 a}$$

$$\delta_- \propto t^{-1/6-\sqrt{2/3}} \simeq t^{-0.98}, \tag{10.5.12 b}$$

which is remarkably close to the correct results given in Equation (10.5.3).

10.6 Newtonian Theory in a Dust Universe

Having mentioned the basic properties of the Jeans instability in the expanding Universe, and given some approximate physical arguments for the results, we should now put more flesh on these bones and go through a systematic translation of the previous chapter into the framework of the expanding universe models. For simplicity, we concentrate upon the case of a dust (zero-pressure) model, and we shall adopt a Newtonian approach as before.

The system of Equations (10.2.1) admits a solution that describes the expansion (or contraction) of a homogeneous and isotropic distribution of matter:

$$\rho = \rho_0 \left(\frac{a_0}{a}\right)^3, \tag{10.6.1 a}$$

$$\boldsymbol{v} = \frac{\dot{a}}{a}\boldsymbol{r}, \tag{10.6.1 b}$$

$$\varphi = \tfrac{2}{3}\pi G \rho r^2, \tag{10.6.1 c}$$

$$p = p(\rho, S), \tag{10.6.1 d}$$

$$s = \text{const.}; \tag{10.6.1 e}$$

$\boldsymbol{r}$ is a physical coordinate, related to the comoving coordinate $\boldsymbol{r}_0$ by the relation

$$\boldsymbol{r} = \boldsymbol{r}_0 \frac{a}{a_0}. \tag{10.6.2}$$

One defect of the solution (10.6.1) is that for $r \to \infty$, both v and φ diverge. Only a relativistic treatment can remedy this problem, so we shall ignore it for the present, making some comments later, in Section 11.12, on the correct analysis.

We proceed by looking for small perturbations $\delta\rho$, $\delta\boldsymbol{v}$, $\delta\varphi$ and δp to the zero-order solution represented by Equations (10.6.1). The equations for the perturbations can then be written

$$\dot{\delta\rho} + 3\frac{\dot{a}}{a}\delta\rho + \frac{\dot{a}}{a}(\boldsymbol{r}\cdot\nabla)\delta\rho + \rho(\nabla\cdot\delta\boldsymbol{v}) = 0, \tag{10.6.3 a}$$

$$\dot{\delta\boldsymbol{v}} + \frac{\dot{a}}{a}\delta\boldsymbol{v} + \frac{\dot{a}}{a}(\boldsymbol{r}\cdot\nabla)\delta\boldsymbol{v} = -\frac{1}{\rho}\nabla\delta p - \nabla\delta\varphi, \tag{10.6.3 b}$$

$$\nabla^2\delta\varphi - 4\pi G\delta\rho = 0, \tag{10.6.3 c}$$

$$\dot{\delta s} + \frac{\dot{a}}{a}(\boldsymbol{r}\cdot\nabla)\delta s = 0, \tag{10.6.3 d}$$

where the dots denote partial derivatives with respect to time. We now neglect the terms in $\boldsymbol{r}\cdot\nabla$ because we make the calculations in a coordinate system where the background velocity $\boldsymbol{v}$ is zero. In fact, this trick does not always work: these terms actually correspond to terms which appear only in the Newtonian framework and they give rise to inconsistencies if there is a non-zero pressure; see Lima *et al.* (1997).

As we did earlier, we now look for solutions in the form of small plane-wave departures from the exact solution represented by (10.6.1):

$$\delta u_i = u_i(t)\exp(\mathrm{i}\boldsymbol{k}\cdot\boldsymbol{r}), \tag{10.6.4}$$

where the variables u_i, for $i = 1, 2, 3, 4$, are related to the quantities D, $\boldsymbol{V}$, Φ, Σ introduced in Section 10.2; their amplitudes here, however, have to depend on time; the perturbation in the pressure is again expressed in terms of $\delta\rho$ and δs. The $u_i(t)$ cannot be functions of the type $u_{0i}\exp(\mathrm{i}\omega t)$, because the coefficients of the equations depend on time. We should also note that the wavevector k corresponds to a wavelength λ which varies with time according to the law (10.6.2), simply because of the expansion of the Universe:

$$k = \frac{2\pi}{\lambda} = \frac{2\pi}{\lambda_0}\frac{a_0}{a} = k_0\frac{a_0}{a}; \tag{10.6.5}$$

for this reason the exponential in (10.6.4) does not depend upon time. One can obtain (after some work!) the perturbation equations corresponding to those given in (10.2.5):

$$\dot{D} + 3\frac{\dot{a}}{a}D + \mathrm{i}\rho\boldsymbol{k}\cdot\boldsymbol{V} = 0, \tag{10.6.6 a}$$

$$\dot{\boldsymbol{V}} + \frac{\dot{a}}{a}\boldsymbol{V} + \mathrm{i}v_s^2\boldsymbol{k}\frac{D}{\rho} + \mathrm{i}\frac{\boldsymbol{k}}{\rho}\left(\frac{\partial p}{\partial s}\right)_\rho \Sigma + \mathrm{i}\boldsymbol{k}\Phi = 0, \tag{10.6.6 b}$$

$$k^2\Phi + 4\pi G D = 0, \tag{10.6.6 c}$$

$$\dot{\Sigma} = 0. \tag{10.6.6 d}$$

This system admits a static (time-independent) solution of entropic type, in which

$$\delta s = \Sigma_0 \exp(\mathrm{i}\boldsymbol{k}\cdot\boldsymbol{r}). \tag{10.6.7}$$

The vortical solutions can be obtained by putting $D = \Phi = \Sigma = 0$ and the condition that $\boldsymbol{V}$ is perpendicular to $\boldsymbol{k}$. From (10.6.6 *b*) we get

$$\dot{\boldsymbol{V}} + \frac{\dot{a}}{a}\boldsymbol{V} = 0, \tag{10.6.8}$$

which has solutions

$$\boldsymbol{V} = \boldsymbol{V}_0\frac{a_0}{a}, \tag{10.6.9}$$

with $\boldsymbol{V}_0$ perpendicular to $\boldsymbol{k}$. The Equation (10.6.9) can be obtained in another way, by applying the law of conservation of angular momentum $\mathcal{L}$, due to the absence of dissipative processes,

$$\mathcal{L} \simeq \rho a^3 V a = \text{const.} \tag{10.6.10}$$

(V is the modulus of $\boldsymbol{V}$).

The solutions with $\Sigma = 0$ and $\boldsymbol{V}$ parallel to $\boldsymbol{k}$ are more interesting from a cosmological point of view. In this case the Equations (10.6.6) become

$$\dot{D} + 3\frac{\dot{a}}{a}D + \mathrm{i}\rho k V = 0, \tag{10.6.11 a}$$

$$\dot{V} + \frac{\dot{a}}{a}V + \mathrm{i}k\left(v_s^2 - \frac{4\pi G\rho}{k^2}\right)\frac{D}{\rho} = 0. \tag{10.6.11 b}$$

Putting $D = \rho\delta$ in (10.6.11 *a*) gives

$$\dot{\delta} + \mathrm{i}kV = 0, \tag{10.6.12}$$

which, upon differentiation, yields

$$\ddot{\delta} + \mathrm{i}k\left(\dot{V} - \frac{\dot{a}}{a}V\right) = 0. \tag{10.6.13}$$

Obtaining V and $\dot{V}$ from (10.6.12) and (10.6.13) and substituting in (10.6.11 *b*) gives

$$\ddot{\delta} + 2\frac{\dot{a}}{a}\dot{\delta} + (v_s^2k^2 - 4\pi G\rho)\delta = 0, \tag{10.6.14}$$

which in the static case and with $\delta \propto \exp(\mathrm{i}\omega t)$ corresponds to the dispersion relation (10.2.7).

As we shall see, for wavelengths λ such that the second term in the parentheses in (10.6.14) is much less than the first, i.e. for $\lambda \ll \lambda_J$, where

$$\lambda_J \simeq v_s\left(\frac{\pi}{G\rho}\right)^{1/2}, \tag{10.6.15}$$

we have two oscillating solutions, while for wavelengths $\lambda \gg \lambda_J$ we have two solutions which involve the phenomenon of gravitational instability.

10.7 Solutions for the Flat Dust Case

The solutions of Equation (10.6.13) depend on the background model relative to which the perturbations are defined. The simplest model we can look at is the flat, matter-dominated Einstein–de Sitter universe which we shall use first to derive some key results. In this model,

$$\rho = \frac{1}{6\pi Gt^2}, \tag{10.7.1 a}$$

$$a = a_0\left(\frac{t}{t_0}\right)^{2/3}, \tag{10.7.1 b}$$

$$\frac{\dot{a}}{a} = \frac{2}{3t}, \tag{10.7.1 c}$$

and the velocity of sound, assuming that the matter comprises monatomic particles of mass m, is given by

$$v_s = \left(\frac{5k_BT_m}{3m}\right)^{1/2} = \left(\frac{5k_BT_{0m}}{3m}\right)^{1/2}\frac{a_0}{a}. \tag{10.7.2}$$

Substituting these results into (10.6.13), one obtains

$$\ddot{\delta} + \frac{4}{3}\frac{\dot{\delta}}{t} - \frac{2}{3t^2}\left(1 - \frac{v_s^2k^2}{4\pi G\rho}\right)\delta = 0. \tag{10.7.3}$$

This equation, for $k \to 0$, is solved with a trial solution of the form $\delta \propto t^n$, with n constant; one gets the exact result that there are two modes, one growing,

$$\delta_+ \propto t^{2/3}, \tag{10.7.4}$$

and one decaying,

$$\delta_- \propto t^{-1}. \tag{10.7.5}$$

One can try to solve Equation (10.7.3) in the case $k \neq 0$ using the same trial solution. We obtain

$$\frac{\delta\rho}{\rho} \propto t^{-[1\pm 5(1-6v_s^2k^2/25\pi G\rho)^{1/2}]/6} \exp(\mathrm{i}\boldsymbol{k}\cdot\boldsymbol{r}). \tag{10.7.6}$$

This power-law solution is, in fact, only correct with constant n for $k \to 0$, but the approximate solution (10.7.6) yields important physical insights. When the expression inside the square root in Equation (10.7.6) is positive, that is for

$$\lambda > \lambda_J' = \frac{\sqrt{24}}{5} v_s \left(\frac{\pi}{G\rho}\right)^{1/2}, \tag{10.7.7}$$

the solutions of (10.7.3) represent the gravitational instability of the system according to which the density fluctuations grow with time. When $\lambda < \lambda_J'$, there are oscillating solutions.

As we mentioned above, the solutions for $\lambda \neq 0$ are approximate because they are derived under the assumption that the index n of the trial power-law solution is constant in time. In general, however, it will depend on time through the behaviour of the ratio λ_J'/λ. We shall discuss this fact in more detail later, in § 10.10. The exponent n does not depend on time if the equation of state is of the form $p \propto \rho^{4/3}$ (i.e. in the plasma epoch with $z < z_{eq}$). In this case the Equation (10.7.4) is exact, and the term in $t^{-1/6}$ which comes from (10.7.7) can be obtained using the theory of adiabatic invariants in the manner discussed in Section 10.6.

It is also worth noting the fact that the Jeans length λ_J is identical to that introduced in Equation (10.2.6) of the previous chapter. In this respect, no new physics is involved when one moves to the expanding (or contracting) case.

10.8 The Growth Factor

The Equation (10.6.13) admits analytic solutions for $\lambda \gg \lambda_J$ also in models where $\Omega_0 \neq 1$. Using the parametric variables ϑ and ψ introduced in Section 2.4 and substituting in (10.6.14) yields the equations

$$(1-\cos\vartheta)\frac{\mathrm{d}^2\delta}{\mathrm{d}\vartheta^2} + \sin\vartheta\frac{\mathrm{d}\delta}{\mathrm{d}\vartheta} - 3\delta = 0, \tag{10.8.1}$$

for $\Omega_0 > 1$, and

$$(\cosh\psi - 1)\frac{d^2\delta}{d\psi^2} + \sinh\psi\frac{d\delta}{\delta\psi} - 3\delta = 0, \tag{10.8.2}$$

for $\Omega_0 < 1$. They have solutions of increasing and decreasing type of the form

$$\delta_+ \propto -\frac{3\vartheta\sin\vartheta}{(1-\cos\vartheta)^2} + \frac{5+\cos\vartheta}{1-\cos\vartheta}, \tag{10.8.3 a}$$

$$\delta_- \propto \frac{\sin\vartheta}{(1-\cos\vartheta)^2}, \tag{10.8.3 b}$$

for $\Omega_0 > 1$, and

$$\delta_+ \propto -\frac{3\psi\sinh\psi}{(\cosh\psi-1)^2} + \frac{5+\cosh\psi}{\cosh\psi-1}, \tag{10.8.4 a}$$

$$\delta_- \propto \frac{\sinh\psi}{(\cosh\psi-1)^2}, \tag{10.8.4 b}$$

for $\Omega_0 < 1$. The relationship between proper time t and the parametric variables ψ and ϑ is given in Section 2.4. In both cases one can verify that, for small values of ϑ or ψ, that is for $t \ll t_0$, one obtains Equation (10.5.3), so that all these cases are identical at early times when the curvature terms in the Friedmann equations are negligible. It is interesting to note that in open universes the growing solution δ_+ remains practically constant for $\cosh\psi \geqslant 5$, which corresponds to a redshift $z \leqslant z^* \simeq \frac{2}{5}\Omega$ if $\Omega \ll 1$; we shall also come across this result later in this section.

Now that we have obtained a number of solutions for different cosmological models, it is helpful to introduce a general notation to describe the growth of fluctuations. The name *growth factor* is given to the relative size of the solution δ_+ as a function of t: thus, the growth factor in the interval (t_i, t_0) is $A_{i0} = \delta_+(t_0)/\delta_+(t_i)$. For reasons which will become clearer later on, the most interesting value of the growth factor will be that relative to $t_i = t_{rec}$. From Equations (10.8.3 *a*), (10.7.3) and (10.8.4 *a*) concerning δ_+ and (2.4.6), (2.2.6 *a*) and (2.4.2) we obtain:

$$A_{r0} = (1+z_{rec})\frac{5[-3\vartheta_0\sin\vartheta_0 + (1-\cos\vartheta_0)(5+\cos\vartheta_0)]}{(1-\cos\vartheta_0)^3}, \tag{10.8.5}$$

for $\Omega_0 > 1$, where $\cos\vartheta_0 = (2\Omega_0^{-1} - 1)$;

$$A_{r0} = 1 + z_{rec}, \tag{10.8.6}$$

for $\Omega_0 = 1$;

$$A_{r0} = (1+z_{rec})\frac{5[-3\psi_0\sinh\psi_0 - (1-\cosh\psi_0)(5+\cosh\psi_0)]}{(\cosh\psi_0-1)^3}, \tag{10.8.7}$$

for $\Omega_0 < 1$, where $\cosh\psi_0 = (2\Omega_0^{-1} - 1)$. The growth factor A_{r0} is an increasing function of the density parameter Ω: it varies from a value of 10 for $\Omega_0 \simeq 10^{-2}$, to a value of order 300 for $\Omega_0 \simeq 10^{-1}$, to 1500 for $\Omega_0 = 1$, and 3000 for $\Omega_0 \simeq 4$.

To give a more succinct summary of the effect of cosmology on the growth of perturbations, it is helpful to introduce the quantity f, defined by

$$f(\Omega_0) \equiv \frac{\mathrm{d}\log\delta_+}{\mathrm{d}\log a}. \tag{10.8.8}$$

This gives the growth factor relative to the Einstein–de Sitter case with the advantage that it does not require a translation between scale factor and time. It is an extremely helpful approximation to take

$$f(\Omega_0) \simeq \Omega^{0.6} \tag{10.8.9}$$

for models with $\Lambda = 0$ (Peebles 1980). If there is a cosmological constant, it actually does not make much difference to f. A better fit in such cases is

$$f \simeq \Omega_0^{0.6} + \frac{\Omega_\Lambda}{70}(1 + \tfrac{1}{2}\Omega_0). \tag{10.8.10}$$

10.9 Solution for Radiation-Dominated Universes

The procedure followed in Section 10.6 for a matter-dominated universe can also be followed, with appropriate modifications, for a universe which is radiation dominated. As we have already noted, in radiation universes the gravitational 'source' in the Einstein equations must include pressure terms, so a Newtonian treatment will not suffice. For pure radiation we have that $\rho + 3p/c^2 = 2\rho$. As well as the equations of energy and momentum conservation, we must also take account of the effect of radiation pressure. One can demonstrate that the relativistic analogues of Equations (10.5.1) can be written in the form

$$\frac{\partial\rho}{\partial t} + \nabla\cdot\left(\rho + \frac{p}{c^2}\right)\boldsymbol{v} = 0, \tag{10.9.1 a}$$

$$\left(\rho + \frac{p}{c^2}\right)\left(\frac{\partial\boldsymbol{v}}{\partial t} + \boldsymbol{v}\cdot\nabla\boldsymbol{v}\right) + \nabla p + \left(\rho + \frac{p}{c^2}\right)\nabla\varphi = 0, \tag{10.9.1 b}$$

$$\nabla^2\varphi - 4\pi G\left(\rho + 3\frac{p}{c^2}\right) = 0; \tag{10.9.1 c}$$

we have not bothered to write down the appropriate law of conservation of entropy, since we shall only be interested from now on in longitudinal adiabatic perturbations. Following the same method as we did in Section 10.6, we arrive at equations which are analogous to Equation (10.6.13):

$$\ddot{\delta} + 2\frac{\dot{a}}{a}\dot{\delta} + (v_s^2k^2 - \tfrac{32}{3}\pi G\rho)\delta = 0, \tag{10.9.2}$$

in which the velocity of sound is now $v_s = c/\sqrt{3}$. Let us concentrate upon finding the solution for a flat universe, which will be a good approximation to our Universe before matter–radiation equivalence. For this model we have

$$\rho = \frac{3}{32\pi G t^2}, \tag{10.9.3 a}$$

$$a = a_{\mathrm{eq}}\left(\frac{t}{t_{\mathrm{eq}}}\right)^{1/2}, \tag{10.9.3 b}$$

$$\frac{\dot{a}}{a} = \frac{1}{2t}, \tag{10.9.3 c}$$

which, upon substitution in (10.9.2), gives

$$\ddot{\delta} + \frac{\dot{\delta}}{t} - \frac{1}{t^2}\left(1 - \frac{3v_{\mathrm{s}}^2 k^2}{32\pi G\rho}\right)\delta = 0. \tag{10.9.4}$$

For $k \to 0$ Equation (10.9.4) is solved by $\delta \propto t^n$, with n constant; one again gets a growing mode, but in the form

$$\delta_+ \propto t, \tag{10.9.5}$$

while the decaying mode is again of the form

$$\delta_- \propto t^{-1}. \tag{10.9.6}$$

Looking for solutions of the power-law form also for $k \neq 0$, one finds similar (non-exact) results to those in Section 10.7, but with λ'_{J} given by

$$\lambda'_{\mathrm{J}} = v_{\mathrm{s}}\left(\frac{3\pi}{8G\rho}\right)^{1/2}. \tag{10.9.7}$$

Going further still, one can extend these analyses to models with a general equation of state of the form $p = w\rho c^2$, with w constant and $v_{\mathrm{s}} \neq 0$. In general one now has $v_{\mathrm{s}} = w^{1/2}c$ for $w > 0$, but for a matter-dominated universe ($w \simeq 0$) the value of v_{s} must be defined in an appropriate manner. For example in the case $w = 0$, which corresponds either to dust or a collisionless fluid, v_{s}^2 is of order the mean square velocity of the particles. In any case, the general result for λ'_{J} can be written

$$\lambda'_{\mathrm{J}} = \frac{\sqrt{24}}{5+9w} v_{\mathrm{s}}\left(\frac{\pi}{G\rho}\right)^{1/2}, \tag{10.9.8}$$

and the increasing and decreasing modes for scale $\lambda \gg \lambda'_{\mathrm{J}}$ are of the exact form

$$\delta_+ \propto t^{2(1+3w)/3(1+w)}, \tag{10.9.9}$$

$$\delta_- \propto t^{-1}. \tag{10.9.10}$$

10.10 The Method of Autosolution

There is another method which can be used to study the evolution of perturbations in the regime with $\lambda \gg \lambda_J$: the method of *autosolution*, pioneered in a paper by Zel'dovich and Barenblatt (1958). This method is based on the property that a spherical perturbation with diameter $\lambda \gg \lambda_J$ evolves in exactly the same manner as a universe model. This is essentially a consequence of Birkhoff's theorem in general relativity, which is the relativistic analogue of Newton's famous Spherical Theorem. In the simplest case of a sphere which is homogeneous and isotropic, the evolution is just that of a Friedmann model with parameters differing slightly from the surrounding (unperturbed) universe. In particular, the density ρ_p inside the perturbation will be different from the density of the universe ρ; the difference between ρ_p and ρ evolves with time because the interior and exterior universe evolve according to different equations.

The Friedmann equations regarding the evolution of a universe comprised of a fluid with equation of state $p = w\rho c^2$ can be written in the form

$$\dot{a}^2 = Aa^{-(1+3w)} + B, \tag{10.10.1}$$

where the constants A and B are given by

$$A = \dot{a}_0^2 \Omega_{0w} a_0^{1+3w}, \tag{10.10.2 a}$$

$$B = \dot{a}_0^2(1 - \Omega_{0w}). \tag{10.10.2 b}$$

It is clear that Equation (10.10.1) just represents conservation of energy. To obtain the evolution of ρ_p we consider perturbations of the total energy or, alternatively, of the time of origin of the expansion of the model described by Equation (10.10.1).

Concerning the energy, we have

$$\dot{a}_p^2 = Aa_p^{-(1+3w)} + B + \epsilon, \tag{10.10.3}$$

where ϵ is such that

$$|\epsilon| \ll |Aa_p^{-(1+3w)} + B|; \tag{10.10.4}$$

this quantity is proportional to the perturbation to the energy. We can easily obtain, from (10.10.1) and (10.10.3), that

$$t = \int_0^a \frac{\mathrm{d}a}{[Aa^{-(1+3w)} + B]^{1/2}} = \int_0^{a_p} \frac{\mathrm{d}a}{[Aa^{-(1+3w)} + B + \epsilon]^{1/2}}, \tag{10.10.5 a}$$

which can be approximated by

$$t \simeq \int_0^{a_p} \frac{\mathrm{d}a}{[Aa^{-(1+3w)} + B]^{1/2}} - \tfrac{1}{2}\epsilon \int_0^{a_p} \frac{\mathrm{d}a}{[Aa^{-(1+3w)} + B]^{3/2}}. \tag{10.10.5 b}$$

Using the fact that $\int_0^{a_p} f(a)\,\mathrm{d}a - \int_0^{a} f(a)\,\mathrm{d}a \simeq (a_p - a)f(a)$, from equation (10.10.5) we find

$$\delta a = a_p - a \simeq \tfrac{1}{2}\epsilon[Aa^{-(1+3w)} + B]^{1/2} \int_0^{a_p} \frac{\mathrm{d}a}{[Aa^{-(1+3w)} + B]^{3/2}}, \tag{10.10.6 a}$$

which gives

$$\delta \simeq \tfrac{1}{2}\epsilon[Aa^{-(1+3w)} + B]^{1/2} \int_0^a \frac{da}{[Aa^{-(1+3w)} + B]^{3/2}}. \tag{10.10.6 b}$$

The evolution of the perturbation $\delta = (\rho_p - \rho)/\rho$ is therefore given by

$$\delta = -3(1+w)\frac{\delta a}{a} \simeq -\tfrac{3}{2}(1+w)\epsilon\frac{[Aa^{-(1+3w)} + B]^{1/2}}{a} \int_0^a \frac{da}{[Aa^{-(1+3w)} + B]^{3/2}}. \tag{10.10.7}$$

The sign of ϵ has the opposite sense to that of δ: an underdense region has an excess of energy compared with the background universe, and vice versa. In the special case of a flat universe the total energy, which is related to B, is exactly zero, and Equation (10.10.7) becomes

$$\delta \simeq -\frac{3(1+w)}{5+9w}\frac{\epsilon}{A}a^{1+3w} \propto t^{2(1+3w)/3(1+w)}, \tag{10.10.8}$$

which coincides with the result given in Equation (10.9.6). In the case of an open universe, for $t \gg t^*$ (see Section 2.3), we have instead that $A \simeq 0$ and, from (10.10.7), we obtain

$$\delta \simeq -\tfrac{3}{2}(1+w)\frac{\epsilon}{B} = \text{const.}, \tag{10.10.9}$$

in accordance with the result found for $w = 0$ in Section 11.4. This result can also be obtained by observing that, for $t \gg t^*$, the characteristic time for the Jeans instability to grow, $\tau_J \simeq (G\rho)^{-1/2}$, is much greater than the characteristic time of the expansion of the universe, $\tau_H = a/\dot{a}$. In fact, one can easily show, using the formulae derived in Section 2.3, that

$$\tau_J \simeq \frac{1}{(G\rho)^{1/2}} \simeq \frac{1}{(G\rho^*)^{1/2}}\left(\frac{t}{t^*}\right)^{3(1+w)/2}, \tag{10.10.10}$$

while we have

$$\tau_H = \frac{a}{\dot{a}} \simeq t^*\left(\frac{t}{t^*}\right) \simeq \frac{1}{(G\rho^*)^{1/2}}\frac{t}{t^*}, \tag{10.10.11}$$

from which

$$\frac{\tau_J}{\tau_H} \simeq \left(\frac{t}{t^*}\right)^{(1+3w)/2} \gg 1. \tag{10.10.12}$$

Equation (10.10.7) represents the solution that increases with respect to time, δ_+. To obtain the decreasing solution δ_-, one must perturb the time at which the expansion begins. We have, respectively, that

$$t = \int_0^a \frac{da}{[Aa^{-(1+3w)} + B]^{1/2}}, \tag{10.10.13 a}$$

$$t - \tau = \int_0^{a_p} \frac{da}{[Aa^{-(1+3w)} + B]^{1/2}}, \tag{10.10.13 b}$$

where the parameter τ represents the time lag (either positive or negative) between the perturbed and unperturbed solutions. From the preceding equations one obtains

$$\tau \simeq -\frac{\delta a}{[Aa^{-(1+3w)} + B]^{1/2}}, \qquad (10.10.14)$$

from which

$$\delta \simeq 3(1+w)\tau \frac{[Aa^{-(1+3w)} + B]^{1/2}}{a}. \qquad (10.10.15)$$

The sign of δ is this time the same as the sign of τ. In the special case of the flat Einstein–de Sitter model we have, in accordance with our previous calculations,

$$\delta \simeq 3(1+w)\tau A^{1/2} a^{-3(1+w)/2} \propto t^{-1}; \qquad (10.10.16)$$

for an open universe with $t \gg t^*$ we obtain

$$\delta \simeq 3(1+w)\tau \frac{B^{1/2}}{a} \propto t^{-1}, \qquad (10.10.17)$$

with a behaviour as a function of time which is in this case independent of w. In general, however, Equation (10.10.15) represents a decreasing perturbation with a behaviour that depends upon w.

10.11 The Meszaros Effect

As we shall see later on, in a universe composed of non-relativistic matter and relativistic particles (radiation, massless neutrinos, etc.), there can exist a mode of perturbation in which the non-relativistic component is perturbed with respect to a homogeneous distribution while the relativistic component remains unperturbed. If the matter component is entirely baryonic, this type of perturbation is often called *isothermal*, and a picture of structure formation based on this type of fluctuation was popular in the 1970s. In the 1980s, alternative scenarios were developed in which an important role is played by various forms of non-baryonic matter (massive neutrinos, axions, photinos, etc.): perturbations which involve this component and not the others (baryons, photons, massless neutrinos) are usually termed *isocurvature* fluctuations, because these fluctuations do not modify the local spatial curvature. It is consequently important to study the evolution of perturbations of a non-relativistic component with density $\rho_{\rm nr}$ in a universe dominated by a fluid of relativistic particles of density $\rho_{\rm r}$. The Universe is dominated by such a fluid at redshifts given by the inequality (5.3.4).

The problem of the evolution of perturbations through $z_{\rm eq}$ has been studied by various authors, the first being Meszaros (1974): one finds that the growing-mode perturbation $\delta_{\rm nr}$ remains 'frozen' until $z_{\rm eq}$ even when $\lambda \gg \lambda_{\rm J}$. This effect of freezing-in of perturbations or 'stagnation' or the *Meszaros effect* is very important for models in which galaxies and clusters of galaxies are formed by the growth of primordial fluctuations in a universe dominated by cold dark matter. We should point out that this effect does not require perturbations of isocurvature form: it is a generic feature of models with a period of domination by relativistic particles. To form structure one requires at the very least that the perturbations to the non-relativistic particle distribution, $\delta_{\rm nr}$, should be of order unity. The time available for fluctuations to grow from a small amplitude up to this is changed if there is an extended period of stagnation. The problem is exacerbated if $\Omega \ll 1$ because of the freezing out of perturbations when the universe becomes dominated by curvature. We shall describe the detailed consequences of this effect later; for the moment let us just describe the basic physics.

Let us begin with a qualitative argument. The characteristic time for a gravitational instability process to boost the perturbations in the non-relativistic component $\delta_{\rm nr}$ is given by the Jeans timescale, $\tau_{\rm J} \simeq (G\rho_{\rm nr})^{-1/2}$, while the characteristic time for the expansion of the universe is given by $\tau_{\rm H} \simeq (G\rho_{\rm r})^{-1/2}$ before $z_{\rm eq}$; the two timescales are similar after $z_{\rm eq}$. Consequently, as long as the Universe is dominated by the relativistic component, the fluctuations in the other component remain frozen; the perturbation can only grow after $z_{\rm eq}$.

We can now study this effect in an analytical manner, restricting ourselves for simplicity to the case of a flat universe and $\lambda \gg \lambda_{\rm J}$. Introducing the variable

$$y = \frac{\rho_{\rm nr}}{\rho_{\rm r}} = \frac{a}{a_{\rm eq}}, \tag{10.11.1}$$

one finds that the equation describing the perturbation in the non-relativistic component $\delta = \delta\rho_{\rm nr}/\rho_{\rm nr}$ becomes

$$\ddot{\delta} + 2\frac{\dot{a}}{a}\dot{\delta} - 4\pi G\rho_{\rm nr}\delta = 0. \tag{10.11.2}$$

One then obtains

$$\frac{{\rm d}^2\delta}{{\rm d}y^2} + \frac{2+3y}{2y(1+y)}\frac{{\rm d}\delta}{{\rm d}y} - \frac{3\delta}{2y(1+y)} = 0, \tag{10.11.3}$$

which has, as usual, two solutions, one increasing and one decreasing. We shall forget about the decaying mode from now on: interested readers can calculate the relevant behaviour for the decaying mode themselves. We have

$$\delta_+ \propto 1 + \tfrac{3}{2}y. \tag{10.11.4}$$

Before $z_{\rm eq}$ ($y < 1$) the growing mode is practically frozen: the total growth in the interval $(0, t_{\rm eq})$ is only

$$\frac{\delta_+(y=1)}{\delta_+(y=0)} = \tfrac{5}{2}; \tag{10.11.5}$$

after $z_{\rm eq}$ the solution rapidly matches the law in a matter-dominated Einstein–de Sitter universe:

$$\delta_+(y \gg 1) \propto y \propto a \propto t^{2/3}. \tag{10.11.6}$$

10.12 Relativistic Solutions

As we have already explained, the solution of the linear evolution of perturbations, i.e. perturbations with $|\delta| \ll 1$, in Friedmann models within the framework of general relativity was studied for the first time by Lifshitz (1946). In the relativistic approach one proceeds in a quite different manner from the Newtonian treatment we have concentrated upon so far. The fundamental object one should treat perturbatively is usually taken to be the metric g_{ij}, to which one adds small perturbations h_{ij}. One problem that arises immediately is to distinguish between real physical perturbations, and those that arise purely from the choice of reference coordinate system. These latter perturbation modes are called 'gauge modes' and one can avoid them by choosing a particular gauge and then finding the gauge modes by hand, or by choosing gauge-invariant combinations of physical variables. In any case, the perturbed metric becomes

$$g'_{ij} = g_{ij} + h_{ij}. \tag{10.12.1}$$

For the energy-momentum tensor one adopts a tensor T'_{ij}, which is perturbed relative to an ideal fluid, so that ρ, p and U_i are perturbed relative to their values in the background Friedmann model. One then writes down the Einstein equations in terms of the (perturbed) metric g'_{ij} and the (perturbed) energy-momentum tensor T'_{ij}. The procedure is complicated from an analytical point of view, so we just summarise the results here. We find there are three perturbation types which can be classified as *tensor*, *vector* and *scalar modes*.

There are in fact two solutions of tensor type, both corresponding to the propagation of gravitational waves. Gravitational waves are described by an equation of state of radiative type and their amplitude h_{ij} varies with time according to

$$h^i{}_j \propto \text{const.}, \qquad h^i{}_j \propto t^{-1} \tag{10.12.2 a}$$

for a matter-dominated Einstein–de Sitter universe, and according to

$$h^i{}_j \propto \text{const.}, \qquad h^i{}_j \propto t^{-1/2} \tag{10.12.2 b}$$

for the analogous radiation-dominated universe. The solutions (10.12.2) correspond to wavelengths $\lambda \gg ct$; for $\lambda \ll ct$ we have instead two oscillating solutions:

$$h_{ij} \propto t^{5/8} J_{\pm 3/2}(3ckt), \qquad h_{ij} \propto t^{3/4} J_{\pm 1/2}(2ckt), \tag{10.12.3}$$

where J are Bessel functions.

While the tensor modes have no Newtonian analogue, the vector modes are similar to phenomena which appear in the Newtonian analysis. They correspond to rotational modes in the velocity field, which have velocity $\boldsymbol{v}$ perpendicular to the wavevector $\boldsymbol{k}$. Their amplitude varies according to

$$v_t \propto [(\rho c^2 + p)a^4]^{-1} \tag{10.12.4}$$

which, in a matter-dominated universe with $p \ll \rho c^2 \propto a^{-3}$, becomes

$$v_t \propto a^{-1}, \tag{10.12.5 a}$$

corresponding to (10.6.8), while for a radiation-dominated universe we have

$$v_t = \text{const.} \tag{10.12.5 b}$$

The Equation (10.12.4) can, in a certain sense, be interpreted as a kind of conservation law for angular momentum $\mathcal{L}$, in which one replaces the matter density by $(\rho + p/c^2)$. Equation (10.12.4) can then be written in the form

$$\mathcal{L} \simeq (\rho + p/c^2)a^3 v_t a \simeq \text{const.}, \tag{10.12.6}$$

which is known as *Loytsianski's theorem*, an extension of Equation (10.6.10).

The final perturbation type, the scalar mode, actually represents the longitudinal compressional density wave we have been discussing in most of this chapter. One finds in the relativistic approach the same results as we have introduced in a Newtonian approximation.

In modern cosmological theories involving inflation the relativistic treatment is extremely important; while we can handle the growth of fluctuations inside the horizon R_c adequately using the Newtonian treatment we have described, fluctuations outside the horizon must be handled using general relativity. In particular, in inflationary theories one must consider the super-horizon evolution of scalar fluctuations, i.e. when $\lambda > R_c$, in a model where the equation of state is of the form $p = w\rho c^2$, with $w < -\frac{1}{3}$. We mention this problem again in Section 13.6.

Bibliographic Notes on Chapter 10

The pioneering works by Silk (1967, 1968), as well as Doroshkevich *et al.* (1967), Peebles and Yu (1970), Weinberg (1971), Chibisov (1972) and Field (1971) are all still worth reading. Weinberg (1972) summarises much of this historical work; see also Zel'dovich (1965). For detailed perturbation theory and alternative formulations of the material we have covered in this chapter, see Efstathiou and Silk (1983), Kodama and Sasaki (1984), Efstathiou (1990) and Peacock (1999).

Problems

1. Calculate the Jeans length for air at room temperature.
2. How is the expression for the Jeans length modified in the presence of a magnetic field?
3. Derive Equations (10.6.6 *a*) and (10.6.6 *b*).
4. Show that the solutions to (10.7.3) for finite $\lambda > \lambda_J'$ have the form given by equation (10.7.6). Thus obtain the correct form in the limit $\lambda \to \infty$, i.e. $\delta_+ \propto t^{2/3}$, $\delta_- \propto t^{-1}$.
5. Derive Equation (10.11.3) and obtain the growing mode solution (10.11.4).

11

Gravitational Instability of Baryonic Matter

11.1 Introduction

In this chapter we shall apply the principle of the Jeans instability to models of the Universe in which the dominant matter component is baryonic. As we shall see, the adoption of a realistic physical fluid brings in many more complications than we found in our previous analyses of gravitational instability in purely dust or radiation universes. The interaction of matter with radiation during the plasma epoch is one such complication which we have not addressed so far. Although the baryon-dominated models are in this sense more realistic than the simple ones we have used in our illustration of the basic physics, we should make it clear at the outset that these models are not successful at explaining the origin of the structure observed in our Universe. In the next chapter we shall explain why this is so and why models including non-baryonic weakly interacting dark matter may be more successful than the baryon-dominated ones. Nevertheless, we feel it is important to study the baryonic situation in some detail. Our primary reason for this is pedagogical. Although it is believed that there is non-baryonic matter, there certainly are baryons in our Universe. Whatever the dominant form of the matter, we must in any case understand the behaviour of baryons in the presence of radiation during the cosmological expansion. The simplest way to understand this behaviour is to study a model which includes only these two ingredients. Once we have understood the physics here, we can go on to study the effect of other components. The baryon-dominated models also provide an interesting insight into the history of the study of large-scale structure, and their analysis is an interesting part of the development of the subject in the late 1960s and in the 1970s. We begin with some comments on the form of perturbations in baryonic models.

11.2 Adiabatic and Isothermal Perturbations

Before recombination, the Universe was composed of a plasma of ionised matter and radiation, interacting via Compton scattering with characteristic times given by $\tau_{e\gamma}$ and $\tau_{\gamma e}$, described in Section 9.2. For simplicity we neglect the presence of helium nuclei in this plasma, and take it to be composed entirely of protons and electrons. We shall also neglect the role of neutrinos in most of this discussion.

As we have seen in Chapter 10, there exist a number of possible perturbation modes in a self-gravitating fluid. There are vortical perturbations (transverse waves) which do not interest us here. There are also perturbations of adiabatic or entropic type, the first time dependent, the second independent of time in the static case studied in Chapter 10. The distinction between these two latter types of perturbation remains when one moves to the cosmological case of an expanding background model.

The entropy per unit mass of a fluid composed of matter and radiation in a volume V has a very high value because of the enormous value of the entropy per baryon σ_r. In other words, the entropy is carried almost entirely by the radiation:

$$S = \tfrac{4}{3}\sigma T^3 V \propto \sigma_{\rm rad} \propto \frac{T^3}{\rho_{\rm m}} \propto \frac{\rho_{\rm r}^{3/4}}{\rho_{\rm m}}. \tag{11.2.1}$$

A perturbation which leaves S invariant – an *adiabatic perturbation* – is made up of perturbations in both the matter density $\rho_{\rm m}$ and the radiation density $\rho_{\rm r}$ (or, equivalently, T, the radiation temperature) such that

$$\frac{\delta S}{S} = \frac{\delta\sigma_{\rm rad}}{\sigma_{\rm rad}} = \frac{3}{4}\frac{\delta\rho_{\rm r}}{\rho_{\rm r}} - \frac{\delta\rho_{\rm m}}{\rho_{\rm m}} = \frac{3\delta T}{T} - \frac{\delta\rho_{\rm m}}{\rho_{\rm m}} = 0; \tag{11.2.2}$$

this means that

$$\delta_{\rm m} \equiv \frac{\delta\rho_{\rm m}}{\rho_{\rm m}} = 3\frac{\delta T}{T} = \frac{3}{4}\frac{\delta\rho_{\rm r}}{\rho_{\rm r}} \equiv \frac{3}{4}\delta_{\rm r}. \tag{11.2.3}$$

As we have seen in Section 7.4, the value of $\sigma_{\rm rad}$ may be explained by microscopic physics involving a GUT or electroweak phase transition. If such a microphysical explanation is correct, one might expect small inhomogeneities to have the same value of $\sigma_{\rm r}$ and therefore be of adiabatic type.

A perturbation of *entropic* type or an *isothermal perturbation* is such that a non-zero perturbation in the matter component $\delta_{\rm m} \neq 0$ is not accompanied by any fluctuation in the radiation component. In other words there is no inhomogeneity in the radiation temperature, hence the word isothermal. This type of fluctuation is closely related, but not identical, to the isocurvature fluctuations discussed in the previous chapter and also in the next one. The physical reason why $\delta T \simeq 0$ rests on the fact that such fluctuations are more or less independent of time; the high thermal conductivity of the cosmological medium allows the temperature to be levelled out by heat conduction. A perturbation with $\delta\rho_{\rm m} \neq 0$ is held frozen and therefore time independent by the strong frictional 'drag' forces between the matter and radiation fluid. An exact treatment of this problem confirms, at least to a first approximation, this division into two main types of perturbation.

After recombination, and the consequent decoupling of matter and radiation, the perturbations $\delta\rho_{\rm m}$ in the total matter density evolve in the same way regardless of whether they were originally of adiabatic or isothermal type. Because there is essentially no interaction between the matter and radiation, and the radiation component is dynamically negligible compared with the matter component, the Universe behaves as a single-fluid dust model.

Before recombination a generic perturbation can be decomposed into a superposition of adiabatic and isothermal modes which evolve independently; the two modes can be thought of as similar to the normal modes of a dynamical system. To understand what is going on it is therefore useful, as a first approximation, to study the behaviour of each mode separately.

11.3 Evolution of the Sound Speed and Jeans Mass

As we have already explained, the distinction between adiabatic and isothermal perturbations only has meaning before recombination. In this period we shall denote the relevant sound speeds for the adiabatic and isothermal modes by $v_{\rm s}^{(\rm a)}$ and $v_{\rm s}^{(\rm i)}$, respectively.

The adiabatic sound speed, $v_{\rm s}^{(\rm a)}$, is that of a plasma with density $\rho = \rho_{\rm m} + \rho_{\rm r}$ and pressure $p = p_{\rm r} + p_{\rm m} \simeq p_{\rm r} \simeq \frac{1}{3}\rho_{\rm r}c^2$. We assume the neutrinos are massless. Recalling Equation (11.2.3), we therefore have

$$v_{\rm s}^{(\rm a)} = \left(\frac{\partial p}{\partial \rho}\right)_S^{1/2} \simeq \frac{c}{\sqrt{3}}\left[1 + \left(\frac{\partial \rho_{\rm m}}{\partial \rho_{\rm r}}\right)_S\right]^{-1/2} = \frac{c}{\sqrt{3}}\left(1 + \frac{3}{4}\frac{\rho_{\rm m}}{\rho_{\rm r}}\right)^{-1/2}. \tag{11.3.1}$$

This equation gives $v_{\rm s}^{(\rm a)} \simeq c/\sqrt{3}$ for $t \ll t_{\rm eq}$, while $v_{\rm s}^{(\rm a)} \simeq 0.76c/\sqrt{3}$ for $t = t_{\rm eq}$ and during the interval $t_{\rm rec} > t \gg t_{\rm eq}$, which exists only if $\Omega_{\rm b}h^2 \geqslant 4 \times 10^{-2}$, we have

$$v_{\rm s}^{(\rm a)} \simeq \frac{c}{\sqrt{3}}\left(\frac{4\rho_{\rm r}}{3\rho_{\rm m}}\right)^{1/2} \simeq \frac{c}{\sqrt{3}}\left(\frac{1+z}{1+z_{\rm eq}}\right)^{1/2} \simeq 2 \times 10^8 \left(\frac{1+z}{1+z_{\rm eq}}\right)^{1/2} \text{ m s}^{-1}. \tag{11.3.2}$$

In the following considerations we assume for simplicity that $v_{\rm s}^{(\rm a)} = c/\sqrt{3}$ for $z \geqslant z_{\rm eq}$ and $v_{\rm s}^{(\rm a)} = (c/\sqrt{3})[(1+z)/(1+z_{\rm eq})]^{1/2}$ for $z \leqslant z_{\rm eq}$. In reality the transition between these two regimes will be much smoother than this.

The isothermal sound speed $v_s^{(i)}$ is that appropriate for a gas of monatomic particles of mass m_p (the proton mass) and temperature $T_m \simeq T_r = T_{0r}(1+z)$, i.e.

$$v_s^{(i)} = \left(\frac{\partial p_m}{\partial \rho_m}\right)_S^{1/2} = \left(\frac{\gamma k_B T}{m_p}\right)^{1/2}, \tag{11.3.3}$$

with $\gamma = \frac{5}{3}$ for hydrogen, which gives

$$v_s^{(i)} \simeq \left(\frac{k_B T_{rec}}{m_p}\right)^{1/2}\left(\frac{1+z}{1+z_{rec}}\right)^{1/2} \simeq 5\times 10^5\left(\frac{1+z}{1+z_{rec}}\right)^{1/2} \text{ m s}^{-1}, \tag{11.3.4}$$

where we have assumed that $T_{rec} = T(z_{rec}) \simeq 4000$ K. The velocity of sound associated with matter perturbations after z_{rec} is given by $v^{(i)}$ and one finds that $T_m \simeq T_r$ in this period only for $z \gtrsim 300$; see Section 9.4. After this, until the moment of reheating, $T_m \propto (1+z)^2$, so that Equation (11.3.4) should be modified. However, as far as the origin of galaxies and clusters is concerned, the value of $v_s^{(i)}$ for $z \ll z_{rec}$ is not important so we shall not discuss it further here.

We have already introduced the Jeans length, λ_J. An alternative way of specifying the physical scale appropriate for gravitational instability is to deal with a mass scale. For this reason, we shall define the *Jeans mass* to be the mass contained in a sphere of radius $\frac{1}{2}\lambda_J$

$$M_J = \tfrac{1}{6}\pi\rho_m\lambda_J^3; \tag{11.3.5}$$

in this expression we have assumed that, for any value of the equation-of-state parameter w, the relation

$$\lambda_J \simeq v_s\left(\frac{\pi}{G\rho}\right)^{1/2} \tag{11.3.6}$$

is a good approximation. More accurate expressions can be found in Section 10.9, but we shall not use them in this order-of-magnitude analysis. It is useful to note the obvious relation between mass and length scales $M \propto \rho\lambda^3$ so that, for example, 1 Mpc corresponds to $10^{11}(\Omega_0 h^2)^{-1}M_\odot$.

Before recombination we must distinguish between adiabatic and isothermal perturbations. We begin with the Jeans mass associated with adiabatic perturbations, $M_J^{(a)}$, for which one must insert the quantity $v_s^{(a)}$ in place of v_s in the Equation (11.4.2). One should also use $\rho = \rho_m + \rho_r$ because the total density is included in the terms describing the self-gravity of the perturbation. For simplicity we can adopt the approximate relations that $\rho \simeq \rho_r$ for $z > z_{eq}$ and $\rho \simeq \rho_m$ for $z < z_{eq}$. Together with the other approximations we have introduced above for $v_s^{(a)}$ we find that, for $z \gtrsim z_{eq}$,

$$M_J^{(a)} = \tfrac{1}{6}\pi\rho_m\left[\frac{c}{\sqrt{3}}\left(\frac{\pi}{G\rho}\right)^{1/2}\right]^3 \simeq M_J^{(a)}(z_{eq})\left(\frac{1+z}{1+z_{eq}}\right)^{-3}, \tag{11.3.7 a}$$

where

$$M_J^{(a)}(z_{eq}) \simeq 3.5\times 10^{15}(\Omega h^2)^{-2}M_\odot, \tag{11.3.7 b}$$

while in the interval $z_{eq} > z > z_{rec}$, if it exists, we have

$$M_J^{(a)} \simeq \tfrac{1}{6}\pi\rho_m\left[\frac{c}{\sqrt{3}}\left(\frac{1+z}{1+z_{eq}}\right)^{1/2}\left(\frac{\pi}{G\rho}\right)^{1/2}\right]^3 \simeq M_J(z_{eq}) \simeq \text{const.} \tag{11.3.8}$$

This is an approximate relation. In reality, if $z_{eq} \gg z_{rec}$, because ρ_r is small at z_{rec}, the value of the Jeans mass at recombination, $M_J^{(a)}(z_{rec})$, will be about a factor three higher than $M_J^{(a)}(z_{eq})$.

Now turning to the isothermal perturbations, we must use the expression given in Equation (11.3.4) for $v_s^{(i)}$ in place of v_s. We then find that, in the interval $z_{eq} > z > z_{rec}$,

$$M_J^{(i)} \simeq \tfrac{1}{6}\pi\rho_m\left(\frac{\pi k_B T_m}{G m_p \rho_m}\right)^{3/2} \simeq \text{const.} \simeq M_{J,rec}^{(i)} \simeq 5\times 10^4(\Omega h^2)^{-1/2}M_\odot. \tag{11.3.9}$$

It is interesting to note that both $M_J^{(a)}$ and $M_J^{(i)}$ remain roughly constant during the interval (if it exists) between equivalence and recombination. After recombination, since we are only interested in the matter perturbations, the Jeans mass M_J can be taken to coincide with $M_J^{(i)}$ while $T_m \simeq T_r$, and then thereafter the behaviour is roughly proportional to $(1+z)^{3/2}$.

11.4 Evolution of the Horizon Mass

An important concept which we have not yet come across in the study of gravitational instability is that of the *cosmological horizon*. Essentially this defines the scale over which different parts of a perturbation can be in causal contact with each other at a particular epoch. We shall not worry too much here about the technical issue of whether we should use the particle horizon, R_H, or the radius of the speed of light sphere, R_c, to characterise the horizon. In the case we are considering here, these differ only by a factor of order unity anyway, so we shall use the radius of the particle horizon, R_H, to define the horizon mass by analogy with the definition of the Jeans mass:

$$M_H = \tfrac{1}{6}\pi\rho R_H^3, \tag{11.4.1}$$

which represents the total mass inside the particle horizon which of course includes the effective mass contributed by the radiation. It is often more interesting to consider only the baryonic part of this mass, since that is the part that will dominate any structures that form after z_{rec}. Thus we have

$$M_{Hb} = \tfrac{1}{6}\pi\rho_m R_H^3. \tag{11.4.2}$$

Before equivalence, the Universe is well described by an Einstein–de Sitter model of pure radiation for which, using results from Chapters 2 and 5 and with the assumption that $\rho \simeq \rho_r$,

$$M_{Hb} \simeq \tfrac{1}{6}\pi\rho_m(2ct)^3 \simeq M_H(z_{eq})\left(\frac{1+z}{1+z_{eq}}\right)^{-3}, \tag{11.4.3 a}$$

where

$$M_{\rm H}(z_{\rm eq}) \simeq 5 \times 10^{14}(\Omega h^2)^{-2} M_{\odot}, \qquad (11.4.3\,b)$$

which is a little less than $M_{\rm J}^{(a)}$. For $z \leqslant z_{\rm eq}$ and $\Omega z \gg 1$, and using the same approximations as the previous expression, we have

$$M_{\rm Hb} \simeq \tfrac{1}{6}\pi\rho_{\rm m}(3ct)^3 \simeq M_{\rm H}(z_{\rm eq})\left(\frac{1+z}{1+z_{\rm eq}}\right)^{-3/2}. \qquad (11.4.4)$$

By analogy with the relations (11.4.1) and (11.4.3) we can obtain before equivalence

$$M_{\rm H} \simeq \tfrac{1}{6}\pi\rho(2ct)^3 \simeq M_{\rm H}(z_{\rm eq})\left(\frac{1+z}{1+z_{\rm eq}}\right)^{-2}, \qquad (11.4.5)$$

while, for $z < z_{\rm eq}$, it becomes

$$M_{\rm H} \simeq M_{\rm Hb} \simeq M_{\rm H}(z_{\rm eq})\left(\frac{1+z}{1+z_{\rm eq}}\right)^{-3/2}. \qquad (11.4.6)$$

We can define the *horizon entry* of a mass scale M to be the time (or, more usefully, redshift) at which the mass scale M coincides with the mass inside the horizon. It is most useful to write this in terms of the baryonic mass given by Equation (11.4.2). The redshift of horizon entry for the mass scale M is denoted $z_{\rm H}(M)$ and is therefore given implicitly by the relation

$$M_{\rm Hb}(z_{\rm H}(M)) = M. \qquad (11.4.7)$$

From Equation (11.4.3) we find that for $M < M_{\rm H}(z_{\rm eq})$

$$z_{\rm H}(M) \simeq z_{\rm eq}\left(\frac{M}{M_{\rm H}(z_{\rm eq})}\right)^{-1/3}, \qquad (11.4.8)$$

with $z_{\rm H}(M) > z_{\rm eq}$, while for $M > M_{\rm H}(z_{\rm eq})$ one obtains, using Equation (11.4.4),

$$z_{\rm H}(M) \simeq z_{\rm eq}\left(\frac{M}{M_{\rm H}(z_{\rm eq})}\right)^{-2/3}, \qquad (11.4.9)$$

with $z_{\rm H}(M) < z_{\rm eq}$ and $z_{\rm H}(M) \gg \Omega^{-1}$. The relations (11.4.8) and (11.4.9) will be useful later in Chapter 14 when we look at the variance of fluctuations as a function of their horizon entry.

11.5 Dissipation of Acoustic Waves

Having established two basic physical scales – the Jeans scale and the horizon scale – which will play a strong role in the evolution of structure, we must now investigate other physical processes which can modify the purely gravitational evolution of perturbations. We shall begin by considering adiabatic fluctuations in some detail.

The most important physical phenomenon we have to deal with is the interaction between matter and radiation during the plasma epoch and the consequent dissipation due to viscosity and thermal conduction. We shall study the basic physics in this section and the more detailed ramifications in Section 12.7. As we shall see, dissipative processes act significantly on sound waves with a wavelength λ, or an effective mass scale $M = \frac{1}{6}\pi\rho_m\lambda^3$, less than a certain characteristic scale λ_D, called the *dissipation scale* whose corresponding mass scale, M_D, is called the *dissipation mass*. During the period in which we are interested (the period before recombination), it turns out that $M_D \ll M_J$ for both adiabatic and isothermal perturbations; however, the dissipation mass for isothermal perturbations has no practical significance for cosmology.

The effect of these dissipative processes upon an adiabatic perturbation is to decrease its amplitude. From a kinetic point of view this is because of the phenomenon of diffusion, which slowly moves particles into the region outside the perturbation. One can assume for all practical purposes that, after a time t, a perturbation of wavelength $\lambda < \lambda_D(t)$, where $\lambda_D(t)$ is the mean diffusion length for particles in a time t, is totally dissipated. Given that the particles travel in an arbitrary direction, the effect is a complete randomisation of the original fluctuation so that it becomes smeared out and dissipated. The distance λ_D is obviously connected with the mean free path $\bar{l}$ of the particles.

On scales $\lambda < \bar{l}$ the fluctuation is dissipated in a time of order the wave period and over a distance of order the wavelength λ. In this case it does not make sense to talk about diffusion, and the role of λ_D is taken by $\bar{l}$. We therefore have *free streaming* of particles, which is important in the models we discuss in the next chapter, which have perturbations in a fluid of collisionless particles. On scales $\lambda \gg \bar{l}$, it is more illuminating to employ a macroscopic model, where dissipation is attributed to the presence of viscosity η and thermal conductivity D_t. Evidently, however, there is a strict connection between the coefficients of viscosity and thermal conductivity on the one hand, and the coefficient of diffusion D and its related length scale λ_D on the other. On scales $\lambda \simeq \bar{l}$ the model for dissipation we must use cannot be a fluid model, but must be based on kinetic theory.

Let us elaborate these concepts in more mathematical detail. The phenomenon of diffusion is described by *Fick's law*:

$$J_m \equiv \rho_m \boldsymbol{v} = -D\nabla\rho_m, \tag{11.5.1}$$

where J_m is the matter flux caused by the density gradient $\nabla\rho_m$ and D is called the coefficient of diffusion. Together with the continuity equation, Equation (11.5.1) furnishes Fick's second law

$$\frac{\partial\rho_m}{\partial t} - D\nabla^2\rho_m = 0. \tag{11.5.2}$$

There is a formal analogy of this relation with the equation of heat conduction

$$\frac{\partial T}{\partial t} - D_t\nabla^2 T = 0 \tag{11.5.3}$$

($D_t = \lambda_t/\rho c_t$ is the coefficient of thermal diffusion; c_t is the specific heat; λ_t is the thermal conductivity), which is obtained easily from the Fourier postulate about conduction, similar to Equation (11.5.1), and from the calorimetric equation.

It is obvious from Equations (11.5.2) and (11.5.3) that the coefficients D and D_t have the same dimensions as each other, and also the same as those of the kinematic viscosity $\nu = \eta/\rho$ which appears in the Navier–Stokes equation

$$\frac{\partial \boldsymbol{v}}{\partial t} + \boldsymbol{v} \cdot \nabla \boldsymbol{v} = -\frac{\nabla p}{\rho} + \nu \nabla^2 \boldsymbol{v}; \tag{11.5.4}$$

according to this formal analogy, one is invited to interpret ν as a sort of coefficient of velocity diffusion. We have that

$$[D] = [D_t] = [\nu] = \mathrm{m^2\ s^{-1}}. \tag{11.5.5}$$

Adopting a kinetic treatment to confirm these relations, one finds that

$$D \simeq D_t \simeq \nu \simeq \tfrac{1}{3}\bar{v}\bar{l} = \frac{1}{3}\frac{\bar{l}^2}{\tau} = \tfrac{1}{3}\bar{v}^2\tau, \tag{11.5.6}$$

where $\bar{v}$ is the mean particle velocity and τ is the mean time between two consecutive particle collisions.

From a dimensional point of view, the mean distance $\bar{d} \gg \bar{l}$ affected after a time $t \gg \tau$ by the three 'diffusion' processes described above are, respectively,

$$\bar{d}_d \simeq (Dt)^{1/2}, \qquad \bar{d}_t \simeq (D_t t)^{1/2}, \qquad \bar{d}_\nu \simeq (\nu t)^{1/2}, \tag{11.5.7}$$

which, by Equation (11.5.6), corresponds to

$$\bar{d} \simeq \bar{l}\left(\frac{t}{\tau}\right)^{1/2}. \tag{11.5.8}$$

This relationship is easy to demonstrate by assuming that all these diffusion processes can be attributed to the diffusion of particles by a simple random walk.

Following on from (11.5.8), the *dissipation scale* (or the diffusion scale) of an acoustic wave at time t is therefore

$$\lambda_D(t) = \bar{l}\left(\frac{t}{\tau}\right)^{1/2} = \bar{v}(t\tau)^{1/2} = (\bar{l}\bar{v}t)^{1/2}. \tag{11.5.9}$$

We define the *dissipation time* of a perturbation of wavelength λ by the quantity

$$\tau_D(\lambda) = \tau\left(\frac{\lambda}{\bar{l}}\right)^2 = \frac{\lambda^2}{\bar{v}^2\tau} = \frac{\lambda^2}{\bar{l}\bar{v}}, \tag{11.5.10}$$

i.e. the time when $\lambda_D[\tau_D(\lambda)] = \lambda$. In particular, the times for dissipation through thermal conduction and viscosity are, respectively,

$$\tau_{D_t}(\lambda) \simeq \frac{\lambda^2}{D_t}, \qquad \tau_\nu(\lambda) \simeq \frac{\lambda^2}{\nu}. \tag{11.5.11}$$

In a situation where both these phenomena are present, the characteristic time for dissipation $\tau_{\rm dis}(\lambda)$ is given by the relation

$$\frac{1}{\tau_{\rm dis}(\lambda)} = \frac{1}{\tau_{\nu}(\lambda)} + \frac{1}{\tau_{D_{\rm t}}(\lambda)}, \tag{11.5.12}$$

characteristic of processes acting in parallel.

The full (non-relativistic) theory of dissipation of acoustic waves through viscosity and thermal conduction yields the following result

$$\frac{1}{\tau_{\rm dis}(\lambda)} \equiv -\frac{\dot{E}}{E} = \frac{4\pi^2}{\lambda^2}\left[\tfrac{4}{3}\nu\left(1 + \frac{3}{4}\frac{\zeta}{\eta}\right) + D_{\rm t}(1-\gamma^{-1})\right], \tag{11.5.13}$$

where E is the mechanical energy transported by the sound wave, ζ is the second viscosity, and γ is the adiabatic index. The Equation (11.5.13) verifies the applicability of Equation (11.5.12).

11.6 Dissipation of Adiabatic Perturbations

We now apply the physics described in the previous section to adiabatic perturbations in the plasma epoch of the expanding Universe described in Chapter 9. In the period prior to recombination, when $\tau_{\rm ep} \ll \tau_{\rm e\gamma}(\tau_{\gamma\rm e})$, one can treat the plasma-photon system as an imperfect radiative fluid, where the effect of dissipation manifests itself as an imperfect thermal equilibrium between matter and radiation. In this situation, the kinematic viscosity and the coefficient of thermal diffusion are given by

$$\nu \simeq \frac{4}{15}\frac{\rho_{\rm r}c^2}{\rho_{\rm r}+\rho_{\rm m}}\tau_{\gamma\rm e} \simeq \tfrac{4}{5}D_{\rm t}. \tag{11.6.1}$$

Equation (11.6.1) cannot be used in Equation (11.5.13), which was obtained in a non-relativistic treatment. There are special processes which modify Equation (11.5.13) in the relativistic limit: for example the thermal conduction is not proportional to ∇T, but to $\nabla T - [T/(p+\rho c^2)]\nabla p$. In particular, Equation (11.5.11) becomes

$$\tau_{D_{\rm t}} = \frac{\lambda^2}{4\pi^2}\left(\frac{\rho_{\rm m}+\frac{4}{3}\rho_{\rm r}}{\rho_{\rm m}}\right)^2\frac{6}{c^2\tau_{\gamma\rm e}}, \tag{11.6.2 a}$$

$$\tau_{\nu} = \frac{\lambda^2}{4\pi^2}\left(\frac{\rho_{\rm m}+\frac{4}{3}\rho_{\rm r}}{\rho_{\rm r}}\right)\frac{45}{8c^2\tau_{\gamma\rm e}} = \frac{15\rho_{\rm m}^2}{16\rho_{\rm r}(\rho_{\rm m}+\frac{4}{3}\rho_{\rm r})}\tau_{D_{\rm t}}. \tag{11.6.2 b}$$

The net dissipation time is, from (11.5.12),

$$\tau_{\rm dis} = \frac{\tau_{D_{\rm t}}\tau_{\nu}}{\tau_{D_{\rm t}}+\tau_{\nu}}. \tag{11.6.3}$$

Before equivalence, when $\rho_r > \rho_m$, we have

$$\tau_\nu \simeq \left(\frac{\rho_m}{\rho_r}\right)^2 \tau_{D_t} < \tau_{D_t}, \tag{11.6.4}$$

from which

$$\tau_{dis} \simeq \tau_\nu \simeq \frac{\lambda^2}{4\pi^2}\frac{15}{2c^2\tau_{\gamma e}}, \tag{11.6.5}$$

while after equivalence, when $\rho_m > \rho_r$, we have

$$\tau_\nu \simeq \frac{\rho_m}{\rho_r}\tau_{D_t} > \tau_{D_t}, \tag{11.6.6}$$

from which

$$\tau_{dis} \simeq \tau_{D_t} \simeq \frac{\lambda^2}{4\pi^2}\frac{6}{c^2\tau_{\gamma e}}. \tag{11.6.7}$$

Thus, before equivalence, the dissipation can be attributed mainly to the effect of radiative viscosity and, after equivalence, it is mainly due to thermal conduction. In any case the quantity τ_{dis} does not change by much between these two epochs: Equations (11.6.5) and (11.6.7) show that, in the final analysis, the dissipation of acoustic waves in the plasma epoch is due to the diffusion of photons.

As we have explained, we must consider dissipation after a time t on scales characterised by a mass $M < M_D(t)$ or by a length $\lambda < \lambda_D(t)$. It is straightforward to verify, within the framework of the approximations introduced above, that the condition $\lambda < \lambda_D(t)$ is identical to the condition that $\tau_{dis}(\lambda) < t$. It therefore emerges that

$$\tau_{dis}(\lambda) = \left(\frac{\lambda}{\lambda_D}\right)^2 t = \left(\frac{M}{M_D}\right)^{2/3} t. \tag{11.6.8}$$

For adiabatic perturbations of mass $M < M_J^{(a)}(z_{eq})$, the time t is the interval of time Δt in which such perturbations evolve like acoustic waves: given that $M_{Hb} \simeq M_J^{(a)}$ before equivalence, this interval is approximated by $\Delta t(M) \simeq t - t(z_H(M)) \simeq t$, where now t stands for cosmological proper time; $t(z_H(M))$ is negligible with respect to t for the range of masses we are interested in.

Before equivalence the dissipation scale for adiabatic perturbations is, from Equations (11.6.8) and (11.6.5),

$$\lambda_D^{(a)} \simeq 2.3c(\tau_{\gamma e}t)^{1/2}, \tag{11.6.9}$$

where t is given by Equation (5.6.7) and $\tau_{\gamma e}$ is given by Equation (9.2.9). The corresponding dissipation mass scale is then given by

$$M_D^{(a)} = \frac{1}{6}\pi\rho_m\lambda_D^{(a)^3} \simeq 0.5\left(\frac{m_p c}{\sigma_T G^{1/2}}\right)^{3/2}(\rho_{0c}^2\rho_{0r}^3\Omega^2)^{-1/4}(1+z)^{-9/2}, \tag{11.6.10 a}$$

which yields

$$M_{\rm D}^{(a)} \simeq 7\times 10^{10}(\Omega h^2)^{-5}\left(\frac{1+z}{1+z_{\rm eq}}\right)^{-9/2} M_\odot. \qquad (11.6.10\,b)$$

If $\Omega h^2 \leqslant 4\times 10^{-2}$, then $z_{\rm rec} \geqslant z_{\rm eq}$ and the mass scale for dissipation at recombination becomes $M_{\rm D}^{(a)}(z_{\rm rec}) \leqslant 10^{17} M_\odot$.

If the Universe is sufficiently dense so that $z_{\rm eq} > z_{\rm rec}$, we can obtain in a similar manner, using Equations (11.6.8), (11.6.7) and (2.2.6b) for the period $z_{\rm eq} > z > z_{\rm rec}$, the result

$$\lambda_{\rm D}^{(a)} \simeq 2.5c(\tau_{\gamma e} t)^{1/2}. \qquad (11.6.11)$$

The dissipation mass scale is then

$$M_{\rm D}^{(a)} \simeq 0.9\left(\frac{m_{\rm p}c}{\sigma_T G^{1/2}}\right)^{3/2}(\rho_{0c}\Omega)^{-5/4}(1+z)^{-15/4}$$

$$\simeq 8\times 10^{7}(\Omega h^2)^{-5}\left(\frac{1+z}{1+z_{\rm eq}}\right)^{-15/4} M_\odot. \qquad (11.6.12)$$

At recombination we have $M_{\rm D}^{(a)}(z_{\rm rec}) \simeq 10^{12}(\Omega h^2)^{-5/4} M_\odot$.

As we shall see, the value of $M_{\rm D}^{(a)}(z_{\rm rec})$ is of great significance for structure formation. Its magnitude depends on the density parameter through the quantity $\Omega_0 h^2$. Approximate numerical values for $4\times 10^{-2} \leqslant \Omega_0 h^2 \leqslant 2$ are $10^{17} M_\odot \geqslant M_{\rm D}^{(a)}(z_{\rm rec}) \geqslant 4\times 10^{11} M_\odot$. The first to calculate the value of $M_{\rm D}^{(a)}(z_{\rm rec})$ was Silk (1967) - for this reason the quantity $M_{\rm D}^{(a)}$ is also known as the *Silk mass*. It is interesting to note that

$$M_{\rm D}^{(a)} \simeq (M_\gamma M_{\rm Hb})^{1/2}, \qquad (11.6.13)$$

where M_γ is the mass contained within a sphere of diameter $l_\gamma = c\tau_{\gamma e}$. The reason for the relation (11.6.13) is implicit in Equation (11.5.9).

In the case where there is a significant amount of non-baryonic matter so that $\Omega \neq \Omega_{\rm b}$, which is the case we shall discuss in the next chapter, Silk damping of course still occurs, but the damping mass scale changes. It is a straightforward exercise to show that, in this case, the corresponding value at $z_{\rm rec}$ can be obtained from the above case if $z_{\rm rec} > z_{\rm eq}$ by changing Ω to $\Omega_{\rm b}$ and, if $z_{\rm rec} < z_{\rm eq}$, by changing Ω to $(\Omega_{\rm b}\Omega^9)^{1/10}$.

The importance of the Silk mass can be explained as follows. Without taking account of dissipative processes, the amplitude of an acoustic wave on a mass scale $M < M_{\rm J}^{(a)}$ would remain constant in time during radiation domination and would decay according to a $t^{-1/6}$ law in the period between equivalence and recombination. The dissipative processes we have considered cause a decrease of the amplitude of such waves, with a rate of attenuation that depends upon M. In fact the energy of the wave $E \propto A^2$ is damped exponentially. The time for a wave to damp away completely is therefore much less than the timescale for the next scale to enter the horizon. The upshot of this is that fluctuations on all scales less than the Silk mass are completely obliterated by photon diffusion almost immediately. No structure will therefore be formed on a mass scale less than this.

11.7 Radiation Drag

We now turn our attention to physical processes which are important for isothermal rather than adiabatic fluctuations. We have already mentioned that isothermal perturbations on a scale $M > M_J^{(i)}$ are frozen-in because of a kind of viscous friction force acting on particles trying to move through a smooth radiation background. This force is essentially due to *radiation drag*. We can show schematically that this freezing-in effect is relevant if the viscous forces on the perturbation F_v per unit mass dominate the self-gravitational force F_g per unit mass. This condition is that

$$\frac{F_v}{m} \simeq \frac{v}{\tau_{e\gamma}} \simeq \frac{\lambda}{t\tau_{e\gamma}} > \frac{F_g}{m} \simeq G\rho_m\lambda \simeq \frac{\lambda}{t^2}, \tag{11.7.1}$$

where we have used the fact that $\rho_m \simeq (Gt^2)^{-1}$, and we are now interested in the period defined by $z_{eq} \gg z \gg \Omega^{-1}$. The inequality (11.7.1) holds for $t > \tau_{e\gamma}$, which is true before recombination. Now let us treat this phenomenon in a more precise way. If a perturbation in the ionised component (plasma) moves with a velocity $v \ll c$ relative to an unperturbed radiation background, any electron encounters a force opposing its motion that has magnitude

$$f_v \simeq \tfrac{4}{3}\sigma_T\rho_r c^2\frac{v}{c} = \tfrac{4}{3}\sigma_T\sigma T^4\frac{v}{c}. \tag{11.7.2}$$

This applies also to electron–proton pairs because for $z > z_{rec}$ the protons are always strictly coupled to the motion of the electrons. In fact, because of the Doppler effect, an electron moving through the radiation background experiences a radiation temperature which varies with the angle ϑ between its velocity and the line of sight:

$$T(\vartheta) = T\left[1 - \left(\frac{v}{c}\right)^2\right]^{1/2}\left(1 - \frac{v}{c}\cos\vartheta\right)^{-1} \simeq T\left(1 + \frac{v}{c}\cos\vartheta\right), \tag{11.7.3}$$

which corresponds to an energy flux in the solid angle $d\Omega$ of

$$d\Phi = i(\vartheta)\, d\Omega = \frac{1}{4\pi}\rho_r(\vartheta)c^3\, d\Omega = \frac{1}{4\pi}\sigma T^4(\vartheta)c\, d\Omega, \tag{11.7.4}$$

and a momentum flux in the direction of the velocity of

$$dP_\vartheta = \frac{1}{c}\cos\vartheta\, d\Phi \simeq \frac{1}{4\pi}\sigma T^4\left(1 + 4\frac{v}{c}\cos\vartheta\right)\cos\vartheta\, d\Omega. \tag{11.7.5}$$

The momentum acquired by an electron per unit time, which is caused by the anisotropic radiation field experienced by it, is therefore

$$f_v = \sigma_T\int_\Omega dP_\vartheta = -\tfrac{4}{3}\sigma_T\sigma T^4\frac{v}{c} = -\frac{m_p v}{\tau_{e\gamma}}; \tag{11.7.6}$$

since the Thomson cross-section of a proton is a factor $(m_p/m_e)^2$ smaller than that of an electron, the force suffered by the protons is negligible. Equation (11.7.6) is a definition, in fact, of the characteristic time $\tau_{e\gamma}$ for the transfer of momentum between proton–electron pairs and photons which we have encountered already in Section 9.2.

Taking account of this frictional force f_v, the equation which governs the gravitational instability of isothermal perturbations, derived according to the methods laid out in Section 11.2, yields

$$\ddot{\delta}_m + \left(2\frac{\dot{a}}{a} + \frac{1}{\tau_{e\gamma}}\right)\dot{\delta}_m + ({v_s^{(i)}}^2 k^2 - 4\pi G\rho_m)\delta_m = 0. \tag{11.7.7}$$

For $M > M_J^{(i)}$ and $z_{eq} > z \geqslant z_{rec}$, Equation (11.7.7) becomes

$$\ddot{\delta}_m + \left(\frac{4}{3t} + \frac{A}{t^{8/3}}\right)\dot{\delta}_m - \frac{2}{3}\frac{\delta_m}{t^2} \simeq 0, \tag{11.7.8}$$

where the constant A is given by

$$A \simeq \frac{4}{3}\frac{\sigma_T \rho_{0r} c}{m_p} t_{0c}^{8/3} (\Omega h^2)^{-4/3}; \tag{11.7.9}$$

the second term in parentheses in (11.7.8) dominates the first if $\tau_{e\gamma} < t$, i.e. before decoupling. In this period, an approximate solution to (11.7.9) is

$$\delta_m \propto \exp\frac{2t^{5/3}}{5A} \simeq \exp[10^5(\Omega h^2)^{1/2}(1+z)^{-5/2}] \simeq \text{const.}: \tag{11.7.10}$$

the perturbation remains practically constant before recombination.

As a final remark in this section, we should make it clear that this freezing-in of perturbations due to radiation drag is not the same as the Meszaros effect discussed in Section 10.11, which is a purely kinematic effect and does not require any collisional interaction between matter and radiation.

11.8 A Two-Fluid Model

In the previous sections of this chapter we have treated the primordial plasma as a single, imperfect fluid of matter and radiation. This model is good enough for $\tau_{\gamma e} \ll \tau_H \simeq t$ and for $\lambda \gg c\tau_{\gamma e} = l_\gamma$; all this is true at times well before recombination and decoupling. A better treatment can be adopted for the period running up to recombination by considering the matter and radiation components as two fluids interacting with each other on characteristic timescales $\tau_{e\gamma}$ and $\tau_{\gamma e}$. We shall see, however, that even this method has its limitations, which we discuss at the end of this section.

Let us indicate the temporal parts of the perturbations to the density and velocity of the matter and radiation components by δ_m, δ_r, V_m and V_r, respectively; the spatial dependence of the perturbations is assumed to be of the form $\exp(i\boldsymbol{k}\cdot\boldsymbol{r})$, as previously. We thus find for longitudinal perturbations in the matter component

$$\dot{\delta}_{\rm m} + {\rm i}kV_{\rm m} = 0, \qquad (11.8.1\,a)$$

$$\dot{V}_{\rm m} + \frac{\dot{a}}{a}V_{\rm m} + \frac{V_{\rm m} - V_{\rm r}}{\tau_{\rm e\gamma}} + {\rm i}kv_{\rm sm}^2\delta_{\rm m} - \frac{\rm i}{k}4\pi G(\rho_{\rm m}\delta_{\rm m} + 2\rho_{\rm r}\delta_{\rm r}) = 0, \qquad (11.8.1\,b)$$

where the terms involving $\tau_{\rm e\gamma}$ take account of the interaction between matter and radiation, and $v_{\rm sm}$ coincides with $v_{\rm s}^{(\rm i)}$. For the radiation component we find, using results from the previous chapter,

$$\dot{\delta}_{\rm r} + \tfrac{4}{3}{\rm i}kV_{\rm r} = 0, \qquad (11.8.2\,a)$$

$$\dot{V}_{\rm r} + \frac{\dot{a}}{a}V_{\rm r} + \frac{V_{\rm r} - V_{\rm m}}{\tau_{\rm e\gamma}} + {\rm i}k\tfrac{3}{4}v_{\rm sr}^2\delta_{\rm r} - \frac{\rm i}{k}4\pi G(\rho_{\rm m}\delta_{\rm m} + 2\rho_{\rm r}\delta_{\rm r}) = 0, \qquad (11.8.2\,b)$$

where the term including $\tau_{\gamma\rm e}$ takes into account the interaction between matter and radiation (the factor $\frac{4}{3}$ is due to pressure), and $v_{\rm sr} = c/\sqrt{3}$. From Equations (11.8.1) and (11.8.2) we obtain, respectively,

$$\ddot{\delta}_{\rm m} + \left(\frac{2\dot{a}}{a} + \frac{1}{\tau_{\rm e\gamma}}\right)\dot{\delta}_{\rm m} - \frac{3\dot{\delta}_{\rm r}}{4\tau_{\rm e\gamma}} + \left[v_{\rm sm}^2k^2 - 4\pi G\rho_{\rm m}\left(1 + \frac{2\delta\rho_{\rm r}}{\delta\rho_{\rm m}}\right)\right]\delta_{\rm m} = 0, \quad (11.8.3\,a)$$

$$\ddot{\delta}_{\rm r} + \left(\frac{2\dot{a}}{a} + \frac{1}{\tau_{\gamma\rm e}}\right)\dot{\delta}_{\rm r} - \frac{4\dot{\delta}_{\rm m}}{3\tau_{\gamma\rm e}} + \left[v_{\rm sr}^2k^2 - \tfrac{32}{3}\pi G\rho_{\rm r}\left(1 + \frac{2\delta\rho_{\rm m}}{\delta\rho_{\rm r}}\right)\right]\delta_{\rm r} = 0. \quad (11.8.3\,b)$$

One can solve the system (11.8.3) by putting

$$\delta_{\rm m} \propto \delta_{\rm r} \propto \exp({\rm i}\omega t), \qquad (11.8.4)$$

where the frequency ω is in general complex and time dependent. One makes the hypothesis at the outset that $\tau_\omega \equiv \omega/\dot{\omega} > t \simeq \tau_{\rm H} = a/\dot{a}$, so that $\dot{\delta}_{m(r)} \simeq \omega\delta_{m(r)}$. Afterwards one must discard the solutions with $\tau_\omega \leqslant \tau_{\rm H}$: one finds that, on the scales of interest (i.e. $M \geqslant M_{\rm D}^{(\rm a)}$), this happens soon after recombination. Putting the result (11.8.4) in (11.8.3) in light of this hypothesis yields a somewhat cumbersome *dispersion relation* in the form

$$\omega^4 + c_3\omega^3 + c_2\omega^2 + c_1\omega + c = 0, \qquad (11.8.5)$$

in which

$$c_3 = {\rm i}\left(4\frac{\dot{a}}{a} + \frac{1}{\tau_{\rm e\gamma}} + \frac{1}{\tau_{\gamma\rm e}}\right), \qquad (11.8.6\,a)$$

$$c_2 = -\left[v_{\rm sr}^2(k^2 - k_{\rm Jr}^2) + v_{\rm sm}^2(k^2 - k_{\rm Jm}^2) + 2\frac{\dot{a}}{a}\left(2\frac{\dot{a}}{a} + \frac{1}{\tau_{\rm e\gamma}} + \frac{1}{\tau_{\gamma\rm e}}\right)\right]\omega^2, \qquad (11.8.6\,b)$$

$$c_1 = -{\rm i}\left[v_{\rm sr}^2(k^2 - k_{\rm Jr}^2)\left(2\frac{\dot{a}}{a} + \frac{1}{\tau_{\rm e\gamma}}\right)\right.$$
$$\left. + v_{\rm sm}^2(k^2 - k_{\rm Jm}^2)\left(2\frac{\dot{a}}{a} + \frac{1}{\tau_{\gamma\rm e}}\right) + \left(\frac{v_{\rm sr}^2k_{\rm Jr}^2}{\tau_{\gamma\rm e}} + \frac{v_{\rm sm}^2k_{Jm}^2}{\tau_{\rm e\gamma}}\right)\right] \qquad (11.8.6\,c)$$

and

$$c_0 = (v_{sr} v_{sm} k)^2 (k^2 - k_{Jr}^2 - k_{Jm}^2), \tag{11.8.6 d}$$

where k_{Jm} and k_{Jr} are the wavenumbers appropriate to the wavelengths given by equations (10.6.15) and (10.9.7). The dispersion relation is of fourth order in ω. For a given k there exist four solutions $\omega_i(k)$, with $i = 1, 2, 3, 4$, and there are also four perturbation modes. Next one puts an expression of the form (11.8.4) in the equations for V_m and V_r, (11.8.1 *b*) and (11.8.2 *b*), with the same restriction on τ_ω. Then substituting in these four equations the solutions $\omega_i(k)$ one obtains the four perturbation modes:

$$\delta_{m(r),i} = D_{m(r)}[k, \omega_i(k)] \exp i[\boldsymbol{k} \cdot \boldsymbol{r} + \omega_i(k)t], \tag{11.8.7 a}$$

$$v_{m(r),i} = V_{m(r)}[k, \omega_i(k)] \exp i[\boldsymbol{k} \cdot \boldsymbol{r} - \omega_i(k)t]. \tag{11.8.7 b}$$

The analytical study of the acoustic modes described by the Equations (11.8.7) is very complicated, except in special cases where one can simplify the dispersion relation to transform it into a cubic equation or, most usefully, a quadratic equation. In general the ith root of (11.8.5) is complex:

$$\omega_i(k) = \mathrm{Re}\,\omega_i(k) + \mathrm{i}\,\mathrm{Im}\,\omega_i(k). \tag{11.8.8}$$

One has wavelike propagation when $\mathrm{Re}\,\omega_i(k) \neq 0$; in this case one can easily see that $\omega_j(k) = -\omega_i^*(k)$ is also a solution: these two solutions represent waves propagating in opposite sense to each other, with phase velocity $v_s(k) = |\mathrm{Re}\,\omega_i(k)|/k$ and amplitude which decreases with time when $\mathrm{Im}\,\omega_i(k) < 0$; the characteristic time for the wave to decay is given by $\tau_i = |\mathrm{Im}\,\omega_i(k)|^{-1}$.

One has gravitational instability when $\mathrm{Re}\,\omega_i(k) = 0$. This instability can be of either increasing or decreasing type according to whether $\mathrm{Im}\,\omega_i(k)$ is greater than or less than zero, and the characteristic time for the evolution of the instability is given by $\tau_i = |\mathrm{Im}\,\omega_i(k)|^{-1}$.

In general, before decoupling there are two modes of approximately adiabatic nature, in the sense that $\delta_r/\delta_m \simeq \frac{4}{3}$. These modes are unstable for $M > M_J^{(a)}$, so that one increases and the other decreases; for $M < M_J^{(a)}$ they evolve like damped acoustic waves with the sound speed $v_s \simeq v_s^{(a)}$. A third mode, again of approximately adiabatic type, also exists but is non-propagating and always damped. The fourth and final mode is approximately isothermal (in the sense that $|\delta_r| \ll |\delta_m|$), so that for $M > M_J^{(i)}$ it is an unstable growing mode, but with a characteristic growth time $\tau > \tau_H$, so it is effectively frozen-in. During decoupling, the last two of these modes gradually transform themselves into two isothermal modes which oscillate like waves for $M < M_J^{(i)}$ with a sound speed $v_s \simeq v_s^{(i)}$, and are unstable (one growing and the other decaying) for $M > M_J^{(i)}$. The first two modes become purely radiative, i.e. $\delta_m \simeq 0$, which are unstable for wavelengths greater than the appropriate Jeans length for radiation $\lambda_J^{(r)}$ and which oscillate like waves propagating at a speed $c/\sqrt{3}$ practically without damping for $\lambda < \lambda_J^{(r)}$. These last two modes are actually spurious, since in reality the radiation after decoupling behaves like a collisionless fluid which cannot be described by an equation of the form (11.8.2). A more exact treatment of the radiation shows that, for $\lambda > \lambda_J^{(r)}$ and after decoupling, there is a rapid damping of these purely radiative perturbations due to the free streaming of photons whose mean free path is $l_\gamma \gg \lambda$.

The analysis of the two-fluid model yields qualitatively similar results to those already noted for $z < z_{rec}$. One novel outcome of this treatment is that, in general, the four modes correspond neither to purely adiabatic nor purely isothermal modes. A generic perturbation must be thought of as a combination of four perturbations, each one in the form of one of these four fundamental modes. Given that each mode evolves differently, the nature of the perturbation must change with time; one can, for example, begin with a perturbation of pure adiabatic type which, in the course of its evolution, assumes a character closer to a mode of isothermal type, and vice versa. One can attribute this phenomenon to the continuous exchange of energy between the various modes.

The two-fluid model furnishes an estimate of $M_D(z_{rec})$ in a different way to that we obtained previously. Let us define $M_D(z_{rec})$ to be the mass scale corresponding to a wavenumber k such that, for the approximately adiabatic modes with $M < M_J^{(a)}$, we have $|\,\mathrm{Im}\,\omega(k)|t_{rec} \simeq 1$. In this way, one finds a value of $M_D(z_{rec})$ which is a little larger than that we found previously.

Now we turn to the limitations of the two-fluid approach to the matter–radiation plasma. There are three main problems. First, the Equations (11.8.1) and (11.8.2) do not take into account all necessary relativistic corrections. One cannot trust the results obtained with these equations on scales comparable with, or greater than, the scale of the cosmological horizon. Secondly, the description of the radiation as a fluid is satisfactory on length scales $\lambda \gg c\tau_{\gamma e}$ and for epochs during which $\tau_{\gamma e}(\tau_{e\gamma}) \ll \tau_H$. On the scales of interest, $M \simeq M_J^{(a)}(z_{rec})$, these conditions are true only for $z \gg z_{rec}$. For later times, or for smaller scales, it is necessary to adopt an approach which is completely kinetic; we shall describe this kind of approach in Section 12.10. The last major problem we should mention, and which we have mentioned before, is that the approximations used to derive the dispersion relation (11.8.5) from the system of Equations (11.8.3) are only acceptable for $z > z_{rec}$.

The numerical solution of the system of fully relativistic equations describing the matter and radiation perturbations (in a kinetic approach), and the perturbations in the spatial geometry (i.e. metric perturbations) is more complex still. Such computations enable one to calculate with great accuracy, given for generic initial conditions at the entry of a baryonic mass scale in the cosmological horizon, the detailed behaviour of $\delta_m(M)$, as well as the perturbations to the radiation component and hence the associated fluctuations in the cosmic microwave background on scales of interest. We shall comment upon this latter topic in the next section.

11.9 The Kinetic Approach

As we have already mentioned, the exact relativistic treatment of the evolution of cosmological perturbations is very complicated. One must keep track not only of perturbations to both the matter and radiation but also of fluctuations in the metric. The Robertson–Walker metric describing the unperturbed background must be replaced by a metric whose components g'_{ik} differ by infinitesimal quantities from the original g_{ik}: the deviations δg_{ik} are connected with the perturbations to the matter and radiation by the Einstein equations. There is also the problem referred to in Section 10.12 concerning the choice of *gauge*. This is a subtle problem which we shall not describe in detail at the moment, although we will return to it briefly in Chapter 17 where we discuss the cosmic microwave background. The simplest approach is to adopt a *synchronous gauge* characterised by the metric

$$ds^2 = (c\,dt)^2 - a^2[\gamma_{\alpha\beta} - h_{\alpha\beta}(\boldsymbol{x}, t)]\,dx^\alpha\,dx^\beta, \tag{11.9.1}$$

where $|h_{\alpha\beta}| \ll 1$. The treatment is considerably simplified if the unperturbed metric is flat so that $\gamma_{\alpha\beta} = \delta_{\alpha\beta}$, where $\delta_{\alpha\beta}$ is the Kronecker symbol: $\delta_{\alpha\beta} = 1$ for $\alpha = \beta$, $\delta_{\alpha\beta} = 0$ for $\alpha \neq \beta$. This is also the case in an approximate sense if the Universe is not flat, but one is looking at scales much less than the curvature radius or at very early times.

The time evolution of the trace h of the tensor $h_{\alpha\beta}$ is related to the evolution of matter and radiation perturbations

$$\ddot{h} + 2\frac{\dot{a}}{a}\dot{h} = 8\pi G(\rho_{\rm m}\delta_{\rm m} + 2\rho_{\rm r}\delta_{\rm r}). \tag{11.9.2}$$

The equations that describe the evolution of the time-dependent parts $\delta_{\rm m}$ and $V_{\rm m}$ of the perturbations in the density and velocity of the matter are

$$\dot{\delta}_{\rm m} + \mathrm{i}kV_{\rm m} = \tfrac{1}{2}\dot{h}, \tag{11.9.3 a}$$

$$\dot{V}_{\rm m} + \frac{\dot{a}}{a}V_{\rm m} + \frac{V_{\rm m} - V_{\rm r}}{\tau_{\rm e\gamma}} = 0; \tag{11.9.3 b}$$

the perturbation in the velocity of the radiation $V_{\rm r}$ will be defined a little later.

As far as the radiation perturbations are concerned, one can demonstrate that their evolution is described by a single equation involving the *brightness function* $\delta^{(\rm r)}(\boldsymbol{x}, t)$, whose Fourier transform can be written

$$\delta_{\rm r}(k, t) = \frac{1}{4\pi}\int \delta_k^{(\rm r)}(\vartheta, \varphi, t)\,d\Omega : \tag{11.9.4}$$

the quantity $\delta_k^{(\rm r)}$ at any point involves contributions from photons with momenta directions specified by the spherical polar angles ϑ and φ. The differential equation which describes the evolution of $\delta_k^{(\rm r)}$, which was first derived from the Liouville equation by Peebles and Yu (1970), is

$$\dot{\delta}_k^{(\rm r)} + \mathrm{i}kc\cos\vartheta\,\delta_k^{(\rm r)} + \frac{1}{\tau_{\gamma\rm e}}\left(\delta_{\rm r} + 4\frac{V_{\rm m}}{c}\cos\vartheta - \delta_k^{(\rm r)}\right) = 2\cos^2\vartheta\,\dot{h}, \tag{11.9.5}$$

where ϑ is the angle between the photon momentum and the wave vector $\boldsymbol{k}$, which we assume to define the polar axis of a local coordinate system. Given the rotational symmetry, one can expand $\delta_k^{(\mathrm{r})}$ in angular moments σ_l defined with respect to the Legendre polynomials

$$\delta_k^{(\mathrm{r})} = \sum_l (2l+1) P_l(\cos\vartheta)\sigma_l(k,t). \tag{11.9.6}$$

The perturbation δ_r coincides with the moment σ_0, while the velocity perturbation V_r which appears in (11.9.3 *b*) is given by $\frac{1}{4}\sigma_1$.

It is comparatively straightforward to show that the evolution of the brightness function is governed by a hierarchy of equations for the moments σ_l:

$$\dot{\sigma}_0 + \mathrm{i}k\sigma_1 = \tfrac{2}{3}\dot{h} \qquad (l = 0), \tag{11.9.7 a}$$

$$\dot{\sigma}_1 + \mathrm{i}k(\tfrac{2}{3}\sigma_2 + \tfrac{1}{3}\sigma_0) = \frac{4}{3}\frac{V_\mathrm{m} - V_\mathrm{r}}{\tau_{\gamma\mathrm{e}}} \qquad (l = 1), \tag{11.9.7 b}$$

$$\dot{\sigma}_2 + \mathrm{i}k(\tfrac{3}{5}\sigma_3 + \tfrac{2}{5}\sigma_1) = \tfrac{4}{15}\dot{h} - \frac{3\sigma_2}{4\tau_{\gamma\mathrm{e}}} \qquad (l = 2), \tag{11.9.7 c}$$

$$\dot{\sigma}_l + \mathrm{i}k\left(\frac{l+1}{2l+1}\sigma_{l+1} + \frac{l}{2l+1}\sigma_{l-1}\right) = -\frac{\sigma_l}{\tau_{\gamma\mathrm{e}}} \qquad (l \geqslant 3). \tag{11.9.7 d}$$

One can verify that the two-fluid approximation practically coincides with the system of Equations (11.9.2)–(11.9.3 *b*) and (11.9.7) if one puts $\sigma_3 = 0$ and neglects $\dot{\sigma}_2$ in Equation (11.9.7 *c*), thus truncating the hierarchy. This approximation is good in the epoch during which $\tau_{\gamma\mathrm{e}} \ll \tau_\mathrm{H}$, which is in practice any time prior to recombination, and on large scales, such that $\lambda \gg c\tau_{\gamma\mathrm{e}}$. In the general situation, both during and after recombination, the system can be solved only by truncating the hierarchy at some suitably high value of l; the number of l-modes one has to take grows steadily as decoupling and recombination proceed. A couple of examples of a full numerical solution of the evolution of perturbations in the matter δ_m and radiation δ_r components in an adiabatic scenario are shown in Figures 11.1 and 11.2. The mass scale in both these calculations is of order $10^{15} M_\odot$. Notice how the matter and radiation perturbations oscillate together in both calculations until recombination, whereafter the radiation perturbation stays roughly constant and the matter perturbation becomes unstable and grows until the present epoch. Figure 11.1 shows a model with $\Omega = 1$ so that the growth after recombination is a pure power law, while Figure 11.2 has $\Omega = 0.1$, so that the effect of the growth factor (Section 11.4) in flattening out the behaviour of the perturbations is clear. In the opposite limit to that of the validity of the two-fluid approach, one has $\tau_{\gamma\mathrm{e}} \gg \tau_\mathrm{H}$, which is much later than recombination or for small scales such that $\lambda \ll c\tau_{\gamma\mathrm{e}}$. In such a case we have

$$\dot{\delta}_k^{(\mathrm{r})} + \mathrm{i}kc\cos\vartheta\,\delta_k^{(\mathrm{r})} = 2\cos^2\vartheta\,\dot{h}, \tag{11.9.8}$$

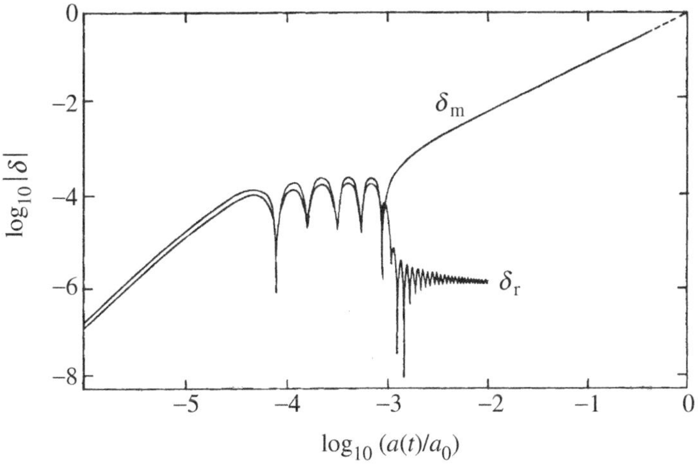

Figure 11.1 Evolution of perturbations, corresponding to a mass scale $10^{15} M_\odot$, in the baryons δ_m and photons δ_r in a Universe with $\Omega = 1$.

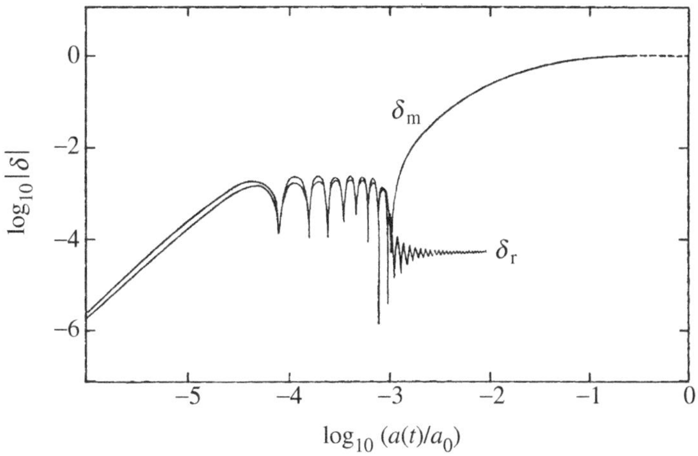

Figure 11.2 Evolution of perturbations, corresponding to a mass scale $10^{15} M_\odot$, in the baryons δ_m and photons δ_r in a Universe with $\Omega = 0.1$.

which is called the *equation of free streaming*. With appropriate approximations, the Equation (11.9.8) can be solved directly.

The value of the brightness function $\delta^{(r)}$ at time t_0 is connected with the fluctuations observed today in the temperature of the cosmic microwave background, but in the latest models of structure formation this method of calculating it is not adequate. In any case our aim in this chapter was to explain the basic physics behind baryonic fluctuations, without trying to create a model we can compare in detail with observations. We shall explain the more complete theory in Chapter 17, together with the observational developments.

11.10 Summary

We have chosen to investigate the behaviour of density perturbations in a baryon-radiation universe in some detail mainly for pedagogical reasons, that is to illustrate the important physics and display the required machinery. In fact, it is not thought possible that structure in the Universe grew in such a scenario. We shall explain why this is so and make some comments about the development of baryon-only models during the 1970s in Chapter 15.

We end by summarising the most important consequences for structure formation of the physics we have discussed in this chapter. First is the effect of the evolution of the characteristic mass scales $M_J^{(a)}$, $M_J^{(i)}$ and $M_D^{(a)}$. The behaviour of an *adiabatic* perturbation depends upon its characteristic mass scale. For perturbations on scales $M > M_J^{(a)}(z_{eq}) \simeq 4 \times 10^{15}(\Omega h^2)^{-2} M_\odot$, i.e. 10^{15}–$10^{18} M_\odot$ for acceptable values of the parameter Ωh^2, we have a wavelength greater than the Jeans length either before decoupling or after, when the Jeans mass drops to $M_J \simeq 10^5 (\Omega h^2)^{-1/2} M_\odot$. Such scales therefore experience uninterrupted growth (we shall neglect the decaying modes in this study). The growth law is

$$\delta_m \simeq \tfrac{3}{4}\delta_r \propto t \propto (1+z)^{-2} \tag{11.10.1}$$

before equivalence, and

$$\delta_m \simeq \tfrac{3}{4}\delta_r \propto t^{2/3} \propto (1+z)^{-1} \tag{11.10.2}$$

in the period, if it exists, between equivalence and recombination. After decoupling, the radiation must be treated like a 'gas' of collisionless particles and the evolution of its perturbations must be handled in a more sophisticated manner than the classical gravitational instability treatment. We described this approach briefly in Section 11.10. As far as δ_m is concerned, the growth law is still given by Equation (11.10.2) for $\Omega = 1$ and also for $\Omega z \gg 1$ if $\Omega < 1$. More precise formulae are given in Section 11.4.

In the case of perturbations with mass in the interval

$$M_J^{(a)}(z_{eq}) > M > M_D^{(a)}(z_{rec}) \simeq 10^{12}\text{–}10^{14} M_\odot, \tag{11.10.3}$$

for acceptable values of Ωh^2, we have the following evolutionary sequence. In the period before their entry into the cosmological horizon defined by $z_H(M)$, the perturbations evolve according to Equation (11.10.1); in the period between $z_H(M)$ and z_{rec} they oscillate like acoustic waves with a sound speed $v_s^{(a)}$ and with constant amplitude for $z > z_{eq}$ and amplitude decreasing as $t^{-1/6}$ between equivalence and recombination; after decoupling they become unstable again and evolve like masses with $M > M_J^{(a)}$. Perturbations with masses $M < M_D^{(a)}(z_{rec})$ evolve as before until the time $t_D(M)$ at which $M = M_D^{(a)}$. After $t_D(M)$ these fluctuations become rapidly dissipated. The bottom line is that only the perturbations with $M > M_D(z_{rec})$ can survive from the plasma epoch until the period after recombination. It is interesting to note that this characteristic scale is similar to that of a rich cluster of galaxies.

As we have seen, *isothermal* perturbations with

$$M > M_J^{(i)}(z_{rec}) \simeq 5 \times 10^4 (\Omega h^2)^{-1/2} M_\odot \tag{11.10.4}$$

are frozen-in until the epoch defined by $z_i = \min(z_{eq}, z_{rec})$. After this time, they are unstable and can grow according to the same law that applies to adiabatic perturbations at late times. We shall not worry about the evolution of perturbations on scales less than $M_J^{(i)}(z_{rec})$, because these have no real cosmological relevance. It is interesting to note that $M_J^{(i)}(z_{rec})$ is of the same order as the mass of a globular cluster.

Bibliographical Notes on Chapter 11

Historically important papers relevant to this chapter are Peebles and Yu (1970), Wilson and Silk (1981) and Wilson (1983). An alternative formulation of the kinetic approach is given by Efstathiou (1990).

Problems

1. What is the energy stored in a primordial acoustic wave? When these waves are dissipated by Silk damping, where does this energy go?
2. Derive the dispersion relation (11.8.5).
3. Derive the Equations (11.9.7) using the definitions given in Section 11.9.

12

Non-baryonic Matter

12.1 Introduction

We shall now extend the analyses of the previous two chapters to study the evolution of perturbations in models of the Universe dominated by dark matter which is not in the form of baryons. As we saw in Section 4.4, dynamical considerations suggest that the value of Ω at the present epoch is around $\Omega_{\rm dyn} \simeq 0.2$ and may well be higher. Given that modern observations of the light-element abundances require $\Omega_{\rm b}h^2 \simeq 0.02$ to be compatible with cosmological nucleosynthesis calculations, at least part of this mass must be in the form of non-baryonic particles (or perhaps primordial black holes which formed before nucleosynthesis and therefore did not participate in it). As we have seen, most examples of the inflationary universe predict flat spatial sections which, in the absence of a cosmological constant, implies Ω very close to unity at the present time. If this is true, then the Universe must be dominated by non-baryonic material to such an extent that the baryons constitute only a fraction of a percent of the total amount of matter.

One of the problems in these models is that we do not know enough about high-energy particle physics to know for sure which kinds of particles can make up the dark matter, nor even what mass many of the predicted particles might be expected to have. Our approach must therefore be to keep an open mind about the particle physics, but to place constraints where appropriate using astrophysical considerations.

We begin by running briefly through the physics of particle production in the early Universe, and then go on to describe the effect of different kinds of particles on the evolution of perturbations. Theories of galaxy formation based on the properties of different kinds of dark matter are then discussed in a qualitative way.

12.2 The Boltzmann Equation for Cosmic Relics

If the Universe is indeed dominated by non-baryonic matter, it is obviously important to figure out the present density of various types of candidate particle expected to be produced in the early stages of the Big Bang. In general, we shall use the suffix X to denote some generic particle species produced in the early Universe; we call such particles *cosmic relics*. We know that relics with a predicted present mass density of $\Omega_X > 1$ are excluded by observations while those with $\Omega_X < 0.1$ at the present time, though possible, would not contribute enough of the matter density to be relevant for structure formation.

We distinguish at the outset between two types of cosmic relics: *thermal* and *non-thermal*. Thermal relics are held in thermal equilibrium with the other components of the Universe until they decouple; a good example of this type of relic is the massless neutrino, although this is of course not a candidate for the gravitating dark matter. One can subdivide this class into *hot* and *cold* relics. The former are relativistic when they decouple, and the latter are non-relativistic. Non-thermal relics are not produced in thermal equilibrium with the rest of the Universe. Examples of this type would be monopoles, axions and cosmic strings. The case of non-thermal relics is much more complicated than the thermal case, and no general prescription exists for calculating their present abundance. We shall concentrate in this chapter on thermal relics, which seem to be based on better-established physics, and for which a general treatment is possible. In practice, it turns out in fact that this approach is also quite accurate for particles like the axion anyway.

The time evolution of the number density n_X of some type of particle species X is generally described by the Boltzmann equation:

$$\frac{\mathrm{d}n_X}{\mathrm{d}t} + 3\frac{\dot{a}}{a}n_X + \langle\sigma_{\mathrm{A}}v\rangle n_X^2 - \psi = 0, \tag{12.2.1}$$

where the term in $\dot{a}/a$ takes account of the expansion of the Universe, $\langle\sigma_{\mathrm{A}}v\rangle n_X^2$ is the rate of collisional annihilation (σ_{A} is the cross-section for annihilation reactions, and v is the mean particle velocity); ψ denotes the rate of creation of particle pairs. If the creation and annihilation processes are negligible, one has the expected solution: $n_{X\mathrm{eq}} \propto a^{-3}$. This solution also holds if the creation and annihilation terms are non-zero, but equal to each other, i.e. if the system is in equilibrium: $\psi = n_{X\mathrm{eq}}^2\langle\sigma_{\mathrm{A}}v\rangle$. Thus, Equation (12.2.1) can be written in the form

$$\frac{\mathrm{d}n_X}{\mathrm{d}t} + 3\frac{\dot{a}}{a}n_X + \langle\sigma_{\mathrm{A}}v\rangle(n_X^2 - n_{X\mathrm{eq}}^2) = 0 \tag{12.2.2}$$

or, introducing the comoving density

$$n_{\mathrm{c}} = n\left(\frac{a}{a_0}\right)^3, \tag{12.2.3}$$

in the form

$$\frac{a}{n_{\mathrm{c,eq}}}\frac{\mathrm{d}n_{\mathrm{c}}}{\mathrm{d}a} = -\frac{\langle\sigma_{\mathrm{A}}v\rangle n_{\mathrm{eq}}}{\dot{a}/a}\left[\left(\frac{n_{\mathrm{c}}}{n_{\mathrm{c,eq}}}\right)^2 - 1\right] = -\frac{\tau_{\mathrm{H}}}{\tau_{\mathrm{coll}}}\left[\left(\frac{n_{\mathrm{c}}}{n_{\mathrm{c,eq}}}\right)^2 - 1\right], \tag{12.2.4}$$

where $\tau_{\rm coll} = 1/\langle\sigma_A v\rangle n_{\rm eq}$ is the mean time between collisions and $\tau_H = a/\dot{a}$ is the characteristic time for the expansion of the Universe; we have dropped the subscript X for clarity. Equation (12.2.4) has the approximate solution

$$n_c \simeq n_{\rm c,eq} \qquad (\tau_{\rm coll} \ll \tau_H), \tag{12.2.5 a}$$

$$n_c \simeq \text{const.} \simeq n_c(t_d) \qquad (\tau_{\rm coll} \gg \tau_H), \tag{12.2.5 b}$$

where t_d is the moment of 'freezing out' of the creation and annihilation reactions, defined by

$$\tau_{\rm coll}(t_d) \simeq \tau_H(t_d). \tag{12.2.6}$$

More exact solutions to Equation (12.2.4) behave in a qualitatively similar way to this approximation.

12.3 Hot Thermal Relics

As we have explained, hot thermal relics are those that decouple while they are still relativistic. Let us assume that the particle species X becomes non-relativistic at some time t_{nX}, such that

$$Ak_B T(t_{nX}) \simeq m_X c^2 \tag{12.3.1}$$

($A \simeq 3.1$ or 2.7 is a statistical-mechanical factor which takes these two values according to whether X is a fermionfermions or a boson). For simplicity we take $A = 3$ to get rough estimates. Hot relics are thus those for which $t_{nX} > t_{dX}$, where t_{dX} is defined by Equation (12.2.6).

Let us denote by g_X the statistical weight of the particle X and by g_X^* the effective number of degrees of freedom of the Universe at t_{dX}. Following the same kind of reasoning as in Chapter 8, based on the conservation of entropy per unit comoving volume, we have

$$g_X^* T_{0X}^3 = 2T_{0r}^3 + \tfrac{7}{8} \times 2 \times N_\nu T_{0\nu}^3 = g_0^* T_{0r}^3, \tag{12.3.2}$$

where T_{0X} is the present value of the effective temperature defined by the mean particle momentum via

$$\bar{p}_X \simeq 3\frac{k_B T_X}{c}, \tag{12.3.3}$$

T_{0r} is the present temperature of the photon background and $T_{0\nu} = (\frac{4}{11})^{1/3} T_{0r}$ takes account of the N_ν neutrino families; $g_0^* \simeq 3.9$ for $N_\nu = 3$. We thus obtain from (12.3.2)

$$T_{0X} = \left(\frac{g_0^*}{g_X^*}\right)^{1/3} T_{0r}. \tag{12.3.4}$$

This equation also applies to neutrinos if one puts

$$g_\nu^* = 2 + \tfrac{7}{8} \times 2 \times N_\nu + \tfrac{7}{8} \times 2 \times 2 \tag{12.3.5}$$

(photons, neutrinos and electrons all contribute to g_ν^*). In this case we obtain the well-known relation

$$T_{0\nu} = (\tfrac{4}{11})^{1/3} T_{0\mathrm{r}} = 0.7 T_{0\mathrm{r}}. \tag{12.3.6}$$

The present number-density of X particles is

$$n_{0X} \simeq 0.5 B g_X \left(\frac{T_{0X}}{T_{0\mathrm{r}}}\right)^3 n_{0\mathrm{r}} \simeq 0.5 B g_X \frac{g_0^*}{g_X^*} n_{0\mathrm{r}}, \tag{12.3.7}$$

where $B = \frac{3}{4}$ or 1 according to whether the particle X is a fermion or a boson. The density parameter corresponding to these particles is then just

$$\Omega_\mathrm{X} = \frac{m_\mathrm{X} n_{0X}}{\rho_{0\mathrm{c}}} \simeq 2 B g_X \frac{g_0^*}{g_X^*} \frac{m_\mathrm{X}}{10^2\ \mathrm{eV} h^2}. \tag{12.3.8}$$

Equations (12.3.7) and (12.3.8) are to be compared with Equations (8.5.5) and (8.5.10). For example, consider hypothetical particles with mass $m_\mathrm{X} \simeq 1$ KeV, which decouple at $T \simeq 10^2$–10^3 MeV when $g_X^* \simeq 10^2$; these have $\Omega_\mathrm{X} \simeq 1$.

Let us now apply Equation (12.3.8) to an example: the case of a single massive neutrino species with $m_\nu \simeq 1$ MeV, which decouples at a temperature of a few MeV when $g_X^* = 10.75$ (taking account of photons, electrons and three types of massless neutrinos). The condition that the cosmic density of such relics should not be much greater than the critical density requires that $m_\nu < 90$ eV: this bound was obtained by Cowsik and McClelland (1972). If, instead, all the neutrino types have mass around 10 eV, then their density will be given by the equation already presented in Section 8.5:

$$\Omega_\nu h^2 \simeq 0.1 N_\nu \frac{\langle m_\nu \rangle}{10\ \mathrm{eV}}. \tag{12.3.9}$$

Equations (12.3.1) and (12.3.4) can be used to calculate the redshift corresponding to $t_{\mathrm{n}X}$:

$$z_{\mathrm{n}X} \simeq 1.43 \times 10^5 \left(\frac{g_X^*}{g_0^*}\right)^{1/3} \frac{m_\mathrm{X}}{10^2\ \mathrm{eV}}. \tag{12.3.10}$$

The moment of equivalence, t_eq, between the relativistic components (photons, massless neutrinos) and the non-relativistic particles (X after $t_{\mathrm{n}X}$ and baryons) is given by

$$z_\mathrm{eq} = \frac{\Omega_\mathrm{X}}{K_0 \Omega_\mathrm{r}} \simeq 2.3 \times 10^4 \frac{\Omega_\mathrm{X} h^2}{K_0}, \tag{12.3.11}$$

if one assumes that $\Omega_\mathrm{X} \gg \Omega_\mathrm{b}$, and neglects the contribution of baryons to Ω. In Equation (12.3.11) we have $K_0 \simeq 1 + 0.227 N_\nu$ taking account of the massless neutrinos. It is clear that we cannot have $z_{\mathrm{n}X} < z_\mathrm{eq}$; in the case where the collisionless component dominates at $t_{\mathrm{n}X}$ one assumes $z_{\mathrm{n}X} = z_\mathrm{eq}$.

Because Ω_X is proportional to m_X by Equation (12.3.8), one can write

$$z_{\mathrm{n}X} \simeq 7 \times 10^4 \frac{1}{g_X} \left(\frac{g_X^*}{g_0^*}\right)^{4/3} \Omega_\mathrm{X} h^2 \tag{12.3.12 a}$$

and

$$z_{\rm eq} \simeq 5 \times 10^4 g_X \left(\frac{g_0^*}{g_X^*}\right) \frac{m_X}{10^2 \text{ eV}}, \tag{12.3.12 b}$$

which complement Equations (12.3.10) and (12.3.11). In particular, if the X particles are massive neutrinos, we can obtain

$$z_{\rm n\nu} \simeq 2 \times 10^4 \frac{\langle m_\nu \rangle}{10 \text{ eV}} \simeq \frac{2 \times 10^5}{N_\nu} \Omega_\nu h^2, \tag{12.3.13 a}$$

and

$$z_{\rm eq} \simeq 4 \times 10^3 N_\nu \frac{\langle m_\nu \rangle}{10 \text{ eV}} \simeq 4 \times 10^4 \Omega_\nu h^2 < z_{\rm n\nu}; \tag{12.3.13 b}$$

$\langle m_\nu \rangle$ is the average neutrino mass.

12.4 Cold Thermal Relics

Calculating the density of cold thermal relics is much more complicated than for hot relics. At the moment of their decoupling the number density of particles in this case is given by a Boltzmann distribution:

$$n(t_{\rm dX}) = g_X \frac{1}{\hbar^3} \left(\frac{m_X k_{\rm B} T_{\rm dX}}{2\pi}\right)^{3/2} \exp\left(-\frac{m_X c^2}{k_{\rm B} T_{\rm dX}}\right). \tag{12.4.1 a}$$

The present density of cold relics is therefore

$$n_{0X} = n(t_{\rm dX}) \left[\frac{a(t_{\rm dX})}{a_0}\right]^3 = n(t_{\rm dX}) \frac{g_0^*}{g_X^*} \left(\frac{T_{\rm 0r}}{T_{\rm dX}}\right)^3. \tag{12.4.1 b}$$

The problem is to find $T_{\rm dX}$, that is to say the temperature at which Equation (12.2.6) is true. The characteristic time for the expansion of the Universe at $t_{\rm dX}$ is

$$\tau_{\rm H}(t_{\rm dX}) \simeq 0.3 \frac{\hbar T_{\rm P}}{g_X^{*1/2} k_{\rm B} T_{\rm dX}^2}, \tag{12.4.2}$$

which is the same as appeared in Equation (7.1.6), while the characteristic time for collisional annihilations is given by

$$\tau_{\rm coll}(t_{\rm dX}) = \left[n(t_{\rm dX}) \sigma_0 \left(\frac{k_{\rm B} T_{\rm dX}}{m_X c^2}\right)^q\right]^{-1}, \tag{12.4.3}$$

where we have made the assumption that

$$\langle \sigma_{\rm A} v \rangle = \sigma_0 \left(\frac{k_{\rm B} T}{m_X c^2}\right)^q : \tag{12.4.4}$$

$q = 0$ or 1 for most kinds of reaction. Introducing the variable $x = m_X c^2/k_B T$, the condition $\tau_{coll}(x) = \tau_H(x)$ is true when $x = x_{dX} = m_X c^2/k_B T_{dX} \gg 1$. The value of x_{dX} must be found by an approximate solution of Equation (12.2.6), which reads

$$x_{dX}^{q-1/2} \exp x_{dX} = 0.038 \frac{g_X}{(g_X^*)^{1/2}} \frac{c}{\hbar^2} m_P m_X \sigma_0 = C, \tag{12.4.5}$$

where m_P is the Planck mass. One therefore obtains

$$x_{dX} \simeq \ln C - (q - 1/2) \ln(\ln C). \tag{12.4.6}$$

The present density of relic particles is then

$$\rho_{0X} \simeq 10 g_X^{*-1/2} \frac{(k_B T_{0r})^3}{\hbar c^4 \sigma_0 m_P} x_{dX}^{n+1}. \tag{12.4.7}$$

As an application of Equation (12.4.4), one can consider the case of a heavy neutrino of mass $m_\nu \gg 1$ MeV. If the neutrino is a Dirac particle (i.e. if the particle and its antiparticle are not equivalent), then the cross-section in the non-relativistic limit varies as v^{-1} corresponding to $q = 0$ in (12.4.4), for which $\sigma_0 = \text{const.} \simeq 0.8 g_{wk}^2 (m_\nu^2 c/\hbar^4)$ (g_{wk} is the weak interaction coupling constant). Putting $g_\nu = 2$ and $g_\nu^* \simeq 60$ one finds that $x_{d\nu} \simeq 15$, corresponding to a temperature $T_{d\nu} \simeq 70\ (m_\nu/\text{GeV})$ MeV. Placing this value of $x_{d\nu}$ in Equation (12.4.7), the condition that $\Omega_\nu h^2 < 1$ implies that $m_\nu > 1$ GeV: this limit was found by Lee and Weinberg (1977), amongst others. If, on the other hand, the neutrino is a Majorana particle (i.e. if the particle and its antiparticle are equivalent), the annihilation rate $\langle \sigma_A v \rangle$ has terms in x^{-q} with $q = 0$ and 1, thus complicating matters considerably. Nevertheless, the limit on m_ν we found above does not change. In fact we find $m_\nu > 5$ GeV. If the neutrino has mass $m_\nu \simeq 100$ GeV, the energy scale of the electroweak phase transition, the cross-section is of the form $\sigma_A \propto T^{-2}$ and all the previous calculations must be modified.

The relations (12.3.10) and (12.3.11) which supply z_{nX} and z_{eq} remain substantially unchanged, except that in the expression for z_{nX} one should replace g_X^* by g_{nX}^*, the value of g^* at t_{nX}.

12.5 The Jeans Mass

In this section we shall study the evolution of the Jeans mass M_{JX} and the free-streaming mass M_{fX} for a fluid of collisionless particles. As we have explained in Section 10.3 and Chapter 11, we need first to determine the behaviour of the mean particle velocity v_X in the various relevant cosmological epochs. These epochs are the two intervals $t < t_{nX}$ and $t > t_{nX}$ for hot relics; the three intervals $t < t_{nX}$, $t_{nX} \leqslant t \leqslant t_{dX}$ and $t > t_{dX}$ for cold relics. In the first case (hot relics) we have, roughly,

$$v_X \simeq \frac{c}{\sqrt{3}} \qquad (z \geqslant z_{nX}), \tag{12.5.1 a}$$

$$v_X \simeq \frac{c}{\sqrt{3}} \frac{1+z}{1+z_{nX}} \qquad (z \leqslant z_{nX}), \tag{12.5.1 b}$$

while for the cold relics we have instead

$$v_X = \frac{c}{\sqrt{3}} \qquad (z \geqslant z_{nX}), \tag{12.5.2 a}$$

$$v_X \simeq \frac{c}{\sqrt{3}}\left(\frac{1+z}{1+z_{nX}}\right)^{1/2} \qquad (z_{nX} \geqslant z \geqslant z_{dX}), \tag{12.5.2 b}$$

$$v_X \simeq \frac{c}{\sqrt{3}}\left(\frac{1+z_{dX}}{1+z_{nX}}\right)^{1/2}\frac{1+z}{1+z_{dX}} \qquad (z \leqslant z_{dX}). \tag{12.5.2 c}$$

One defines the *Jeans mass* for the collisionless component to be the quantity

$$M_{JX} = \tfrac{1}{6}\pi m_X n_X \lambda_{JX}^3; \tag{12.5.3}$$

the Jeans length λ_{JX} is given by Equation (10.3.11) where one replaces v_* by v_X from above:

$$\lambda_{JX} = v_X\left(\frac{\pi}{G\rho}\right)^{1/2}. \tag{12.5.4}$$

The total density ρ includes contributions from a relativistic component ρ_r (photons and massless neutrinos), the collisionless component ρ_X and the baryonic component ρ_b which, in the first approximation, can be neglected. One can put $\rho \simeq \rho_r$ for $z > z_{eq}$ and $\rho \simeq \rho_X$ for $z < z_{eq}$.

Now let us consider the case of *hot thermal relics.* Assuming that $z_{nX} > z_{eq}$ we easily obtain

$$M_{JX} \simeq \tfrac{1}{6}\pi\rho_{0c}\left(\frac{c}{\sqrt{3}}\right)^3\left(\frac{\pi}{G\rho_{0r}}\right)^{3/2}(1+z)^{-3}\Omega_X \simeq M_{JX}(z_{nX})\left(\frac{1+z}{1+z_{nX}}\right)^{-3} \tag{12.5.5}$$

for $z \geqslant z_{nX}$, where

$$M_{JX}(z_{nX}) \simeq 3.5\times 10^{15}\left(\frac{1+z_{eq}}{1+z_{nX}}\right)^3(\Omega_X h^2)^{-2}M_\odot; \tag{12.5.6 a}$$

$$M_{JX} \simeq \text{const.} \simeq M_{JX}(z_{nX}) = M_{JX,\max} \tag{12.5.6 b}$$

for $z_{nX} \geqslant z \geqslant z_{eq}$; and

$$M_{JX} \simeq M_{JX}(z_{nX})\left(\frac{1+z}{1+z_{eq}}\right)^{3/2} \tag{12.5.7}$$

for $z \leqslant z_{eq}$. The mass $M_{JX}(z_{nX})$ represents the maximum value of M_{JX}. Its value depends on the type of collisionless particle. The highest value of this mass is obtained for particles having $z_{nX} \simeq z_{eq}$, such as neutrinos with a mass around $\langle m_\nu\rangle \simeq 10$ eV. In this case we have

$$M_{J\nu,\max} \simeq 3.5\times 10^{15}(\Omega_\nu h^2)^{-2}M_\odot, \tag{12.5.8 a}$$

which corresponds to a length scale

$$\lambda_{J\nu,\max} \simeq 6(\Omega_\nu h^2)^{-1}\ \text{Mpc}, \tag{12.5.8 b}$$

so that, using equation (12.3.9), we have

$$\lambda_{J\nu,\max} \simeq \frac{60}{N_\nu}\left(\frac{\langle m_\nu\rangle}{10\ \mathrm{eV}}\right)^{-1}. \tag{12.5.8c}$$

More accurate expressions from full numerical calculations are given in Chapter 15.

Before $z_{n\nu} \simeq z_{eq}$ the Jeans mass $M_{J\nu}$ practically coincides with $M_J^{(a)}$, the Jeans mass corresponding to adiabatic perturbations in a plasma of baryons and radiation. As we have seen above, $M_J^{(a)}$ grows after z_{eq} and reaches a maximum value at z_{rec}. In cases in which $z_{nX} > z_{eq}$, the difference between $M_{JX,\max}$ and $M_J^{(a)}(z_{rec})$ is large.

Now we turn to *cold thermal relics*. One can show that

$$M_{JX} \simeq M_{JX}(z_{nX})\left(\frac{1+z}{1+z_{nX}}\right)^{-3}, \tag{12.5.9a}$$

$$M_{JX} \simeq M_{JX}(z_{nX})\left(\frac{1+z}{1+z_{nX}}\right)^{-3/2}, \tag{12.5.9b}$$

$$M_{JX} \simeq \text{const.} \simeq M_{JX}(z_{dX}) = M_{JX,\max}, \tag{12.5.9c}$$

$$M_{JX} \simeq M_{JX}(z_{dX})\left(\frac{1+z}{1+z_{eq}}\right)^{3/2} \tag{12.5.9d}$$

in the four redshift intervals $z \geqslant z_{nX}$, $z_{nX} \geqslant z \geqslant z_{dX}$, $z_{dX} \geqslant z \geqslant z_{eq}$ and $z \leqslant z_{eq}$, respectively. The maximum value of the Jeans mass for typical cold-dark-matter particles is too small to be of interest in cosmology.

As we have already explained, in a collisionless fluid perturbations on scales less than the Jeans mass do not just oscillate but can be damped by two physical processes: in the ultrarelativistic regime, when the particle velocities are all of order $v \simeq c$, the amplitude of a perturbation decays because particles move with a large 'directional' dispersion from overdense to underdense regions, and vice versa; in the non-relativistic regime there is also a considerable spread in the particle velocities which tends to smear out the perturbation. This second damping mechanism is similar to the *Landau damping* that occurs in plasma physics, and is also known as *phase mixing*. In either case, to order of magnitude, after a time t perturbations are dissipated on a scale $\lambda \simeq \lambda_{fX}$, with

$$\lambda_{fX} \simeq a(t)\int_0^t \frac{v_X}{a(t')}\,dt'. \tag{12.5.10}$$

The scale λ_{fX} is called the *free-streaming scale*. We introduce here the *free-streaming mass*:

$$M_{fX} = \tfrac{1}{6}\pi m_X n_X \lambda_{fX}^3. \tag{12.5.11}$$

Let us again turn to the case of *hot thermal relics* with $z_{nX} > z_{eq}$. In this case we find

$$M_{fX}(t) \simeq 0.6 M_{JX}. \tag{12.5.12}$$

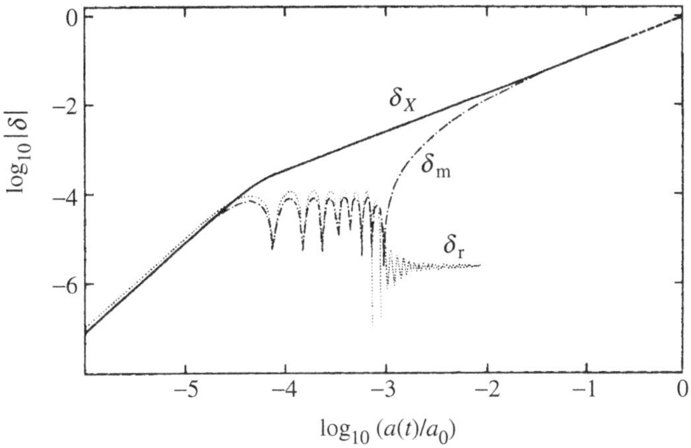

Figure 12.1 Evolution of perturbations on a scale $M \simeq 10^{15}M_\odot$ for the cold component δ_X, baryonic component δ_m and photons δ_r in a model dominated by CDM ($\Omega = 1$, $h = 0.5$). This scale enters the horizon after radiation domination, so the stagnation effect is not seen.

Soon after z_{eq} the curve of M_{fX} intersects the curve for M_{JX}, as can be seen in Figure 12.1. One can therefore assume that all perturbations in the collisionless component δ_X corresponding to masses $M < M_{JX,max}$ will be completely obliterated by free streaming. A more detailed treatment for the neutrinos, using the kinetic approach described in Chapter 11, shows that

$$\delta_\nu \simeq \delta_0 \left[1 + \left(\frac{M_{J\nu,max}}{M}\right)^{2/3}\right]^{-4}, \tag{12.5.13}$$

where δ_0 is the amplitude of perturbations when $M = M_{J\nu}$ and δ_ν is the amplitude remaining when M again becomes larger than $M_{J\nu}$. Perturbations with $M < 0.5M_{J\nu}(z_{nX})$ are in practice dissipated completely. Analogous considerations lead one to conclude that for *cold thermal relics*, the phenomenon of free streaming erodes all perturbations with masses $M < M_{JX,max}$.

Non-thermal cosmic relic particles, because they are not in equilibrium with the other components of the Universe, have a mean velocity v_X which is negligible compared even with that of cold relics. The maximum values of the Jeans mass and the free-streaming mass are therefore very low. In this case, perturbations on all the scales of interest can grow uninterrupted by damping processes. They do, however, suffer stagnation through the Meszaros effect before z_{eq}. After recombination they can give rise to fluctuations in the baryonic counterpart on scales of order $M \simeq M_J^{(i)}(z_{rec}) \simeq 10^5 M_\odot$ or larger.

12.6 Implications

Having established the relevant physics, and shown how important mass scales vary with cosmic epoch, we now briefly discuss the principal implications for models of structure formation with collisionless relic particles. Historically, there have been two important scenarios involving: *hot dark matter* (HDM) in which the collisionless dark matter takes the form of a hot thermal relic; and *cold dark matter* (CDM) in which the dark matter is either a cold thermal relic, or perhaps a non-thermal relic such as an axion.

12.6.1 Hot Dark Matter

Recall that hot dark matter corresponds to thermal relics with $z_{\mathrm{n}X} \simeq z_{\mathrm{eq}}$ and therefore with a maximum value of $M_{\mathrm{J}X}$ of order $10^{14} M_{\odot}$ or greater. A typical HDM candidate particle is a neutrino species with mass of the order of 10 eV. When a perturbation enters the cosmic horizon in a universe dominated by such particles it will have $\delta_{\mathrm{r}} \simeq \delta_{\mathrm{m}} \simeq \delta_X$. Fluctuations in the relic component δ_X with $M > M_{\mathrm{J}X}(z_{\mathrm{n}X})$ can enjoy a period of uninterrupted growth (apart from a brief interval of stagnation due to the action of the Meszaros effect ending at z_{eq}). If the primordial spectrum of perturbations has an amplitude decreasing with scale, as we shall explain in the next chapter, one will first form structure in the collisionless component on the scale $M \simeq M_{\mathrm{J}X}(z_{\mathrm{n}X})$. The first structures to form are called pancakes, as in the adiabatic baryon model. In the range of scales between $M_{\mathrm{J}X}(z_{\mathrm{n}X})$ and $M_{\mathrm{J}}^{(\mathrm{a})}(z_{\mathrm{rec}})$ the fluctuations in the matter component undergo oscillations like acoustic waves until recombination. At z_{rec}, in this range of scales, we therefore have

$$\delta_{\mathrm{r}} \simeq \delta_{\mathrm{m}} \simeq A_X(M)^{-1}\delta_X, \tag{12.6.1}$$

with $A_X(M) \geqslant 1$. The factor A_X, of order unity for $M \simeq M_{\mathrm{J}}^{(\mathrm{a})}(z_{\mathrm{rec}})$, has a maximum value

$$A_{X,\mathrm{max}} \simeq \frac{z_{\mathrm{n}X}}{z_{\mathrm{rec}}} \geqslant \frac{z_{\mathrm{eq}}}{z_{\mathrm{rec}}} \simeq 10 \tag{12.6.2}$$

for the scale $M \simeq M_{\mathrm{J}X}(z_{\mathrm{n}X})$. After recombination the perturbations in the baryonic matter component again become unstable and begin to grow like the perturbations δ_X. The latter fluctuations, being more than an order of magnitude larger than δ_{m}, dominate the self-gravity of the system so that after recombination the baryonic material follows the behaviour of the dark matter: $\delta_{\mathrm{m}} \simeq \delta_X$. This happens very quickly, as the following argument demonstrates. If there is more than one matter component, then equation (10.6.14) becomes

$$\ddot{\delta}_i + 2\frac{\dot{a}}{a}\dot{\delta}_i + v_{\mathrm{s}}^2 k^2 \delta_{\mathrm{i}} = 4\pi G \sum_j \rho_j \delta_j, \tag{12.6.3}$$

where the sum is taken over all the matter components; see also equations (11.9.3 *a*) and (11.9.3 *b*). This can be derived from a two-fluid model ignoring the factors of $\frac{4}{3}$ and 2 corresponding to radiation pressure and the gravitational effect of pressure, respectively, and letting $\tau_{\mathrm{e}\gamma} = \tau_{\gamma\mathrm{e}} \to \infty$. In this case the two fluids are baryons, b, and dark matter, X, and the initial conditions are such that $\delta_X \gg \delta_{\mathrm{b}}$ at

$t_{\rm rec}$. In an Einstein–de Sitter model Equation (12.6.3) for the baryonic component can be written

$$\ddot{\delta}_{\rm b} + 2\frac{\dot{a}}{a}\dot{\delta}_{\rm b} + v_{\rm s}^2 k^2 \delta_{\rm b} = 4\pi G(\rho_{\rm b}\delta_{\rm b} + \rho_X \delta_X) \simeq 4\pi G \rho_X \delta_X. \qquad (12.6.4)$$

This equation is easily solved, since we know that $\delta_X \propto t^{2/3}$, by the ansatz $\delta_{\rm b} = At^p$. One thus finds that

$$\delta_{\rm b}(M) \simeq \frac{\delta_X}{1 + [M_{\rm J}^{(\rm i)}(z_{\rm rec})/M]^{2/3}} \propto t^{2/3}, \qquad (12.6.5)$$

so that the baryonic fluctuations catch up the dark matter virtually instantaneously.

12.6.2 Cold Dark Matter

Particles of cold dark matter correspond to cold thermal relics (or non-thermal relics such as axions), with $z_{nX} \gg z_{\rm eq}$. For such particles the maximum value of $M_{{\rm J}X}$ is quite small compared with scales of cosmological interest. Perturbations in the collisionless component δ_X are frozen-in by the Meszaros effect until $z_{\rm eq}$, but enjoy uninterrupted growth on scales $M > M_{{\rm J}X}$ after $z_{\rm eq}$. In this case, assuming as before that the spectrum of initial fluctuations decreases with mass scale, as discussed in the next chapter, the first structure to form has a mass of order $M \simeq M_{\rm J}^{(\rm i)}(z_{\rm rec}) \simeq 10^5 M_\odot$; the limit here is essentially provided by the pressure of the baryons after recombination. Although fluctuations are not dissipated in this model on small scales, the stagnation effect does suppress their growth compared with large scales, so the spectrum of fluctuations is severely modified: see Chapter 15, where we discuss these effects in detail. More detailed computations, based on kinetic theory, have shown that in both the CDM and HDM models, the residual fluctuations in the microwave radiation background are much smaller than those in the adiabatic baryon picture. This result can be understood from a qualitative point of view, by simply recognising that fluctuations on the scales $M_{{\rm J}X,\rm max} < M < M_{\rm J}^{(\rm a)}(z_{\rm rec})$ are roughly a factor A_X smaller in this case than in the old adiabatic picture. As an example, in Figure 12.1 we show the results of a full numerical computation of the evolution of the perturbations δ_X, $\delta_{\rm m}$ and $\delta_{\rm r}$ corresponding to a mass scale $M \simeq 10^{15} M_\odot$ for a CDM model with a Hubble parameter $h = 0.5$. One can compare this result with the similar computations shown in the previous chapter for baryonic models. The CDM model in particular produces rather low fluctuations in the CMB radiation. Until relatively recently, this was considered an asset, but with the COBE discovery of the radiation it seems to be a weakness: COBE seems to have detected larger fluctuations than CDM would predict, as discussed in Chapter 17.

12.6.3 Summary

By a relatively simple consideration of time and length scales, we have shown in this chapter how the presence of a significant component of non-baryonic material alters the growth rate of perturbations under gravitational instability. It has not been our aim in this chapter to develop complete models of structure formation based on this idea, but simply to explain the physical origin of the difference with respect to models with baryons only. The two main points to remember are that

1. models with non-baryonic dark matter typically induce smaller fluctuations in the radiation background than those with only baryons;
2. structure can survive on scales less than the Silk mass in a cold-dark-matter universe (because fluctuations in the dark-matter component are not affected by photon diffusion);
3. structure is destroyed on small scales in a hot-dark-matter universe because of the free streaming of the non-baryonic component.

In Chapter 15 we will explain how these ingredients manifest themselves in more complete models of structure formation.

Bibliographic Notes on Chapter 12

The standard manifesto for structure formation within CDM models is Blumenthal *et al.* (1984), while the first detailed numerical computations were by Davis *et al.* (1985). This basic model has been developed much further; see, for example, Frenk *et al.* (1988). A detailed account of the evolution of CDM perturbations is given by Liddle and Lyth (1993). Neutrino-dominated universes are discussed by, for example, White *et al.* (1983). This general material is covered well by Padmanabhan (1993) and Peacock (1999). The possibility of directly detecting dark-matter candidates is discussed, for example, in Klapdor-Kleingrothaus and Zuber (1997).

Problems

1. Derive the approximate solutions (12.2.5 *a*) and (12.2.5 *b*).
2. Derive the approximate solution (12.4.5).
3. Compare the solutions obtained in Questions 1 and 2 with numerical solutions of Equations (12.2.4) and (12.2.6).

13

Cosmological Perturbations

13.1 Introduction

In the previous chapters we have studied the linear evolution of a perturbation described as a plane wave with corresponding wave vector $\boldsymbol{k}$. This representation is useful because a generic perturbation can be represented as a superposition of such plane waves (by the Fourier representation theorem) which, while they are evolving linearly, evolve independently of each other. In general we expect fluctuations to exist on a variety of mass or length scales and the final structure forming will depend on the growth of perturbations on different scales relative to each other. In this chapter we shall therefore look at perturbations in terms of their spectral composition and explain how the various spectral properties might arise.

A particularly important problem connected with the primordial spectrum of perturbations is to understand its origin. In the 1970s the form of the spectrum was generally assumed in an *ad hoc* fashion to have the properties which seemed to be required to explain the origin of structure in either the adiabatic or isothermal scenario. A particular spectrum, suggested independently by Peebles and Yu (1970), Harrison (1970) and Zel'dovich (1972), but now usually known as the *Harrison–Zel'dovich* or *scale-invariant spectrum*, was taken to be the most 'natural' choice for initial fluctuations according to various physical arguments. Further motivation for this choice arrived in 1982 in the form of inflationary models, which, as we shall see in Section 13.6, usually predict a spectrum of the scale-invariant form. The details of these fluctuations, which are generated by quantum oscillations of the scalar field driving the inflationary epoch, were first worked out by Guth and Pi (1982), Hawking (1982) and Starobinsky (1982). This result was very important, because it represented the first time that any particular choice of the spectrum of initial perturbations has been strongly motivated by physics.

As far as the evolution of the perturbation spectra is concerned, it is clear that the theory must depend on the nature of the particles which dominate the Universe, baryonic or non-baryonic, hot or cold, and on the nature of the fluctuations themselves, adiabatic or isothermal, curvature or isocurvature. We shall explain how these factors alter or 'modulate' the primordial spectrum later in this chapter. Because the fluctuations are, in some sense, 'random' in origin, we shall also need to introduce some statistical properties which can be used to describe density fluctuations, namely the power spectrum, variance, probability distribution and correlation functions.

13.2 The Perturbation Spectrum

To describe the distribution of matter in the Universe at a given time and its subsequent evolution one might try to divide it into volumes which initially evolve independently of each other. Fairly soon, however, this independence would no longer hold as the gravitational forces between one cell and its neighbours become strong. It is therefore not a good idea to think of a generic perturbation as a sum of spatial components. It is a much better idea to think of the perturbation as a superposition of plane waves which have the advantage that they evolve independently while the fluctuations are still linear. This effectively means that one represents the distribution as independent components not in real space, but in Fourier transform space, or reciprocal space, in terms of the wavevectors of each component $\boldsymbol{k}$.

Let us consider a volume V_{u}, for example a cube of side $L \gg l_{\mathrm{s}}$, where l_{s} is the maximum scale at which there is significant structure due to the perturbations; V_{u} can be thought of as a 'fair sample' of the Universe if this is the case. It is possible therefore to construct, formally, a 'realisation' of the Universe by dividing it into cells of volume V_{u} with periodic boundary conditions at the faces of each cube. This device will be convenient for many applications but should not be taken too literally. Indeed, one can take the limit $V_{\mathrm{u}} \to \infty$ in most cases, as we shall see later.

Let us denote by $\langle\rho\rangle$ the mean density in a volume V_{u} and $\rho(\boldsymbol{x})$ to be the density at a point specified by the position vector $\boldsymbol{x}$ with respect to some arbitrary origin. As usual we define the fluctuation $\delta(\boldsymbol{x}) = [\rho(\boldsymbol{x}) - \langle\rho\rangle]/\langle\rho\rangle$. In light of the above comments we take this to be expressible as a Fourier series:

$$\delta(\boldsymbol{x}) = \sum_{\boldsymbol{k}} \delta_{\boldsymbol{k}} \exp(\mathrm{i}\boldsymbol{k}\cdot\boldsymbol{x}) = \sum_{\boldsymbol{k}} \delta_{\boldsymbol{k}}^{*} \exp(-\mathrm{i}\boldsymbol{k}\cdot\boldsymbol{x}), \tag{13.2.1}$$

where the assumption of periodic boundary conditions $\delta(L, y, z) = \delta(0, y, z)$, etc., requires that the wavevector $\boldsymbol{k}$ has components

$$k_x = n_x \frac{2\pi}{L}, \qquad k_y = n_y \frac{2\pi}{L}, \qquad k_z = n_z \frac{2\pi}{L}, \tag{13.2.2}$$

with n_x, n_y and n_z integers. The Fourier coefficients $\delta_{\boldsymbol{k}}$ are complex quantities given, as it is straightforward to see, by

$$\delta_{\boldsymbol{k}} = \frac{1}{V_u} \int_{V_u} \delta(\boldsymbol{x}) \exp(-i\boldsymbol{k} \cdot \boldsymbol{x}) \, d\boldsymbol{x}; \tag{13.2.3}$$

because of conservation of mass in V_u we have $\delta_{\boldsymbol{k}=\boldsymbol{0}} = 0$; because of the reality of $\delta(\boldsymbol{x})$ we have $\delta_{\boldsymbol{k}}^* = \delta_{-\boldsymbol{k}}$.

If, instead of the volume V_u, we had chosen a different volume V_u', the perturbation within the new volume would again be represented by a series of the form (13.2.1), but with different coefficients $\delta_{\boldsymbol{k}}$. If one imagines a large number N of such volumes, i.e. a large number of 'realisations' of the Universe, one will find that $\delta_{\boldsymbol{k}}$ varies from one to the other in both amplitude and phase. If the phases are random, not only across the ensemble of realisations, but also from node to node within each realisation, then the density field has Gaussian statistics which we shall discuss in detail in Section 13.7. For the moment, however, it suffices to note the following property. Although the mean value of the perturbation $\delta(\boldsymbol{x}) \equiv \delta$ across the statistical ensemble is identically zero by definition, its mean square value, i.e. its *variance* σ^2, is not. It is straightforward to show that

$$\sigma^2 \equiv \langle \delta^2 \rangle = \sum_{\boldsymbol{k}} \langle |\delta_{\boldsymbol{k}}|^2 \rangle = \frac{1}{V_u} \sum_{\boldsymbol{k}} \delta_k^2, \tag{13.2.4}$$

where the average is taken over an ensemble of realisations. The quantity δ_k is defined by the relation (13.2.4) and its meaning will become clearer later, in Section 13.8. One can see from Equation (13.2.4) that $\langle |\delta_{\boldsymbol{k}}|^2 \rangle$ is the contribution to the variance due to waves of wavenumber $\boldsymbol{k}$. If we now take the limit $V_u \to \infty$ and assume that the density field is statistically homogeneous and isotropic, so that there is no dependence on the direction of $\boldsymbol{k}$ but only on $k = |\boldsymbol{k}|$, we find

$$\sigma^2 = \frac{1}{V_u} \sum_{\boldsymbol{k}} \delta_k^2 \to \frac{1}{2\pi^2} \int_0^\infty P(k) k^2 \, dk, \tag{13.2.5}$$

where we have, for simplicity, put $\delta_k^2 = P(k)$ in the limit $V_u \to \infty$. The quantity $P(k)$ is called the power spectral density function of the field δ or, more loosely, the *power spectrum*. The variance does not depend on spatial position but on time, because the perturbation amplitudes $\delta_{\boldsymbol{k}}$ evolve. The quantity σ^2 therefore tells us about the amplitude of perturbations, but does not carry information about their spatial structure.

As we shall see, it is usual to assume that the perturbation power spectrum $P(k)$, at least within a certain interval in k, is given by a power law

$$P(k) = Ak^n; \tag{13.2.6}$$

the exponent n is usually called the *spectral index*. The exponent need not be constant over the entire range of wave numbers: the convergence of the variance in (13.2.5) requires that $n > -3$ for $k \to 0$ and $n < -3$ for $k \to \infty$.

Equation (13.2.5) can also be written in the form

$$\sigma^2 = \frac{1}{2\pi^2}\int_0^\infty P(k)k^2\,\mathrm{d}k = \int_{-\infty}^{+\infty} \Delta(k)\,\mathrm{d}\ln k, \tag{13.2.7}$$

where the dimensionless quantity

$$\Delta(k) = \frac{1}{2\pi^2}P(k)k^3 \tag{13.2.8}$$

represents the contribution to the variance per unit logarithmic interval in k. We shall find this quantity useful to compare with observations of galaxy clustering on large scales in Section 16.6. If $\Delta(k)$ has only one pronounced maximum at k_{max}, then the variance is given approximately by

$$\sigma^2 \simeq \Delta(k_{\mathrm{max}}) = \frac{1}{2\pi^2}P(k_{\mathrm{max}})k_{\mathrm{max}}^3. \tag{13.2.9}$$

Some other useful properties of the spectrum $P(k)$ are its *spectral moments*

$$\sigma_l^2 = \frac{1}{2\pi^2}\int_0^\infty P(k)k^{2(l+1)}\,\mathrm{d}k, \tag{13.2.10}$$

where the index l (which is an integer) is the order; the zeroth-order moment is just the variance σ^2. Typically, such as for power-law spectra, these moments do not converge and it is necessary to filter the spectrum to get meaningful results; we discuss this in Section 13.3 and thereafter. Higher-order moments of the (filtered) spectrum contain information about the shape of $P(k)$ just as moments of a probability distribution contain information about its shape. As we shall see in Section 14.8, many interesting properties of the fluctuation field $\delta(\boldsymbol{x})$ can be expressed in terms of the spectral moments or combinations of them such as

$$\gamma = \frac{\sigma_1^2}{\sigma_2\sigma_0}, \qquad R_* = \sqrt{3}\frac{\sigma_1}{\sigma_2}, \tag{13.2.11}$$

where γ and R_* are usually called the *spectral parameters.*

13.3 The Mass Variance

13.3.1 Mass scales and filtering

The problem with the variance σ^2 is that it contains no information about the relative contribution to the fluctuations from different $\boldsymbol{k}$ modes. It may also be formally infinite, if the integral in Equation (13.2.5) does not converge. It is convenient therefore to construct a statistical description of the fluctuation field as a function of some 'resolution' scale R. Let $\langle M\rangle$ be the mean mass found inside a spherical volume V of radius R:

$$\langle M\rangle = \langle\rho\rangle V = \tfrac{4}{3}\pi\langle\rho\rangle R^3. \tag{13.3.1}$$

One defines the *mass variance* inside the volume V to be the quantity σ_{M}^2 given by

$$\sigma_{\mathrm{M}}^2 = \frac{\langle (M - \langle M \rangle)^2 \rangle}{\langle M \rangle^2} = \frac{\langle \delta M^2 \rangle}{\langle M \rangle^2}, \tag{13.3.2}$$

where the average is made over all spatial volumes V; σ_M is the *RMS* (root mean square) *mass fluctuation*. Using the Fourier decomposition of Equation (13.2.1), Equation (13.3.2) becomes

$$\sigma_{\mathrm{M}}^2 = \frac{1}{V^2} \left\langle \int_V \int_V \sum_{\boldsymbol{k}} \delta_{\boldsymbol{k}} \exp(\mathrm{i}\boldsymbol{k} \cdot \boldsymbol{x}) \sum_{\boldsymbol{k}'} \delta_{\boldsymbol{k}'} \exp(\mathrm{i}\boldsymbol{k}' \cdot \boldsymbol{x}')\, \mathrm{d}\boldsymbol{x}\, \mathrm{d}\boldsymbol{x}' \right\rangle, \tag{13.3.3 a}$$

which can be written

$$\sigma_{\mathrm{M}}^2 = \frac{1}{V^2} \left\langle \sum_{\boldsymbol{k},\boldsymbol{k}'} \delta_{\boldsymbol{k}} \delta_{\boldsymbol{k}'}^* \int_V \exp(\mathrm{i}\boldsymbol{k} \cdot \boldsymbol{x})\, \mathrm{d}\boldsymbol{x} \int_V \exp(-\mathrm{i}\boldsymbol{k}' \cdot \boldsymbol{x}')\, \mathrm{d}\boldsymbol{x}' \right\rangle \tag{13.3.3 b}$$

and then as

$$\sigma_{\mathrm{M}}^2 = \frac{1}{V^2} \left\langle \sum_{\boldsymbol{k},\boldsymbol{k}'} \delta_{\boldsymbol{k}} \delta_{\boldsymbol{k}'}^* \exp[\mathrm{i}(\boldsymbol{k} - \boldsymbol{k}') \cdot \boldsymbol{x}_0] \times I_1 \times I_2 \right\rangle, \tag{13.3.3 c}$$

where

$$I_1 = \int_V \exp[\mathrm{i}\boldsymbol{k} \cdot (\boldsymbol{x} - \boldsymbol{x}_0)]\, \mathrm{d}(\boldsymbol{x} - \boldsymbol{x}_0) \tag{13.3.3 d}$$

and

$$I_2 = \int_V \exp[-\mathrm{i}\boldsymbol{k}' \cdot (\boldsymbol{x}' - \boldsymbol{x}_0)]\, \mathrm{d}(\boldsymbol{x}' - \boldsymbol{x}_0). \tag{13.3.3 e}$$

This can then be seen to give

$$\sigma_{\mathrm{M}}^2 = \sum_{\boldsymbol{k}} \langle |\delta_{\boldsymbol{k}}|^2 \rangle \left[\frac{1}{V} \int_V \exp(\mathrm{i}\boldsymbol{k} \cdot \boldsymbol{y})\, \mathrm{d}\boldsymbol{y} \right]^2 = \sum_{\boldsymbol{k}} \langle |\delta_{\boldsymbol{k}}|^2 \rangle I^2 = \frac{1}{V_{\mathrm{u}}} \sum_{\boldsymbol{k}} \delta_k^2 W^2(kR). \tag{13.3.3 f}$$

In the above equations $\boldsymbol{x}_0$ is the centre of a sphere of volume V, and a mean is taken over all such spheres, i.e. over all positions $\boldsymbol{x}_0$. We have used the relationship

$$\langle \exp[\mathrm{i}(\boldsymbol{k} - \boldsymbol{k}') \cdot \boldsymbol{x}_0] \rangle = \delta_{\boldsymbol{k}\boldsymbol{k}'}^{\mathrm{D}}, \tag{13.3.4}$$

where $\delta_{\boldsymbol{k}\boldsymbol{k}'}^{\mathrm{D}}$ is the Kronecker delta function, which is more usually written $\delta^{\mathrm{D}}(\boldsymbol{k} - \boldsymbol{k}')$ and is not to be confused with $\delta_{\boldsymbol{k}}$, such that $\delta_{\boldsymbol{k}\boldsymbol{k}'}^{\mathrm{D}} = 0$ if $\boldsymbol{k} \neq \boldsymbol{k}'$ and $\delta_{\boldsymbol{k}\boldsymbol{k}'}^{\mathrm{D}} = 1$, if $\boldsymbol{k} = \boldsymbol{k}'$. The function $W(kR)$ in Equation (13.3.3) is called the *window function*; an expression for this can be found by developing $\exp(\mathrm{i}\boldsymbol{k} \cdot \boldsymbol{y})$ in spherical harmonics, given the symmetry of the system around the point $\boldsymbol{x}_0$:

$$\exp(\mathrm{i}\boldsymbol{k} \cdot \boldsymbol{y}) = \sum_{l,m} j_l(kr) i^l (2l + 1) P_l^{|m|}(\cos\vartheta) \exp(\mathrm{i}m\varphi), \tag{13.3.5}$$

where j_l are spherical Bessel functions, $P_l^{|m|}$ are the associated Legendre polynomials, and r, ϑ and φ are spherical polar coordinates. The integral I in Equation (13.3.3 f) then becomes

$$I = \sum_{l,m} i^l(2l+1) \int_0^{2\pi} \exp(im\varphi)\,d\varphi \int_0^{\pi} P_l^{|m|}(\cos\vartheta)\sin\vartheta\,d\vartheta \int_0^R j_l(kr)r^2\,dr \tag{13.3.6 a}$$

or, alternatively,

$$I = 4\pi \int_0^R j_0(kr)r^2\,dr = \frac{4\pi}{k^3}(\sin kR - kR\cos kR) \tag{13.3.6 b}$$

(the integrals over ϑ and φ are zero unless $m = l = 0$); in this way the window function is just

$$W(kR) = \frac{3(\sin kR - kR\cos kR)}{(kR)^3}; \tag{13.3.7}$$

its behaviour is such that $W(x) \simeq 1$ for $x \leqslant 1$ and $|W(x)| \leqslant x^{-2}$ for $x \gg 1$.

Passing to a continuous distribution of plane waves, i.e. in the limit expressed by Equation (13.2.5), the mass variance is

$$\sigma_M^2 = \frac{1}{2\pi^2} \int_0^\infty P(k)W^2(kR)k^2\,dk < \sigma^2, \tag{13.3.8}$$

which, as it must be, is a function of R and therefore of M.

The significance of the window function is the following: the dominant contribution to σ_M^2 is from perturbation components with wavelength $\lambda \simeq k^{-1} > R$, because those with higher frequencies tend to be averaged out within the window volume; we have tacitly assumed that the spectrum is falling with decreasing k, so waves with much larger λ contribute only a small amount. We will return to this point in Section 14.4, where we discuss effects occurring at the edge of the window.

13.3.2 Properties of the filtered field

One can think of the result expressed by Equation (13.3.8) also as a special case of a more general situation. It is often interesting to think of the fluctuation field as being 'filtered' with a low-pass filter. The filtered field, $\delta(\boldsymbol{x}; R_f)$, may be obtained by convolution of the 'raw' density field with some function F having a characteristic scale R_f:

$$\delta(\boldsymbol{x}; R_f) = \int \delta(\boldsymbol{x}')F(|\boldsymbol{x} - \boldsymbol{x}'|; R_f)\,d\boldsymbol{x}'. \tag{13.3.9}$$

The filter F has the following properties: $F = \text{const.} \simeq R_f^{-3}$ if $|\boldsymbol{x} - \boldsymbol{x}'| \ll R_f$, $F \simeq 0$ if $|\boldsymbol{x} - \boldsymbol{x}'| \gg R_f$, $\int F(\boldsymbol{y}; R_f)\,d\boldsymbol{y} = 1$. For example, the *'top-hat' filter*, with a sharp cut off, is defined by the relation

$$F_{TH}(|\boldsymbol{x} - \boldsymbol{x}'|; R_{TH}) = \frac{3}{4\pi R_{TH}^3}\Theta\left(1 - \frac{|\boldsymbol{x} - \boldsymbol{x}'|}{R_{TH}}\right), \tag{13.3.10}$$

where Θ is the Heaviside step function ($\Theta(y) = 0$ for $y \leqslant 0$, $\Theta(y) = 1$ for $y > 0$). Another commonly used filter is the *Gaussian filter*:

$$F_{\mathrm{G}}(|\boldsymbol{x}-\boldsymbol{x}'|; R_{\mathrm{G}}) = \frac{1}{(2\pi R_{\mathrm{G}}^2)^{3/2}} \exp\left(-\frac{|\boldsymbol{x}-\boldsymbol{x}'|^2}{2R_{\mathrm{G}}^2}\right). \tag{13.3.11}$$

The mass contained in a volume of radius R_{TH} is equal to that contained in a Gaussian 'ball', cf. Equation (13.3.16), if $R_{\mathrm{G}} = 0.64R_{\mathrm{TH}}$.

Using the concept of the filtered field we can repeat all considerations we made in Section 14.2 concerning the variance. In place of σ^2 we have the variance of the field $\delta(\boldsymbol{x}; R_{\mathrm{f}})$

$$\sigma^2(R_{\mathrm{f}}) = \frac{1}{2\pi^2}\int_0^\infty P(k; R_{\mathrm{f}})k^2\,\mathrm{d}k = \frac{1}{2\pi^2}\int_0^\infty P(k)W_F^2(kR_{\mathrm{f}})k^2\,\mathrm{d}k, \tag{13.3.12}$$

where $W_F(kR_{\mathrm{f}})$ is now the Fourier transform of the filter F. The spectrum of the filtered field is given by

$$P(k; R_{\mathrm{f}}) = W_F^2(kR_{\mathrm{f}})P(k). \tag{13.3.13}$$

In the top-hat case we have

$$W_{\mathrm{TH}}(kR_{\mathrm{TH}}) = \frac{3(\sin kR_{\mathrm{TH}} - kR_{\mathrm{TH}}\cos kR_{\mathrm{TH}})}{(kR_{\mathrm{TH}})^3}, \tag{13.3.14}$$

which coincides with (13.3.7) with $R = R_{\mathrm{TH}}$; this result is due to the definition of the mass in Equation (13.3.1) as the mass contained in a sphere of radius R. The window function for a Gaussian filter is

$$W_{\mathrm{G}}(kR_{\mathrm{G}}) = \exp[-\tfrac{1}{2}(kR_{\mathrm{G}})^2], \tag{13.3.15}$$

which can be thought of as similar to the mass-in-sphere calculation, but with a sphere having blurred edges

$$\langle M\rangle = 4\pi\langle\rho\rangle\int_0^\infty \exp\left(-\frac{r^2}{2R^2}\right)r^2\,\mathrm{d}r. \tag{13.3.16}$$

By analogy with this expression for the generic mass M, one can find a mass variance using a window function of the form (13.3.15). In general, therefore, the mass variance of a density field $\delta(\boldsymbol{x})$ is given by the relation

$$\sigma_{\mathrm{M}}^2 = \frac{1}{2\pi^2}\int_0^\infty P(k)W_F^2(kR)k^2\,\mathrm{d}k, \tag{13.3.17}$$

where the expression for the window function depends on whichever filter, or effective mass, is used.

13.3.3 Problems with filters

One of the reasons why one might prefer a Gaussian filter over the apparently simpler top hat is illustrated by applying Equation (13.3.17) to a power-law spectrum of the form (13.2.6). As we have said, in order for σ^2 to converge, the spectrum $P(k)$ must have an asymptotic behaviour as $k \to \infty$ of the form k^{n_∞}, with $n_\infty < -3$. For this reason we can only take Equation (13.2.6) to be valid for wavenumbers smaller than a certain value k_∞, after which the spectral index either changes slope to n_∞ or there is a rapid cut-off in $P(k)$. The convergence for small k, however, requires that $n > -3$. If one puts Equation (13.2.6) directly into (13.3.17) and assumes a top-hat filter, so that $W(kR) = 1$ for $k \leqslant 1/R \equiv k_{\mathrm{M}}$, $|W(kR)| \simeq (k/k_{\mathrm{M}})^{-2}$ for $k_{\mathrm{M}} \leqslant k \leqslant k_\infty$, and $P(k) = 0$ for $k > k_\infty$, one obtains, for the interval $-3 < n < 1$,

$$\sigma_{\mathrm{M}}^2 \simeq \frac{A}{2\pi^2}\left\{\frac{4k_{\mathrm{M}}^{n+3}}{(1-n)(3+n)}\left[1 - \frac{n+3}{4}\left(\frac{k_{\mathrm{M}}}{k_\infty}\right)^{1-n}\right]\right\}, \tag{13.3.18 a}$$

which becomes

$$\sigma_{\mathrm{M}}^2 \simeq \frac{2Ak_{\mathrm{M}}^{n+3}}{\pi^2(1-n)(3+n)} \propto R^{-(n+3)} : \tag{13.3.18 b}$$

the mass variance σ_M depends on the spectral index n according to

$$\sigma_M \propto M^{-(3+n)/6} \equiv M^{-\alpha}; \tag{13.3.19}$$

we call the exponent $\alpha = \frac{1}{6}(3+n)$ the *mass index*. For values $n > 1$ one finds, however, that

$$\sigma_{\mathrm{M}}^2 \simeq \frac{A}{2\pi^2}\left\{\frac{k_\infty^{n-1}k_{\mathrm{M}}^4}{n-1}\left[1 - \frac{4}{n+3}\left(\frac{k_{\mathrm{M}}}{k_\infty}\right)^{n-1}\right]\right\}, \tag{13.3.20 a}$$

which is

$$\sigma_{\mathrm{M}}^2 \simeq \frac{Ak_\infty^{n-1}k_{\mathrm{M}}^4}{2\pi^2(n-1)} \propto R^{-4} \propto M^{-4/3}, \tag{13.3.20 b}$$

and therefore

$$\sigma_M \propto M^{-2/3} : \tag{13.3.21}$$

the mass index does not depend on the original spectral index. The result (13.3.21) is also obtained if $n = 1$, apart from a logarithmic term. The reason for this result is that we have taken for the definition of σ_M the variance of fluctuations inside a sphere with sharp edges. This corresponds to an extended window function in Fourier space. When $n \geqslant 1$ the spectral components which enter the integral at the edges of the window function become significant contributors to the variance: σ_M^2 defined by Equation (13.3.17) is no longer a useful measure of the mass fluctuations on a particular scale R, but is dominated by edge effects which are sensitive to fluctuations on a much smaller scale than R. These effects are a form of surface noise which depends on the number of 'particles' at the boundary; a statistical fluctuation arises according to whether a particle happens to lie just inside, on or just outside the boundary. If the expected number of particles on a surface of area S is N_S, then we clearly have

$$\delta N_S \propto N_S^{1/2} \propto S^{1/2} \propto M^{1/3}, \tag{13.3.22}$$

so that

$$\sigma_M \simeq \frac{\delta M}{M} \propto \frac{\delta N_S}{M} \propto M^{-2/3}, \tag{13.3.23}$$

in accordance with Equation (13.3.21). This misleading result can be corrected if one makes a more realistic definition of the volume corresponding to the mass scale M. If one smears out the edges of the sphere such as, for example, via a Gaussian filter (13.3.11), one obtains

$$\sigma_{\mathrm{M}}^2 = \frac{1}{2\pi^2}\int_0^\infty P(k)\exp(-k^2R^2)k^2\,\mathrm{d}k; \tag{13.3.24}$$

the new window function passes sharply from a value of order unity, for $k < 1/R = k_{\mathrm{M}}$, to a vanishingly small value for $k > k_{\mathrm{M}}$: the blurring out of the sphere has therefore made the window function sharper. With the new definition one finds, for any n,

$$\sigma_{\mathrm{M}}^2 = \frac{A}{4\pi^2}\Gamma(\tfrac{1}{2}(n+3))R^{-(n+3)} \tag{13.3.25}$$

(Γ is the Euler gamma function), which has a dependence on R which is now in accord with Equation (13.3.18). The behaviour of σ_M is therefore generally valid if one uses a Gaussian filter function.

13.4 Types of Primordial Spectra

Having established the description of a primordial stochastic density field in terms of its power spectrum and related quantities, we should now indicate some possibilities for the form of this spectrum. It is also important to develop some kind of intuitive understanding of what the spectrum means physically.

It is the usual practice to suppose that some mechanism, perhaps inflation, lays down the initial spectrum of perturbations at some very early time, say $t = t_{\mathrm{p}}$, which one is tempted to identify with the earliest possible physical timescale, the Planck time. The cosmological horizon at this time will be very small, so the fluctuations on scales relevant to structure formation will be outside the horizon. As time goes on, perturbations on larger and larger scales will enter the horizon as they grow by gravitational instability, become modified by the various damping and stagnation processes discussed in the previous chapters and, eventually, after recombination, give rise to galaxies and larger structures. The final structures which form will therefore depend upon the primordial spectrum to a large extent, but also upon the cosmological parameters and the form of any dark matter. It is common to assume a primordial spectrum of a power-law form:

$$P(k; t_p) = A_p k^{n_p}. \tag{13.4.1}$$

In general, one would expect the amplitude A_p and the spectral index n_p to depend on k so that Equation (13.4.1) defines the effective amplitude and index for a given k. In most models, however, n_p is effectively constant over the entire range of scales relevant to the observable Universe. The mass variance corresponding to Equation (13.4.1) is

$$\sigma_M(t_p) = K_p \left(\frac{M}{M_H(t_p)} \right)^{-(3+n_p)/6} \propto M^{-\alpha_p}, \tag{13.4.2}$$

where $M_H(t_p)$ is some reference mass scale which, for convenience, we take to be the horizon mass at time t_p.

Clearly the discussion in Section 13.2 demonstrates that a perfectly homogeneous distribution of mass in which $\delta(\boldsymbol{x}) = 0$ has a power spectrum which is identically zero for all k and therefore has zero mass variance on any scale. To interpret other behaviours of σ_M^2 it is perhaps helpful to think of the mass distribution as being composed of point particles with identical mass m. If these particles are distributed completely randomly throughout space, then the fluctuations in a volume V - which contains on average N particles and, therefore, on average, a mass $M = mN$ - will be due simply to statistical fluctuations in the number of particles from volume to volume. For random (Poisson) distributions this means that $\langle \delta N^2 \rangle^{1/2} \simeq N^{1/2}$, so that the RMS mass fluctuation is given by

$$\sigma_M = \frac{\delta N}{N} \simeq N^{-1/2} \propto M^{-1/2}, \tag{13.4.3}$$

corresponding, by Equation (13.4.2), to a value of the mass index $\alpha = \frac{1}{2}$ and therefore to a spectral index $n = 0$. Since $P(k)$ is independent of k this is usually called a *white-noise spectrum*.

Alternatively, if the distribution of particles is not random throughout space but is instead random over spherical 'bubbles' with sharp edges, the RMS mass fluctuations becomes

$$\sigma_M \simeq \frac{N_S^{1/2}}{N} \simeq (4\pi)^{1/2} \left(\frac{3}{4\pi} \right)^{1/3} \frac{N^{1/3}}{N} \propto N^{-2/3} \propto M^{-2/3}, \tag{13.4.4}$$

as we have mentioned above; the mass fluctuation expressed by Equation (13.4.4) corresponds to a mass index $\alpha = \frac{2}{3}$ and to a spectral index $n = 1$. If the edges of the spheres are blurred, then the 'surface effect' is radically modified and it is then possible to show that

$$\sigma_M \propto N^{-5/6} \propto M^{-5/6}, \tag{13.4.5}$$

corresponding to a mass index $\alpha = \frac{5}{6}$ and a spectral index $n = 2$. Equation (13.4.5) can be found if one assumes that one can create the perturbed distribution from a homogeneous distribution by some rearrangement of the matter which conserves mass. It would be reasonable to infer that this rearrangement can only take place over scales less than the horizon scale when the fluctuations were laid down, which gives a natural scale to the 'bubbles' we mentioned above. From Equation (13.2.3) one obtains

$$\delta_{\boldsymbol{k}} = \frac{1}{V_{\mathrm{u}}}\left[\int_{V_{\mathrm{u}}} \delta(\boldsymbol{x})\,\mathrm{d}\boldsymbol{x} - \mathrm{i}\boldsymbol{k}\cdot\int_{V_{\mathrm{u}}} \boldsymbol{x}\delta(\boldsymbol{x})\,\mathrm{d}\boldsymbol{x} - \tfrac{1}{2}k^2\cdots+\cdots\right]. \tag{13.4.6}$$

In calculating the mass variance σ_M^2, as we have explained, one counts only the waves with $k < R^{-1}$, for which the term $\boldsymbol{k}\cdot\boldsymbol{x}$ is small: in the series (13.4.6) the higher and higher terms are smaller and smaller. Conservation of mass requires that the first term is zero, or that $\delta_{\boldsymbol{k}} \propto k$ and therefore $\sigma_M \propto M^{-5/6}$. If one also requires that linear momentum is conserved or, in other words, that the centre of mass of the system does not move, then the second term in (13.4.6) is also zero and we obtain $\delta_{\boldsymbol{k}} \propto k^2$, corresponding to a spectral index $n = 4$ and therefore to a mass index $\alpha = \frac{7}{6}$:

$$\sigma_M \propto M^{-7/6}. \tag{13.4.7}$$

It is tempting to imagine that fluctuations in the number of particles inside the horizon might lead to a 'natural' form for the initial spectrum. Such a spectrum has some severe problems, however. If one takes the time t_{p} to be the Planck time, for example, the horizon contains on average only one 'Planck particle' and one cannot think of the spatial distribution within this scale as random in the sense required above. Moreover, the white-noise spectrum actually predicts a very chaotic cosmology in which a galactic-scale perturbation would arrive at the nonlinear growth phase (Chapter 14) much before t_{eq}. Let us consider a perturbation with a typical galaxy mass, $10^{11}M_{\odot}$, which contains $N_{\mathrm{b}} \simeq 10^{69}$ baryons corresponding to $N \simeq N_{\mathrm{b}}\sigma_{0\mathrm{r}} \simeq 10^{78}$ particles and therefore characterised by $\sigma_M \simeq N^{-1/2} \simeq 10^{-39}$. This perturbation would arrive at the nonlinear regime at a time t_{c} given, approximately, by

$$\sigma_M(t_{\mathrm{p}})\frac{t_{\mathrm{c}}}{t_{\mathrm{p}}} \simeq \sigma_M(t_{\mathrm{p}})\left(\frac{T_{\mathrm{p}}}{T_{\mathrm{c}}}\right)^2 \simeq 1; \tag{13.4.8}$$

in Equation (13.4.8) we have supposed that $t_{\mathrm{c}} < t_{\mathrm{eq}}$, and this is confirmed *a posteriori* by the result $T_{\mathrm{c}} \simeq 10^{12}$ K. Such collapses would have a drastic effect on the isotropy and spectrum of the microwave background radiation and on nucleosynthesis, so would consequently not furnish an acceptable theory of galaxy formation.

The spectrum (13.4.5), often called the *particles-in-boxes spectrum*, also has problems. It only makes sense to treat the perturbations from a statistical point of view when the horizon contains a reasonably large number of particles, say $N_{\mathrm{i}} \simeq 100$. This happens at a time t_{i} corresponding to a temperature $T_{\mathrm{i}} \simeq 2\times10^{18}$ GeV. A fluctuation on a scale M of the order of the horizon mass at $T_{\rm i}$ has $\sigma_M(t_{\rm i}) \simeq N_{\rm i}^{-1/2}$ if the particles are distributed randomly, but, as we have explained above, the 'surface effect' might produce an RMS mass fluctuation of the form

$$\sigma_M(t_{\rm i}) = B_{\rm p} N^{-5/6}, \tag{13.4.9}$$

for $N > N_{\rm i}$. The constant $B_{\rm p}$ is obtained in a first approximation by putting $\sigma_M(N = N_{\rm i}) = N_{\rm i}^{-1/2}$; one thus finds $B_{\rm p} \simeq 5$. However, even in this case, the variance on a scale $M \simeq 10^{11} M_\odot$ yields a completely unsatisfactory result. Taking, as in the previous case, $N \simeq 10^{78}$ and allowing the perturbation to grow uninterruptedly ($\sigma_M \propto t$, for $t < t_{\rm eq}$, and $\sigma_M \propto t^{2/3}$, for $t > t_{\rm eq}$), i.e. without taking account of periods of damping or oscillation, one finds

$$\sigma_M(t_0) \simeq \sigma_M(t_{\rm i}) \frac{t_{\rm eq}}{t_{\rm i}} \left(\frac{t_0}{t_{\rm eq}} \right)^{2/3} = \sigma_M(t_{\rm i}) \left(\frac{T_{\rm i}}{T_{\rm eq}} \right)^2 \frac{T_{\rm eq}}{T_{0r}} \simeq 10^{-7} : \tag{13.4.10}$$

the fluctuation would not yet have arrived at the nonlinear regime and could not therefore have formed structure. Equation (13.4.10) is valid for $\Omega = 1$ and things get worse if $\Omega < 1$. On the scales of galaxies the amplitude of the white-noise spectrum, $n_{\rm p} = 0$, is too high, while that of the particles-in-boxes spectrum, $n_{\rm p} = 2$, is much too low.

The problems arising from spectra obtained by reshuffling matter within a horizon volume have led most cosmologists to abandon such an origin and appeal to some process which occurs apparently outside the horizon to lay down some appropriate spectrum. As already mentioned, in the early 1970s, Peebles and Yu (1970), Harrison (1970) and Zel'dovich (1970), working independently, suggested a spectrum with $n_{\rm p} = 1$, corresponding to

$$\sigma_M(t_{\rm p}) = K_{\rm p} \left(\frac{M}{M_{H,p}} \right)^{-2/3} \tag{13.4.11}$$

(the value of $K_{\rm p}$ proposed by Zel'dovich was of the order of 10^{-4}, so as to produce fluctuations in the cosmic microwave background at a lower level than the observational limits of that time, while still allowing galaxy formation by the present epoch). This spectrum, called the *Harrison-Zel'dovich spectrum*, is of the same form as Equation (13.4.4), but is not interpreted as a surface effect. One of its properties is that fluctuations in the gravitational potential, $\delta\varphi$, or, in relativistic terms, in the metric, are independent of length scale r. In fact

$$\delta\varphi(r) \simeq \frac{G\delta M}{r} \simeq G\delta\rho(r) r^2 \simeq G\rho\sigma_M r^2 \propto \sigma_M M^{2/3} = \text{const.}, \tag{13.4.12}$$

if Equation (13.4.11) holds. The Equation (13.4.11) therefore characterises a spectrum which has a metric containing 'wrinkles' with an amplitude independent of scale. As we shall see in Section 14.5, fluctuations of this form enter the cosmological horizon with a constant value of the variance, equal to $K_{\rm p}^2$. For these reasons this spectrum is often called the *scale-invariant spectrum*. We shall see in

Section 14.6 that a spectrum of density fluctuations close to this form is in fact a common feature of inflationary models.

As a final remark in this section, we should mention that the spectrum of the density perturbation δ can also be used to construct the spectrum of the perturbations to the gravitational potential, $\delta\varphi$, and to the velocity field $\boldsymbol{v}$ in linear theory. The results are particularly simple. Since $\nabla^2\delta\varphi \propto \delta$, one has $k^2\varphi_k \propto \delta_k$, where φ_k is the Fourier transform of $\delta\varphi$, so that $P_\varphi(k) \propto P(k)k^{-4}$. For a density fluctuation spectrum with spectral index n one therefore has $n_\varphi = n - 4$ so that, for $n = 1$, one has $n_\varphi = -3$. This spectrum is generally, i.e. whether it refers to a potential, velocity or density field, called the *flicker-noise spectrum*, and the associated variance has a logarithmic divergence at small k. The velocity field is the gradient of a velocity potential which is just proportional to the gravitational potential so that $\boldsymbol{v}_k \propto \boldsymbol{k}\varphi_k$ and $P_{\mathrm{v}}(k) \propto P(k)k^{-2}$. We discuss velocity and potential perturbations in more detail in Chapter 18, where the exact expressions for the appropriate power spectra are also given.

13.5 Spectra at Horizon Crossing

In Section 11.5 we defined the time at which a perturbation of mass M enters the horizon; we found that, for $M \leqslant M_{\mathrm{H}}(z_{\mathrm{eq}}) \simeq 5 \times 10^{15}(\Omega h^2)^{-2}M_\odot$, this moment corresponds to a redshift

$$z_{\mathrm{H}}(M) \simeq z_{\mathrm{eq}}\left(\frac{M}{M_{\mathrm{H}}(z_{\mathrm{eq}})}\right)^{-1/3} \geqslant z_{\mathrm{eq}}, \tag{13.5.1}$$

while, for $M \geqslant M_{\mathrm{H}}(z_{\mathrm{eq}})$, we have

$$z_{\mathrm{H}}(M) \simeq z_{\mathrm{eq}}\left(\frac{M}{M_{\mathrm{H}}(z_{\mathrm{eq}})}\right)^{-2/3} \leqslant z_{\mathrm{eq}}; \tag{13.5.2}$$

this relation is valid for a flat universe or an open universe for $z \gg \Omega^{-1}$; in this section we shall assume the simplest case of $\Omega = 1$.

We propose to calculate the variance σ_M^2 corresponding to a scale M at the time defined by $z_{\mathrm{H}}(M)$ if the primordial fluctuation spectrum is of the power-law form (13.4.2). The perturbation grows without interruption from the moment of its origin, which we called t_{p}, to the time in which it enters the cosmological horizon, with a law $\sigma_M \propto t \propto (1+z)^{-2}$ before equivalence and $\sigma_M \propto t^{2/3} \propto (1+z)^{-1}$ after equivalence. If $z_{\mathrm{H}}(M) > z_{\mathrm{eq}}$, we therefore have

$$\sigma_M[z_{\mathrm{H}}(M)] \simeq \sigma_M(t_{\mathrm{p}})\left(\frac{1+z_{\mathrm{p}}}{1+z_{\mathrm{H}}(M)}\right)^2 = \sigma_M(t_{\mathrm{p}})\left(\frac{M}{M_{\mathrm{H}}(z_{\mathrm{p}})}\right)^{2/3} = K_{\mathrm{p}}\left(\frac{M}{M_{\mathrm{H}}(z_{\mathrm{p}})}\right)^{-\alpha_{\mathrm{H}}}, \tag{13.5.3}$$

where $\alpha_{\mathrm{H}} = \alpha_{\mathrm{p}} - \frac{2}{3}$. If, on the other hand, $z_{\mathrm{H}}(M) < z_{\mathrm{eq}}$, we have

$$\sigma_M(z_{\mathrm{H}}(M)) \simeq \sigma_M(t_{\mathrm{p}})\left(\frac{1+z_{\mathrm{p}}}{1+z_{\mathrm{eq}}}\right)^2 \frac{1+z_{\mathrm{eq}}}{1+z_{\mathrm{H}}(M)} = K_{\mathrm{p}}\left(\frac{M}{M_{\mathrm{H}}(z_{\mathrm{p}})}\right)^{-\alpha_{\mathrm{H}}}, \tag{13.5.4}$$

again identical to (13.5.3). The index α_H is the mass index of fluctuations at their entry into the cosmological horizon. This has a corresponding spectral index n_H, in accord with (13.4.1), which one finds from

$$\alpha_H = \alpha_p - \tfrac{2}{3} = \tfrac{1}{2} + \tfrac{1}{6}n_p - \tfrac{2}{3} = \tfrac{1}{2} + \tfrac{1}{6}(n_p - 4) = \tfrac{1}{2} + \tfrac{1}{6}n_H; \tag{13.5.5}$$

one therefore has

$$n_H = n_p - 4. \tag{13.5.6}$$

The Equation (13.5.5) indicates that the Harrison–Zel'dovich scale-invariant spectrum with $n_p = 1$ arrives at the cosmological horizon with a mass variance which is independent of M and equal to K_p^2. Steeper spectra ($n_p > 1$, $\alpha_p > \frac{2}{3}$) have a variance which decreases with increasing M at horizon entry; shallower spectra ($n_p < 1$, $\alpha_p < \frac{2}{3}$) have variance increasing with M. For this latter type, there is the problem that, on sufficiently large scales, one has a universe with extremely large fluctuations which would include separate closed mini-universes. There is clearly then a strong motivation for having a spectrum which, whatever its origin, produces a mass index $\alpha_p \geqslant \frac{2}{3}$ on the very largest scales. As a final comment, notice that the spectral index of fluctuations at horizon entry (13.5.6) is precisely the same as the spectral index for fluctuations in the gravitational potential field, defined in Section 13.4.

13.6 Fluctuations from Inflation

We have already mentioned that one of the virtues of the inflationary cosmology is that it predicts a spectrum of perturbations which might be adequate for the purposes of structure formation. The source for these fluctuations is the quantum field Φ which drives inflation in the manner described in Section 7.10. A full treatment of the origin of these fluctuations is outside the scope of this book since it requires advanced techniques from quantum field theory. Here we shall merely give an outline; Brandenberger (1985) gives a nice review. In this section we use units where $\hbar = c = k_B = 1$.

Suppose that the expectation value of the scalar field $\Phi(\boldsymbol{x}, t)$ is homogeneous in space, i.e. $\langle \Phi(\boldsymbol{x}, t) \rangle = \Phi(t)$. It then follows an equation of motion of the form

$$\ddot{\Phi} + 3H\dot{\Phi} + V'(\Phi) = 0, \tag{13.6.1}$$

cf. Equation (7.10.5), where V is the effective potential and the prime denotes a derivative with respect to Φ. As we mentioned in Section 7.10, most inflationary models satisfy the 'slow-rolling' conditions which we shall assume here because these simplify the calculations. Let us introduce these conditions again in a more quantitative way. In the slow-rolling approach the motion of the field is damped so that the force V' is balanced by the viscosity term $3H\dot{\Phi}$: $\dot{\Phi} \simeq -V'/3H$. This is the first slow-rolling condition. The second slow-rolling condition in fact corresponds to two requirements: firstly that the parameter ϵ, defined by

$$\epsilon \equiv \frac{m_P^2}{16\pi}\left(\frac{V'}{V}\right)^2, \tag{13.6.2}$$

should be small, i.e.

$$\epsilon \ll 1, \tag{13.6.3}$$

which effectively means that $V \gg \dot{\Phi}^2$, the condition for inflation to occur; secondly that

$$H^2 \simeq \frac{8\pi V}{3m_{\rm P}^2}, \tag{13.6.4}$$

which, together with (13.6.3), implies that the scale factor is evolving approximately exponentially: $a \propto \exp(Ht)$. The third condition is that η, defined by

$$\eta \equiv \frac{m_{\rm P}^2 V''}{8\pi V}, \tag{13.6.5}$$

should satisfy

$$|\eta| \ll 1, \tag{13.6.6}$$

which can be thought of as a consistency requirement on the other two conditions, since it can be obtained from them by differentiation.

We now have to understand what happens when we perturb the equation (13.6.1). Assuming, as always, that the spatial fluctuations in the Φ field, $\delta\Phi = \phi$, can be decomposed into Fourier modes $\phi_{\boldsymbol{k}}$ by analogy with (13.2.1), we obtain

$$\ddot{\phi}_{\boldsymbol{k}} + 3H\dot{\phi}_{\boldsymbol{k}} + \left[\left(\frac{k}{a}\right)^2 + V''\right]\phi_{\boldsymbol{k}} = 0. \tag{13.6.7}$$

It turns out, for reasons we shall not go into, that the V'' term in Equation (13.6.7) is negligible when a given fluctuation scale is pushed out beyond the horizon. The resulting equation then looks just like a damped harmonic oscillator for any particular k mode. Applying some quantum theory, it is possible to calculate the expected fluctuations in each 'mode' of this system in much the same way as one calculates the ground-state oscillations in any system of quantum oscillators. One finds the solution

$$\langle|\phi_{\boldsymbol{k}}|^2\rangle = \frac{H^2}{2k^3}. \tag{13.6.8}$$

One can think of this effect as similar to the Hawking radiation from the event horizon of a black hole: there is an event horizon in de Sitter space and one therefore sees a thermal background at a temperature $T_{\rm H} = H/2\pi$ which corresponds to fluctuations in the Φ field in the same manner as the thermal fluctuations at the Planck epoch we discussed in Chapter 6.

From (13.6.8) we can define a quantity $\Delta_\phi(k)$ by (13.2.8) so that $\Delta_\phi = \text{const.} \propto H$. These fluctuations are therefore of the same amplitude (in an appropriately defined sense), i.e. independent of scale as long as H is constant.

These considerations establish the form of the spectrum appropriate to the fluctuations in Φ but we have not yet arrived at the spectrum of the density perturbations themselves. The resolution of this step requires some technicalities concerning gauge choices which we shall skip in this case. What we are interested in at the end is the amplitude of the fluctuations when they enter the cosmological horizon after inflation has finished. If we define $\Delta^2_{\rm H}(k)$ to be the value of $\Delta^2(k)$ for the fluctuations in the density at scale k when they reenter the horizon after inflation, one can find

$$\Delta^2_{\rm H}(k) \simeq \frac{V_*}{m_{\rm P}^4 \epsilon_*}, \tag{13.6.9}$$

where the '$*$' denotes the value of V or ϵ at the time when the perturbation left the horizon during inflation. One therefore sees the fluctuation on reentry which was determined by the conditions just as it left, which is physically reasonable. One does not know the values of these parameters *a priori*, however, so they cannot be used to predict the spectral amplitude. In an exactly exponential inflationary epoch V_* and ϵ_* are constant so that $\Delta^2_{\rm H}(k)$ is constant. Since $\Delta^2 \propto k^3 P(k)$, and $P_{\rm H}(k) \propto P(k)k^{-4}$ from (13.5.6), we therefore have $P(k) \propto k$, which is the Harrison–Zel'dovich spectrum we mentioned before in Section 13.4.

In fact, the generic inflationary prediction is not for a pure de Sitter expansion, so that the quantity $\Delta^2_{\rm H}$ is not exactly independent of scale. It is straightforward to show that the actual spectral index is related to the slow-roll parameters ϵ_* (13.6.2) and η_* (13.6.5) when the perturbation scale k leaves the horizon via

$$n = 1 + 2\eta_* - 6\epsilon_*, \tag{13.6.10}$$

which gives $n = 1$ in the slow-rolling limit, as expected.

The quantum oscillations in Φ also lead to the generation of a stochastic background of gravitational waves with a spectrum and amplitude which depends on a different combination of slow-roll parameters from the scalar density fluctuation spectrum (in fact, the gravitational wave spectrum depends only on ϵ). The relative amplitudes of the gravitational waves and scalar perturbations also depend on the shape of the potential. Since gravitational waves are of no direct relevance to structure formation, we shall not discuss them in more detail here. Gravitational waves can, in principle, also generate temperature fluctuations in the cosmic microwave background, so we shall discuss them briefly in Section 17.4 and they may ultimately be detectable, a possibility we discuss in Chapter 21.

We should also mention that the quantum fluctuations in ϕ_k have random phases and therefore should be Gaussian (see Section 14.7) in virtually all realistic inflationary models (except perhaps those with multiple scalar fields or where the field evolution is nonlinear). This is because one usually assumes the field Φ to be in its ground state: zero point fluctuations are then those of a ground-state harmonic oscillator in quantum mechanics, i.e. Gaussian. Along with the computational advantages we shall mention later, this is a strong motivation for assuming that $\delta(\boldsymbol{x})$ is a Gaussian random field.

13.7 Gaussian Density Perturbations

In Section 13.2 we defined the power spectrum $P(k)$ of density perturbations, which measures the amplitude of the fluctuations as a function of wavenumber k or, equivalently, mass scale M. For some purposes, however, it is necessary to know not only the spectrum, that is the mean square fluctuation of a given wavenumber, but also the (probability) distribution of the fluctuations in either real space or Fourier space. Returning to the discussion we made in Section 13.2, consider a (large) number N of realisations of our periodic volume and label these realisations by $V_{u1}, V_{u2}, V_{u3}, \dots, V_{uN}$. It is meaningful to consider the probability distribution $\mathcal{P}(\delta_{\boldsymbol{k}})$ of the relevant coefficients

$$\delta_{\boldsymbol{k}} = |\delta_{\boldsymbol{k}}| \exp(\mathrm{i}\vartheta_{\boldsymbol{k}}) = \mathrm{Re}\,\delta_{\boldsymbol{k}} + \mathrm{i}\,\mathrm{Im}\,\delta_{\boldsymbol{k}} \tag{13.7.1}$$

from realisation to realisation across this ensemble. Let us assume that the distribution is statistically homogeneous and isotropic (as it must be if the Cosmological Principle holds), and that the real and imaginary parts have a Gaussian distribution and are mutually independent, so that

$$\mathcal{P}(w) = \frac{V_u^{1/2}}{(2\pi\alpha_k^2)^{1/2}} \exp\left(-\frac{w^2 V_u}{2\alpha_k^2}\right), \tag{13.7.2}$$

where w stands for either the real part or the imaginary part of $\delta_{\boldsymbol{k}}$ and $\alpha_k^2 = \delta_k^2/2$; δ_k^2 is the spectrum (see Section 13.2). This is the same as the assumption that the phases $\vartheta_{\boldsymbol{k}}$ in Equation (13.7.1) are mutually independent and randomly distributed over the interval between $\vartheta = 0$ and $\vartheta = 2\pi$. In this case the moduli of the Fourier amplitudes have a Rayleigh distribution:

$$\mathcal{P}(|\delta_{\boldsymbol{k}}|, \vartheta_{\boldsymbol{k}})\,\mathrm{d}|\delta_{\boldsymbol{k}}|\,\mathrm{d}\vartheta_{\boldsymbol{k}} = \frac{|\delta_{\boldsymbol{k}}| V_u}{2\pi\delta_k^2} \exp\left(-\frac{|\delta_{\boldsymbol{k}}|^2 V_u}{2\delta_k^2}\right) \mathrm{d}|\delta_{\boldsymbol{k}}|\,\mathrm{d}\vartheta_{\boldsymbol{k}}. \tag{13.7.3}$$

Because of the assumption of statistical homogeneity and isotropy of the Universe, the quantity $\delta_{\boldsymbol{k}}$ depends only on the modulus of the wavevector $\boldsymbol{k}$, denoted k, and not on its direction. It is fairly simple to show that, if the Fourier quantities $|\delta_{\boldsymbol{k}}|$ have the Rayleigh distribution, then the probability distribution $\mathcal{P}(\delta)$ of $\delta = \delta(\boldsymbol{x})$ in real space is Gaussian, so that

$$\mathcal{P}(\delta)\,\mathrm{d}\delta = \frac{1}{(2\pi\sigma^2)^{1/2}} \exp\left(-\frac{\delta^2}{2\sigma^2}\right) \mathrm{d}\delta. \tag{13.7.4}$$

In fact, Gaussian statistics in real space do not require the distribution (13.7.3) for the Fourier component amplitudes. One can see that $\delta(\boldsymbol{x})$ is simply a sum over a large number of Fourier modes. If the phases of each of these modes are random, then the central limit theorem will guarantee that the resulting superposition will be close to a Gaussian distribution if the number of modes is large. While (13.7.3) provides the formal definition of a Gaussian random field, the main requirement in practice is simply that the phases are random. As we explained in Section 14.6,

Gaussian fields are strongly motivated by inflation. This class of field is the generic prediction of inflationary models where the density fluctuations are generated by quantum fluctuations in a scalar field during the inflationary phase.

For a Gaussian field δ, not only can the distribution function of values of δ at individual spatial positions be written in the form (13.7.4), but also the N-variate joint distribution of a set of $\delta_{\rm i} \equiv \delta(\boldsymbol{x}_i)$ can be written as a multivariate Gaussian distribution:

$$\mathcal{P}_N(\delta_1, \dots, \delta_N) = \frac{\|\boldsymbol{M}\|^{1/2}}{(2\pi)^{N/2}} \exp(-\tfrac{1}{2}\boldsymbol{V}^{\rm T} \cdot \boldsymbol{M} \cdot \boldsymbol{V}), \tag{13.7.5}$$

where $\boldsymbol{M}$ is the inverse of the correlation matrix $\boldsymbol{C} = \langle \delta_i \delta_j \rangle$, $\boldsymbol{V}$ is a column vector made from the $\delta_{\rm i}$, and $\boldsymbol{V}^{\rm T}$ is its transpose. An example for $N = 2$ will be given in equation (14.8.2). This expression (13.7.5) is considerably simplified by the fact that $\langle \delta_{\rm i} \rangle = 0$ by construction. The expectation value $\langle \delta_{\rm i} \delta_j \rangle$ can be expressed in terms of the *covariance function*, $\xi(r_{ij})$,

$$\langle \delta(\boldsymbol{x}_i)\delta(\boldsymbol{x}_j) \rangle = \xi(|\boldsymbol{x}_i - \boldsymbol{x}_j|) = \xi(r_{ij}), \tag{13.7.6}$$

where the averages are taken over all spatial positions with $|\boldsymbol{x}_i - \boldsymbol{x}_j| = r_{ij}$, and the second equality follows from the assumption of statistical homogeneity and isotropy. We shall see in the next section that $\xi(r)$ is intimately related to the power spectrum, $P(k)$. This means that the power spectrum or, equivalently, the covariance function of the density field is a particularly important statistic because it provides a complete statistical characterisation of the density field as long as it is Gaussian.

The ability to construct not only the N-dimensional joint distribution of values of δ, but also joint distributions of spatial derivatives of δ of arbitrary order, $\partial^n \delta / \partial x_i^n$, all of the form (13.7.5), but which involve spectral moments (13.2.10), is what makes Gaussian random fields so useful from an analytical point of view. The properties of Gaussian random fields are also interesting in the framework of biased galaxy-formation theories, which we discuss in Section 15.7. In this context one is particularly interested in regions of particularly high density which one might associate with galaxies. For example, one can show that the number of peaks of the density field per unit volume with height $\delta(\boldsymbol{x})/\sigma_0$ in the range ν to $\nu + {\rm d}\nu$, with $\nu \gg 1$, is

$$\mathcal{N}_{pk}(\nu)\,{\rm d}\nu \simeq \frac{1}{(2\pi)^2} \frac{\gamma}{R_*^3} (\nu^3 - 3\nu) \exp(-\tfrac{1}{2}\nu^2)\,{\rm d}\nu, \tag{13.7.7}$$

while the total number of peaks per unit volume with height exceeding $\nu\sigma$ is

$$n_{pk}(\nu) \simeq \frac{1}{(2\pi)^2} \frac{\gamma}{R_*^3} (\nu^2 - 1) \exp(-\tfrac{1}{2}\nu^2); \tag{13.7.8}$$

the quantities R_* and γ are defined by Equation (13.2.11). The mean distance between peaks of any height is of order $4R_*$. The ratio $R_0 = \sigma_0/\sigma_1 \simeq R_*/\gamma$ represents the order of magnitude of the coherence length of the field, i.e. the value of r at which the covariance function $\xi(r)$ becomes zero.

13.8 Covariance Functions

It is now appropriate to discuss the statistical properties of spatial fluctuations in ρ. We shall have recourse to much of this material in Chapter 16, when we discuss the comparison of galaxy-clustering data with quantities related to the density fluctuation, δ. Let us define the covariance function, introduced in the previous section by Equation (13.7.6), in terms of the density field $\rho(\boldsymbol{x})$ by

$$\xi(r) = \frac{\langle[\rho(\boldsymbol{x}) - \langle\rho\rangle][\rho(\boldsymbol{x}+\boldsymbol{r}) - \langle\rho\rangle]\rangle}{\langle\rho\rangle^2} = \langle\delta(\boldsymbol{x})\delta(\boldsymbol{x}+\boldsymbol{r})\rangle, \tag{13.8.1}$$

where the mean is taken over all points $\boldsymbol{x}$ in a representative volume V_u of the Universe in the manner of Section 13.2. From Equation (13.2.1) we have

$$\xi(\boldsymbol{r}) = \frac{1}{V_\mathrm{u}}\int_{V_\mathrm{u}}\sum_{\boldsymbol{k}}\delta_{\boldsymbol{k}}\exp(\mathrm{i}\boldsymbol{k}\cdot\boldsymbol{x})\sum_{\boldsymbol{k}'}\delta^*_{\boldsymbol{k}'}\exp[-\mathrm{i}\boldsymbol{k}'\cdot(\boldsymbol{x}+\boldsymbol{r})]\,\mathrm{d}\boldsymbol{x}, \tag{13.8.2 a}$$

which becomes

$$\xi(\boldsymbol{r}) = \sum_{\boldsymbol{k}}\langle|\delta_{\boldsymbol{k}}|^2\rangle\exp(-\mathrm{i}\boldsymbol{k}\cdot\boldsymbol{r}). \tag{13.8.2 b}$$

Passing to the limit $V_\mathrm{u} \to \infty$, equation (13.8.2 b) becomes

$$\xi(\boldsymbol{r}) = \frac{1}{(2\pi)^3}\int P(k)\exp(-\mathrm{i}\boldsymbol{k}\cdot\boldsymbol{r})\,\mathrm{d}\boldsymbol{k}. \tag{13.8.3}$$

One can also find the inverse relation quite easily:

$$\langle|\delta_{\boldsymbol{k}}|^2\rangle = \frac{1}{V_\mathrm{u}}\int\xi(\boldsymbol{r})\exp(\mathrm{i}\boldsymbol{k}\cdot\boldsymbol{r})\,\mathrm{d}\boldsymbol{r}. \tag{13.8.4}$$

Passing to the limit $V_\mathrm{u} \to \infty$, the preceding relation can be shown to be

$$P(k) = \int\xi(\boldsymbol{r})\exp(\mathrm{i}\boldsymbol{k}\cdot\boldsymbol{r})\,\mathrm{d}\boldsymbol{r}: \tag{13.8.5}$$

the power spectrum is just the Fourier transform of the covariance function, a result known as the *Wiener-Khintchine theorem*. If μ is the cosine of the angle between $\boldsymbol{k}$ and $\boldsymbol{r}$, the integral over all directions of $\boldsymbol{r}$ gives

$$\int_\Omega \exp(-\mathrm{i}kr\mu)\,\mathrm{d}\Omega = \int_0^{2\pi}\mathrm{d}\phi\int_{-1}^{+1}\exp(-\mathrm{i}kr\mu)\,\mathrm{d}\mu = 4\pi\frac{\sin kr}{kr}. \tag{13.8.6}$$

It turns out therefore that

$$\xi(r) = \frac{1}{2\pi^2}\int_0^\infty P(k)\frac{\sin kr}{kr}k^2\,\mathrm{d}k, \tag{13.8.7}$$

which has inverse

$$P(k) = 4\pi\int_0^\infty \xi(r)\frac{\sin kr}{kr}r^2\,\mathrm{d}r. \tag{13.8.8}$$

Averaging equation (13.8.2 *b*) over $\boldsymbol{r}$ gives

$$\langle \xi(\boldsymbol{r}) \rangle_{\boldsymbol{r}} = \frac{1}{V_{\mathrm{u}}} \sum_{\boldsymbol{k}} \langle |\delta_{\boldsymbol{k}}|^2 \rangle \int \exp(-\mathrm{i}\boldsymbol{k} \cdot \boldsymbol{r}) \, \mathrm{d}\boldsymbol{r} = 0. \tag{13.8.9}$$

In a homogeneous and isotropic universe the function $\xi(\boldsymbol{r})$ does not depend on either the origin or the direction of $\boldsymbol{r}$, but only on its modulus; the result (13.8.9) implies therefore that

$$\lim_{r \to \infty} \frac{1}{r^3} \int_0^r \xi(r') r'^2 \, \mathrm{d}r' = 0 : \tag{13.8.10}$$

in general the covariance function must change sign – from positive at the origin, at which (13.8.1) guarantees $\xi(0) = \langle \sigma^2 \rangle \geqslant 0$, to negative at some r – to make the overall integral (13.8.10) converge in the correct way. A perfectly homogeneous distribution would have $P(k) \equiv 0$ and $\xi(r)$ would be identically zero for all r.

The meaning of the function $\xi(r)$ can be illustrated by the following example. Imagine that the material in the Universe is distributed in regions of the same size r_0 with density fluctuations $\delta > 0$ and $\delta < 0$. In this case the product $\delta(\boldsymbol{x})\delta(\boldsymbol{x}+\boldsymbol{r})$ will be, on average, positive for distances $r < r_0$ and negative for $r > r_0$. This means that the function $\xi(r)$ reaches zero at a value $r \simeq r_0$, which represents the mean size of regions and therefore the coherence length of the fluctuation field. Inside the regions themselves, where $\xi(r) > 0$, there is correlation, while, outside the regions, where $\xi(r) < 0$, there is anticorrelation.

The function $\xi(r)$ is the two-point covariance function. In an analogous manner it is possible to define spatial covariance functions for $N > 2$ points. For example, the three-point covariance function is

$$\zeta(r, s, t) = \frac{\langle [\rho(\boldsymbol{x}) - \langle \rho \rangle][\rho(\boldsymbol{x} + \boldsymbol{r}) - \langle \rho \rangle][\rho(\boldsymbol{x} + \boldsymbol{s}) - \langle \rho \rangle] \rangle}{\langle \rho \rangle^3}, \tag{13.8.11}$$

which gives

$$\zeta(r, s, t) = \langle \delta(\boldsymbol{x})\delta(\boldsymbol{x} + \boldsymbol{r})\delta(\boldsymbol{x} + \boldsymbol{s}) \rangle, \tag{13.8.12}$$

where the mean is taken over all the points $\boldsymbol{x}$ and over all directions of $\boldsymbol{r}$ and $\boldsymbol{s}$ such that $|\boldsymbol{r} - \boldsymbol{s}| = t$: in other words, over all points defining a triangle with sides r, s and t.

The generalisation of (13.8.12) to $N > 3$ is obvious. It is convenient to define quantities related to the N-point covariance functions called the *cumulants*, κ_N, which are constructed from the moments of order up to and including N. The cumulants are defined as the part of the expectation value $\langle \delta_1 \dots \delta_N \rangle$ ($\delta_1 \equiv \delta(\boldsymbol{x}_1)$, etc.), of which (13.8.12) is the special case for $N = 3$, which cannot be expressed in terms of expectation values of lower order. Cumulants are also sometimes called the *connected part* of the corresponding covariance function. To determine them in terms of $\langle \delta_1 \delta_2 \dots \delta_N \rangle$ for any order, one simply expresses the required expectation value as a sum over all distinct possible partitions of the set $\{1, \dots, N\}$, ignoring the ordering of the components of the set; the cumulant is just the part of this sum which corresponds to the unpartitioned set. This definition makes use of the *cluster expansion*. For example, the possible partitions of the set $\{1, 2, 3\}$ are $(\{1\}, \{2, 3\})$, $(\{2\}, \{1, 3\})$ $(\{3\}, \{1, 2\})$, $(\{1\}, \{2\}, \{3\})$ and the unpartitioned set $(\{1, 2, 3\})$. This means that the expectation value can be written

$$\begin{aligned}\langle\delta_1\delta_2\delta_3\rangle &= \langle\delta_1\rangle_c\langle\delta_2\delta_3\rangle_c + \langle\delta_2\rangle_c\langle\delta_1\delta_3\rangle_c \\ &\quad + \langle\delta_3\rangle_c\langle\delta_1\delta_2\rangle_c + \langle\delta_1\rangle_c\langle\delta_2\rangle_c\langle\delta_3\rangle_c + \langle\delta_1\delta_2\delta_3\rangle_c. \quad (13.8.13)\end{aligned}$$

The cumulants are $\kappa_3 \equiv \langle\delta_1\delta_2\delta_3\rangle_c$, $\kappa_2 = \langle\delta_1\delta_2\rangle_c$, etc. Since $\langle\delta\rangle = 0$ by construction, $\kappa_1 = \langle\delta_1\rangle_c = \langle\delta_1\rangle = 0$. Moreover, $\kappa_2 = \langle\delta_1\delta_2\rangle_c = \langle\delta_1\delta_2\rangle$. The second- and third-order cumulants are simply the same as the covariance functions. The fourth- and higher-order quantities are different, however. The particularly useful aspect of the cumulants which motivates their use is that all κ_N for $N > 2$ are zero for a Gaussian random field; for such a field the odd N expectation values are all zero, and the even ones can be expressed as combinations of $\langle\delta_i\delta_j\rangle$ in such a way that the connected part is zero.

It is possible to define $\xi(r)$ also in terms of a discrete distribution of masses rather than a continuous density field. Formally one can write the density field $\rho(\boldsymbol{x}) = \sum_i m_i\delta^{\mathrm{D}}(\boldsymbol{x} - \boldsymbol{x}_i)$, where the sum is taken over all the mass points labelled by i and found at position $\boldsymbol{x}_i$; δ^{D} is the Dirac function. If all the $m_i = m$, the mean density is $\langle\rho\rangle = n_V m$. The probability of finding a mass point in a randomly chosen volume δV at $\boldsymbol{x}$ is therefore $\delta P = m^{-1}\rho(\boldsymbol{x})\delta V$; the joint probability of finding a point in δV_1 and a point in δV_2 separated by a distance r is

$$\begin{aligned}\delta^2 P_2 &= \frac{\langle\rho(\boldsymbol{x})\rho(\boldsymbol{x}+\boldsymbol{r})\rangle}{m^2}\delta V_1\delta V_2 \\ &= n_V^2\frac{\langle\rho(\boldsymbol{x})\rho(\boldsymbol{x}+\boldsymbol{r})\rangle}{\langle\rho\rangle^2}\delta V_1\delta V_2 \\ &= n_V^2[1 + \xi(r)]\delta V_1\delta V_2, \quad (13.8.14)\end{aligned}$$

which defines $\xi(r)$ to be the *two-point correlation function* of the mass points. The same result holds if we take the probability of finding a point in a small volume δV, where the density is ρ, to be proportional to ρ. This forms the so-called *Poisson clustering model* which we shall use later, in Section 16.6.

One can also extend the (discrete) correlations to orders $N > 2$ by a straightforward generalisation of equation (13.8.14):

$$\delta^N P_N = n_V^N[1 + \xi^{(N)}(\boldsymbol{r})]\delta V_1 \ldots \delta V_N, \quad (13.8.15)$$

where $\boldsymbol{r}$ stands for all the r_{ij} separating the N points. However, the function $\xi^{(N)}(\boldsymbol{r})$, which is called the total N-point correlation function, contains contributions from correlations of orders less than N. For example, the number of triplets is larger than a random distribution partly because there are more pairs than in a random distribution:

$$\delta^3 P_3 = n_V^3[1 + \xi_{23} + \xi_{13} + \xi_{12} + \zeta_{123}]\delta V_1\delta V_2\delta V_3. \quad (13.8.16)$$

The part of $\xi^{(3)}$ which does not depend on ξ_{ij}, usually written ζ_{123}, is called the *irreducible* or *connected* three-point function. The four-point correlation function $\xi^{(4)}$ will contain terms in ζ_{ijk}, $\xi_{ij}\xi_{kl}$ and ξ_{ij}, which must be subtracted to give the connected four-point function η_{1234}. The connected correlation functions are analogous to the cumulants defined above for continuous variables, and are constructed from the same cluster expansion. The only difference is that, for discrete distributions, one interprets single partitions (e.g. $\langle\delta_1\rangle_c$) as having the value unity rather than zero. For the two-point function there are only two partitions, $(\{1\},\{2\})$ and $(\{1,2\})$. The first term would correspond to $\langle\delta_1\rangle\langle\delta_2\rangle = 0$ in the continuous variable case because $\langle\delta\rangle = 0$, but the two expectation values are each assigned a value of unity in the discrete variable case, so that $\delta^2 P_2 \propto 1 + \xi(r)$ and $\xi^{(2)}(r) = \xi(r)$, as expected. For the three-point function, the right-hand side of Equation (13.8.12) has, first, three terms corresponding to the three terms in ξ_{ij} in Equation (13.8.16), then a product of three single-partitions each with the value unity, and finally a triplet which corresponds to the connected part ζ_{123}. This reconciles the forms of (13.8.16) and (13.8.12) and shows that $\xi^{(3)} = \xi_{23} + \xi_{13} + \xi_{12} + \zeta_{123}$. This procedure can be generalised straightforwardly to higher N.

13.9 Non-Gaussian Fluctuations?

As we have explained, the power spectrum of density fluctuations scales in the linear regime in such a way that each mode evolves independently according to the growth law. This means, for example, that $\sigma_M \propto t^{2/3}$ in an Einstein–de Sitter model. Since each mode evolves independently, the random-phase hypothesis of Section 13.7 continues to hold as the perturbations evolve linearly and the distribution of δ should therefore remain Gaussian.

Notice, however, that δ is constrained to have a value $\delta \geqslant -1$, otherwise the energy density ρ would be negative. The Gaussian distribution (13.7.3) always assigns a non-zero probability to regions with $\delta < -1$. The error in doing this is negligible when σ_M is small because the probability of $\delta < -1$ is then very small, but, as fluctuations enter the nonlinear regime with $\sigma_M \simeq 1$, the error must increase to a point where the Gaussian distribution is a very poor approximation to the true distribution function. What happens is that, as the fluctuations evolve into this regime, mode-coupling effects cause the initial distribution to skew, generating a long tail at high δ while they are also bounded at $\delta = -1$. Notice, however, that if the mass distribution is smoothed on a scale M, one should recover the regime where $\sigma_M \ll 1$, where the field will still be Gaussian. Large scales therefore continue to evolve linearly, even when small scales have undergone nonlinear collapse in the manner described in the next chapter.

The generation of non-Gaussian features as a result of the nonlinear evolution of initially Gaussian perturbations is well known and can be probed using numerical simulations or analytical approximations. We shall not say much about this question here, except to remark that, on scales where such effects are important, the power spectrum, or, equivalently, the covariance function, does not furnish a complete statistical description of the properties of the density field δ.

Despite the strong motivation for the Gaussian scenario from inflationary models we should at least mention the possibility that either the primordial fluctuations are not Gaussian or that some later mechanism, apart from gravity, induces non-Gaussian behaviour during their evolution.

Attempts to construct inflationary models with non-Gaussian fluctuations due to oscillations in Φ have largely been unsuccessful. It is necessary to have some kind of feature in the potential $V(\Phi)$ or to have more than one scalar field. There are, however, some other possibilities. First, as we mentioned briefly in Section 7.6, it is possible that some form of topological defect might survive a phase transition in the early Universe. These defects comprise regions of trapped energy density which could act as seeds for structure formation. However, in such pictures the seeds are very different from quantum fluctuations induced during inflation and would be decidedly non-Gaussian at very early times. One of the early favourites for a theory based on this idea was the *cosmic-string scenario* in which one-dimensional string-like defects act as seeds. The behaviour of a network of cosmic strings is difficult to handle even with numerical methods and this scenario did not live up to its early promise. The original idea was that the evolving network would form loops of string which shrink and produce gravitational waves; as they do so they accrete matter. More accurate simulations, however, showed that this does not happen and that small loops cannot be responsible for structure formation. A revised version of this theory has been suggested more recently, in which long pieces of string, moving relativistically, produce 'wakes' which can give rise to sheet-like inhomogeneities. Another possibility is that three-dimensional defects called *textures*, rather than one-dimensional strings, might be the required seed. Perhaps primordial black holes could also act as a form of zero-dimensional seed. These pictures do not seem as compelling as the 'inflationary paradigm' we have mentioned above, but they are not ruled out by present observations.

The second possibility is that some astrophysical mechanism might induce non-Gaussian behaviour. A possible example is that some kind of *cosmic explosion*, perhaps associated with early formation of very massive objects, could form a blast wave which would push material around into a bubbly or cellular pattern at early times (e.g. Ostriker and Cowie 1981). This would be non-Gaussian and would subsequently evolve under its own gravity to form a distribution very dissimilar to that which would form in an inflationary model. Unfortunately, this model seems to be ruled out by the lack of any distortions in the spectrum of the microwave background radiation; see Chapter 19.

Although there is no strongly compelling physical motivation for non-Gaussian fluctuations, one should be sure to test the Gaussian assumption as rigorously as possible. One can do this in many ways, using the microwave background and galaxy-clustering statistics. Until non-Gaussian models are shown to be excluded by the observations, there is always the possibility that some physics we do not yet understand created initial fluctuations of a very different form to those predicted by inflation.

Bibliographic Notes on Chapter 13

An interesting discussion of the properties of primordial power spectra is given by Gott (1980). Adler (1981) and Vanmarcke (1983) are useful texts on the general mathematical properties of Gaussian random fields; application of Gaussian random fields in a cosmological context are discussed by Bardeen *et al.* (1986), a famous paper known to the community as BBKS. Non-Gaussian perturbations are discussed by Brandenberger (1990) and Coulson *et al.* (1994).

Problems

1. Show that if a perturbation field has a power spectrum of the form $k\exp(-\lambda_0 k)$, then the covariance function crosses zero at $r = \lambda_0\sqrt{3}$. Give a physical interpretation of this result.

2. Calculate the spectral parameters (13.2.11) for the power spectrum defined in Question 1.

3. A lognormal field $Y(\boldsymbol{r})$ is defined by $Y(\boldsymbol{r}) = \exp[X(\boldsymbol{r})]$, where X is a Gaussian random field. Calculate the two-point covariance function of Y in terms of the covariance function of X.

4. For the lognormal field Y defined in Question 3 calculate the three-point function (a) in terms of the two-point function of X, and (b) in terms of the two-point function of Y.

5. Repeat Questions 3 and 4 for the χ^2 field defined by $Z = X^2$, where X is a Gaussian random field.

14

Nonlinear Evolution

After recombination, fluctuations in the matter component δ on a scale $M > M_J^{(i)}(z_{rec}) \simeq 10^5 M_\odot$ grow according to the theory developed in Chapters 10–12 while $|\delta| \ll 1$. This is obviously a start, but it cannot be used to follow the evolution of structure into the strongly nonlinear regime where overdensities can exist with $\delta \gg 1$. A cluster of galaxies, for example, corresponds to a value of δ of order several hundred or more. To account for structure formation we therefore need to develop techniques for studying the nonlinear evolution of perturbations. This is a much harder problem than the linear case, and exact solutions are difficult to achieve. We shall mention some analytical and numerical approaches in this chapter.

14.1 The Spherical 'Top-Hat' Collapse

The simplest approach to nonlinear evolution is to follow an inhomogeneity which has some particularly simple form. This is not directly relevant to interesting cosmological models, because the real fluctuations are expected to be highly irregular and random. Considering cases of special geometry can nevertheless lead to important insights. In this spirit let us consider a spherical perturbation with constant density inside it which, at an initial time $t_i \simeq t_{rec}$, has an amplitude $\delta_i > 0$ and $|\delta_i| \ll 1$. This sphere is taken to be expanding with the background universe in such a way that the initial peculiar velocity at the edge, V_i, is zero. As we have mentioned before, the symmetry of this situation means that we can treat the perturbation as a separate universe and, for simplicity, we assume that the background universe at t_i is described by an Einstein–de Sitter model; in this case we get

$$\delta = \delta_+(t_i)\left(\frac{t}{t_i}\right)^{2/3} + \delta_-(t_i)\left(\frac{t}{t_i}\right)^{-1}, \tag{14.1.1 a}$$

$$V = i\frac{\dot{\delta}}{k} = \frac{i}{k_i t_i}\left[\tfrac{2}{3}\delta_+(t_i)\left(\frac{t}{t_i}\right)^{-1/3} - \delta_-(t_i)\left(\frac{t}{t_i}\right)^{-4/3}\right] \tag{14.1.1 b}$$

(as usual, the symbol '+' indicates the growing mode, while '−' denotes the decaying mode). The combination of growing and decreasing modes in Equations (14.1.1) is necessary to satisfy the correct boundary condition on the velocity: $V_i = 0$ requires that $\delta_+(t_i) = \frac{3}{5}\delta_i$. One can assume that, after a short time, the decaying mode will become negligible and the perturbation remaining will just be $\delta \simeq \delta_+(t_i)$. Let us take the initial value of the Hubble expansion parameter to be H_i. Assuming that pressure gradients are negligible, the sphere representing the perturbation evolves like a Friedmann model whose initial density parameter is given by

$$\Omega_p(t_i) = \frac{\rho(t_i)(1+\delta_i)}{\rho_c(t_i)} = \Omega(t_i)(1+\delta_i), \tag{14.1.2}$$

where the suffix 'p' denotes the quantity relevant for the perturbation, while $\rho(t_i)$ and $\Omega(t_i)$ refer to the unperturbed background universe within which the perturbation resides. Structure will be formed if, at some time t_m, the spherical region ceases to expand with the background universe and instead begins to collapse. This will happen to any perturbation with $\Omega_p(t_i) > 1$. From Equations (14.1.2) and (2.6.4) this condition can easily be seen to be equivalent to

$$\delta_+(t_i) = \tfrac{3}{5}\delta_i > \frac{3}{5}\frac{1-\Omega(t_i)}{\Omega(t_i)} = \frac{3}{5}\frac{1-\Omega}{\Omega(1+z_i)}, \tag{14.1.3}$$

where Ω is the present value of the density parameter. In universes with $\Omega < 1$, however, the fluctuation must exceed the critical value $(1-\Omega)/\Omega(1+z_i)$; it is interesting to note that in this case the condition (14.1.3) implies that the growing perturbation reaches the nonlinear regime before the time t^* at which the universe becomes curvature dominated and therefore enters a phase of undecelerated free expansion. For $\Omega \geqslant 1$, on the other hand, there is no problem.

The expansion of the perturbation is described by the equation

$$\left(\frac{\dot{a}}{a_i}\right)^2 = H_i^2\left[\Omega_p(t_i)\frac{a_i}{a} + 1 - \Omega_p(t_i)\right], \tag{14.1.4}$$

from which we easily obtain that the density of the perturbation at time t_m is

$$\rho_p(t_m) = \rho_c(t_i)\Omega_p(t_i)\left[\frac{\Omega_p(t_i)-1}{\Omega_p(t_i)}\right]^3; \tag{14.1.5}$$

the value of t_m, from Equation (2.4.9) (where t_0 is replaced by t_i) and Equation (14.1.5), is just

$$t_m = \frac{\pi}{2H_i}\frac{\Omega_p(t_i)}{[\Omega_p(t_i)-1]^{3/2}} = \frac{\pi}{2H_i}\left[\frac{\rho_c(t_i)}{\rho_p(t_m)}\right]^{1/2} = \left[\frac{3\pi}{32G\rho_p(t_m)}\right]^{1/2}. \tag{14.1.6}$$

In an Einstein–de Sitter universe the ratio χ between the background density, $\rho(t_\mathrm{m})$, and the density inside the perturbation, $\rho_\mathrm{p}(t_\mathrm{m})$, is obtained from the previous equation and from

$$\rho(t_\mathrm{m}) = \frac{1}{6\pi G t_\mathrm{m}^2}; \tag{14.1.7}$$

it follows that

$$\chi = \frac{\rho_\mathrm{p}(t_\mathrm{m})}{\rho(t_\mathrm{m})} = \left(\frac{3\pi}{4}\right)^2 \simeq 5.6, \tag{14.1.8}$$

which corresponds to a perturbation $\delta_+(t_\mathrm{m}) \simeq 4.6$; the extrapolation of the linear growth law, $\delta_+ \propto t^{2/3}$, would have yielded, from (14.1.6),

$$\delta_+(t_\mathrm{m}) = \delta_+(t_\mathrm{i})\left(\frac{t_\mathrm{m}}{t_\mathrm{i}}\right)^{2/3} = \delta_+(t_\mathrm{i})(\tfrac{3}{4}\pi)^{2/3}\frac{\Omega_\mathrm{p}(t_\mathrm{i})^{2/3}}{\delta_\mathrm{i}} \simeq \tfrac{3}{5}(\tfrac{3}{4}\pi)^{2/3} \simeq 1.07, \tag{14.1.9}$$

corresponding to the approximate value $\rho_\mathrm{p}(t_\mathrm{m})/\rho(t_\mathrm{m}) \simeq 1 + \delta_+(t_\mathrm{m}) \simeq 2.07$. The perturbation will subsequently collapse and, if one can still ignore pressure effects and the configuration remains spherically symmetric, in a time t_c of order $2t_\mathrm{m}$, one will find an infinite density at the centre. In fact, when the density is high, slight departures from this symmetry will result in the formation of shocks and considerable pressure gradients. Heating of the material will occur due to the dissipation of shocks which converts some of the kinetic energy of the collapse into heat, i.e. random thermal motions. The end result will therefore be a final equilibrium state which is not a singular point but some extended configuration with radius R_vir and mass M. From the virial theorem the total energy of the fluctuation is

$$E_\mathrm{vir} = -\frac{1}{2}\frac{3GM^2}{5R_\mathrm{vir}}. \tag{14.1.10}$$

If in the collapsing phase we can ignore the possible loss of mass from the system due to effects connected with shocks, and possible loss of energy by thermal radiation, the energy and mass in (14.1.10) are the same as the fluctuation had at time t_m,

$$E_\mathrm{m} = -\frac{3}{5}\frac{GM^2}{R_\mathrm{m}}, \tag{14.1.11}$$

where R_m is the radius of the sphere at the moment of maximum expansion. Having assumed that the pressure is zero, in Equation (14.1.11) no account is taken of the contribution of thermal energy; the kinetic energy due to the expansion is zero by definition at this point. From Equations (14.1.10) and (14.1.11) we therefore have $R_\mathrm{m} = 2R_\mathrm{vir}$, so that the density in the equilibrium state is $\rho_\mathrm{p}(t_\mathrm{vir}) = 8\rho_\mathrm{p}(t_\mathrm{m})$. One usually assumes that at t_c, the time of maximum compression, the density is of order $\rho_\mathrm{p}(t_\mathrm{vir})$. Numerical simulations of the collapse allow an estimate to be made of the time taken to reach equilibrium: one finds that $t_\mathrm{vir} \simeq 3t_\mathrm{m}$. If at times t_c and t_vir the universe is still described by an Einstein–de Sitter model, the ratios between the density in the perturbation and the mean density of the universe at these times are

$$\frac{\rho_{\rm p}(t_{\rm c})}{\rho(t_{\rm c})} = 2^2 8\chi \simeq 180, \tag{14.1.12 a}$$

$$\frac{\rho_{\rm p}(t_{\rm vir})}{\rho(t_{\rm vir})} = 3^2 8\chi \simeq 400, \tag{14.1.12 b}$$

respectively. An extrapolation of linear perturbation theory would give

$$\delta_+(t_{\rm c}) \simeq \tfrac{3}{5}(\tfrac{3}{4}\pi)^{2/3}2^{2/3} \simeq 1.68, \tag{14.1.13 a}$$

$$\delta_+(t_{\rm vir}) \simeq \tfrac{3}{5}(\tfrac{3}{4}\pi)^{2/3}3^{2/3} \simeq 2.20, \tag{14.1.13 b}$$

which correspond to values of 2.68 and 3.20 for the ratio of the densities, in place of the exact values given by Equations (14.1.12 *a*) and (14.1.12 *b*).

14.2 The Zel'dovich Approximation

The model discussed in the previous section, though very instructive in its conclusions, suffers from some notable defects. Above all, reasonable models of structure formation do not contain primordial fluctuations at $t_{\rm i} \simeq t_{\rm rec}$, which are organised into neat homogeneous spherical regions with zero peculiar velocity at their edge. Moreover, even if this were the case at the beginning, such a symmetrical configuration is strongly unstable with respect to the growth of non-radial motions during the expansion and collapse phases of the inhomogeneity. In fact, the classic work of Lin *et al.* (1965) showed that, for a generic triaxial perturbation, the collapse is expected to occur not to a point, but to a flattened structure of quasi-two-dimensional nature. The usual descriptive term for such features is *pancakes.*

The spherical top-hat model is only reasonably realistic for perturbations on scales just a little larger than $M_{\rm J}^{(\rm i)}(z_{\rm rec})$. In this case, however, pressure is not negligible and dissipation can be significant during the collapse. Presumably what form in such a situation are more or less spherical protoobjects in which gravity is balanced by pressure forces.

It is more complicated to study the development of perturbations on scales $M \geqslant M_{\rm D}^{(\rm a)}(z_{\rm rec})$. Of course, one could simply resort to numerical methods like those we shall discuss in Section 15.5. However, some simplifying assumptions are possible. For example, in this situation, pressure would be effectively zero and the fluid can be treated like dust. Under this assumption it is in fact possible to understand the growth of structure analytically using a clever approximation devised by Zel'dovich (1970). This approximation actually predicts that the density in certain regions - called *caustics* - should become infinite, but the gravitational acceleration caused by these regions remains finite. Of course, in any case one cannot justify ignoring pressure when the density becomes very high, for much the same reason as we discussed in Section 15.1 in the context of spherical collapse: one forms shock waves which compress infalling material. At a certain point the process of accretion onto the caustic will stop: the condensed matter is contained by gravity within the final structure, while the matter which has not passed through the shock wave is held up by pressure. It has been calculated that about half the material inside the original fluctuation is reheated and compressed by the shock wave. An important property of the structures which thus form is that they are strongly unstable to fragmentation. In principle, therefore, one can generate structure on smaller scales than the pancake.

Let us now describe the Zel'dovich approximation in more detail, and show how it can follow the evolution of perturbations until the formation of pancakes. Imagine that we begin with a set of particles which are uniformly distributed in space. Let the initial (i.e. Lagrangian) coordinate of a particle in this unperturbed distribution be $\boldsymbol{q}$. Now each particle is subjected to a displacement corresponding to a density perturbation. In the Zel'dovich approximation the Eulerian coordinate of the particle at time t is

$$\boldsymbol{r}(t, \boldsymbol{q}) = a(t)[\boldsymbol{q} - b(t)\nabla_q \Phi_0(\boldsymbol{q})], \tag{14.2.1}$$

where $\boldsymbol{r} = a(t)\boldsymbol{x}$, with $\boldsymbol{x}$ a comoving coordinate, and we have made $a(t)$ dimensionless by dividing throughout by $a(t_{\mathrm{i}})$, where t_{i} is some reference time which we take to be the initial time. The derivative on the right-hand side is taken with respect to the Lagrangian coordinates. The dimensionless function $b(t)$ describes the evolution of a perturbation in the linear regime, with the condition $b(t_{\mathrm{i}}) = 0$, and therefore solves the equation

$$\ddot{b} + 2\frac{\dot{a}}{a}\dot{b} - 4\pi G\rho b = 0. \tag{14.2.2}$$

This equation corresponds to (10.6.14), with vanishing pressure term, which describes the gravitational instability of a matter-dominated universe. For a flat matter-dominated universe we have $b \propto t^{2/3}$ as usual. The quantity $\Phi_0(\boldsymbol{q})$ is proportional to a velocity potential, i.e. a quantity of which the velocity field is the gradient, because, from Equation (14.2.1),

$$\boldsymbol{V} = \frac{\mathrm{d}\boldsymbol{r}}{\mathrm{d}t} - H\boldsymbol{r} = a\frac{\mathrm{d}\boldsymbol{x}}{\mathrm{d}t} = -a\dot{b}\nabla_q \Phi_0(\boldsymbol{q}); \tag{14.2.3}$$

this means that the velocity field is irrotational. The quantity $\Phi_0(\boldsymbol{q})$ is related to the density perturbation in the linear regime by the relation

$$\delta = b\nabla_q^2 \Phi_0, \tag{14.2.4}$$

which is a simple consequence of Poisson's equation.

The Zel'dovich approximation is therefore simply a linear approximation with respect to the particle displacements rather than the density, as was the linear solution we derived above. It is conventional to describe the Zel'dovich approximation as a first-order Lagrangian perturbation theory, while what we have dealt with so far for $\delta(t)$ is a first-order Eulerian theory. It is also clear that Equation (14.2.1) involves the assumption that the position and time dependence of the displacement between initial and final positions can be separated. Notice that particles in the Zel'dovich approximation execute a kind of inertial motion on straight line trajectories.

The Zel'dovich approximation, though simple, has a number of interesting properties. First, it is exact for the case of one-dimensional perturbations up to the moment of shell crossing. As we have mentioned above, it also incorporates irrotational motion, which is required to be the case if it is generated only by the action of gravity (due to the Kelvin circulation theorem). For small displacements between $\boldsymbol{r}$ and $a(t)\boldsymbol{q}$, one recovers the usual (Eulerian) linear regime: in fact, Equation (14.2.1) defines a unique mapping between the coordinates $\boldsymbol{q}$ and $\boldsymbol{r}$ (as long as trajectories do not cross); this means that $\rho(\boldsymbol{r},t)\,\mathrm{d}^3r = \langle\rho(t_\mathrm{i})\rangle\,\mathrm{d}^3q$ or

$$\rho(\boldsymbol{r},t) = \frac{\langle\rho(t)\rangle}{|J(\boldsymbol{r},t)|}, \tag{14.2.5}$$

where $|J(\boldsymbol{r},t)|$ is the determinant of the Jacobian of the mapping between $\boldsymbol{q}$ and $\boldsymbol{r}$: $\partial\boldsymbol{r}/\partial\boldsymbol{q}$. Since the flow is irrotational, the matrix J is symmetric and can therefore be locally diagonalised. Hence

$$\rho(\boldsymbol{r},t) = \langle\rho(t)\rangle\prod_{i=1}^{3}[1 + b(t)\alpha_\mathrm{i}(\boldsymbol{q})]^{-1}: \tag{14.2.6}$$

the quantities $1 + b(t)\alpha_\mathrm{i}$ are the eigenvalues of the matrix J (the α_i are the eigenvalues of the deformation tensor). For times close to t_i, when $|b(t)\alpha_\mathrm{i}| \ll 1$, Equation (14.2.6) yields

$$\delta \simeq -(\alpha_1 + \alpha_2 + \alpha_3)b(t), \tag{14.2.7}$$

which is the law of perturbation growth in the linear regime.

Equation (14.2.6) indicates that at some time t_sc, when $b(t_\mathrm{sc}) = -1/\alpha_j$, an event called *shell-crossing* occurs such that a singularity appears and the density becomes formally infinite in a region where at least one of the eigenvalues (in this case α_j) is negative. This condition corresponds to the situation where two points with different Lagrangian coordinates end up at the same Eulerian coordinate. In other words, particle trajectories have crossed and the mapping (14.2.1) is no longer unique. A region where the shell-crossing occurs is called a caustic. For a fluid element to be collapsing, at least one of the α_j must be negative. If more than one is negative, then collapse will occur first along the axis corresponding to the most negative eigenvalue. If there is no special symmetry, one therefore expects collapse to be generically one dimensional, i.e. to a sheet or 'pancake'. Only if two (or three) negative eigenvalues, very improbably, are equal in magnitude can the collapse occur to a filament (or point). One therefore expects 'pancake' formation to be the generic result of structure collapse.

The Zel'dovich approximation matches very well the evolution of density perturbations in full N-body calculations until the point where shell crossing occurs

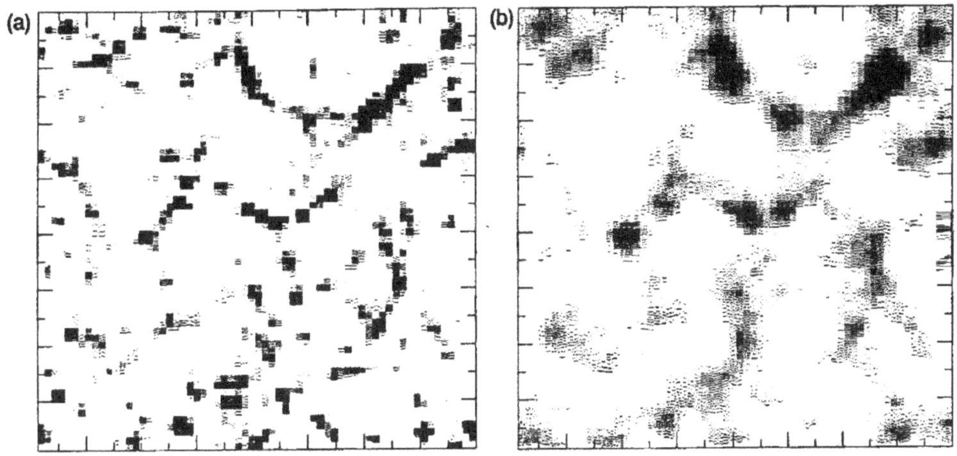

Figure 14.1 Comparison of the Zel'dovich approximation (b) and an N-body experiment (a) for the same initial conditions. Agreement is good, except for the 'fuzzy' appearance of the pancake regions which is due to the motion of particles after shell-crossing.

(Coles *et al.* 1993a); we shall discuss N-body methods later on. After this, the approximation breaks down completely. According to Equation (14.2.1) particles continue to move through the caustic in the same direction as they did before. Particles entering a pancake from either side merely sail through it and pass out the opposite side. The pancake therefore appears only instantaneously and is rapidly smeared out. In reality, the matter in the caustic would feel the strong gravity there and be pulled back towards it before it could escape through the other side. Since the Zel'dovich approximation is only kinematic it does not account for these close-range forces and the behaviour in the strongly nonlinear regime is therefore described very poorly. Furthermore, this approximation cannot describe the formation of shocks and phenomena associated with pressure. The problem of shell-crossing is inevitable in the Zel'dovich approximation. In order to prevent this from interfering too much in calculations, one can filter out the small-scale fluctuations from the initial conditions which give rise to shell-crossing. If the power spectrum is a decreasing function of mass, then the large scales can be evolving in the quasilinear regime (i.e. before shell-crossing) even when a higher resolution would reveal considerable small-scale caustics. By smoothing the density field one removes these small-scale events but does not alter the kinematical evolution of the large-scale field. The best way to implement this idea appears to be to filter the initial power spectrum according to

$$P(k) \to P(k) \exp(-k^2/k_G^2), \tag{14.2.8}$$

where $k_{\rm nl} < k_G < 1.5k_{\rm nl}$ and $k_{\rm nl}$ is the characteristic nonlinear wavenumber given approximately by

$$\frac{1}{2\pi^2} \int_0^{k_{\rm nl}} P(k)k^2\, dk = 1, \tag{14.2.9}$$

so that the RMS density fluctuation σ_M on a scale $R \simeq 2\pi/k_{\rm nl}$ is of order unity. The performance of the Zel'dovich approximation, the 'smoothed' Zel'dovich approximation and a full N-body simulation from a realisation of Gaussian initial conditions is shown in Figure 14.1.

14.3 The Adhesion Model

The smoothed Zel'dovich approximation merely ignores the problem of shell-crossing. If one is forced to deal with it, in other words if one wants to study the mass distribution on scales where $\sigma_M > 1$, then one must come up with some other approach. One relatively straightforward way to extend the Zel'dovich approximation is through the so-called *adhesion model*.

In the adhesion model one assumes that the particles stick to each other when they enter a caustic region because of an artificial viscosity which is intended to simulate the action of strong gravitational effects inside the overdensity forming there. This 'sticking' results in a cancellation of the component of the velocity of the particle perpendicular to the caustic. If the caustic is two dimensional, the particles will move in its plane until they reach a one-dimensional interface between two such planes. This would then form a filament. Motion perpendicular to the filament would be cancelled, and the particles will flow along it until a point where two or more filaments intersect, thus forming a node. The smaller the viscosity term is, the thinner the sheets and filaments will be, and the more point-like the nodes will be. Outside these structures, the Zel'dovich approximation is still valid to high accuracy. Comparing simulations made within this approximation with full N-body calculations shows that it is quite accurate for overdensities up to $\delta \simeq 10$.

Let us begin by rewriting the Euler and continuity equations, together with the Poisson equation (all ignoring the effects of pressure), in a slightly altered form

$$\frac{\partial \boldsymbol{V}}{\partial t} + \frac{\dot{a}}{a}\boldsymbol{V} + \frac{1}{a}(\boldsymbol{V} \cdot \nabla_{\boldsymbol{x}})\boldsymbol{V} = -\frac{1}{a}\nabla_{\boldsymbol{x}}\varphi, \tag{14.3.1 a}$$

$$\frac{\partial \rho}{\partial t} + 3\frac{\dot{a}}{a}\rho + \frac{1}{a}\nabla_{\boldsymbol{x}} \cdot \rho\boldsymbol{V} = 0, \tag{14.3.1 b}$$

$$\nabla^2\varphi = 4\pi G a^2 \rho, \tag{14.3.1 c}$$

which are Equations (10.2.1 *b*), (10.2.1 *a*) and (10.2.1 *c*), with $\boldsymbol{v} = \boldsymbol{r}\dot{a}/a + \boldsymbol{V}$, $\boldsymbol{V} = a\dot{\boldsymbol{x}}$ and $\boldsymbol{r} = a(t)\boldsymbol{x}$; $\boldsymbol{x}$ is a comoving coordinate. The Equation (14.3.1 *c*) is not needed in this section, but we have included it here for the sake of completeness. The Zel'dovich approximation is equivalent to putting the right-hand side of (14.3.1 *a*) equal to $(2\dot{a}/a + \ddot{b}/\dot{b})\boldsymbol{V}$. In this case, with the substitution $\eta = a^3\rho$ and $\boldsymbol{U} = \boldsymbol{V}/a\dot{b} = \mathrm{d}\boldsymbol{x}/\mathrm{d}b$, the first two of the preceding equations become

$$\frac{\partial \eta}{\partial b} + \nabla_{\boldsymbol{x}} \cdot \eta\boldsymbol{U} = 0 \tag{14.3.2 a}$$

$$\frac{\partial \boldsymbol{U}}{\partial b} + (\boldsymbol{U} \cdot \nabla_{\boldsymbol{x}})\boldsymbol{U} = 0. \tag{14.3.2 b}$$

The adhesion model involves modifying the Equation (14.3.2 *b*) by introducing a viscosity term ν, which allows the particles to stick together:

$$\frac{\partial \boldsymbol{U}}{\partial b} + (\boldsymbol{U} \cdot \nabla_{\boldsymbol{x}})\boldsymbol{U} = \nu \nabla_{\boldsymbol{x}}^2 \boldsymbol{U}. \tag{14.3.3}$$

The effect of this term is to make the particles 'feel' the inside of collapsed structures. It remains negligible outside these regions. The viscosity ν has the dimensions of 'length squared' in this representation because our 'time' coordinate is actually dimensionless, so the model basically requires that $d \simeq \sqrt{\nu}$ should be much less than the typical dimension of the structures forming. Equation (14.3.3) is well known in the mathematical literature as the *Burgers equation*. In many cases, and this is true in our case, this equation has an exact solution. With the so-called Hopf–Cole substitution,

$$\boldsymbol{U} = -2\nu \nabla_{\boldsymbol{x}} \ln W, \tag{14.3.4}$$

Equation (14.3.3) becomes the diffusion equation

$$\frac{\partial W}{\partial b} = \nu \nabla_{\boldsymbol{x}}^2 W, \tag{14.3.5}$$

which, in the original variables, has the solution

$$\boldsymbol{U}(\boldsymbol{x}, t) = \frac{\int b(t)^{-1}(\boldsymbol{x} - \boldsymbol{q}) \exp[(2\nu)^{-1} G(\boldsymbol{x}, \boldsymbol{q}, b)]\, \mathrm{d}^3 q}{\int \exp[(2\nu)^{-1} G(\boldsymbol{x}, \boldsymbol{q}, b)]\, \mathrm{d}^3 q}, \tag{14.3.6}$$

where

$$G(\boldsymbol{x}, \boldsymbol{q}, b) = \Phi_0(\boldsymbol{q}) - \frac{(\boldsymbol{x} - \boldsymbol{q})^2}{2b}. \tag{14.3.7}$$

For small values of ν the main contribution to the integral in Equation (14.3.6) comes from regions where the function G has a maximum. This property allows a simplified treatment of the problem. The Eulerian position of the particle can be found by solving the integral equation

$$\boldsymbol{x}(\boldsymbol{q}, t) = \boldsymbol{q} + \int_0^{b(t)} \boldsymbol{U}[\boldsymbol{x}(\boldsymbol{q}, b'), b']\, \mathrm{d}b'. \tag{14.3.8}$$

The adhesion model furnishes results in accord with the Zel'dovich approximation at distances $l \gg d$ from the structure, but allows one to follow the formation of structure insofar as it prevents structure from being erased by shell-crossing. It also allows one to avoid the singularities which occur in the usual Zel'dovich approximation. In many simple cases the solution (14.3.6) does indeed allow one to study the formation of structure to high accuracy even in a highly advanced phase of nonlinearity.

The spatial distribution of particles obtained by letting the parameter ν tend to zero represents a sort of 'skeleton' of the real structure: nonlinear evolution generically leads to the formation of a quasicellular structure, which is similar to a 'tessellation' of irregular polyhedra having pancakes for faces, filaments for edges and nodes at the vertices. This skeleton, however, evolves continuously as structures merge and disrupt each other through tidal forces; gradually, as evolution proceeds, the characteristic scale of the structures increases. In order to interpret the observations we have already described in Chapter 4, one can think of the giant 'voids' as being the regions internal to the cells, while the cell nodes correspond to giant clusters of galaxies. While analytical methods, such as the adhesion model, are useful for mapping out the skeleton of structure formed during the nonlinear phase, they are not adequate for describing the highly nonlinear evolution within the densest clusters and superclusters. In particular, the adhesion model cannot be used to treat the process of merging and fragmentation of pancakes and filaments due to their own (local) gravitational instabilities.

14.4 Self-similar Evolution

A possible way to treat highly nonlinear evolution in the framework of 'bottom-up' scenarios is to introduce the concept of *self-similarity* or *hierarchical clustering*. As we have already explained, in the isothermal baryon model or in the more modern CDM model, the first structures to enter the nonlinear regime are expected to be on a mass scale of order $M_J^{(i)}(z_{rec})$. Galaxies and larger structures then form by merging of such objects into objects of higher mass. This process is qualitatively different from that described by the Zel'dovich and adhesion approximations, which are more likely to be accurate on scales relevant to clusters and superclusters, while we need something else to describe the formation of structure on scales up to this.

14.4.1 A simple model

To illustrate some of these ideas, let us assume that the Universe is well-described by an Einstein–de Sitter model. A perturbation with mass $M > M_J$, which we use from now on to mean $M_J^{(i)}(z_{rec})$, arrives in the nonlinear regime, approximately, at a time t_M such that

$$\sigma_M(t_{rec})\left(\frac{t_M}{t_{rec}}\right)^{2/3} \simeq 1, \tag{14.4.1}$$

where $\sigma_M(t_{rec})$ is the RMS mass fluctuation on the scale M at $t = t_{rec}$. One therefore has the relationship

$$t_M \simeq t_{rec}\sigma_M(t_{rec})^{-3/2} = t_J\left(\frac{M}{M_J}\right)^{3\alpha_{rec}/2}, \tag{14.4.2}$$

where the quantity α_{rec} is defined in Section 13.4. From Equation (14.4.2) it follows that

$$M \simeq M_J\left(\frac{t_M}{t_J}\right)^{2/3\alpha_{rec}}, \tag{14.4.3}$$

where $t_J = t_M$ for $M \simeq M_J$. As we explained in Section 14.1, if we think of the perturbation as a spherical 'blob', then the time t_M practically coincides with the moment at which the perturbation ceases to expand with the background Universe and begins to collapse. In the general case expressed by (14.4.2), one can apply the simple scheme described in Section 14.1: one can easily obtain from Equations (14.1.6) and (14.4.2) that, at virial equilibrium, the perturbation has a density

$$\rho_M \simeq \frac{3\pi}{32Gt_M^2} \simeq \rho_J\left(\frac{M}{M_J}\right)^{-3\alpha_{\rm rec}}, \tag{14.4.4}$$

where we have put $\rho_M(M_J) = \rho_J$. If r_M is the radius of a (collapsed) perturbation of mass M, from (14.4.4) and from the fact that $M \simeq \rho_M r_M^3$, one finds

$$\rho_M = \rho_J\left(\frac{r_M}{r_J}\right)^{-\gamma_{\rm vir}}, \tag{14.4.5}$$

where the meaning of r_J is clear; the exponent $\gamma_{\rm vir}$ is given by the relation

$$\gamma_{\rm vir} = \frac{9\alpha_{\rm rec}}{3\alpha_{\rm rec}+1} = \frac{3(n_{\rm rec}+3)}{5+n_{\rm rec}}. \tag{14.4.6}$$

From Equations (14.4.2) and (14.4.5) we obtain

$$r_M = r_J\left(\frac{t_{\rm vir}}{t_J}\right)^{2/\gamma_{\rm vir}}. \tag{14.4.7}$$

We can also relate the mass M to the virial velocities generated by it, V_M, in this model. The result is

$$M \propto V_M^{12/(1-n_{\rm rec})}. \tag{14.4.8}$$

If $n_{\rm rec} = -2$, then this can explain the $M \propto V_M^4$ relationship implied by the observed correlation between L and V for galaxies, known as the Tully–Fisher relationship, Equation (4.3.2).

A simple interpretation of the model just described, which is called the *hierarchical clustering model*, is the following. The Universe at time t_{M_*} on a scale $r < r_{M_*}$ contains condensed objects of various masses M and corresponding sizes r_M according to a hierarchical arrangement, in which the objects of one scale are the building blocks from which objects on higher scales are made. This arrangement holds up to the scale M_* which is the largest mass scale to have reached virial equilibrium. For masses greater than M_*, fluctuations are small and still evolving in the linear regime so that, for $r > r_{M_*}$, we have $\delta\rho_{\rm m}(r) \propto \sigma_M \propto M^{-\alpha_{\rm rec}} \propto r^{-3\alpha_{\rm rec}} = r^{-(3+n_{\rm rec})/2}$. These small fluctuations will grow and, when $t > t_{M_*}$, objects on a higher mass scale than M_* will collapse and form a higher level of the hierarchy. Simple though it is, this description seems to provide a fairly accurate representation of the behaviour of N-body simulations of hierarchical clustering in the highly nonlinear phase.

We can take this formulation further and model the behaviour of the two-point correlation of the matter fluctuations. Let us divide the possible range of masses at time t_0 into three intervals: (a) scales corresponding to masses still in the linear regime, i.e. those with $t_M > t_0$ or, equivalently, $M > M(t_0) = M_0$; (b) scales which have reached their radius of maximum expansion but have not yet reached virial equilibrium - for these scales $t_0 > t_M > t_0/3$; and (c) scales which have reached virial equilibrium, i.e. those with $t_M < t_0/3$.

The relationship between M and r for scales in the first interval is just

$$M = \tfrac{4}{3}\pi[\rho_{0\mathrm{m}} + \delta\rho_{\mathrm{m}}(r)]r^3 \simeq \tfrac{4}{3}\pi\rho_{0\mathrm{m}}r^3, \tag{14.4.9}$$

while for the second and the third we have

$$M = \tfrac{4}{3}\pi\rho_{\mathrm{c}M}r^3, \tag{14.4.10}$$

where $\rho_{\mathrm{c}M}$ is the density of the condensation of mass M which coincides with ρ_{M} given in (14.4.5) for those condensations already virialised. Because $\rho_{\mathrm{c}M} \gg \langle\rho\rangle$ for scales of interest in this context we have, from Section 13.7,

$$\xi(r) \simeq \left\langle\frac{\rho_{\mathrm{c}M}(r)}{\rho}\right\rangle - 1 \simeq \left\langle\frac{\rho_{\mathrm{c}M}(r)}{\rho}\right\rangle. \tag{14.4.11}$$

For the scales which are still in the linear regime we have

$$\xi(r) \simeq \sigma_{\mathrm{M}}^2 \propto r^{-(n_{\mathrm{rec}}+3)}. \tag{14.4.12}$$

From Equations (14.4.5) and (14.4.11) one can obtain, for the third interval,

$$\xi(r) \simeq (72\chi - 1)\left(\frac{r}{r_{\mathrm{vir}}}\right)^{-\gamma_{\mathrm{vir}}}, \tag{14.4.13}$$

where r_{vir} is the scale which has just reached virial equilibrium and which corresponds to a mass scale M_{vir}.

In the second interval we cannot write an exact expression for $\xi(r)$ for any value of r. For the scale r_{M_0}, which has just reached maximum expansion, we have $\xi(r_{M_0}) \simeq \chi - 1$. For scales $r_{\mathrm{vir}} \leqslant r \leqslant r_{M_0}$ one can introduce a covariance function which is approximated by a power law, by analogy with Equations (14.4.12) and (14.4.13), so that it matches the exact values at r_{vir} and r_{M_0}:

$$\xi(r) \simeq (72\chi - 1)\left(\frac{r}{r_{\mathrm{vir}}}\right)^{-\bar{\gamma}} \simeq (\chi - 1)\left(\frac{r}{r_{M_0}}\right)^{-\bar{\gamma}}, \tag{14.4.14}$$

with exponent $\bar{\gamma}$ given by

$$\bar{\gamma} = \frac{\ln[(72\chi - 1)/(\chi - 1)]}{\ln(r_{M_0}/r_{\mathrm{vir}})}. \tag{14.4.15}$$

Let us recall that, from (14.4.3), we have

$$M_0 = \tfrac{4}{3}\pi r_{M_0}^3 \chi \rho_{0\mathrm{m}} = M_\mathrm{J}\left(\frac{t_0}{t_\mathrm{J}}\right)^{2/3\alpha_\mathrm{rec}}, \tag{14.4.16}$$

$$M_\mathrm{vir} = \tfrac{4}{3}\pi r_\mathrm{vir}^3 72\chi \rho_{0\mathrm{m}} = M_\mathrm{J}\left(\frac{t_0}{3t_\mathrm{J}}\right)^{2/3\alpha_\mathrm{rec}}, \tag{14.4.17}$$

so that

$$\bar{\gamma} = \frac{3\ln[(72\chi - 1)/(\chi - 1)]}{\ln 72 + \ln 81/(3 + n_\mathrm{rec})} \simeq \frac{3.18}{1 + 1.03/(3 + n_\mathrm{rec})}. \tag{14.4.18}$$

One can show that for $\Omega \neq 1$ one has $\chi' = \pi^2/[4\Omega(H_0 t_0)^2]$ instead of $\chi = (\frac{3}{4}\pi)^2 \simeq 5.6$; for $\Omega = 0.1$, for example, this yields $\chi' \simeq 30.6$ and Equation (14.4.18) gives $\gamma' = 3.03/[1 + 0.349/(3 + n_\mathrm{rec})]$.

In this way, in the case $\Omega = 1$, one obtains practically the complete behaviour of $\xi(r)$ for a given n_rec; the only part not covered is that in which $\chi - 1 \simeq 5 \geqslant \xi(r) \geqslant 1$, where the correlation function passes gradually between the behaviour described by Equations (14.4.12) and (14.4.14). In the case $\Omega = 0.1$ the missing range is larger, $\chi' - 1 \simeq 30 \geqslant \xi(r) \geqslant 1$. In any case these results can probably only be interpreted meaningfully in the regime where $\xi \gg 1$. It is interesting to note that, with a spectral index at recombination given by $n_\mathrm{rec} \simeq 0$, we have $\gamma_\mathrm{vir} \simeq 1.8$.

14.4.2 Stable clustering

An alternative approach to self-similar evolution that makes a closer contact with dynamics of clustering evolution is to proceed from the power spectrum. Consider the behaviour of the linear power spectrum smoothed on a scale R_f; this is defined in Equation (13.3.12). At any time there will be a characteristic *comoving* scale R^* such that the spectrum smoothed on that scale has unit variance. If we assume a flat Friedmann model so that the linear density fluctuations grow as $t^{2/3}$ and an initial power-law spectrum of the form $P(k) = Ak^n$, then this characteristic scale varies as

$$R^*(t) \propto t^{4/(3n+9)}. \tag{14.4.19}$$

This, in turn corresponds to a characteristic mass scale M^* that varies as

$$M^* \propto t^{4/(n+3)}. \tag{14.4.20}$$

The assumption that there is self-similar evolution corresponds to the assumption that the two-point correlation function in the nonlinear regime $\xi(x,t)$ is a function of a single similarity variable $s = x/t^\alpha$, where the value of α is fixed by Equation (14.4.19) if the nonlinear behaviour matches onto the growth in the linear regime.

This idea can be connected with the behaviour of velocities by writing an equation for the conservation of pairs of particles:

$$\frac{\partial \xi(x,t)}{\partial t} + \frac{1}{ax^2}\frac{\partial}{\partial x}[x^2 \langle v_{21}(x,t)\rangle (1+\xi(x,t))] = 0 \tag{14.4.21}$$

where $\langle v_{21}(x,t)\rangle$ is the mean relative velocity of particles with separation x at time t (Davis and Peebles 1977; Peebles 1980). Under the similarity transformation mentioned above this equation assumes the form

$$-\alpha s\frac{\mathrm{d}\xi}{\mathrm{d}s} + \frac{1}{s^2}\frac{\mathrm{d}}{\mathrm{d}s}[s^2\langle v_{21}(s)/at^{\alpha-1}\rangle(1+\xi)] = 0. \tag{14.4.22}$$

Now for very small separations it seems to be a reasonable *ansatz* to assume the clumps of matter are stable so that on average there is no net change in separation, i.e.

$$\langle \dot{\boldsymbol{r}}_{12}\rangle = \dot{a}\boldsymbol{x}_{12} + a\langle \dot{\boldsymbol{x}}_{12}\rangle = 0. \tag{14.4.23}$$

This is called the *stable clustering* limit. Putting (14.4.23) into (14.4.22) and solving for ξ yields

$$\xi(s) \propto s^{-\gamma}, \tag{14.4.24}$$

where γ turns out to be the same as γ_{vir} given in (14.4.6).

14.4.3 Scaling of the power spectrum

The idea that some form of self-similarity might apply to the evolution of clustering into the nonlinear regime led Hamilton *et al.* (1991) to construct an ingenious model for how the power spectrum itself might evolve. In the linear regime $P(k)$ retains its initial shape, once clustering becomes strong its shape will change.

The basic idea is as follows. Let r_0 be a Lagrangian comoving coordinate defined by

$$r_0^3 = \int_0^r (1+\xi)\,\mathrm{d}^3\boldsymbol{r} = r^3(1+\bar{\xi}), \tag{14.4.25}$$

where $\bar{\xi}$ is the mean correlation function interior to some radius r. The Lagrangian radius r_0 can be thought of as the size of a patch of the initial conditions that collapses to a size r when the structure goes nonlinear. At early times r and r_0 coincide but as time passes r shrinks relative to r_0. In the linear regime $\bar{\xi} \ll 1$ simply grows as the square of the linear growth law, i.e. if $\Omega_0 = 1$ it grows as $t^{4/3}$ or, alternatively, as a^2. If there is a stable clustering regime for $\bar{\xi} \gg 1$, then the growth law must be $\bar{\xi} \propto a^3$ since the structures are fixed in physical coordinates.

These two limits motivate the suggestion that, anywhere between the two limiting cases of linear and stable clustering, the evolution of $\bar{\xi}$ might be described by a kind of universal function of the initial mean correlation $\bar{\xi}_0(r_0)$ and a, i.e.

$$\bar{\xi} = F[a^2\bar{\xi}_0(r_0)], \tag{14.4.26}$$

where $F[x]$ is unity for small x and proportional to $x^{3/2}$ for large x. Hamilton *et al.* (1991) compare this idea with the results of full numerical computations. They find that it works reasonably well, and provide a fitting formula for F that works in the intermediate regime. A subsequent study by Jain *et al.* (1995) refined and extended this approach.

14.4.4 Comments

Although this analysis is very simplified, it does give results which agree, at least qualitatively, with full N-body simulations of hierarchical clustering. It is possible to extend the ideas of self-similarity further, to the analysis of higher-order correlations. Although this latter approach yields what is called the hierarchical model for reduced N-point correlation functions, which is described in Section 16.4, this should not be thought of as a logical consequence of the highly approximate model we have described in this section. This general picture of self-similar clustering is also the motivation behind attempts to calculate the mass function of condensed objects, which we describe in the next section.

14.5 The Mass Function

The *mass function* $n(M)$, also called the *multiplicity function*, of cosmic structures such as galaxies is defined by the relation

$$dN = n(M)\,dM, \tag{14.5.1}$$

which gives the number of the structures in question per unit volume with mass contained in the interval between M and $M + dM$. It is clear that the mass function and the luminosity function, defined in Section 4.5, contain the same information as long as one knows the value of the ratio M/L for the objects because

$$\Phi(L) = n(M)\frac{dM}{dL} \simeq n(M)\left\langle\frac{M}{L}\right\rangle. \tag{14.5.2}$$

This ratio, as we have mentioned in Chapter 4, is not known with any great certainty: for example, it seems to have values of order 10, 100 and 400 in solar units for galaxies, groups of galaxies and clusters, respectively. It is in practice impossible to recover the mass function from the observed luminosity function. On the other hand, in many cosmological problems, above all in those involving counts of objects at various distances, it is important to have an analytic expression for the mass function. This must therefore be calculated by some appropriate theoretical model. For this reason, Press and Schechter (1974) proposed a simple analytical model to calculate $n(M)$. This method is still used today and, despite simplicity and several obvious shortcomings, is still the most reliable method available for calculating this function analytically.

In the Press–Schechter approach one considers a density fluctuation field $\delta(\boldsymbol{x}; R) \equiv \delta_{\mathrm{M}}$, filtered on a spatial scale R corresponding to a mass M. In particular, if the density field possesses Gaussian statistics (see Section 13.7), the distribution of fluctuations is given by

$$\mathcal{P}(\delta_{\mathrm{M}})\,\mathrm{d}\delta_{\mathrm{M}} = \frac{1}{(2\pi\sigma_M^2)^{1/2}} \exp\left(-\frac{\delta_{\mathrm{M}}^2}{2\sigma_M^2}\right) \mathrm{d}\delta_{\mathrm{M}}. \tag{14.5.3}$$

The probability that at some point the fluctuation δ_{M} exceeds some critical value δ_{c} is expressed by the relation

$$P_{>\delta_{\mathrm{c}}}(M) = \int_{\delta_{\mathrm{c}}}^{\infty} \mathcal{P}(\delta_{\mathrm{M}})\,\mathrm{d}\delta_{\mathrm{M}}; \tag{14.5.4}$$

this quantity depends on the filter mass M and, through the time-dependence of σ_M, on the redshift (or epoch). The probability $P_{>\delta_{\mathrm{c}}}$ is also proportional to the number of cosmic structures characterised by a density perturbation greater than δ_{c}, whether these are isolated or contained within denser structures which collapse with them. For example, in the spherical collapse approximation of Section 14.1, the value $\delta_{\mathrm{c}} \simeq 1.68$, obtained by extrapolating linear theory, represents structures which, having passed the phase of maximum expansion, have collapsed and reached their maximum density. To find the number of regions with mass M which are isolated, in other words surrounded by underdense regions, one must subtract from $P_{>\delta_{\mathrm{c}}}(M)$ the quantity $P_{>\delta_{\mathrm{c}}}(M + \mathrm{d}M)$, proportional to the number of objects entering the nonlinear regime characterised by δ_{c} on the appropriate mass scale. In making this assumption we have completely ignored the so-called *cloud-in-cloud problem*, which is the possibility that at a given instant some object, which is nonlinear on a scale M, can be later contained within another object, on a larger mass scale. It is necessary effectively to take the probability in Equation (14.5.4) to be proportional to the probability that a given point has ever been contained in a collapsed object on some scale greater than M or, in other words, that the only objects which exist on a given scale are those which have just collapsed. If an object has $\delta > \delta_{\mathrm{c}}$ when smoothed on a scale R, it will have $\delta = \delta_{\mathrm{c}}$ when smoothed on some larger scale R' and will therefore be counted again as part of a higher level of the hierarchy. Another problem of this assumption is also obvious: it cannot treat underdense regions properly and therefore, by symmetry, half the mass is not accounted for. In the Press–Schechter analysis this is corrected by multiplying throughout by a factor 2, with the vague understanding that this represents accretion from the underdense regions onto the dense ones. The result is therefore that

$$n(M)M\,\mathrm{d}M = 2\rho_{\mathrm{m}}[P_{>\delta_{\mathrm{c}}}(M) - P_{>\delta_{\mathrm{c}}}(M+\mathrm{d}M)] = 2\rho_{\mathrm{m}}\left|\frac{\mathrm{d}P_{>\delta_{\mathrm{c}}}}{\mathrm{d}\sigma_M}\right|\left|\frac{\mathrm{d}\sigma_M}{\mathrm{d}M}\right|\mathrm{d}M. \tag{14.5.5}$$

The formula (14.5.5) becomes very simple in the case where the RMS mass fluctuation is expressed by a power law:

$$\sigma_M = \left(\frac{M}{M_0}\right)^{-\alpha} \tag{14.5.6}$$

(the preceding relation is also approximately valid if one does not have a pure power law but if α is interpreted as the effective index over the mass scale of interest). In this case we obtain, from Equations (14.5.3), (14.5.4) and (14.5.5), that

$$n(M) = \sqrt{\frac{2}{\pi}} \frac{\delta_{\rm c}\alpha}{\sigma_M} \frac{\rho_{\rm m}}{M^2} \exp\left(-\frac{\delta_{\rm c}^2}{2\sigma_M^2}\right) = \frac{2}{\sqrt{\pi}} \frac{\rho_{\rm m}\alpha}{M_*^2} \left(\frac{M}{M_*}\right)^{\alpha-2} \exp\left[-\left(\frac{M}{M_*}\right)^{2\alpha}\right]. \tag{14.5.7}$$

The mass function thus has a power-law behaviour with an exponential cut-off at the scale

$$M_* = \left(\frac{2}{\delta_{\rm c}^2}\right)^{1/2\alpha} M_0. \tag{14.5.8}$$

It is interesting to note that, for a constant value of the ratio M/L in Equations (14.5.2) and (14.5.7), one can obtain a functional form for the luminosity function $\Phi(L)$ similar to that of the Schechter function introduced in Chapter 4; to match exactly requires $\alpha = \frac{1}{2}$, in other words a white-noise spectrum.

From Equation (14.5.7) it is also possible to derive the time-evolution of an appropriately defined characteristic mass $M_{\rm c}(t)$. In the kinetic theory of fragmentation and coagulation, one often assumes

$$M_{\rm c}(t) = \frac{\int_0^\infty n(M;t)M^2\,{\rm d}M}{\int_0^\infty n(M;t)M\,{\rm d}M}; \tag{14.5.9}$$

the time-dependence comes from the evolution of σ_M. In the simplest case in which σ_M is given by Equation (14.5.6) and is growing in the linear regime one finds that, in an Einstein–de Sitter universe,

$$M_{\rm c}(t) = \pi^{-1/2}\Gamma\left(\frac{1+\alpha}{2\alpha}\right)M_*(t_0)\left(\frac{t}{t_0}\right)^{2/3\alpha} \tag{14.5.10}$$

(Γ is the Gamma function), in accordance with Equation (14.4.3), as one would expect.

The Press–Schechter theory has been very successful and influential because it seems to describe rather well the behaviour of N-body simulations. Nevertheless, there are various assumptions made in this analysis which are extremely hard to justify. First there is the assumption that bound structures essentially form at peaks of the linear density field. While this must be some approximation to the real state of affairs, it can hardly be exact, because matter moved significantly from its initial Lagrangian position during nonlinear evolution as clearly demonstrated by the Zel'dovich approximation. In fact, the problem here is that the Press–Schechter approach does not really deal with localised objects at all but is merely a recipe for labelling points in the primordial density field. It is also quite clear that the device of multiplying the probability (14.5.4) by a factor 2 to obtain Equation (14.5.6) cannot be justified. Some more sophisticated analyses, intended to tackle the cloud-in-cloud problem explicitly, have clarified aspects of the problem. In particular, recent studies have elucidated the real nature of the

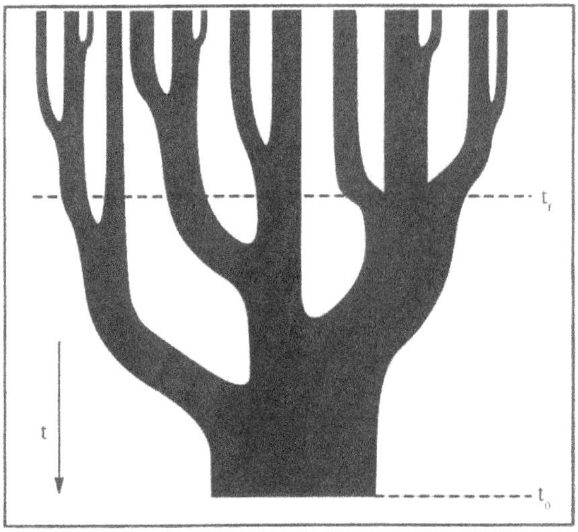

Figure 14.2 Example of a merger tree. The trunk of the tree represents the final mass of a halo and the branches show the various progenitors, with thickness representing the mass of the merging object. Picture courtesy of Sean Cole.

factor 2 as an artefact of overcounting due to cloud-in-cloud effects (Bond *et al.* 1991).

The Press-Schechter model, despite all its failings, is well verified by comparison with N-body simulations and is therefore a useful predictive tool in many circumstances. Its greatest failing however is that it is inherently statistical: mass points are merely labels and no attempt is made to follow the detailed evolution of individual objects. To put this another way, two objects with the same mass M at some time t may have built up through an entirely different series of mergers of smaller objects, sometimes through dramatic encounters of two objects with roughly equal masses, and sometimes through one object steadily consuming much smaller ones. It is likely that these different merger histories give rise to different kinds of object. This approach, pioneered by Lacey and Cole (1993) is illustrated in Figure 14.2.

14.6 N-Body Simulations

The complexity of the physical behaviour of fluctuations in the nonlinear regime makes it impossible to study the details exactly using analytical methods. The methods we have described in Sections 15.1–15.5 are valuable for providing us with a physical understanding of the processes involved, but they do not allow us to make very detailed predictions to test against observations. For this task one must resort to numerical simulation methods.

It is possible to represent part of the expanding Universe as a 'box' containing a large number N of point masses interacting through their mutual gravity. This box, typically a cube, must be at least as large as the scale at which the Universe becomes homogeneous if it is to provide a 'fair sample' which is representative of the Universe as a whole. It is common practice to take the cube as having periodic boundary conditions in all directions, which also assists in some of the computational techniques by allowing Fourier methods to be employed in summing the N-body forces. A number of numerical techniques are available at the present time; they differ, for the most part, only in the way the forces on each particle are calculated. We describe some of the most popular methods here.

14.6.1 Direct summation

The simplest way to compute the nonlinear evolution of a cosmological fluid is to represent it as a discrete set of particles, and then sum the (pairwise) interactions between them directly to calculate the Newtonian forces, as mentioned above. Such calculations are often called particle–particle, or PP, computations. With the adoption of a (small) timestep, one can use the resulting acceleration to update the particle velocity and then its position. New positions can then be used to recalculate the interparticle forces, and so on.

One should note at the outset that these techniques are not intended to represent the motion of a discrete set of particles. The particle configuration is itself an approximation to a fluid. There is also a numerical problem with summation of the forces: the Newtonian gravitational force between two particles increases as the particles approach each other and it is therefore necessary to choose an extremely small timestep to resolve the large velocity changes this induces. A very small timestep would require the consumption of enormous amounts of CPU time and, in any case, computers cannot handle the formally divergent force terms when the particles are arbitrarily close to each other. One usually avoids these problems by treating each particle not as a point mass, but as an extended body. The practical upshot of this is that one modifies the Newtonian force between particles by putting

$$\boldsymbol{F}_{ij} = \frac{Gm^2(\boldsymbol{x}_j - \boldsymbol{x}_i)}{(\epsilon^2 + |\boldsymbol{x}_i - \boldsymbol{x}_j|^2)^{3/2}}, \tag{14.6.1}$$

where the particles are at positions $\boldsymbol{x}_i$ and $\boldsymbol{x}_j$ and they all have the same mass m; the form of this equation avoids infinite forces at zero separations. The parameter ϵ in Equation (14.6.1) is usually called the *softening length* and it acts to suppress two-body forces on small scales. This is equivalent to replacing point masses by extended bodies with a size of order ϵ. Since we are not supposed to be dealing with the behaviour of a set of point masses anyway, the introduction of a softening length is quite reasonable but it means one cannot trust the distribution of matter on scales of order ϵ or less.

If we suppose our simulation contains N particles, then the direct summation of all the $(N-1)$ interactions to compute the acceleration of each particle requires a total of $N(N-1)/2$ evaluations of (14.6.1) at each timestep. This is the crucial limitation of these methods: they tend to be very slow, with the computational time required scaling roughly as N^2. The maximum number of particles for which it is practical to use direct summation is of order 10^4, which is not sufficient for realistic simulations of large-scale structure formation.

14.6.2 Particle–mesh techniques

The usual method for improving upon direct N-body summation for computing inter-particle forces is some form of 'particle–mesh' (PM) scheme. In this scheme the forces are solved by assigning mass points to a regular grid and then solving Poisson's equation on it. The use of a regular grid with periodic boundary conditions allows one to use Fast Fourier Transform (FFT) methods to recover the potential, which leads to a considerable increase in speed. The basic steps in a PM calculation are as follows.

In the following, $\boldsymbol{n}$ is a vector representing a grid position (the three components of $\boldsymbol{n}$ are integers); $\boldsymbol{x}_i$ is the location of the ith particle in the simulation volume; for simplicity we adopt a notation such that the Newtonian gravitational constant $G \equiv 1$, the length of the side of the simulation cube is unity and the total mass is also unity; M will be the number of mesh-cells along one side of the simulation cube, the total number of cells being N; the vector $\boldsymbol{q}$ is $\boldsymbol{n}/M$. First we calculate the density on the grid:

$$\rho(\boldsymbol{q}) = \frac{M^3}{N} \sum_{i=1}^{N} W(\boldsymbol{x}_i - \boldsymbol{q}), \tag{14.6.2}$$

where W defines a weighting scheme designed to assign mass to the mesh. We then calculate the potential by summing over the mesh

$$\varphi(\boldsymbol{q}) = \frac{1}{M^3} \sum_{\boldsymbol{q}'} \mathcal{G}(\boldsymbol{q} - \boldsymbol{q}')\rho(\boldsymbol{q}') \tag{14.6.3}$$

(where $\mathcal{G}$ is an appropriate Green's function for the Poisson equation), compute the resulting forces at the grid points,

$$\boldsymbol{F}(\boldsymbol{q}) = -\frac{1}{N}\boldsymbol{D}\varphi, \tag{14.6.4}$$

and then interpolate to find the forces on each particle,

$$\boldsymbol{F}(\boldsymbol{x}_i) = \sum_{\boldsymbol{q}} W(\boldsymbol{x}_i - \boldsymbol{q})\boldsymbol{F}(\boldsymbol{q}). \tag{14.6.5}$$

In Equation (14.6.4), $\boldsymbol{D}$ is a finite differencing scheme used to derive the forces from the potential. We shall not go into the various possible choices of weighting function W in this brief treatment: possibilities include 'nearest gridpoint' (NGP), 'cloud-in-cell' (CIC) and 'triangular-shaped clouds' (TSC).

We have written the computation of φ as a convolution but the most important advantage of the PM method is that it allows a much faster calculation of the potential than this. The usual approach is to Fourier transform the density field ρ, which allows the transform of φ to be expressed as a product of transforms of the two terms in (14.6.3) rather than a convolution; the periodic boundary conditions allow FFTs to be used to transform backwards and forwards, and this saves a considerable amount of computer time. The potential on the grid is thus written

$$\varphi(l,m,n) = \sum_{p,q,r} \hat{\mathcal{G}}(p,q,r)\hat{\rho}(p,q,r)\exp\left[\mathrm{i}\frac{\pi}{M}(pl+qm+rn)\right], \qquad (14.6.6)$$

where the 'hats' denote Fourier transforms of the relevant mesh quantities. There are different possibilities for the transformed Green's function $\hat{\mathcal{G}}$, the most straightforward being simply

$$\hat{\mathcal{G}}(p,q,r) = \frac{-1}{\pi(p^2+q^2+r^2)}, \qquad (14.6.7)$$

unless $p = q = r = 0$, in which case $\hat{\mathcal{G}} = 0$. Equation (14.6.6) represents a sum, rather than the convolution in Equation (14.6.3), and its evaluation can therefore be performed much more quickly. The calculation of the forces in Equation (14.6.5) can also be speeded up by computing them in Fourier space. An FFT is basically of order $N \log N$ in the number of grid points and this represents a substantial improvement for large N over the direct particle–particle summation technique. The price to be paid for this is that the Fourier summation method implicitly requires that the simulation box has periodic boundary conditions: this is probably the most reasonable choice for simulating a 'representative' part of the Universe, so this does not seem to be too high a price.

The potential weakness of this method is the comparatively poor force resolution on small scales because of the finite spatial size of the mesh. A substantial increase in spatial resolution can be achieved by using instead a hybrid 'particle-particle–particle–mesh' method, which solves the short range forces directly (PP) but uses the mesh to compute those of longer range (PM); hence PP + PM = $\mathrm{P}^3\mathrm{M}$, the usual name of such codes. Here, the short-range resolution of the algorithm is improved by adding a correction to the mesh force. This contribution is obtained by summing directly all the forces from neighbours within some fixed distance r_{s} of each particle. A typical choice for r_{s} will be around three grid units. Alternatively, one can use a modified force law on these small scales to assign any particular density profile to the particles, similar to the softening procedure demonstrated in Equation (14.6.1). This part of the force calculation may well be quite slow, so it is advantageous merely to calculate the short-range force at the start for a large number of points spaced linearly in radius, and then find the actual force by simple interpolation. The long-range part of the force calculation is done by a variant of the PM method described earlier.

Variants of the PM and $\mathrm{P}^3\mathrm{M}$ technique are now the standard workhorses for cosmological clustering studies. Different workers have slightly different interpolation schemes and choices of softening length. Whether one should use PM

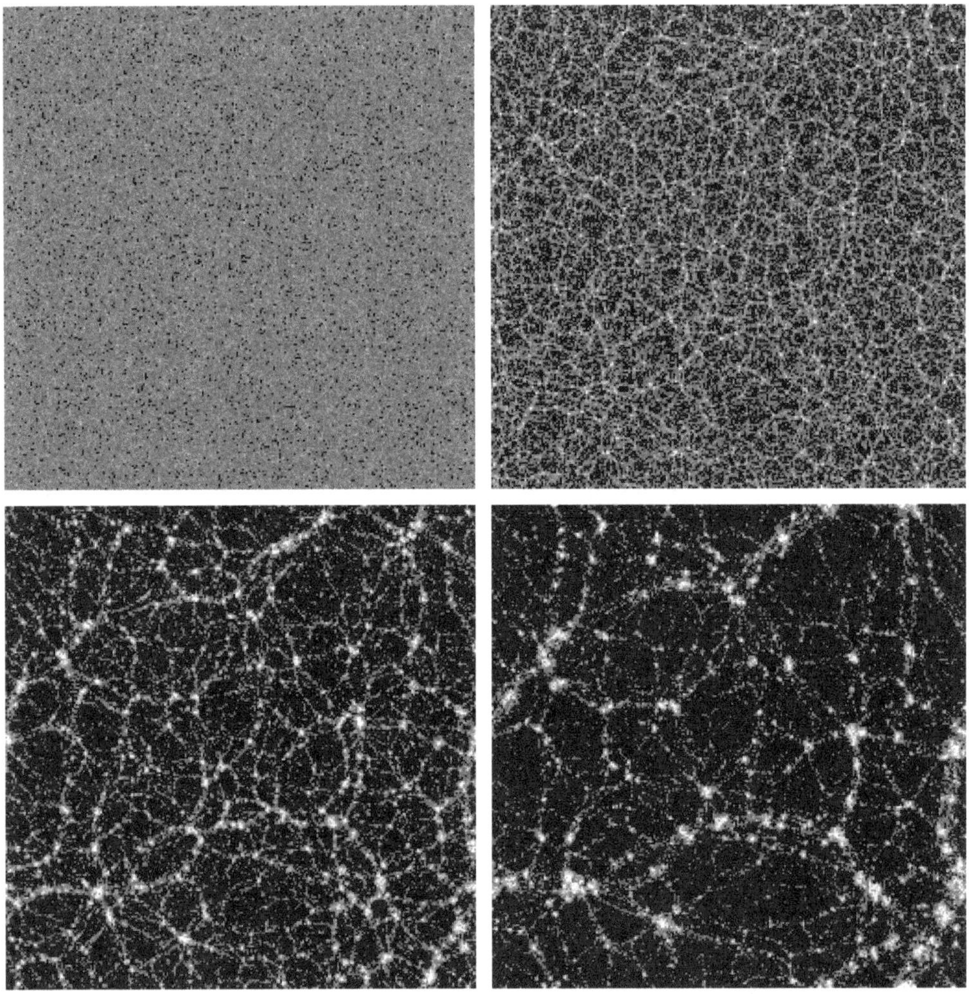

Figure 14.3 Numerical simulations from scale-free initial conditions with spectral index $n = 0$. The time sequence runs from left to right and top to bottom. The development of a filament–cluster–void network with an increasing characteristic size is clearly seen.

or P^3M in general depends upon the degree of clustering one wishes to probe. Strongly nonlinear clustering in dense environments probably requires the force resolution of P^3M. For larger-scale structure analyses, where one does not attempt to probe the inner structure of highly condensed objects, PM is probably good enough. One should, however, recognise that the short-range forces are not computed exactly, even in P^3M, so the apparent extra resolution may not necessarily be saying anything physical.

Some simulations of structure formation in models with scale-free (i.e. $n =$ const.) initial conditions are shown in Figure 14.3. One can see that not only does one form isolated 'blobs' which resemble those handled by the hierarchical model, the appearance of pancakes and filaments is also generic. In the CDM and HDM models, which are not scale free, the behaviour is rather simpler than the scale-free simulations which can be analysed with the techniques of Section 14.4 and 14.5. In the HDM model, where the initial spectrum is cut off on small scales, Zel'dovich pancakes form readily on supercluster scales, but that nonlinear processes do not create galaxy-size fluctuations rapidly enough to agree with the observations. The structure in a CDM model is much more clumpy on small scales but smoother on large scales.

14.6.3 Tree codes

An alternative procedure for enhancing the force resolution of a particle code whilst keeping the necessary demand on computational time within reasonable limits is to adopt a hierarchical subdivision procedure. The generic name given to this kind of technique is 'tree code'. The basic idea is to treat distant clumps of particles as single massive pseudo-particles. The usual algorithm involves a mesh which is divided into cells hierarchically in such a way that every cell which contains more than one particle is divided into 2^3 sub-cells. If any of the resulting sub-cells contains more than one particle, that cell is subdivided again. There are some subtleties involved with communicating particle positions up and down the resulting 'tree', but it is basically quite straightforward to treat the distant forces using the coarsely grained distribution contained in the high level of the tree, while short-range forces use the finer grid. The greatest problem with such codes is that, although they run quite quickly in comparison with particle–mesh methods with the same resolution, they do require considerable memory resources. Their use in cosmological contexts has so far therefore been quite limited, one of the problems being the difficulty of implementing periodic boundary conditions in such algorithms.

14.6.4 Initial conditions and boundary effects

To complete this section, we make a few brief remarks about starting conditions for N-body simulations, and the effect of boundaries and resolution on the final results.

Firstly, one needs to be able to set up the initial conditions for a numerical simulation in a manner appropriate to the cosmological scenario under consideration. For most models this means making a random-phase realisation of the power spectrum - see Section 14.8. This is usually achieved by setting up particles initially exactly on the grid positions, then using the Zel'dovich approximation, Equation (14.2.1), to move them such as to create a density field with the required spectrum and statistics. The initial velocity field is likewise obtained from the primordial gravitational potential. One should beware, however, the effects of the poor $\boldsymbol{k}$-space resolution at long wavelengths. The assignment of $\boldsymbol{k}$-space amplitudes requires a random amplitude for each wave vector contained in the reciprocal-space version of the initial grid. As the wave number decreases, the discrete nature of the grid becomes apparent. For example, there are only three (orthogonal) wave vectors associated with the fundamental mode of the box. When amplitudes are assigned via some random-number generator, one must take care that the statistically poor sampling of $\boldsymbol{k}$-space does not lead to spurious features in the initial conditions. One should use a simulation box which is rather larger than the maximum scale at which there is significant power in the initial spectrum.

At the other extreme, there arises the question of the finite spacing of the grid. This puts an upper limit, known as the *Nyquist frequency*, on the wavenumbers k which can be resolved, which is defined by $k_N = 2\pi/d$, where d is the mesh spacing. Clearly, one should not trust structure on scales smaller than k_N^{-1}.

One is therefore warned that, although numerical methods such as these are the standard way to follow the later nonlinear phases of gravitational evolution, they are not themselves 'exact' solutions of the equations of motion and results obtained from them can be misleading if one does not choose the resolution appropriately.

14.7 Gas Physics

So far we have dealt exclusively with the behaviour of matter under its self-gravity. We have ignored pressure gradient terms in the equation of motion of the matter at all times after recombination. While this is probably a good approximation in the linear and quasilinear regimes, when the Jeans mass is much smaller than scales of cosmological interest, it is probably a very poor representation of the late nonlinear phase of structure formation. As we shall see, hydrodynamical effects are clearly important in determining the behaviour of the baryonic part of galaxies, even if the baryons are only a small fraction of the total mass. Nonlinear hydrodynamical effects connected with the formation of shocks are also very important in determining how a collapsing structure reaches virial equilibrium.

14.7.1 Cooling

One of the important things to explain in hierarchical clustering scenarios is the existence of a characteristic scale of $\sim 10^{11} M_\odot$ in the mass spectrum of galaxies. Because gravity itself does not pick out any scale, some other physical mechanism must be responsible. Since only the baryonic part of the galaxy can be seen, and it is only this part which is known to possess characteristic properties, it is natural to think that gas processes might be involved. A good candidate for such a process is the cooling of the gas forming the galaxy.

Following Rees and Ostriker (1977), let us consider a simple model of a galaxy as a spherical gas cloud (i.e. no non-baryonic material) in the manner of Section 14.1. After collapse and violent relaxation (the process which converts the radial collapse motion into random 'thermal' motions) this cloud will be supported in equilibrium at its virial radius R and will have a temperature $T \propto GM\mu/R$, where μ is the mean molecular weight. If this temperature is high, as it will be for interesting mass scales, the cloud will be radiating and therefore cooling. The balance between pressure support and gravity which determines the size of the object depends on two characteristic timescales: the *cooling time*

$$t_{\rm cool} = -\frac{E}{\dot{E}} \simeq \frac{3\rho k_{\rm B}T}{2\mu\Lambda(T)}, \tag{14.7.1}$$

and the *dynamical time*, defined to be the free-fall collapse time for a sphere of mass M and radius R,

$$t_{\rm dyn} = \frac{\pi}{2}\left(\frac{R^3}{2GM}\right)^{1/2}, \tag{14.7.2}$$

where ρ is the mean baryon density and $\Lambda(T)$ in Equation (14.7.1) is the cooling rate (energy loss rate per unit volume per unit time) for a gas at temperature T (Λ is tabulated in standard physics texts for different kinds of gas). There are three main contributions to cooling in a hydrogen–helium plasma which is what we expect to have in the case of galaxy formation: free–free (bremsstrahlung) radiation, recombination radiation from H and He, and Compton cooling via the cosmic microwave background. This last one is efficient only if $z > 10$ or so. Since it is not known whether galaxy formation might have taken place at such high redshifts, this may play a role but for simplicity we shall ignore it here.

The two timescales $t_{\rm dyn}$ and $t_{\rm cool}$, together with the expansion timescale $\tau_{\rm H} = H^{-1}$, determine how the protogalaxy cools as it collapses. If $t_{\rm cool} > \tau_{\rm H}$, then cooling cannot have been important and the cloud will have scarcely evolved since its formation. If $\tau_{\rm H} > t_{\rm cool} > t_{\rm dyn}$, then the gas can cool on a cosmological timescale, but the fact that it does so more slowly than the dynamical characteristic time means that the cloud can adjust its pressure distribution to maintain the support of the cooling matter. There is thus a relatively quiescent quasi-static collapse on a timescale $t_{\rm cool}$. The last possibility is that $t_{\rm cool} < t_{\rm dyn}$. Now the cloud cools so quickly that dynamical processes are unable to adjust the pressure distribution in time: pressure support will be lost and the gas undergoes a rapid collapse on the free-fall timescale, accompanied by fragmentation on smaller and smaller scales as instabilities develop in the cloud which is behaving isothermally.

It is thought that the condition $t_{\rm cool} < t_{\rm dyn}$ is what determines the characteristic mass scale for galaxies. Only when this criterion is satisfied can the gas cloud collapse by a large factor and fragment into stars which allow the cloud to be identified as a galaxy. Furthermore, if structure formation proceeds hierarchically, the gas must cool on a timescale at least as small as $t_{\rm dyn}$, otherwise it will not be confined in a bound structure on some particular scale but will instead be disrupted as the next level of the hierarchy forms.

Let us now add non-baryonic matter into this discussion. What changes here is that the dynamical timescale for a collapsing cloud will be dominated by the dark matter while cooling is enjoyed only by the gas. Let us assume a spherical collapse model again. Notice that the dynamical timescale (14.7.2) is essentially the time taken for a perturbation to collapse from its maximum extent which can be identified as the turnaround radius $R_{\rm m}$ in Section 15.1. Putting in some numbers one finds that

$$t_{\rm dyn} \simeq 1.5 \times 10^9 \left(\frac{M}{M_\odot}\right)^{-1/2} \left(\frac{R_{\rm m}}{200\ {\rm kpc}}\right)^{3/2} \ {\rm years}. \tag{14.7.3}$$

One can estimate the cooling timescale by assuming that gas makes up a fraction $X_{\rm b}$ of the total mass M and that it is uniformly distributed within the virial radius which will be $R_{\rm m}/2$. We then take the gas temperature to be the same as the virial temperature of the collapsed object: $T \simeq 2GM\mu/5k_{\rm B}R_{\rm m}$. We also assume that the gas has not been contaminated by metals from an early phase of star formation (metals can increase the cooling rate and thus lower the cooling time considerably), and therefore adopt the appropriate value of $\Lambda(T)$ for a pure hydrogen plasma at temperature T. Using Equation (14.7.1) we find that

$$t_{\rm cool} \simeq 2.4 \times 10^8 X_{\rm b}^{-1} \left(\frac{M}{M_\odot}\right)^{1/2} \left(\frac{R_{\rm m}}{200\ {\rm kpc}}\right)^{3/2} \ {\rm years}, \tag{14.7.4}$$

so that the cooling criterion is satisfied when

$$M < M_* \simeq 6.4 \times 10^{12} X_{\rm b}^{-1} M_\odot, \tag{14.7.5}$$

which, for $X_{\rm b} \simeq 0.05$, gives $M_* \simeq 3 \times 10^{11} M_\odot$. While this theory therefore gives a plausible account of the characteristic mass scale for galaxies, it is obviously extremely simplified. Hydrodynamical effects may be important in many other contexts, such as cluster formation, the collapse of pancakes and also the feedback of energy from star formation into the intergalactic medium. A detailed theory of the origin of structure including gas dynamics, dissipation and star formation is, however, still a long way from being realised.

14.7.2 Numerical hydrodynamics

In the above we discussed an example where gas pressure forces are important in the formation of cosmic structure. Understanding of these effects is highly qualitative and applicable only to simple models. In an ideal world, one would like to understand the influence of gas pressure and star formation in a general context. Effectively, this means solving the Euler equation, including the relevant pressure terms, self-consistently. The appropriate equation is

$$\frac{\partial \boldsymbol{V}}{\partial t} + \frac{\dot{a}}{a}\boldsymbol{V} + \frac{1}{a}(\boldsymbol{V} \cdot \nabla_x)\boldsymbol{V} = -\frac{1}{a}\nabla_x \varphi - \frac{1}{a\rho}\nabla_x p. \tag{14.7.6}$$

The field of cosmological hydrodynamics is very much in its infancy, and it is fair to say that there are no analytic approximations that can be implemented with any confidence in this kind of analysis. The only realistic hope for progress in the near future lies with numerical methods, so we describe some of the popular techniques here.

In *smoothed-particle hydrodynamics* (SPH) one typically represents the fluid as a set of particles in the same way as in the N-body gravitational simulations described in Section 14.6. Densities and gas forces at particle locations are thus calculated by summing pairwise forces between particles. Since pressure forces are expected to fall off rapidly with separation, above some smoothing scale h (see below), it is reasonable to insert the gas dynamics into the part of a particle code that details the short-range forces such as the particle–particle part of a P^3M code. It is, however, possible to include SPH dynamics also in other types of simulation, including tree codes.

One technique used to insert SPH dynamics into a P^3M code is to determine local densities and pressure gradients by a process known as kernel estimation. This is essentially equivalent to convolving a field $f(\boldsymbol{x})$ with a filter function W to produce a smoothed version of the field:

$$f_s(\boldsymbol{r}) = \int f(\boldsymbol{x})W(\boldsymbol{x}-\boldsymbol{r})\,\mathrm{d}^3\boldsymbol{x}, \tag{14.7.7}$$

where W contains some implicit smoothing scale; one possible choice of W is a Gaussian. If $f(\boldsymbol{x})$ is just the density field arising from the discrete distribution of particles, then it can be represented simply as the sum of delta-function contributions at each particle location $\boldsymbol{x}_i$ and one recovers Equation (14.6.2). We need to represent the pressure forces in the Euler equation: this is done by specifying the equation of state of the fluid $p = (\gamma - 1)\epsilon\rho$, where ϵ is the thermal energy, ρ the local density and p the pressure. Now one can write the pressure force term in Equation (14.7.6) as

$$-\frac{\nabla p}{\rho} = -\nabla\left(\frac{p}{\rho}\right) - \frac{p}{\rho^2}\nabla\rho. \tag{14.7.8}$$

The gradient of the smoothed function f_s can be written

$$\nabla f_s(\boldsymbol{r}) = \int f(\boldsymbol{x})\nabla W(\boldsymbol{x}-\boldsymbol{r})\,\mathrm{d}^3\boldsymbol{x}, \tag{14.7.9}$$

so that the gas forces can be obtained in the form

$$\boldsymbol{F}_i^{\mathrm{gas}} = -\left(\frac{\nabla p}{\rho}\right)_i \propto -\sum_j \left(\frac{p_i}{\rho_{\mathrm{i}}^2} + \frac{p_j}{\rho_j^2}\right)\nabla W(\boldsymbol{r}_{ij}). \tag{14.7.10}$$

The form of Equation (14.7.10) guarantees conservation of linear and angular momentum when a spherically symmetric kernel W is used. The adiabatic change in the internal energy of the gas can similarly be calculated:

$$\frac{\mathrm{d}\epsilon_i}{\mathrm{d}t} \propto \frac{P_i}{\rho_{\mathrm{i}}^2}\sum_j \nabla W(\boldsymbol{r}_{ij}) \cdot \boldsymbol{v}_{ij}, \tag{14.7.11}$$

where $\boldsymbol{v}_{ij}$ is the relative velocity between particles. For collisions at a high Mach number, defined as the ratio of any systematic velocity to the thermal random velocity, thermal pressure will not prevent the particles from streaming freely, but in real gases there is molecular viscosity which prevents interpenetration of gas clouds. This is modelled in the simulations by introducing a numerical viscosity, the optimal form of which depends upon the nature of the simulation being attempted.

The advantage of particle-based methods is that they are Lagrangian and consequently follow the motion of the fluid. In practical terms, this means that most of the computing effort is directed towards places where most of the particles are and, therefore, where most resolution is required. As mentioned above, particle methods are the standard numerical tool for cosmological simulations. Classical fluid dynamics, on the other hand, has usually followed an *Eulerian* approach where one uses a fixed (or perhaps adaptive) mesh. Codes have been developed which conserve flux and which integrate the Eulerian equations of motion rapidly and accurately using various finite-difference approximation schemes. It has even proved possible to introduce methods for tracking the behaviour of shocks accurately – something which particle codes struggle to achieve. Typically, these codes can treat many more cells than an SPH code can treat particles, but the resolution is usually not so good in some regions because the cells will usually be equally spaced rather than being concentrated in the interesting high-density regions.

An extensive comparison between Eulerian and Lagrangian hydrodynamical methods has recently been performed, which we recommend to anyone thinking of applying these techniques in a cosmological context. Each has its advantages and disadvantages. For example, density resolution is better in the state-of-the-art Lagrangian codes, and the thermal accuracy better in the Eulerian codes. Conversely, Lagrangian methods have poor accuracy in low-density regions, presumably due to statistical effects, while the Eulerian codes usually fail to resolve the temperatures correctly in high-density regions due to the artificially high numerical viscosity in them.

14.8 Biased Galaxy Formation

It should be obvious by now that the complexities of nonlinear gravitational evolution, together with the possible influence of gas-dynamical processes on galaxy formation, mean that a full theory of the formation of these objects is by no means fully developed. Structure on larger scales is less strongly nonlinear, and therefore is less prone to hydrodynamical effects, so may be treated fairly accurately using linear theory as long as $\sigma_M \ll 1$ or, better still, using approximation methods such as the Zel'dovich and adhesion approximations. The problem is that, when one seeks observational data with which to compare theoretical predictions, these data invariably involve the identification of galaxies. Even if we give up on the task of understanding the details of the galaxy-formation process, we still need to know how to relate observations of the large-scale distribution of galaxies to that of the mass.

In Section 13.9 we discussed the Poisson clustering model, which is a statistical statement of the form 'galaxies trace the mass'. In this model the two-point correlation function of galaxies is equal to the covariance function of the underlying density field. In recent years, however, it has become clear that this is probably not a good representation of reality. In the spirit of the spherical collapse model one might imagine that galaxies should form not randomly sprinkled around according to the local density of matter, but at specific locations where collapse, cooling and star formation can occur. Obvious sites for protostructures would therefore be peaks of the density field, rather than randomly chosen sites. This simple idea, together with the assumption that the large-scale cosmological density field is Gaussian (see Section 14.8), led Kaiser (1984) (in a slightly different context; see Section 16.5) to suggest a *biased galaxy formation*, so that the galaxy correlation function and the matter autocovariance function are no longer equivalent. The way such a bias might come about is as follows. Suppose the density field δ_{M}, smoothed on some appropriate mass scale M to define a galaxy, is Gaussian and has variance σ_M^2. The covariance function $\xi(r)$ of δ_{M} is

$$\xi(r) = \langle \delta_{\mathrm{M}}(\boldsymbol{x})\delta_{\mathrm{M}}(\boldsymbol{x}')\rangle, \qquad (14.8.1)$$

where the average is taken over all spatial positions $\boldsymbol{x}$ and $\boldsymbol{x}'$ such that $|\boldsymbol{x}-\boldsymbol{x}'| = r$. If galaxies trace the mass, then the two-point correlation function of galaxies $\xi_{\mathrm{gg}}(r)$ coincides with $\xi(r)$. If galaxies do not trace the mass, this equality need not hold. In particular, imagine a scenario where galaxies only form from high-density regions above some threshold $\delta_{\mathrm{c}} = \nu\sigma_M$, where ν is a dimensionless threshold. The existence of such a threshold is qualitatively motivated by the spherical model of collapse, described in Section 14.1, within which a linear value of $\delta_{\mathrm{c}} \simeq 1.68$ would seem to be required for structure formation. To proceed we need to recall that, for such a Gaussian field, all the statistical information required to specify its properties is contained in the autocovariance function $\xi(r)$. It is straightforward to calculate the correlation function of points exceeding δ_{c} using the Gaussian prescription because the probability of finding two regions separated by a distance r both above the threshold will be just

$$Q_2 = \int_{\delta_{\mathrm{c}}}^{\infty}\int_{\delta_{\mathrm{c}}}^{\infty} \mathcal{P}_2(\delta_1, \delta_2)\,\mathrm{d}\delta_1\,\mathrm{d}\delta_2. \qquad (14.8.2)$$

Now, as explained in Section 13.7, the N-variate joint distribution of a set of δ_{i} can be written as a multivariate Gaussian distribution: for the case where $N = 2$, which is needed in Equation (14.8.2), using the substitution $\delta_{\mathrm{i}} = \nu_i\sigma$ and $w(r) = \xi(r)/\sigma^2$, we find

$$\mathcal{P}_2(\nu_1, \nu_2) = \frac{1}{2\pi}\frac{1}{\sqrt{1 - w^2(r)}}\exp\left(-\frac{\nu_1^2 + \nu_2^2 - 2w(r)\nu_1\nu_2}{2[1 - w^2(r)]}\right). \qquad (14.8.3)$$

The two-point correlation function for points exceeding $\nu_{\mathrm{c}} = \delta_{\mathrm{c}}/\sigma$ is then

$$\xi_{\nu_{\mathrm{c}}} = \frac{Q_2}{Q_1^2} - 1, \qquad (14.8.4)$$

where $Q_1 = P_{>\delta_c}$; see Equation (14.5.4). The exact calculation of the integrals in this equation is difficult but various approximate relations have been obtained. For large ν_c and small w we have

$$\xi_{\nu_c} \simeq \nu_c^2 w(r), \tag{14.8.5}$$

while another expression, valid when w is not necessarily small, is

$$\xi_{\nu_c} \simeq \exp[\nu_c^2 w(r)] - 1. \tag{14.8.6}$$

Kaiser initially introduced this model to explain the enhanced correlations of Abell clusters compared with those of galaxies; see Section 16.5. Here the field δ is initially smoothed with a filter of radius several Mpc to pick out structure on the appropriate scale. If galaxies trace the mass, and so have $\xi_{gg}(r) \simeq \xi(r)$, then the simple relation (14.8.5) explains qualitatively why cluster correlations might have the same slope, but a higher amplitude than the galaxy correlations. This enhancement is natural because rich clusters are defined as structures within which the density of matter exceeds the average density by some fairly well-defined factor in very much the way assumed in this calculation.

This simple argument spawned more detailed analyses of the statistics of Gaussian random fields, culminating in the famous 'BBKS' paper of Bardeen *et al.* (1986), which have refined and extended, while qualitatively confirming, the above calculations. The interest in most of these studies was the idea that galaxies themselves might form only at peaks of the linear density field (this time smoothed with a smaller filtering radius). If galaxies only form from large upward fluctuations in the linear density field, then they too should display enhanced correlations with respect to the matter. This seemed to be the kind of bias required to reconcile the standard CDM model with observations of galaxy-peculiar motions and also the cause of the apparent discrepancy between dynamical estimates of the mass density of the Universe of around $\Omega_0 \simeq 0.2$ when the theoretically favoured value is $\Omega_0 \simeq 1$. We shall discuss the question of velocities in detail in Chapter 18 and we have referred to it also in Chapter 4. Nevertheless, some comments here are appropriate. The velocity argument can be stated simply in terms of a sort of *cosmic virial theorem*. If galaxies trace the mass, and have correlation function $\xi(r)$ and mean pairwise velocity dispersion at a separation r equal to $v^2(r)$, then this theorem states that

$$\Omega \propto \xi(r)(v/r)^2, \tag{14.8.7}$$

with a calculable constant of proportionality; see Section 18.5 for details.

There are problems with this theorem in the context of standard CDM. First, if one runs a numerical simulation of CDM to the point when the correlation function of the mass has the right slope compared with that of the observations, then the accompanying velocities v are far too high. A low-density CDM seems to be a much better bet in this respect, but this may be because the slope of the correlation function is not a very good way to determine the present epoch in a simulation. The same thing, however, seems to happen in our Universe, where the observed correlation function and the observed pairwise peculiar motions give $\Omega \simeq 0.2$. One way out of this, indeed the obvious way out apart from the fact that it appears to contradict inflation, is to have $\Omega \simeq 0.2$ and leave it at that. There is another way out, however, which involves bias of the sort discussed above. Taking (14.8.6) as a qualitative model, one might argue that in fact $\xi(r)$ is wrong by a factor ν^2/σ_M^2 and, if this bias is large, one can reconcile a given v with $\Omega = 1$. A bias factor b, defined by

$$\xi(r)_{\text{galaxies}} = b^2 \xi(r)_{\text{mass}}, \tag{14.8.8}$$

of around $b \simeq 1.5$–3 seems to be required to match small-scale clustering and peculiar velocity data with the standard CDM model. Notice also that true density fluctuations are smaller than the apparent fluctuations in counts of galaxies, so that fluctuations in the microwave background are smaller by a factor $\sim 1/b$ in this picture than they would be if galaxies trace the mass.

The parameter b often arises in the cosmological literature to represent the possible difference between mass statistics and the statistics of galaxy clustering. The usual definition is not (14.8.8) but rather

$$b^2 = \frac{\sigma_8^2(\text{galaxies})}{\sigma_8^2(\text{mass})}, \tag{14.8.9}$$

where σ_8^2 represents the dimensionless variance in either galaxy counts or mass in spheres of radius $8h^{-1}$ Mpc. This choice is motivated by the observational result that the variance of counts of galaxies in spheres of this size is of order unity, so that $b \simeq 1/\sigma_8(\text{mass})$. Unless stated otherwise, this is what we shall mean by b in the rest of this book. Many authors use different definitions, e.g.

$$\frac{\delta N}{N} = b\frac{\delta \rho}{\rho}, \tag{14.8.10}$$

which is called the *linear bias model*. While a relation of the form (14.8.10) clearly entails (14.8.9) and (14.8.8), it does not follow from them, so these definitions are not equivalent. While there is little motivation, other than simplicity, for supposing the bias parameter to be a simple constant multiplier on small scales, it can be shown that, as long as the bias acts as a local function of the density, the form (14.8.8) should hold on large scales, even if the biasing relationship is complicated (Coles 1993).

Alternatives to (14.8.10), which are not equivalent, include the high-peak model and the various local-bias models (Coles 1993). Non-local biases are possible, but it is rather harder to construct such models (Bower *et al.* 1993). If one is prepared to accept an *ansatz* of the form (14.8.10), then one can use linear theory on large scales to relate galaxy-clustering statistics to those of the density fluctuations, e.g.

$$P_{\text{gal}}(k) = b^2 P(k), \tag{14.8.11}$$

as well as the form (14.8.8). This approach is the one most frequently adopted in practice, but the community is becoming increasingly aware of its limitations. A simple model of this kind simply cannot hope to describe realistically the relationship between galaxy formation and environment (Dekel and Lahav 1999).

One should say, however, that there is no compelling reason *a priori* to believe that galaxy formation should be restricted to peaks of particularly high initial density. It is true that peaks collapsing later might produce objects with a lower final density than peaks collapsing earlier, but these could (and perhaps should) still correspond to galaxies. Some astrophysical mechanism must be introduced which will inhibit galaxy formation in the lower peaks. Many mechanisms have been suggested, such as the possibility that star formation may produce strong winds capable of blowing the gas out of shallow potential wells, thus suppressing star formation, but none of these are particularly compelling. We discuss briefly how such a mechanism might also explain the morphological difference between elliptical and spiral galaxies in the next section. It is even possible that some large-scale modulation of the efficiency of galaxy formation might be achieved, perhaps by cosmic explosions or photoionisation due to quasars. Such a modulation would not be local in the sense discussed above and may well lead to a nonlinear bias parameter on large scales. We shall see later, however, in Chapter 17 that the latest clustering observations and the COBE microwave background fluctuations do not seem to support the idea of a strong bias, at least not in a CDM model.

At the present time b has a somewhat dubious status in the field of structure formation. The best way to think of b is not as describing some specific way of relating galaxies to mass, such as in (14.8.10), but as a way of parametrising our ignorance of galaxy formation in much the same way as one should interpret the mixing-length parameter in the theory of stellar convection. As we have mentioned already, to understand how this occurs we need to understand not only gravitational clustering but also star formation and gas dynamics. All this complicated physics is supposed to be contained in the parameter b.

14.9 Galaxy Formation

As we mentioned in Chapter 4, galaxies possess angular momentum. Its amount depends on the morphological type: it is maximum for spirals and S0 galaxies, and minimum for ellipticals. The angular momentum of our Galaxy, a fairly typical spiral galaxy of mass $M \simeq 10^{11} M_{\odot}$, is $J \simeq 1.4 \times 10^{74}$ cm^2 g s^{-1}. The conventional parametrisation of galactic angular momenta is in terms of the ratio between the observed angular velocity, ω, and the angular velocity which would be required to support the galaxy by rotation alone, ω_0:

$$\lambda \equiv \frac{\omega}{\omega_0} \simeq \frac{J/(MR^2)}{(GM/R^3)^{1/2}}, \tag{14.9.1}$$

where the dimensionless angular momentum parameter λ is typically as high as $\lambda \simeq 0.4$ for spirals, but only $\lambda \simeq 0.05$ for ellipticals. It is also probable that clusters of galaxies have some kind of rotation, large for the irregular open clusters like Virgo and smaller for the compact rich clusters like Coma.

The Kelvin circulation theorem guarantees that, in the absence of dissipative processes, an initially irrotational velocity field must remain so. The gravitational force can only create velocity fields in the form of potential flows which have zero curl. For a long time, therefore, the idea was held that the vorticity one appears to see now in galaxies must have been present in the early universe. This idea was developed much further in the theory of galaxy formation by cosmic turbulence which was at its most popular in 1970; this theory, however, predicted very high fluctuations in the temperature of the cosmic microwave background and some additional implausible assumptions were made. For this reason this scenario was rapidly abandoned and we mention it now only out of historical interest.

The origin of the rotation of galaxies within the framework of the theory of gravitational instability is described by a model, the first version of which was actually created by Hoyle (1949) and which has been subsequently modified by various authors and adapted to the various cosmogonical scenarios in fashion over the years (e.g. Efstathiou and Jones 1979). This model attributes the acquisition of angular momentum by a galaxy to the tidal action of protogalactic objects around it, at the epoch when the protogalaxy is just about to form a galaxy. At this epoch, protogalaxies have relatively large size (they will be close to their maximum expansion scale) and have a relatively small spatial separation compared with their size. Analytic calculations and N-body experiments show that this mechanism does indeed give a plausible account of the distribution of angular momentum observed in galactic systems.

This theory is valid in both top-down and bottom-up scenarios of structure formation. There is also another possibility: the circulation theorem is not valid in the presence of dissipative processes such as those accompanying the formation and propagation of a shock wave after the collapse of a pancake; the potential motion of the gas can become rotational after the gas has been compressed by a shock wave. This mechanism has not yet been analysed in great detail partly because of the difficulty in dealing with nonlinear hydrodynamics and partly because of the apparent success of the alternative, simpler scenario based on tidal forces.

In the tidal action model the acquisition of angular momentum by a galaxy takes place in two phases. The first phase commences at the moment a fluctuation begins to grow after recombination and ends when it reaches its maximum expansion, at $t_{\rm m}$; the second phase lasts from then until the present epoch. This second phase is thought to be when the galaxy acquires its own individuality beginning at the stage it collapses, undergoes violent relaxation and reaches virial equilibrium. It can be shown that in the first phase the angular momentum of the perturbation grows roughly like $t^{5/3}$, due to the effects of deviations from the Hubble flow caused by the various sub-condensations which make up the protostructure in question. In the second phase the protogalaxy, which will not in general be spherical, is subject to a torque due to other protogalaxies in its vicinity. One finds that this tidal effect, due to all the surrounding objects, increases the angular momentum of the galaxy according to $\dot{J} \propto t^{-2}$, decreasing with time because the expansion of the Universe carries the protogalaxies away from each other.

The question of the angular momentum of galaxies is intimately related to the origin of the morphological types, discussed in Chapter 4. A full theory of the formation of galaxies is complicated by gas pressure effects, as outlined in Section 14.6, and is yet to be elucidated. Possible answers to both the angular momentum and morphology questions may, however, come from the idea that dissipation is important for spiral galaxies but not for ellipticals. One can connect this to the problem of angular momentum as follows. The tidal action model can generate a value of $\lambda \simeq 0.05$–0.1, not quite large enough to account for spiral galaxies but comfortable for ellipticals. It seems clear for spirals that dissipation must be important to explain why the luminous matter in a galaxy is concentrated in the middle of its dark halo. If the gas collapses through cooling, as described in Section 14.7, then its binding energy will increase while the mass and angular momentum are conserved. If the binding energy of a spherical cloud is $E \simeq GM^2/R$, as usual, then $E \propto 1/R$ as the gas cools and shrinks. This means that $\lambda \propto R^{-1/2}$, so cooling can increase the λ parameter. The problem with this is that, if the galaxy is all baryonic, the rate of increase is rather slow. If, however, there is a dominant dark halo, one can get a much more rapid increase in λ and a value of $\simeq 0.4$–0.5 is reasonable.

The problem of formation of elliptical galaxies is less well understood. The value of their angular momentum seems to be accounted for by the tidal action model if there is no significant dissipation, but how can it be arranged for spirals and ellipticals to be thus separated? A possible explanation for this is that ellipticals formed earlier, when the Universe was denser and star formation (perhaps) more efficient. One might therefore be motivated towards an extension of the idea of biased galaxy formation (Section 14.8) in which the very highest density peaks, which collapse soonest, become ellipticals, while the smaller peaks become spirals. The detailed physics of the dividing line between these two morphologies, which we have supposed may be crudely delineated by the efficiency of dissipation, is still very unclear. An alternative idea is that perhaps all galaxies form like spiral galaxies, but that ellipticals are made from merging of spirals. This would seem to be plausible, given that ellipticals occur predominantly in dense regions. There are also problems with this picture. It is not clear whether ellipticals have the correct density profiles for them to be consistent with mergers of disc galaxies if the mergers are dissipationless. This aspect would have to be explored using numerical simulations.

The difficulty of understanding the complex effects of heating, dissipation and star formation within a continuously evolving clustering hierarchy has spawned the field of semi-analytic galaxy formation. This approach encodes the complex physics of galaxy formation in a set of relatively simple rules applied within a merger-tree description of the formation and merging of dark-matter haloes. The basic picture described in this model is that gas falls into the haloes whereupon it is shock-heated up to the virial temperature of the halo. It then undergoes radiative cooling. The cold gas component thus formed collapses into a rotationally supported disc and provides a reservoir of material that forms stars. The stars thus formed inject energy into the gas through supernova explosions, which also add a sprinkling of heavy elements to the mix. Crucial to this scenario is the assumption that the basic galaxy unit is disc. Elliptical and spheroidal galaxies are made through 'major mergers' of discs as suggested above. See Baugh *et al.* (1998) for a view of the state of this particular art.

14.10 Comments

It is clear that this chapter leaves many questions unanswered. We have shown that, while it is possible to use analytical methods and numerical simulations to understand the behaviour of density perturbations in the nonlinear regime, the complications of gas pressure, dissipation and star formation are still not fully understood. This means that we do not have an entirely satisfactory way of identifying sites of galaxy formation and every attempt to compare calculations with observations must take account of this difficulty. The semianalytic approach has been a major advance in this area but it is still not clear how fully it can account for the observed properties of galaxies of different types.

We also have the problem that, in order to run an N-body simulation or perform an analytical calculation, one needs to normalise the spectrum appropriately. In the past this was done by matching properties of the density fluctuation field to properties of galaxy counts. In more recent times, after the COBE result, the usual approach has become to normalise models to the microwave background anisotropy they predict. Even this latter method still carries some uncertainty, as we shall see in Chapter 17. To this one can add the problem of not knowing the form and quantity of any dark matter, which alters the primordial spectrum before the nonlinear phase is reached. Clearly there is an enormous parameter space to be explored and the tools we have to probe it theoretically are relatively crude.

Nevertheless, there has been substantial progress in recent years in the field of structure formation, and there is considerable cause to be optimistic about the future. Numerical techniques are being refined, the computational power available is steadily increasing and powerful analytical extensions of those we have discussed in this chapter have also been developed. On the observational side, tens of thousands of galaxy redshifts have been compiled over the last three decades. These allow us to probe the distribution of luminous matter on larger and larger scales; models for the bias are used to translate this into the mass distribution. New methods we shall describe in the following chapters have been devised to minimise the bias-dependence of tests of structure-formation scenarios. And finally, the microwave background fluctuations on small angular scales may allow us to test these theoretical ideas in a much more rigorous way than has hitherto been possible.

Bibliographic Notes on Chapter 14

Analytic nonlinear methods for large-scale structure are reviewed by Shandarin and Zel'dovich (1989) and Sahni and Coles (1995). The Burgers equation is discussed by Gurbatov *et al.* (1989). The basics of N-body simulation are discussed by Hockney and Eastwood (1988) in a general context. Numerical N-body techniques in cosmology are discussed by Efstathiou *et al.* (1985) and Bertschinger and Gelb (1991), while SPH variants are covered by Evrard (1988). For a discussion of Eulerian hydrodynamics, see Cen (1992).

Problems

1. For a Universe with $\Omega_0 \neq 1$, show that the generalisation of Equation (14.1.8) is

$$\chi(\Omega_0) = \frac{\pi^2}{4\Omega_0 (H_0 t_0)^2}.$$

2. Show that the Zel'dovich approximation is an exact solution of the one-dimensional gravitational clustering problem provided no trajectories have crossed. (**Hint**: substitute the Zel'dovich trajectories into the Euler equation for the problem and show that the potential gradients implied are consistent with the Poisson equation.)

3. Find the Zel'dovich displacement field corresponding to a spherical 'top-hat' density perturbation like that discussed in Section 14.1. Show that the Zel'dovich approximation predicts the formation of a singularity (i.e. that $\delta \to \infty$ at a finite time).

4. Prove the relation (14.4.19).

5. The self-similar evolution described in Section 14.4.2 requires that very large- and very small-scale velocities give convergent contributions to the peculiar velocity field. What restriction does this place on the spectral index, n, of the density fluctuations?

6. Derive the approximate results (14.8.5) and (14.8.6).

15

Models of Structure Formation

15.1 Introduction

In the preceding four chapters we have laid out the basic ingredients of the theory of cosmological structure formation according to the standard paradigm. The essential components of this recipe are primordial density perturbations, gravitational instability and dark matter, but many variations on this basic theme are viable. Despite the great progress that has undoubtedly been made, further steps are difficult because of uncertainties in the cosmological parameters, in the modelling of relevant physical processes involved in galaxy formation, and in the uncertain relationship between galaxies and the underlying distribution of matter.

Our aim in this chapter is to explain how the various components we have described come together in 'models' of structure formation that can be tested against observations. This will involve taking stock, and reducing the rather detailed physical discussion we have followed so far to a few key ideas and model parameters. Our role is not to advocate one particular mix of ingredients over another, but to point out how these different ingredients might be constrained or ruled out.

For example, as we have seen in Chapter 10, the expansion of the Universe renders the cosmological version of gravitational instability very slow, a power law in time rather than the exponential growth that develops in a static background. This slow rate has the important consequence that the evolved distribution of mass still retains significant memory of the initial state. If the perturbations were to grow exponentially, all memory of the initial conditions would be rapidly erased. This, in turn, has two consequences for theories of structure formation. One is that a detailed model must entail a complete prescription for the form of the initial conditions, and the other is that observations made at the present epoch allow us to probe the form of the primordial fluctuations and thus test the theory.

15.2 Historical Prelude

Progress in the field of structure formation during the 1970s was characterised by the construction of scenarios for the origin of cosmic protostructure in two-component models containing baryonic material and radiation. (As we shall see, the cosmological neutrino background does not greatly influence the evolution of perturbations in matter and radiation, as long as the neutrinos are massless.) There can exist two fundamental modes of perturbations in such a two-component system: *adiabatic perturbations*, in which the matter fluctuations, $\delta_{\rm m} = \delta\rho_{\rm m}/\rho_{\rm m}$, and radiation fluctuations, $\delta_{\rm r} = \delta\rho_{\rm r}/\rho_{\rm r}$, are coupled together so that $4\delta_{\rm m} = 3\delta_{\rm r}$; and *isothermal perturbations*, which involve only fluctuations in the matter component, i.e. $\delta_{\rm r} = 0$. These two kinds of perturbation led to two distinct scenarios for galaxy formation.

In the *adiabatic scenario* the first structures to form are on a large scale, $M \simeq 10^{12}$–$10^{14} M_\odot$, corresponding to clusters or superclusters of galaxies. Galaxies then form by successive processes of fragmentation of these large objects. For this reason the adiabatic scenario is also called a '*top-down*' scenario.

On the other hand, in the *isothermal scenario* the first structures, protoclouds, are formed on a much smaller mass scale, $M \simeq 10^5$–$10^6 M_\odot$, and then structure on larger scales is formed by the successive effect of gravitational instability, a process known as *hierarchical clustering*. For this reason, the isothermal scenario is described as '*bottom-up*'.

The adiabatic and isothermal scenarios were in direct competition with each other during the 1970s. One aspect of this confrontation was that the adiabatic scenario was chiefly championed by the great school of Russian astrophysicists led by Zel'dovich in Moscow, and the isothermal model was primarily an American affair, advocated in particular by Peebles and the Princeton group. In fact, neither of these adversaries actually won the battle: because of several intrinsic difficulties, the baryonic models were overtaken in the 1980s by models involving non-baryonic dark matter.

The main difficulty of the adiabatic scenario was that it predicted rather large angular fluctuations in the temperature of the microwave background, which were in excess of the observational limits. We can illustrate the problem in a simple qualitative manner to bypass the complications of the kinetic approach described above. In a universe made only of baryons with $\Omega_{\rm b} \simeq 1$, photons and massless neutrinos, the density fluctuation $\delta_{\rm m}(z_{\rm rec})M > M_{\rm D}^{(\rm a)}(z_{\rm rec})$ must have amplitude greater than the growth factor between recombination and t_0, which we called $A_{\rm r0}$. From Section 11.4, one can see that, if $\Omega \simeq 1$, then $A_{\rm r0} \simeq z_{\rm rec} \simeq 10^3$; if we are going to produce nonlinear structure by the present epoch, the density fluctuations must have amplitude at least unity by now. Thus, one requires $\delta_m(z_{rec}) \simeq 10^{-3}$ or higher. But these fluctuations in the matter are also accompanied in the adiabatic picture by fluctuations in the radiation which lead to fluctuations in the microwave background temperature $\delta_r \simeq 3\delta T/T \simeq 10^{-3}$, greater than the observational limits on the appropriate scale by more than two orders of magnitude. Moreover, if one recalls the calculations of primordial nucleosynthesis in the standard model, one cannot have Ω_b as large as this, and a (generous) upper bound is given by $\Omega \simeq \Omega_b \simeq 0.1$. This makes things even worse: in an open universe the growth factor is lower than a flat universe: $A_{r0} \simeq z_{rec}/z(t_*) \simeq 10^3\Omega \simeq 10^2$. In such a case the brightness fluctuations on the surface of last scattering exceed the observational limits by more than three orders of magnitude.

There is a possible escape from the limits on microwave background fluctuations provided by the possible existence of a period of reheating after z_{rec}, perhaps caused by the energy liberated during pregalactic stellar evolution, which smooths out some of the fluctuations in the microwave background. There are problems with this escape route, however, as we shall see later in Chapter 19.

The isothermal scenario does not suffer from the same difficulties with the microwave background, chiefly because $\delta_r \simeq 0$ for the isothermal fluctuations, and in any case the mass scale of the crucial first generation of clouds is so small. The major difficulty in this case is that isothermal perturbations are 'unnatural': only very special processes can create primordial fluctuations in the matter component while leaving the radiation component undisturbed. One possibility we should mention is that inflation, which generically produces fluctuations of adiabatic type, can produce isocurvature fluctuations if the scalar field responsible for generating the fluctuations is not the same as the field - the inflaton - that drives the inflation. Isocurvature perturbations are, as we have mentioned, similar to isothermal perturbations but not identical. Indeed a variation of the old isothermal model has been advocated in recent years by Peebles (1987). His *Primordial Isocurvature Baryon Model* (PIB model) circumvents many of the problems of the old isothermal baryon model, but has difficulties of its own.

Difficulties with the adiabatic and isothermal pictures, chiefly the large-amplitude fluctuations they predicted in the cosmic microwave background, opened the way for the theories of the 1980s. These theories were built around the hypothesis that the Universe is dominated by *non-baryonic dark matter*, in the form of weakly interacting (collisionless) particles, perhaps neutrinos with mass $m_\nu \simeq 10$ eV or some other 'exotic' particles (gravitinos, photinos, axions, etc.) predicted by some theories of high-energy particle physics. There are various possible models; the simplest is one of three components: baryonic material, non-baryonic material made of a single type of particle, and radiation (also in this case, the addition of a component of massless neutrinos does not have much effect upon the evolution of perturbations). In this three-component system there are two fundamental perturbation modes again, similar to the two-component system mentioned above. These two modes are *curvature perturbations* (adiabatic modes) and *isocurvature perturbations*. In the first mode, all three components are perturbed ($\delta_m \simeq \delta_r \simeq \delta_i$, where i denotes the 'exotic' component); there is, therefore, a net perturbation in the energy-density and hence a perturbation in the curvature of space–time. In the second type of perturbation, however, the net energy-density is constant, so there is no perturbation to the spatial curvature.

The fashionable models of the 1980s can also be divided into two categories along the lines of the top-down/bottom-up labels we mentioned above. Here the discriminating factor is not the type of initial perturbation, which is usually assumed to be adiabatic in each case, but the form of the dark matter, as we shall discuss in Chapter 13.

In the *hot-dark-matter* (HDM) *scenario*, which is similar in broad outline to the old adiabatic baryon picture, the Universe is dominated by collisionless particles with a very large velocity dispersion (hence the name 'hot'), by virtue of it decoupling from the other components when it is still relativistic. A typical example is a neutrino with mass $m_\nu \simeq 10$ eV.

The *cold-dark-matter* (CDM) *scenario* has certain similarities to the old isothermal picture. This is characterised by the assumption that the Universe is dominated again by collisionless particles, but this time with a very small velocity dispersion (hence the term 'cold'). This can occur if the particles decouple when they are no longer relativistic (typical examples are supersymmetric particles such as gravitinos and photinos) or have never been in thermal equilibrium with the other components (e.g. the axion).

The rapid explosion in the quantity and quality of galaxy-clustering data (Chapters 16 and 18) and the discovery by the COBE team in 1992 of fluctuations in the temperature of the cosmic microwave background on the sky (Chapter 17) have placed strong constraints on these theories. Nevertheless, the general picture that Jeans instability produces galaxies and large-scale structure from small initial fluctuations seems to hold together extremely well. It remains to be seen whether the remaining questions can be resolved, or are symptomatic of a fundamental flaw in the model.

15.3 Gravitational Instability in Brief

In order to focus our attention on the various possible models, let us now recapitulate the essentials of the gravitational instability model.

In order to understand how structures form we need to consider the difficult problem of dealing with the evolution of inhomogeneities in the expanding Universe. We are helped in this task by the fact that we expect such inhomogeneities to be of very small amplitude early on so we can adopt a kind of perturbative approach, at least for the early stages of the problem. If the length scale of the perturbations is smaller than the effective cosmological horizon $d_H = c/H_0$, a Newtonian treatment of the subject is expected to be valid. If the mean free path of a particle is small, matter can be treated as an ideal fluid and the Newtonian equations governing the motion of gravitating particles in an expanding universe that we used in Chapters 10–12 can be used.

From these equations the essential point is that, if one ignores pressure forces, one obtains a simple equation for the evolution of δ:

$$\ddot{\delta} + 2H\dot{\delta} - \tfrac{3}{2}\Omega H^2 \delta = 0. \tag{15.3.1}$$

For a spatially flat universe dominated by pressureless matter, $\rho_0(t) = \frac{1}{6}\pi G t^2$ and Equation (15.3.1) admits two linearly independent power law solutions $\delta(\boldsymbol{x}, t) = D_{\pm}(t)\delta(\boldsymbol{x})$, where $D_{+}(t) \propto a(t) \propto t^{2/3}$ is the growing mode and $D_{-}(t) \propto t^{-1}$ is the decaying mode.

15.4 Primordial Density Fluctuations

The above considerations apply to the evolution of a single Fourier mode of the density field $\delta(\boldsymbol{x}, t) = D_{+}(t)\delta(\boldsymbol{x})$. What is more likely to be relevant, however, is the case of a superposition of waves, resulting from some kind of stochastic process in which the density field consists of a superposition of such modes with different amplitudes. A statistical description of the initial perturbations is therefore required, and any comparison between theory and observations will also have to be statistical.

The spatial Fourier transform of $\delta(\boldsymbol{x})$ is

$$\tilde{\delta}(\boldsymbol{k}) = \frac{1}{(2\pi)^3} \int \mathrm{d}^3\boldsymbol{x}\, \mathrm{e}^{-\mathrm{i}\boldsymbol{k}\cdot\boldsymbol{x}} \delta(\boldsymbol{x}). \tag{15.4.1}$$

It is useful to specify the properties of δ in terms of $\tilde{\delta}$. We can define the *power spectrum* of the field to be (essentially) the variance of the amplitudes at a given value of $\boldsymbol{k}$:

$$\langle \tilde{\delta}(\boldsymbol{k}_1)\tilde{\delta}(\boldsymbol{k}_2)\rangle = P(k_1)\delta^{\mathrm{D}}(\boldsymbol{k}_1 + \boldsymbol{k}_2), \tag{15.4.2}$$

where δ^{D} is the Dirac delta function; this rather cumbersome definition takes account of the translation symmetry and reality requirements for $P(k)$; isotropy is expressed by $P(\boldsymbol{k}) = P(k)$. The analogous quantity in real space is called the two-point correlation function, or, more correctly, the autocovariance function, of $\delta(\boldsymbol{x})$:

$$\langle \delta(\boldsymbol{x}_1)\delta(\boldsymbol{x}_2)\rangle = \xi(|\boldsymbol{x}_1 - \boldsymbol{x}_2|) = \xi(\boldsymbol{r}) = \xi(r), \tag{15.4.3}$$

which is itself related to the power spectrum via a Fourier transform. The shape of the initial fluctuation spectrum is assumed to be imprinted on the universe at some arbitrarily early time. As we have explained, many versions of the inflationary scenario for the very early universe (Guth 1981; Guth and Pi 1982) produce a power-law form

$$P(k) = Ak^n, \tag{15.4.4}$$

with a preference in some cases for the Harrison–Zel'dovich form with $n = 1$ (Harrison 1970; Zel'dovich 1972). Even if inflation is not the origin of density fluctuations, the form (15.4.4) is a useful phenomenological model for the fluctuation spectrum.

These considerations specify the shape of the fluctuation spectrum, but not its amplitude. The discovery of temperature fluctuations in the CMB by COBE has plugged that gap. We discuss the COBE normalisation in Chapter 17 but it is also worth mentioning that the abundance of galaxy clusters also provides a viable method for fixing the primordial amplitude; see, for example, Viana and Liddle (1996).

The power spectrum is particularly important because it provides a complete statistical characterisation of a particular kind of stochastic process: a *Gaussian random field*. This class of field is the generic prediction of inflationary models, in which the density perturbations are generated by Gaussian quantum fluctuations in a scalar field during the inflationary epoch (Guth and Pi 1982; Brandenberger 1985).

15.5 The Transfer Function

We have hitherto assumed that the effects of pressure and other astrophysical processes on the gravitational evolution of perturbations are negligible. In fact, depending on the form of any dark matter, and the parameters of the background cosmology, the growth of perturbations on particular length scales can be suppressed relative to the growth laws discussed above.

We need first to specify the fluctuation mode. In cosmology, the two relevant alternatives are *adiabatic* and *isocurvature*. The former involve coupled fluctuations in the matter and radiation component in such a way that the entropy does not vary spatially; the latter have zero net fluctuation in the energy density and involve entropy fluctuations. Adiabatic fluctuations are the generic prediction from inflation and form the basis of most currently fashionable models.

In the classical Jeans instability, pressure inhibits the growth of structure on scales smaller than the distance traversed by an acoustic wave during the free-fall collapse time of a perturbation. If there are collisionless particles of hot dark matter, they can travel rapidly through the background and this free streaming can damp away perturbations completely. Radiation and relativistic particles may also cause kinematic suppression of growth. The imperfect coupling of photons and baryons can also cause dissipation of perturbations in the baryonic component. The net effect of these processes, for the case of statistically homogeneous initial Gaussian fluctuations, is to change the shape of the original power spectrum in a manner described by a simple function of wave-number - the transfer function $T(k)$ - which relates the processed power spectrum $P(k)$ to its primordial form $P_0(k)$ via $P(k) = P_0(k) \times T^2(k)$. The results of full numerical calculations of all the physical processes we have discussed can be encoded in the transfer function of a particular model (Bardeen *et al.* 1986; Holtzmann 1989). For example, fast-moving or 'hot' dark-matter (HDM) particles erase structure on small scales by the free-streaming effects mentioned above so that $T(k) \to 0$ exponentially for large k; slow-moving or 'cold' dark matter (CDM) does not suffer such strong dissipation, but there is a kinematic suppression of growth on small scales (to be more

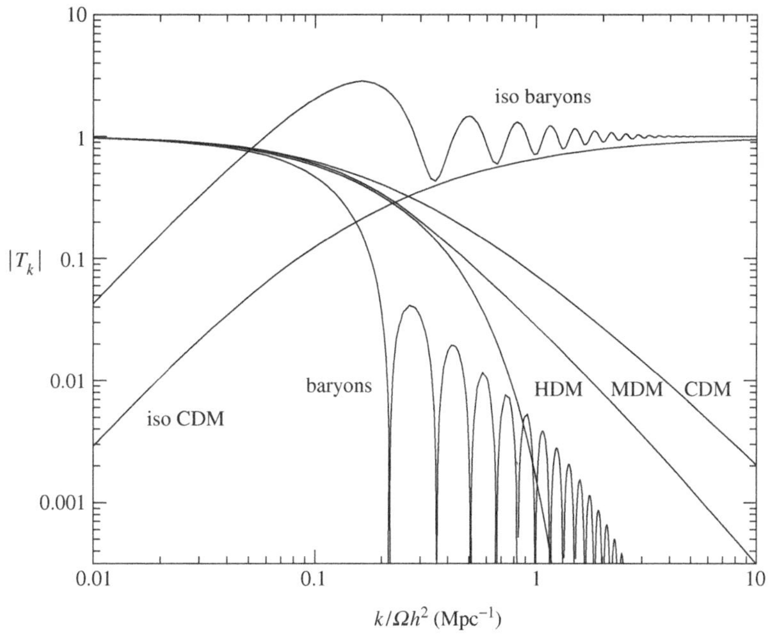

Figure 15.1 Examples of adiabatic transfer functions for baryons, hot dark matter (HDM), cold dark matter (CDM) and mixed dark matter (MDM; also known as CHDM). Isocurvature modes are also shown. Picture courtesy of John Peacock.

precise, on scales less than the horizon size at matter–radiation equality); significant small-scale power nevertheless survives in the latter case. These two alternatives thus furnish two very different scenarios for the late stages of structure formation: the 'top-down' picture exemplified by HDM first produces superclusters, which subsequently fragment to form galaxies; CDM is a 'bottom-up' model because small-scale structures form first and then merge to form larger ones. The general picture that emerges is that, while the amplitude of each Fourier mode remains small, i.e. $\delta(\boldsymbol{k}) \ll 1$, linear theory applies. In this regime, each Fourier mode evolves independently and the power spectrum therefore just scales as

$$P(k,t) = P(k,t_1)\frac{D_+^2(k,t)}{D_+^2(k,t_1)} = P_0(k)T^2(k)\frac{D_+^2(k,t)}{D_+^2(k,t_1)}.$$

For scales larger than the Jeans length, this means that $D_+(k,t) = D_+(t)$ only, so that the shape of the power spectrum is preserved during linear evolution on large scales. The quantity $D_+(t)$ is then just the growth factor δ_+ we discussed in Chapter 10.

Examples of transfer functions are shown in Figure 15.1. Note that the adiabatic transfer functions for CDM and HDM are all smooth, while the baryonic version has strong oscillations. The latter are produced by the acoustic oscillations we remarked upon in Chapter 11. Waves with different modes have different temporal phases which result in the waves arriving at recombination at different stages of their cycle. At recombination the restoring force for the oscillations supplied by pressure disappears and the waves become stranded with an amplitude that depends on wavelength. Since both HDM and CDM are collisionless, there is never any restoring force. Acoustic oscillations therefore do not occur.

The HDM transfer function shows a rapid cut-off at high k caused by free streaming, while CDM displays a graceful 'knee' produced by the Meszaros-suppression of fluctuations inside the horizon prior to matter–radiation equivalence. A characteristic scale for this knee is supplied by $\Omega_0 h^2$: the lower the value of Ω_0 the later the time of matter–radiation equivalence, the bigger the horizon at that point and the larger the scale of the knee.

15.6 Beyond Linear Theory

The linearised equations of motion provide an excellent description of gravitational instability at very early times when density fluctuations are still small ($\delta \ll 1$). The linear regime of gravitational instability breaks down when δ becomes comparable with unity, marking the commencement of the *quasilinear* (or weakly nonlinear) regime. During this regime the density contrast may remain small ($\delta < 1$), but the phases of the Fourier components $\delta_{\boldsymbol{k}}$ become substantially different from their initial values resulting in the gradual development of a non-Gaussian distribution function if the primordial density field was Gaussian. In this regime the shape of the power spectrum changes by virtue of a complicated cross-talk between different wave-modes. The usual approach is to use N-body experiments for strongly nonlinear analyses (Davis *et al.* 1985; Jenkins *et al.* 1998).

Further into the nonlinear regime, bound structures form. The baryonic content of these objects may then become important dynamically: hydrodynamical effects (e.g. shocks), star formation and heating and cooling of gas all come into play. The spatial distribution of galaxies may therefore be very different from the distribution of the (dark) matter, even on large scales. Attempts are only just being made to model some of these processes with cosmological hydrodynamics codes, but it is some measure of the difficulty of understanding the formation of galaxies and clusters that most studies have only just begun to attempt to include modelling the detailed physics of galaxy formation. In the front rank of theoretical efforts in this area are the so-called semi-analytical models, which encode simple rules for the formation of stars within a framework of merger trees that allow the hierarchical nature of gravitational instability to be explicitly taken into account (Baugh *et al.* 1998).

The usual approach is instead simply to assume that the point-like distribution of galaxies, galaxy clusters or whatever,

$$n(\boldsymbol{r}) = \sum_i \delta^{\mathrm{D}}(\boldsymbol{r} - \boldsymbol{r}_i), \qquad (15.6.1)$$

bears a simple functional relationship to the underlying $\delta(\boldsymbol{r})$. An assumption often invoked is that relative fluctuations in the object number-counts and matter density fluctuations are proportional to each other, at least within sufficiently large volumes, according to the *linear biasing* prescription:

$$\frac{\delta n(\boldsymbol{r})}{\bar{n}} = b\frac{\delta\rho(\boldsymbol{r})}{\bar{\rho}}, \tag{15.6.2}$$

where b is what is usually called the biasing parameter. For more detailed discussion see Section 14.8.

15.7 Recipes for Structure Formation

It should now be clear that models of structure formation involve many ingredients which may interact in a complicated way. In the following list, notice that most of these ingredients involve at least one assumption that may well turn out not to be true.

1. A background cosmology. This basically means a choice of Ω_0, H_0 and Λ, assuming we are prepared to stick with the Robertson–Walker metric and the Einstein equations.
2. An initial fluctuation spectrum. This is usually taken to be a power law, but may not be. The most common choice is $n = 1$.
3. A choice of fluctuation mode: usually adiabatic.
4. A statistical distribution of the initial fluctuations. This is often assumed to be Gaussian.
5. A normalisation of the power spectrum, usually taken to be the COBE microwave background measurements but there are other possibilities, such as requiring the abundance of clusters produced by the model to match observations.
6. The transfer function, which requires knowledge of the relevant proportions of 'hot', 'cold' and baryonic material as well as the number of relativistic particle species.
7. A 'machine' for handling nonlinear evolution, so that the distribution of galaxies and other structures can be predicted. This could be an N-body or hydrodynamics code, an approximated dynamical calculation or simply, with fingers crossed, linear theory.
8. A prescription for relating fluctuations in mass to fluctuations in light, frequently the linear bias model.

Historically speaking, the first model incorporating non-baryonic dark matter to be seriously considered was the HDM scenario, in which the universe is dominated by a massive neutrino with mass around 10–30 eV. This scenario has fallen into disrepute because the copious free streaming it produces smooths the matter fluctuations on small scales and means that galaxies form very late. The favoured alternative for most of the 1980s was the CDM model in which the dark-matter particles undergo negligible free streaming owing to their higher mass or non-thermal behaviour. A 'standard' CDM model (SCDM) then emerged in which the cosmological parameters were fixed at $\Omega_0 = 1$ and $h = 0.5$, the spectrum was of the Harrison–Zel'dovich form with $n = 1$ and a significant bias, $b = 1.5$–2.5, was required to fit the observations (Davis *et al.* 1985).

The SCDM model was ruled out by a combination of the COBE-inferred amplitude of primordial density fluctuations, galaxy-clustering power-spectrum estimates on large scales, rich cluster abundances and small-scale velocity dispersions (e.g. Peacock and Dodds 1996). It seems that the standard version of this theory simply has a transfer function with the wrong shape to accommodate all the available data with an $n = 1$ initial spectrum. Nevertheless, because CDM is such a successful first approximation and seems to have gone a long way to providing an answer to the puzzle of structure formation, the response of the community has not been to abandon it entirely, but to seek ways of relaxing the constituent assumptions in order to get a better agreement with observations. Various possibilities have been suggested.

If the total density is reduced to $\Omega_0 \simeq 0.3$, which is favoured by many arguments, then the size of the horizon at matter–radiation equivalence increases compared with SCDM and much more large-scale clustering is generated. This is called the open CDM model, or OCDM for short. The simplest way to describe this effect is to look at the shape of the CDM transfer function shown in Figure 15.1. This shows that position of the 'knee' scales with Ωh if k is measured in Mpc/h. This means that the knee pushes to lower physical wavenumbers, i.e. to larger scales, for low-density models. This is usually taken to define a shape parameter $\Gamma = \Omega_0 h$ so that the SCDM model has $\Gamma = 0.5$ and the OCDM version might have a shape parameter more like 0.2. The scaling with Ω is not quite exact, however: it is broken by the presence of baryons (Peacock and Dodds 1994).

Those unwilling to dispense with the inflationary predilection for flat spatial sections have invoked $\Omega_0 = 0.2$ and a positive cosmological constant (Efstathiou *et al.* 1990) to ensure that $k = 0$; this can be called ΛCDM and is apparently also favoured by the observations of distant supernovae we have mentioned previously (Riess *et al.* 1998; Perlmutter *et al.* 1999). Much the same effect on the power spectrum may be obtained in $\Omega = 1$ CDM models if matter–radiation equivalence is delayed, such as by the addition of an additional relativistic particle species. The resulting models are usually called τCDM (White *et al.* 1995).

Another alternative to SCDM involves a mixture of hot and cold dark matter (CHDM), having perhaps $\Omega_{\text{hot}} = 0.3$ for the fractional density contributed by the hot particles. For a fixed large-scale normalisation, adding a hot component has the effect of suppressing the power-spectrum amplitude at small wavelengths (Davis *et al.* 1992; Klypin *et al.* 1993). A variation on this theme would be to invoke a 'volatile' rather than 'hot' component of matter produced by the decay of a heavier particle (Pierpaoli *et al.* 1996). The non-thermal character of the decay products results in subtle differences in the shape of the transfer function in the CVDM model compared with the CHDM version. Another possi-

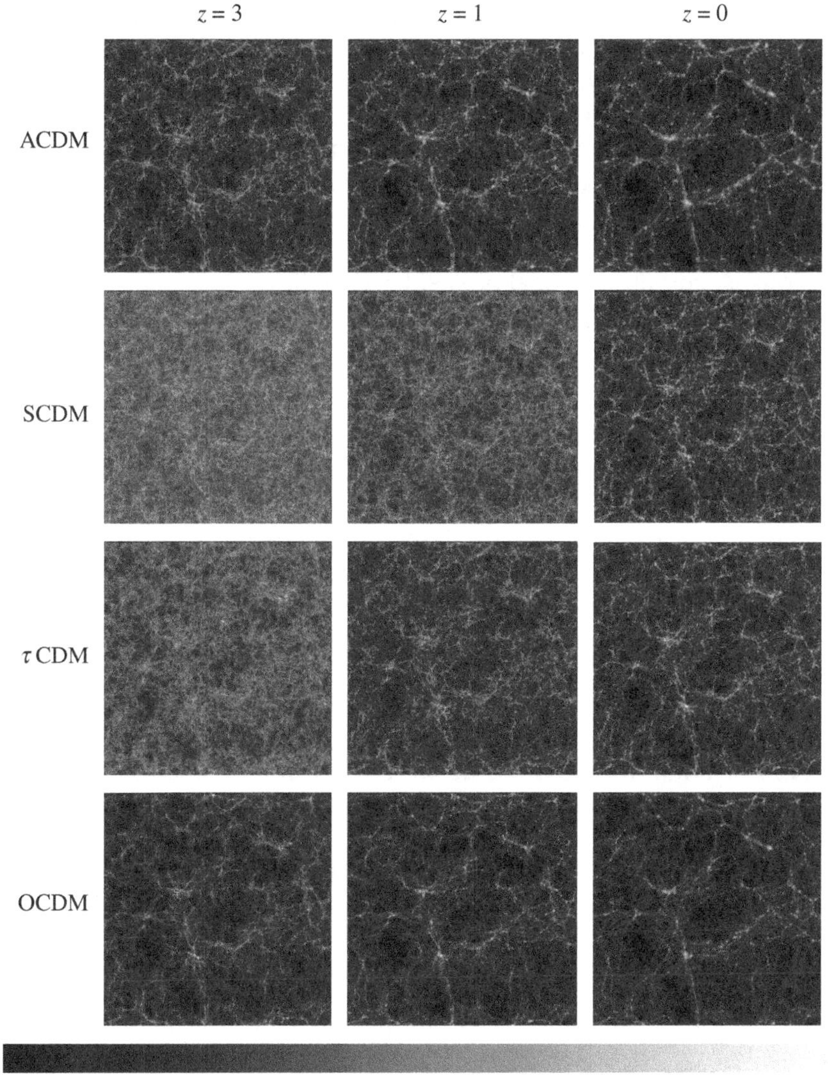

Figure 15.2 Some of the candidate models described in the text, as simulated by the Virgo consortium. Notice that SCDM shows very different structure at $z = 0$ than the three alternatives shown. The models also differ significantly at different epochs. These simulations show the distribution of dark matter only. Picture courtesy of the Virgo Consortium.

bility is to invoke non-flat initial fluctuation spectra, while keeping everything else in SCDM fixed. The resulting 'tilted' models (TCDM) usually have $n < 1$ power-law spectra for extra large-scale power and, perhaps, a significant fraction of tensor perturbations (Lidsey and Coles 1992). Models have also been constructed in which non-power-law behaviour is invoked to produce the required extra power: these are the broken scale-invariance (BSI) models (Gottlober *et al.* 1994).

But diverse though this collection of alternatives may seem, it does not include any models in which the assumption of Gaussian statistics is relaxed. This is at least as important as the other ingredients which have been varied in some of the above models. The reason for this is that fully specified non-Gaussian models are hard to construct, even if they are based on purely phenomenological considerations (Weinberg and Cole 1992; Coles *et al.* 1993b). Models based on topological defects rather than inflation generally produce non-Gaussian features but are computationally challenging (Avelino *et al.* 1998). A notable exception is the ingenious isocurvature model of Peebles (1999).

15.8 Comments

The models we have described in this chapter are not the only possible constructions of the basic gravitational instability scenario, but the list includes most of the current front runners. Our purpose was however not to try guessing the precise combination of parameters describing our universe but instead to set up a set of plausible models so that we can see in Part 4 how the differences between them might be probed.

It is interesting how the appealing simplicity of the standard cold dark matter has been superseded by a collection of apparently more complex third-generation models, all of which have extra free parameters to cover the basic deficiencies of SCDM. There is something very similar to Ptolemy's epicycles in this development and it would be somewhat depressing were it not for the fact that the field has entered a period not only of dramatic observational breakthroughs but of intense interplay between theory and observation.

Bibliographic Notes on Chapter 15

An excellent account of the field of structure-formation theory is given in Peacock (1999) and, with an emphasis on inflation models, by Liddle and Lyth (2000).

Problems

1. Account for the behaviour of the CDM isocurvature transfer function shown in Figure 15.1.

2. Calculate the radius of a sphere within which the average mass corresponds to that of a rich cluster $M_C \simeq 10^{14} M_\odot$. Use this radius within the Press–Schechter formalism described in the previous chapter to derive an expression for the number-density of clusters of mass exceeding M_C and investigate how this number varies with power-spectral index and Ω_0.

3. Rich clusters of galaxies have velocity dispersions of order 1000 km s^{-1} Mpc^{-1} or larger. Show that these objects correspond to metric perturbations of order 10^{-5}.

PART 4

Observational Tests

16

Statistics of Galaxy Clustering

16.1 Introduction

We now turn to the question of how to test theories of structure formation using observations of galaxy clustering. As we have seen, a theory for the origin of galaxies and clusters contains several ingredients which interact in a complicated way to produce the final structure. First, there is the background cosmological model which, in 'standard' theories, will be a Friedmann model specified by two parameters H_0 and Ω. Then we need to know the breakdown of the global mass density into baryons and non-baryonic matter. If the latter exists, we need to know whether it is hot or cold, or a mixture of the two. These two sets of information allow us to supply the transfer function (Section 14.7). If we then assume a spectrum for the primordial fluctuations, either in an *ad hoc* manner or by appealing to an inflationary model, we can use the transfer function to predict the shape of the fluctuation spectrum in the linear regime. But, importantly, we have no way to calculate *a priori* the *normalisation*, or amplitude, of the spectrum.

There are two ways one can attempt to normalise the power spectrum. One is to compare the properties of mass fluctuations predicted within the framework of the model using either linear theory (on sufficiently large scales) or N-body simulations. There are several problems with these approaches. One problem with linear theory is that one cannot be sure how accurate it will be for fluctuations of finite (i.e. measurable) amplitude. One therefore needs to be very careful to choose the appropriate statistical measure of fluctuations to compare the theory with the observations. Moreover, the linear approximation is only expected to be accurate on large scales where, because of the assumption of statistical homogeneity implicit in the Cosmological Principle, the fluctuation level will be small and therefore difficult to measure above sampling noise (statistical uncertainty due to finite survey size). Secondly, one needs to be sure that the sample of galaxies one uses to 'measure' clustering in our observed Universe is large enough to be, in some sense, representative of the Universe as a whole. If one extracts a statistical measure of clustering from a finite sample, then the value of the statistic would be different if one took a sample of the same size at a different place in the Universe. This effect is generally known as '*cosmic variance*', although this is not a particularly good term for the phenomenon it purports to describe. Important though these problems are, they are overshadowed by the obstacle presented by the existence of a bias, as described in Section 14.8. This means that, however accurately one can predict mass fluctuations analytically and however robustly one can measure galaxy fluctuations observationally, one cannot compare the two without assuming some *ad hoc* relationship between galaxies and mass like the linear bias model.

As we shall see, bias complicates all galaxy-clustering studies. If the bias is of the linear form described by Equation (14.8.10), then there is a simple constant multiplier between the 'mass' statistic and the 'galaxies' statistic so that, for example, the shape of the galaxy–galaxy correlation function and the shape of the matter autocovariance function are the same, but the amplitudes are different. In this case, knowing the multiplier b essentially eliminates the problem. On the other hand, the linear bias model is only expected to be applicable on very large scales (and perhaps not even then). Indeed, it is possible to imagine an extreme kind of bias which has the effect that there is very little correlation between the positions of galaxies and concentrations of mass. This is especially the case in scenarios where the bulk of the matter of the Universe is in the form of non-baryonic and therefore non-luminous material. Fortunately, however, there are ways to circumvent the bias problem to achieve a normalisation of the power spectrum or, at least, constrain it.

One way is to look not just at the positions of galaxies, but also at their peculiar motions. These motions are generated by gravity which, in turn, is generated by the whole mass distribution, not just by the luminous part. As we discussed in Section 4.6, the existence of peculiar motions means that the Hubble law is not exactly correct and consequently that a galaxy's redshift is not directly proportional to its distance from the observer. Galaxy redshift surveys generally supply only the redshifts, which are tacitly assumed to translate directly into distances via the Hubble law. Statistical measurements based on redshift surveys are therefore 'distorted' by deviations from the Hubble flow. The direct use of measured peculiar velocities and the indirect use of redshift-space distortions are both discussed in detail in Chapter 18; in the present chapter we shall generally assume that we can measure the statistical quantities in question in real space without worrying about redshift space.

The other way to normalise the spectrum only recently became possible with the COBE discovery of fluctuations in the CMB temperature in 1992. These are generally thought to be due to the influence of primordial fluctuations at $t \simeq t_{\rm rec}$, long before galaxy formation commenced. Knowing the amplitude of these fluctuations allows one, in principle, to compute the amplitude of the power spectrum at the present time without worrying about bias at all. We discuss this, and other issues connected to the CMB, in Chapter 17.

In the present chapter we shall concentrate on the statistical study of the clustering properties of galaxies and galaxy clusters and the relationship between observed statistical properties and theory. We shall use some of the tools introduced in Chapter 14 but will also introduce many new ones including, for example, techniques based on ideas from topology, dynamical systems and condensed matter physics. Different statistical descriptors measure different aspects of the clustering pattern revealed by a survey. Some quantities, such as the two-point correlation function (Section 16.2), the cell-count variance (Section 16.6) and the galaxy power spectrum (Section 16.7) are directly related to, and can therefore constrain, the fluctuation power spectrum. Other approaches, such as percolation analysis (Section 16.9) and topology (Section 16.10), test the morphology of the large-scale galaxy distribution and may therefore be sensitive to the existence of sheets and filaments predicted in the nonlinear phase of perturbation evolution or to features, such as bubbles, which may be connected with some form of non-Gaussian perturbation (Section 14.10). These methods therefore constrain a different set of 'ingredients' of structure-formation models. Other methods, such as higher-order correlations (Section 16.4), can shed light on whether self-similarity is important in the origin of the observed structure. We shall also take the opportunity in this chapter to show specific examples of how recent analyses of the 2dF Galaxy Redshift Survey and Sloan data using these statistical tools have yielded important constraints on models of structure formation. We shall, however, try to place an emphasis on methods rather than existing results, since we anticipate that new data will add much to our understanding of galaxy clustering in the next few years.

16.2 Correlation Functions

We begin our study of statistical cosmology by describing the correlation functions which have, for many years, been the standard way of describing the clustering of galaxies and galaxy clusters in cosmology. The use of these functions was first suggested in the 1960s by Totsuji and Kihara (1969), but their most influential advocate has been Peebles, who, along with several colleagues in the 1970s, carried out a program to extract estimates of these functions from the Lick galaxy catalogue and other data sets; see Peebles (1980) and references therein for details.

The correlation functions furnish a description of the clustering properties of a set of points distributed in space. The space can be three dimensional, but useful results are also obtainable for two-dimensional distributions of positions on the celestial sphere; see Section 16.3. We shall assume in this section that our 'points' are galaxies but this need not be the case. Indeed, this technique has been applied not only to various different kinds of galaxies (optical, infrared, radio) but also to quasars and clusters of galaxies; these latter objects are particularly important, for reasons we shall describe in Section 16.5. We shall also see that the correlation functions are closely related to the functions we described in Section 13.9 as the *covariance functions*, the difference between covariance and correlation functions being that the former describe properties of a continuous density field while the latter describe properties of a clustered set of points.

We have met the simplest correlation function already, in Section 13.9, but we give a more complete definition here. The joint probability $\delta^2 P_2$ of finding one galaxy in a small volume δV_1 and another in the volume δV_2, separated by a vector $\boldsymbol{r}_{12}$, if one chooses the two volumes randomly within a large (representative) volume of the Universe, is given by

$$\delta^2 P_2 = n_V^2 [1 + \xi(r_{12})] \delta V_1 \delta V_2, \tag{16.2.1}$$

where n_V is the mean number-density of galaxies and the function $\xi(r)$ is called the *two-point galaxy-galaxy spatial correlation function*. Because of statistical homogeneity and isotropy, ξ depends only on the modulus of the vector $\boldsymbol{r}_{12}$ (which we have written r_{12} in the equation) and not on its direction. If the galaxies are sprinkled completely randomly in space, then it is clear that $\xi(r_{12}) \equiv 0$; this means that ξ represents the excess probability, compared with a uniform random distribution, of finding another galaxy at a distance r_{12} from a given galaxy. If $\xi(r) > 0$, then galaxies are clustered, and if $\xi(r) < 0$, they tend to avoid each other. For reasons we explained in Section 14.9, if the correlation function is positive at $r_{12} \simeq 0$, it must change sign at large r_{12} so that its volume integral over all r_{12} does not diverge. Equation (16.2.1) implies, for example, that the mean number of galaxies within a distance r of a given galaxy is

$$\langle N \rangle_r = \tfrac{4}{3}\pi n_V r^3 + 4\pi n_V \int_0^r \xi(r'_{12}) r'^2_{12} \, \mathrm{d}r'_{12} : \tag{16.2.2}$$

the second term on the right-hand side of this equation represents the excess number compared with a uniform random distribution.

The two-point correlation function of a self-gravitating distribution of matter evolves rapidly in the nonlinear regime. This means that the shape of $\xi(r)$ in the regime where $\xi \simeq 1$ or greater will be very different from that of the primordial correlation function, and the amplitude will be different from that expected from linear theory. For this reason one cannot expect to use observations of $\xi(r)$ directly to normalise the spectrum. Notice, however, that the second term on the right-hand side of Equation (16.2.2) is an integral over ξ which is weighted to large r, and hence to regions of small $\xi(r)$. This motivates the use of the quantity J_3, defined by

$$J_3(R) \equiv \int_0^R \xi(r) r^2 \, \mathrm{d}r = \tfrac{1}{3} R^3 \int W_{\mathrm{TH}}(kR) P(k) \, \mathrm{d}^3 \boldsymbol{k}, \tag{16.2.3}$$

with R up to several tens of Mpc, to obtain the normalisation; $W_{\rm TH}$ is the top-hat window function introduced in Section 13.3. This kind of normalisation was used frequently before the discovery of CMB temperature fluctuations.

Let us stress again that $\xi(r)$ measures the correlations between galaxies, not the correlations of the mass distribution. These might be equal if galaxies trace the mass, but if galaxy formation is biased they will differ. In the linear bias model – equation (14.8.10) – the galaxy–galaxy correlations will be a factor b^2 higher than the mass correlations.

If one only has a two-dimensional (projected) catalogue, then one can define the *two-point galaxy–galaxy angular correlation function*, $w(\vartheta)$, by

$$\delta^2 P_2 = n_\Omega^2 [1 + w(\vartheta_{12})]\delta\Omega_1 \delta\Omega_2, \tag{16.2.4}$$

which, in analogy with (16.2.1), is just the probability of finding two galaxies in small elements of solid angle $\delta\Omega_1$ and $\delta\Omega_2$, separated by an angle ϑ_{12} on the celestial sphere; n_Ω is the mean number of galaxies per unit solid angle on the sky.

In an analogous manner one can define the correlation functions for $N > 2$ points; we mentioned this in Section 13.9. The definition proceeds from equation (13.8.15), which gives the probability of finding N galaxies in the N (disjoint) volumes δV_i in terms of the total N-point correlation function $\xi^{(N)}$. This function, however, contains contributions from correlations of lower order than N and a more useful statistic is the *reduced* or *connected correlation function*, which is simply that part of $\xi^{(N)}$ which does not depend on correlations of lower order; we shall use $\xi_{(N)}$ for the connected part of $\xi^{(N)}$. One can illustrate the way to extract the reduced correlation function simply using the three-point function as an example. Using the cluster expansion in the form given by equation (13.8.13) and, as instructed in Section 13.9, interpreting the single partitions $\langle \delta_{\rm i} \rangle$ as having the value of unity for point distributions rather than the zero value one uses in the case for continuous fields, we find

$$\delta^3 P_3 = n_V^3 [1 + \xi(r_{12}) + \xi(r_{23}) + \xi(r_{31}) + \zeta(r_{12}, r_{23}, r_{31})]\delta V_1 \delta V_2 \delta V_3, \tag{16.2.5}$$

where $\zeta \equiv \xi_{(3)}$ is the reduced three-point function. The terms $\xi(r_{ij})$ represent the excess number of triplets one gets compared with a random distribution (described by the '1') just by virtue of having more pairs than in a random distribution; the term ζ is the number of triplets above that expected for a distribution with a given two-point correlation function. From now on we shall drop the term 'connected' or 'reduced'; whenever we use an N-point correlation function, it will be assumed to be the reduced one. The three-point angular correlation function z is defined in an analogous manner:

$$\delta^3 P_3 = n_\Omega^3 [1 + w(\vartheta_{12}) + w(\vartheta_{23}) + w(\vartheta_{31}) + z(\vartheta_{12}, \vartheta_{23}, \vartheta_{31})]\delta\Omega_1 \delta\Omega_2 \delta\Omega_3, \tag{16.2.6}$$

which is the probability of finding galaxies in the three solid-angle elements $\delta\Omega_1$, $\delta\Omega_2$ and $\delta\Omega_3$, separated by angles ϑ_{12}, ϑ_{23} and ϑ_{31} on the celestial sphere. For

$N = 4$ the spatial correlation function $\eta \equiv \xi_{(4)}$ is defined by

$$\begin{aligned}\delta^4 P_4 = n_V^4[1 &+ \xi(r_{12}) + \xi(r_{13}) + \xi(r_{14}) + \xi(r_{23}) + \xi(r_{24}) + \xi(r_{34}) \\ &+ \xi(r_{12})\xi(r_{34}) + \xi(r_{13})\xi(r_{24}) + \xi(r_{14})\xi(r_{23}) \\ &+ \zeta(r_{12}, r_{23}, r_{31}) + \zeta(r_{12}, r_{24}, r_{41}) + \zeta(r_{13}, r_{34}, r_{41}) \\ &+ \zeta(r_{23}, r_{34}, r_{42}) + \eta(r_{12}, r_{13}, r_{14}, r_{23}, r_{24}, r_{34})]\delta V_1 \delta V_2 \delta V_3 \delta V_4\end{aligned} \tag{16.2.7}$$

in an obvious notation; one can also define the four-point angular function u in an appropriate manner. The usual notation for the five-point spatial function is $\tau \equiv \xi_{(5)}$ and, for its angular version, t.

16.3 The Limber Equation

One of the most useful aspects of the correlation functions, particularly the two-point correlation function, is that its spatial and angular versions have a relatively simple relationship between them. This allows one to extract an estimate of the spatial function from the angular version. In Section 4.5 we introduced the luminosity function $\Phi(L)$. Let us convert this into a function of magnitude M, as described in Section 1.8, via $\Psi(M) = \Phi(L)|\mathrm{d}L/\mathrm{d}M|$. This allows us to write

$$\delta^2 P = \Psi(M)\delta M \delta V, \tag{16.3.1}$$

which is the probability of finding a galaxy with absolute magnitude between M and $M + \delta M$ in the volume δV. By analogy with Equation (16.2.1) we can also write the joint probability of finding two galaxies, one in δV_1 with magnitude between M_1 and $M_1 + \delta M_1$ and the other in δV_2 with magnitude between M_2 and $M_2 + \delta M_2$, separated by a distance r_{12}, as

$$\delta^4 P = [\Psi(M_1)\Psi(M_2) + G(M_1, M_2, r_{12})]\delta M_1 \delta M_2 \delta V_1 \delta V_2, \tag{16.3.2}$$

where the function G takes account of the correlations between the galaxies. We now suppose that the absolute magnitude of a galaxy is statistically independent of its position with respect to other galaxies, that is to say that $\Psi(M)$ is independent of the strength of clustering. This hypothesis, called the *Limber hypothesis*, seems to be verified by observations but is actually quite a strong assumption: it means, for example, that there is no variation of the luminosity properties of galaxies with the density of their environment. We then write

$$G(M_1, M_2, r_{12}) = \Psi(M_1)\Psi(M_2)\xi(r_{12}). \tag{16.3.3}$$

Projected catalogues generally collect the positions of galaxies brighter than a certain apparent magnitude limit m_0 within some well-defined region on the celestial sphere. To take account of systematic observational errors concerning the objects with apparent magnitude $m \simeq m_0$, one introduces a selection function $f(m - m_0)$

which is the probability that an observer includes a galaxy with apparent magnitude m in the catalogue. The function f should be equal to unity for $m \ll m_0$ (galaxies much brighter than m_0), and practically zero for $m \gg m_0$. A good catalogue will also have a sharp cut-off at $m \simeq m_0$, though this is not always realised in practice. The luminosity function of galaxies has a characteristic magnitude at $M^* \simeq -19.5 + 5\log h$ and tends rapidly to zero for $M < M^*$. Let us assume that the typical distance from the observer of galaxies in the catalogue is D^*, the distance at which a galaxy with absolute magnitude M^* is seen with an apparent magnitude m_0; from Equation (1.8.3) we have

$$D^* = 10^{0.2(m_0-M^*)-5}\ \mathrm{Mpc}. \tag{16.3.4}$$

The number of galaxies in a certain catalogue per unit solid angle, from Equations (16.3.1) and (16.3.4), is given by

$$n_\Omega = D^{*3}\int_0^\infty x^2\,\mathrm{d}x\int_{-\infty}^{+\infty}\Psi(M)f(M-M^*+5\log x)\,\mathrm{d}M = D^{*3}\int_0^\infty \psi(x)x^2\,\mathrm{d}x, \tag{16.3.5}$$

where $x = r/D^*$ and

$$\psi(x) = \int_{-\infty}^{+\infty}\Psi(M)f(M-M^*+5\log x)\,\mathrm{d}M. \tag{16.3.6}$$

The function $\psi(x)$ represents the number of galaxies per unit volume, at a distance given by $r = xD^*$, belonging to the catalogue. This function is given to a good approximation by

$$\psi(x) = n_V x^{-5\beta} \qquad (\beta = 0.25;\ x < 1), \tag{16.3.7 a}$$

$$\psi(x) = n_V x^{-5\alpha} \qquad (\alpha = 0.75;\ 1 < x < x_0), \tag{16.3.7 b}$$

$$\psi(x) = 0 \qquad (x > x_0 \simeq 10^{2/5\alpha} = 10^{8/15}). \tag{16.3.7 c}$$

From Equations (16.3.2) and (16.3.3) one can recover Equation (16.2.4):

$$\begin{aligned}\delta^2 P_2 &= n_\Omega^2[1 + w(\vartheta_{12})]\delta\Omega_1\delta\Omega_2\\ &= D^{*6}\int_0^\infty \psi(x_1)x_1^2\,\mathrm{d}x_1\int_0^\infty \psi(x_2)x_2^2[1+\xi(r_{12})]\,\mathrm{d}x_2\delta\Omega_1\delta\Omega_2,\end{aligned} \tag{16.3.8}$$

where

$$r_{12}^2 = D^{*2}(x_1^2 + x_2^2 - 2x_1x_2\cos\vartheta_{12}). \tag{16.3.9}$$

It is helpful to move to new variables:

$$x = \tfrac{1}{2}(x_1 + x_2), \qquad y = \frac{x_1 - x_2}{x\vartheta_{12}}. \tag{16.3.10}$$

Because the catalogue is assumed to be a 'fair' sample of the Universe, the typical length scale of correlations must be much less than D^*. For this reason the main contribution to the integral over $\xi(r_{12})$ in (16.3.8) comes from points with $x_1 \simeq x_2 \simeq 1$, separated by a small angle ϑ_{12}. For this reason (16.3.9) becomes

$$r_{12}^2 \simeq D^{*2} x^2 \vartheta_{12}^2 (1 + y^2) \tag{16.3.11}$$

and the Equations (16.3.8) and (16.3.11) furnish the relation

$$w(\vartheta_{12}) \simeq \frac{\vartheta_{12} \int_0^\infty \psi^2(x) x^5 \, dx \int_{-\infty}^{+\infty} \xi[D^* x \vartheta_{12} (1 + y^2)^{1/2}] \, dy}{[\int_0^\infty \psi(x) x^2 \, dx]^2}, \tag{16.3.12}$$

called the *Limber equation* (obtained by Limber (1953, 1954) to analyse the correlations of stars in our Galaxy). This relationship has the interesting scaling property that

$$w'\left(\vartheta'_{12} = \frac{D^*}{D^{*'}} \vartheta_{12}\right) = \frac{D^*}{D^{*'}} w(\vartheta_{12}), \tag{16.3.13}$$

where w and w' are the correlation functions corresponding to two catalogues with characteristic distances D^* and $D^{*'}$, respectively.

One can extend the Limber equation to higher-order correlations $N > 2$, still assuming the Limber hypothesis. It is thus possible to relate the angular and spatial N-point functions for $N > 2$. We shall spare the reader the details, but just mention some of the results in the next section.

16.4 Correlation Functions: Results

16.4.1 Two-point correlations

The analysis of two-dimensional catalogues of the projected positions of galaxies on the sky (chiefly the Lick map and, more recently, the APM and COSMOS surveys) has shown that, over a suitable interval of angles ϑ, the angular two-point correlation function $w(\vartheta)$ is well approximated by a power law

$$w(\vartheta) \simeq A^* \vartheta^{-\delta} \quad (\vartheta_{\min} \leqslant \vartheta \leqslant \vartheta_{\max};\ \delta \simeq 0.8), \tag{16.4.1}$$

where the amplitude A^* depends on the characteristic distance D^* of the galaxies in the catalogue, and the angular interval over which the relationship (16.4.1) holds corresponds to a spatial separation $0.1h^{-1}$ Mpc $\leqslant r \leqslant 10h^{-1}$ Mpc at this distance. One can use the scaling relation (16.3.13) to compare the correlation functions of catalogues with different values of D^* and so check the assumptions upon which the analysis is based. Beyond the power-law regime the angular correlation function breaks and rapidly falls to zero.

If one makes the assumption that, over a certain interval of scale, the two-point spatial correlation function is given by

$$\xi(r) = B r^{-\gamma}, \tag{16.4.2}$$

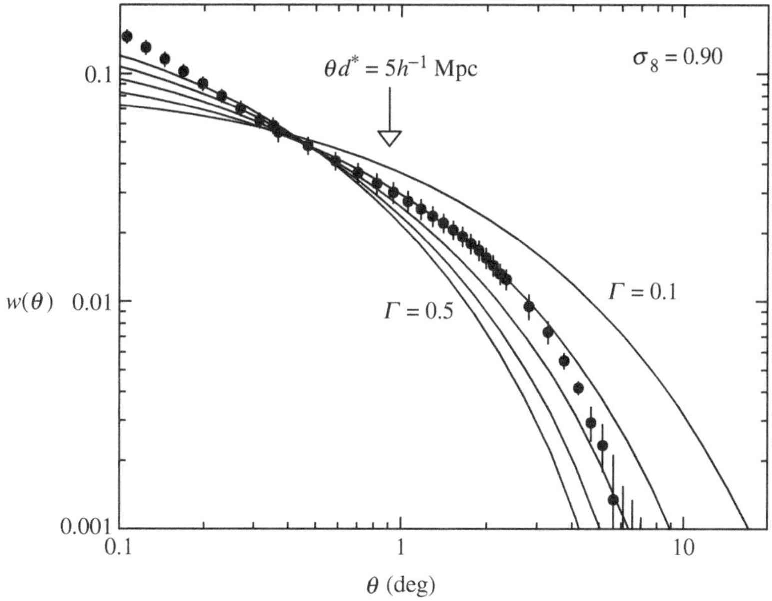

Figure 16.1 The dots with error bars show determinations of $w(\vartheta)$ from the APM survey, while the solid lines show a family of CDM models labelled by the shape parameter Γ. Figure courtesy of Steve Maddox.

then one can recover from Equation (16.3.12) that

$$w(\vartheta) = A\vartheta^{1-\gamma} = A\vartheta^{-\delta}, \tag{16.4.3}$$

where the constants A and B are related by

$$\frac{A}{B} = \frac{\Gamma(1/2)\Gamma[(\gamma-1)/2]}{\Gamma(\gamma/2)} \frac{\int_0^\infty x^{5-\gamma}\psi^2(x)\,\mathrm{d}x}{[\int_0^\infty x^2\psi(x)\,\mathrm{d}x]^2} D^{*-\gamma} \tag{16.4.4}$$

(Γ is the Euler gamma function). The assumption (16.4.2) therefore appears consistent with the angular correlation function (16.4.1) if

$$\xi(r) \simeq \left(\frac{r}{r_{0\mathrm{g}}}\right)^{-\gamma}, \tag{16.4.5}$$

with $r_{0\mathrm{g}} \simeq 5h^{-1}$ Mpc and $\gamma \simeq 1.8$ in the range $0.1h^{-1}$ Mpc $\leqslant r \leqslant 10h^{-1}$ Mpc (e.g. Shanks *et al.* 1989); on larger scales the correlation function tends rapidly towards zero and is difficult to measure above statistical noise. The form of $\xi(r)$ given in (16.4.5) is confirmed by direct, i.e. three-dimensional, determinations from galaxy surveys, as shown in Figure 16.2. The quantity $r_{0\mathrm{g}}$, where $\xi = 1$, is often called the *correlation length* of the galaxy distribution; it marks, roughly speaking, the transition between linear and nonlinear regimes.

The usual method for estimating $\xi(r)$, or $w(\vartheta)$, employs a random Poisson point process generated with the same sample boundary and selection function as the real data; one can then estimate ξ straightforwardly according to

$$1 + \hat{\xi}(r) \simeq \frac{n_{\mathrm{DD}}(r)}{n_{\mathrm{RR}}(r)} \tag{16.4.6}$$

or, more robustly, using either

$$1 + \hat{\xi}(r) \simeq \frac{n_{\mathrm{DD}}(r)}{n_{\mathrm{DR}}(r)} \tag{16.4.7 a}$$

or

$$1 + \hat{\xi}(r) \simeq \frac{n_{\mathrm{DD}}(r) n_{\mathrm{RR}}(r)}{n_{\mathrm{DR}}^2(r)}, \tag{16.4.7 b}$$

where $n_{\mathrm{DD}}(r)$, $n_{\mathrm{RR}}(r)$ and $n_{\mathrm{DR}}(r)$ are the number of pairs with separation r in the actual data catalogue, in the random catalogue and with one member in the data and one in the random catalogue, respectively. In Equations (16.4.6) and (16.4.7) we have assumed, for simplicity, that the real and random catalogues have the same number of points (which they need not). The second of these estimators is more robust to boundary effects (e.g. if a cluster lies near the edge of the survey region), but they both give the same result for large samples.

16.5 The Hierarchical Model

The problem with the higher-order correlation functions $\xi_{(N)}$ is that they are functions of all the distances separating the N points and are consequently much more difficult to interpret than $\xi = \xi_{(2)}$, which is a function of only one variable. It therefore helps to have a model for the higher-order correlations which one can use to interpret the results. The fact that the two-point correlation function has a power-law behaviour suggests that one might look for a *hierarchical model*, i.e. for a self-similar behaviour of the $\xi_{(N)}$ in which the Nth function is related to the $(N-1)$th function and thence all the way down to the two-point function, according to some simple scaling rule. Notice that this assumption is conceptually distinct from the simplified treatment of hierarchical clustering we presented in Section 14.4, i.e. the hierarchical model for correlations does not automatically follow from that discussion. In fact, the hierarchical model here rests on the assumption of scale invariance, i.e. that the higher-order correlations possess no characteristic scale. The appropriate model for the three-point function is

$$\zeta(r_{12}, r_{23}, r_{31}) = \xi_{(3)}(r_{12}, r_{23}, r_{31}) = Q(\xi_{12}\xi_{23} + \xi_{23}\xi_{31} + \xi_{31}\xi_{12}), \tag{16.5.1}$$

where Q is a constant. This form does indeed appear to fit observations fairly well, with a value $Q \simeq 1$ over the range $50h^{-1}$ kpc $< r < 5h^{-1}$ Mpc. The appropriate generalisation of Equation (16.5.1) to $N > 3$ is more complicated, and involves a bit of combinatorial analysis:

$$\xi_{(N)} = \sum_{\text{topologies}} Q_{N,t} \sum_{\text{relabellings}} \prod_{\text{edges}} \xi_{ij}. \tag{16.5.2}$$

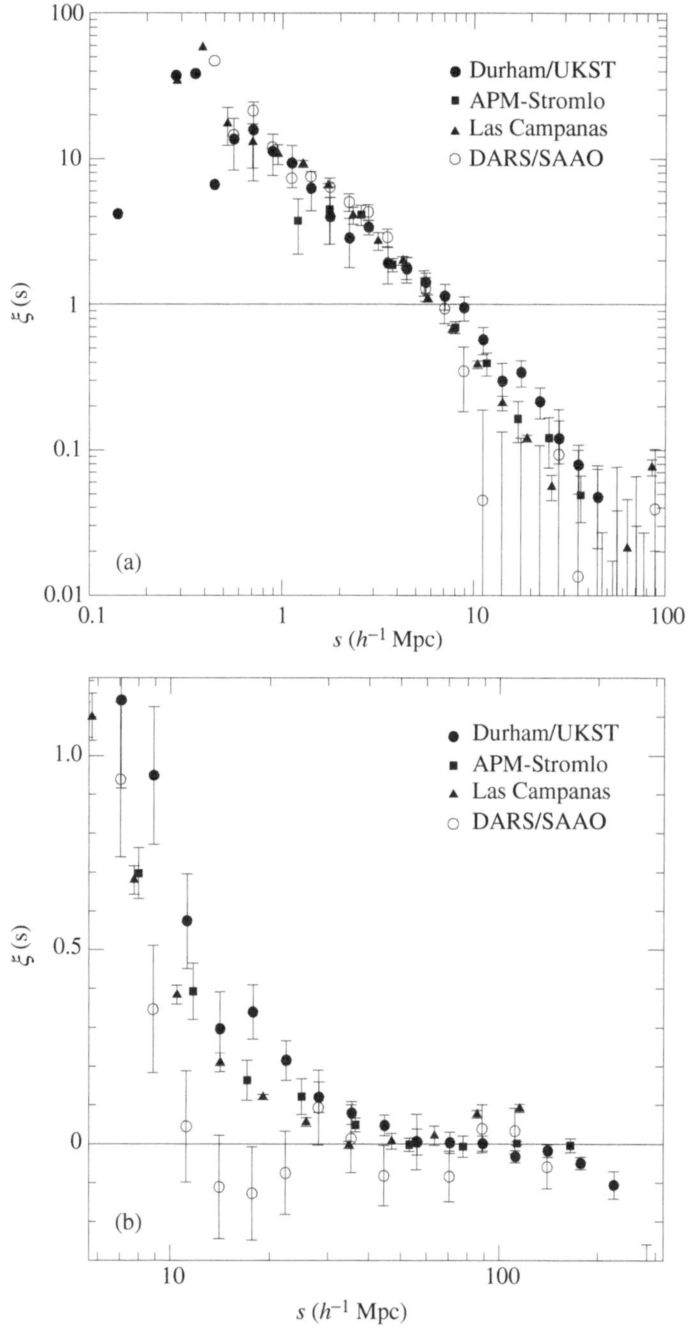

Figure 16.2 Estimates of $\xi(r)$ from different redshift surveys, including the Las Campanas Redshift Survey shown in Figure 4.6. The variable s is shown instead of r to denote determination in redshift space, rather than real space; see Section 18.5. Figure courtesy of Tom Shanks.

The notation here means a product over the $(N - 1)$ edges linking N objects, summed over all relabellings of the objects (l) and summed again over all distinct N-tree graphs with a given topology t weighted by a coefficient $Q_{N,t}$. The four-point term must therefore include two coefficients, one for 'snake' connections and the other for 'star' graphs, as illustrated in Figure 16.3. For $N = 2$ and $N = 3$, the different graphs connecting the points are topologically equivalent, but for $N = 4$ there are two distinct topologies. The topological difference can be seen by considering the result of cutting one edge in the graph. The first 'snake' topology is such that connections can be cut to leave either two pairs, or one pair and a triplet. The second cannot be cut in such a way as to leave two pairs; this is a 'star' topology. There are twelve possible relabellings of the snake and four of the star. For the $N = 5$ function, there are three distinct topologies, illustrated in the figure with 5, 60 and 60 relabellings, respectively. We leave it as an exercise for the reader to show that $N = 6$ has six different topologies, and a total of 1296 different relabellings.

The Lick and Zwicky catalogues have also supplied a rather uncertain estimate of the four-point correlation function, which is given by the approximate relation

$$\eta = \xi_{(4)} \simeq R_a[\xi(r_{12})\xi(r_{23})\xi(r_{34}) + 11 \text{ others}] \\ + R_b[\xi(r_{12})\xi(r_{13})\xi(r_{14}) + 3 \text{ others}], \qquad (16.5.3)$$

where the function η depends on the six independent interparticle distances as in Equation (16.2.7); the first twelve terms correspond to 'snake' topologies and the second four to 'stars'; the quantities R_a and R_b correspond to $Q_{N,t}$ of Equation (16.5.2) for each of the two topologies; from observations, $R_a \simeq 2.5$ and $R_b \simeq 4.3$. This again seems to confirm the hierarchical model. Indeed, as far as one can tell within the statistical errors, all the correlation functions up to $N \simeq 8$ seem to follow a roughly hierarchical pattern. The success of this model is intriguing, particularly as the analysis of galaxy counts in cells seems to confirm that it extends to larger scales than can be probed directly by the correlation functions. A sound theoretical understanding of this success now seems to be emerging: the strongly nonlinear behaviour (16.5.2) is consistent with our understanding of the statistical mechanics of self-gravitating systems through a hierarchy of equations studied first by Born, Bogoliubov, Green, Kirkwood and Yvon, which is known as the BBGKY hierarchy. The behaviour in the weakly nonlinear regime can be understood by perturbation theory.

16.5.1 Comments

The extraction of estimates of $\xi_{(N)}$ from galaxy samples has involved a huge investment of computer power over the last two decades. These functions have yielded important insights into both the statistical properties and possible dynamical origin of the clustering pattern. An important aspect of this is a connection, which we have no space to explore here, between the correlation functions and a dynamical description of self-gravitating systems in terms of the set of equations that make up the BBGKY hierarchy (Davis and Peebles 1977).

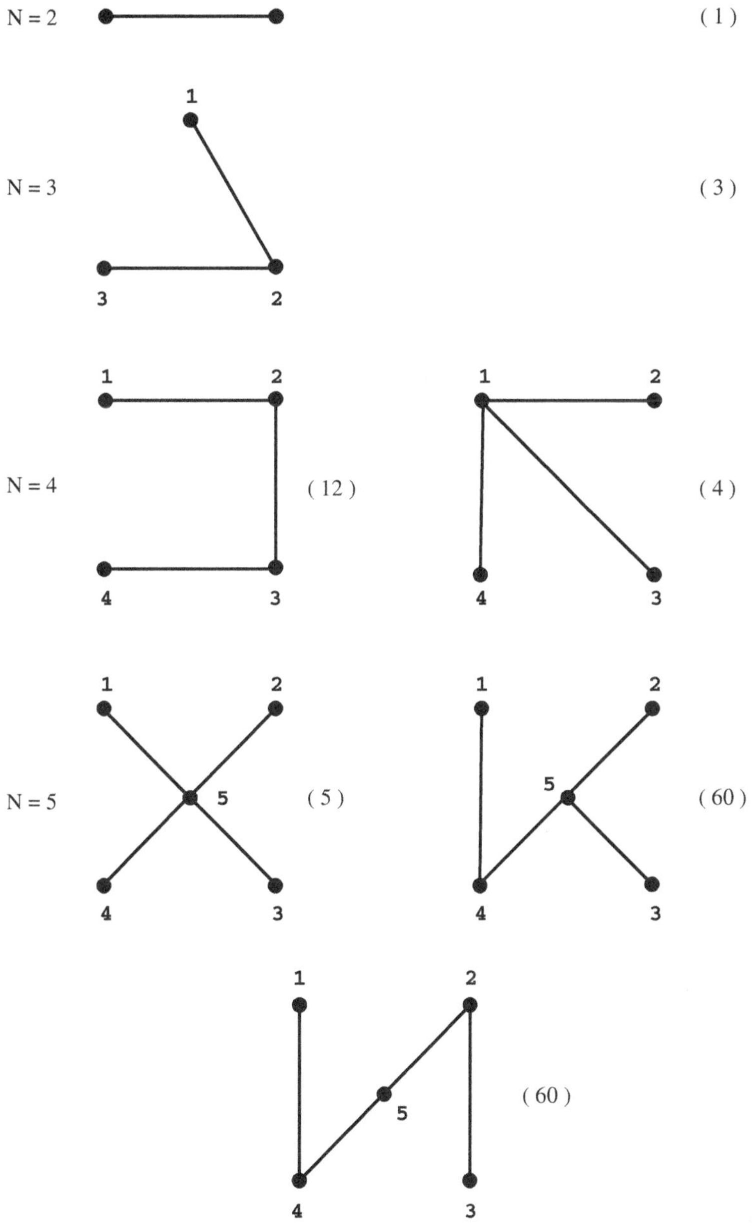

Figure 16.3 Different topologies of graphs connecting the N points for computing correlation functions in the hierarchical model; graphs for $N = 3$, 4 and 5 are shown.

Nevertheless, the statistical information contained in these functions is limited. In order to have a complete statistical description of the properties of a point distribution we need to know all the finite-order correlation functions. Given the computational labour required to extract even the low-order functions from a large sample, this is unlikely to be achieved in practice. This problem is exacerbated by the fact that the correlation functions, even the two-point function, are very difficult to determine from observations on large scales where the evolution of ξ is close to linear and analytical theory is consequently most reliable. For this reason, and the difficulty of disentangling effects of bias from dynamical evolution, it is necessary to look for other statistical descriptions; we shall describe some of these in Sections 16.6–16.10.

16.6 Cluster Correlations and Biasing

As we mentioned above, the correlation function analysis can be applied to other kinds of distributions, including quasars and radio galaxies. In this section we shall concentrate on rich clusters of galaxies; we shall also restrict ourselves to the two-point correlation properties of these objects since the sizes of these samples make it difficult to obtain accurate estimates of higher-order functions. The two-point correlation function for Abell clusters (those containing at least 65 galaxies within the 'Abell radius' of around $1.5h^{-1}$ Mpc) is found to be

$$\xi_c(r) \simeq \left(\frac{r}{r_{0c}}\right)^{-\gamma}, \tag{16.6.1}$$

where $5h^{-1}$ Mpc $\leqslant r \leqslant 75h^{-1}$ Mpc, $r_{0c} \simeq 12\text{–}25h^{-1}$ Mpc and $\gamma \simeq 1.8$. The similarity in shape between (16.6.1) and the galaxy version (16.4.5) is interesting. There is, however, considerable uncertainty about the correct value of the correlation length r_{0c} for these objects because of the possible, indeed probable, existence of systematic errors accumulated during the compilation of the Abell catalogue. Cluster catalogues recently compiled using automated plate-measuring devices suggest values towards the lower end of the quoted range, while the richest Abell clusters (those with more than 105 galaxies inside an Abell radius) may have a correlation length as large as $50h^{-1}$ Mpc. There is indeed some evidence that the correlation length scales with the richness (i.e. density) of the clusters and is consequently higher for the denser, and hence rarer, clusters. It has been suggested that this correlation can be expressed by the relationship

$$\xi_i(r) \simeq \left(\frac{r}{r_{0c,i}}\right)^{-\gamma} \simeq C_i\left(\frac{r}{l_i}\right)^{-\gamma} \quad \left[C_i \simeq \left(\frac{r_{0c,i}}{l_i}\right)^{\gamma} \simeq \text{const.} \simeq 0.4\right] \tag{16.6.2}$$

between the correlation length $r_{0c,i}$ and the mean separation l_i of subsamples selected according to a given richness threshold. The self-similar form of (16.6.2) can be interpreted intuitively as a kind of fractal structure.

The self-similar properties that seem to be implied by both observations and the theory described above lead one naturally to a description of the mass distribution in the language of fractal sets. The prevalence of techniques based on fractal geometry in fields such as condensed matter physics has given rise to a considerable interest in applying these methods to the cosmological context.

To get a rough idea of the fractal description consider the mass contained in a small sphere of radius r around a given galaxy, denoted $M(r)$. In the case where $\xi(r) \gg 1$ we have

$$M(r) \propto \xi(r)r^3 \propto r^{D_2}, \tag{16.6.3}$$

with $D_2 = 3 - \gamma$: since $\xi(r)$ has a power-law form with a slope of around $\gamma \simeq 1.8$, then we have $M(r) \propto r^{1.2}$. In the language of fractals, this corresponds to a *correlation dimension* of $D_2 \simeq 1.2$. One can interpret this very simply by noting that, if the mass is distributed along one-dimensional structures (filaments), then $M(r) \propto r$; two-dimensional sheets would have $M \propto r^2$ and a space-filling homogeneous distribution would have $M \propto r^3$. A fractional dimension like that observed indicates a *fractal structure.*

The first convincing explanation of the relationship between (16.6.1) and (16.4.5) was given by Kaiser (1984). He supposed that galaxy and cluster formation proceeded hierarchically from Gaussian initial conditions in the manner outlined in Section 14.4. If this is the case, then clusters, on mass scales of order $10^{15} M_\odot$, must have formed relatively recently. Moreover, rich clusters are extremely rare objects, with a mean separation of order $60h^{-1}$ Mpc. It is natural therefore to interpret rich clusters as representing the high peaks of a density field which is still basically evolving linearly: the collapse of the highest peaks will not alter the properties of the 'average' density regions significantly. Applying the spherical 'top-hat' collapse model of Section 15.1, the collapse to a bound structure occurs when, roughly speaking, the linearly evolved value of the density perturbation, δ, on the relevant scale reaches a value $\delta_c \simeq 1.68$. If $\Omega \simeq 1$, which we assume for simplicity, then the collapse time t_{coll} will be given by

$$t_{\text{coll}} \simeq t_0 \left(\frac{1.68}{\nu\sigma}\right)^{3/2}, \tag{16.6.4}$$

where t_0 is the present epoch, σ is the RMS mass fluctuation on the scale of clusters and $\delta = \nu\sigma$ is the value of δ obtained from linear theory. The final overdensity of the collapsed structure with respect to the background universe will be, at collapse (see Section 15.1),

$$\delta_f \simeq 180 \left(\frac{t_0}{t_{\text{coll}}}\right)^2, \tag{16.6.5}$$

so that structures which collapse earlier have a higher final density. For $t_0 \geqslant t_{\text{coll}} \geqslant t_0/2$ we have $1.7 \leqslant \nu\sigma \leqslant 2.4$ and $180 \leqslant \delta_f \leqslant 720$. A small difference in collapse time and, therefore, a small difference in ν produces objects with very different final density. For this reason it is reasonable to interpret clusters as being density 'peaks', i.e. as regions where δ exceeds some sharp threshold. On large scales we can use the high-peak biasing formalism described in Section 14.8; the relationship between the correlation function of the 'peaks' and the covariance function of the underlying matter distribution is therefore given by Equation (14.8.5).

For simplicity we assume that galaxies trace the mass, so that equation (14.8.5) becomes

$$\xi_c(r) \simeq \left(\frac{\nu}{\sigma}\right)^2 \xi_g(r), \tag{16.6.6}$$

which, for appropriate choices of ν and σ, can reconcile (16.4.5) with (16.6.1). The model also explains how one might get an increased correlation length with richness: higher peaks have higher ν and correspond to denser systems.

This elucidation of the reason why clusters should have stronger correlations than galaxies is natural because clusters are, by definition, objects with exceptionally high density on some well-defined scale. Kaiser's calculation was, however, subsequently used as the basis for the first models of biased galaxy formation described in Section 14.8. For it to apply to galaxies, however, one has to think of a good reason why galaxies should only form at particularly dense peaks of the matter distribution: some mechanism must be invoked to suppress galaxy formation in 'typical' fluctuations. One should therefore take care to distinguish between the apparent biasing of clusters relative to galaxies and the biasing of galaxies relative to mass; the former is well-motivated physically, the latter, at least with our present understanding of galaxy formation, is not.

In any event, one of the advantages of the cluster distribution is that it can be used to measure correlations on scales where the galaxy–galaxy correlation function vanishes into statistical noise. The cluster–cluster correlation function seems to be positive out to at least $50h^{-1}$ Mpc, while the galaxy–galaxy function is very small, and perhaps negative, for $r \simeq 10h^{-1}$ Mpc.

16.7 Counts in Cells

A simple but useful way of measuring the correlations of galaxies on large scales which does not suffer from the problems of the correlation functions is by looking at the distribution of counts of galaxies in cells, $P_n(V)$. This is defined as the probability of finding n objects in a randomly placed volume V, or the low-order moments of this distribution such as the variance σ^2 and skewness γ which we define below; do not confuse γ with the slope of the two-point correlation function in Equation (16.4.2) or with the spectral parameter in equation (13.2.11). Indeed some of the earliest quantitative analyses of galaxy clustering by Hubble adopted the counts-in-cells approach.

Using only the moments of the cell-count distribution does result in a loss of information compared with the use of the full distribution function, but the advantage is a simple relationship between the moments and the correlation functions, e.g.

$$\sigma^2 \equiv \left\langle \left(\frac{\Delta n}{\bar{n}}\right)^2 \right\rangle = \frac{1}{\bar{n}} + \frac{1}{V^2} \iint \xi_{(2)}(r_{12})\, \mathrm{d}V_1\, \mathrm{d}V_2, \tag{16.7.1}$$

where $\bar{n}$ is the mean number of galaxies in a cell of volume V, i.e. $\bar{n} = n_V V$ (n_V is the mean number-density of galaxies). The derivation of this formula for the variance is quite straightforward. Consider a set of n points (galaxies) distributed in a cell of volume V. Divide the cell into infinitesimal sub-cells dV_k and let each contain n_k galaxies. If the dV_k are small enough, then n_k can only be 0 or 1. Clearly $n = \sum n_k$. The expected number of galaxies in the cell is

$$\langle n \rangle = \bar{n} = \sum \langle n_k \rangle = \int_V n \, dV = n_V V. \tag{16.7.2}$$

The mean squared value of n is

$$\langle n^2 \rangle = \sum \langle n_k^2 \rangle + \sum_{k \neq l} \langle n_k n_l \rangle. \tag{16.7.3}$$

Because n_k is only either 0 or 1, the first term must be the same as $\sum \langle n_k \rangle$; the second term is obviously just $n_V^2 \, dV_1 \, dV_2 (1 + \xi_{12})$, so that

$$\langle n^2 \rangle = n_V V + (n_V V)^2 + n_V^2 \int \xi_{12} \, dV_1 \, dV_2. \tag{16.7.4}$$

The form (16.7.1) then follows when the result is expressed in terms of

$$\left\langle \left(\frac{n - \bar{n}}{\bar{n}} \right)^2 \right\rangle = \left\langle \left(\frac{\Delta n}{\bar{n}} \right)^2 \right\rangle. \tag{16.7.5}$$

The $1/\bar{n}$ term in Equation (16.7.1) is due to Poisson fluctuations: it is a discreteness effect. Apart from this, the second-order moment is simply an integral of the two-point correlation function over the volume V, and is therefore related to the mass variance defined by Equation (13.3.8) for a sharp window function. The same is true for higher-order moments, but the discreteness terms are more complicated and the integrals must be taken over the cumulants. For example, following a similar derivation to that above, the *skewness* γ can be written

$$\gamma \equiv \left\langle \left(\frac{\Delta n}{\bar{n}} \right)^3 \right\rangle = -\frac{2}{\bar{n}^2} + \frac{3\sigma^2}{\bar{n}} + \frac{1}{V^3} \iiint \xi_{(3)} \, dV_1 \, dV_2 \, dV_3. \tag{16.7.6}$$

Equation (16.7.1) provides a good way of measuring the two-point correlation function on large scales. Use of the skewness and higher-order moments descriptors is now also possible. The usual formulation is to write the ratio of the Nth-order moment to the $(N-1)$th power of the variance as S_N. For example, in terms of γ and σ^2, the hierarchical parameter S_3 is just γ/σ^4. In the hierarchical model the S_N should be constant, independent of the cell volume. For the simple hierarchical distribution (16.5.1) we have $S_3 = 3Q$, which seems to be in reasonable agreement with measured skewnesses. There should be some scale dependence of clustering properties if the initial power spectrum is not completely scale free, so one would not expect S_3 to be accurately constant on all scales in, for example, the CDM model. It is, however, a very slowly varying quantity. Within the considerable errors, there seems to be a roughly hierarchical behaviour of the clustering data consistent with most gravitational instability models of structure formation.

This is further confirmation of the comments we made in Section 16.4 about the success of the hierarchical model.

Although it is encouraging that these different approximations do agree with each other to a reasonable degree and also seem to behave in roughly the same way as the data, it is advisable to be cautious here. The skewness is a relatively crude statistical descriptor and many different non-Gaussian distributions have the same skewness, but very different higher-order moments. One could proceed by measuring higher and higher order moments from the data, but this is probably not a very efficient way to proceed. It is perhaps better to focus instead upon the distribution function of cell counts, $P_n(V)$, rather than its moments. The problem is that, except for a few special cases, it is not possible to derive the distribution function analytically even in the limit of large V.

The distribution function of galaxy counts leads naturally on to the *void probability function* (VPF), the probability that a randomly selected volume V is completely empty. Properties of voids are also appealing for intuitive reasons: these are the features that stand out most strikingly in the visual appearance of the galaxy distribution. The *generating function* of the count probabilities, defined by

$$\mathcal{P}(\lambda) \equiv \sum_{N=0}^{\infty} \lambda^N P_N(V), \tag{16.7.7}$$

can be shown to be a sum over the 'averaged' connected correlation functions of all orders,

$$\log \mathcal{P}(\lambda) = \sum_{N=1}^{\infty} \frac{(\lambda - 1)^N}{N!} (\bar{n})^N \bar{\xi}_{(N)}, \tag{16.7.8}$$

(White 1979), where

$$\bar{\xi}_{(N)} \equiv \frac{1}{V^N} \int \cdots \int \xi_{(N)}(r_{ij})\, \mathrm{d}V_1 \cdots \mathrm{d}V_N. \tag{16.7.9}$$

Setting $\lambda = 0$ in Equation (16.7.8), we obtain

$$\log P_0(V) = \sum_{N=1}^{\infty} \frac{(-\bar{n})^N}{N!} \bar{\xi}_{(N)} \tag{16.7.10}$$

as long as this sum converges. The VPF is quite easy to extract from simulations or real data and depends strongly upon correlations of all orders; it is therefore a potentially useful diagnostic of the clustering. Studies of the VPF again seem to support the view that clustering on scales immediately accessible to observations is roughly hierarchical in form.

Although the VPF is unquestionably a useful statistic, it pays no attention to the geometry of the voids, or their topology. Typically one uses a spherical test volume, so a flat or filamentary void will not register in the VPF with a V corresponding to its real volume. Moreover, because the voids which seem most obvious to the eye are not actually completely empty: these do not get counted at all in the VPF statistic. The search for a better statistic for describing void probabilities is under way and is an important task.

16.8 The Power Spectrum

There are many advantages, particularly on large scales, in not measuring the two-point correlation function directly, but through its Fourier transform. The Wiener–Khintchine theorem (13.8.5) shows that, for a statistically homogeneous random field, the two-point covariance function is the Fourier transform of the *power spectrum*. One might expect therefore that one can define a useful power spectrum for galaxy clustering which is the inverse of the two-point correlation function. For power-law primordial spectra $P(k) \propto k^n$, one can show that $\xi(r) \propto -\sin(\pi n/2) r^{-(3+n)}$ $(n > -3)$, which can be used to deduce the power spectrum from a knowledge of ξ in regions where it can be represented as a power law. On the other hand, one would imagine that a better procedure is to estimate $P(k)$ directly from the data without worrying about $\xi(r)$, particularly on large scales. This is indeed the case. There are some subtleties, however, because the discreteness of the galaxy counts induces a 'white-noise' contamination into the power spectrum which must be removed.

For a discrete distribution of N points (galaxies) we can define the Fourier transform as

$$\delta(\boldsymbol{k}) = \frac{1}{N} \sum \exp(\mathrm{i}\boldsymbol{k} \cdot \boldsymbol{x}), \tag{16.8.1}$$

where the sum is taken over all galaxy positions $\boldsymbol{x}$. If the distribution were random, the coefficients $\delta(\boldsymbol{k})$ would be generated by a random walk in the complex plane. It is then straightforward to show that the variance of the modulus of $\delta(\boldsymbol{k})$ is given by

$$\langle |\delta(\boldsymbol{k})|^2 \rangle = \frac{1}{N}. \tag{16.8.2}$$

In principle, one can therefore just subtract the quantity $1/N$ from the quantity $|\delta(\boldsymbol{k})|^2$ determined by (16.8.1). In fact, the power spectrum is estimated over a region of $\boldsymbol{k}$-space which defines an interval in the modulus of $\boldsymbol{k}$, denoted k. One therefore needs to subtract off the 'shot-noise' contribution for each $\boldsymbol{k}$ which enters this estimate, so that

$$P(k) \simeq \sum_{\boldsymbol{k}} |\delta(\boldsymbol{k})|^2 - \frac{n_k}{N}, \tag{16.8.3}$$

where n_k is the number of $\boldsymbol{k}$ modes involved in the sum.

Even this does not work, however, unless we have a cubic sample volume (which is unlikely to be the case). It is necessary, in fact, to think of the observed sample as being a modulation of the real density field by some selection function $f(\boldsymbol{x})$, which can also take account of the fact that some galaxies will be missed at larger distances from the observer in a survey limited by apparent magnitude. To account for this, one therefore has to subtract off from $\delta(\boldsymbol{k})$ the Fourier transform of $f(\boldsymbol{x})$ before doing the subtraction in (16.8.3). One also has to correct for the effect of f at modulating the Fourier coefficients of δ. It turns out that the observed power spectrum is just a convolution of the 'true' power spectrum with the function $|f_k|^2$, the squared modulus of the Fourier transform of $f(\boldsymbol{x})$. This also induces an error in n_k, since the number of $\boldsymbol{k}$ modes depends on the volume after modulation, rather than on the idealised cubic volume mentioned above. Correcting for all these effects requires some care.

To be precise, $P(k)$ is actually a spectral density function, and should have units of volume. To avoid the possible dependence of $P(k)$ upon the sample volume it is more useful to deal for comparison purposes with a dimensionless power spectrum $\Delta^2(k) \simeq k^3P(k)$ in the manner of Equation (14.2.8). The power spectrum of galaxy clustering has been analysed for a number of different samples and the results are reasonably well fitted by the functional form:

$$\Delta^2(k) = \frac{(k/k_0)^{1.6}}{1 + (k/k_c)^{-2.4}}. \tag{16.8.4}$$

The best-fitting value for the parameters are $k_c \simeq 0.015$–$0.025h$ Mpc^{-1} and $k_0 \simeq 0.19h$ Mpc^{-1}, but k_0 depends quite sensitively upon the accuracy of the various selection functions. This form, on large scales, is similar to a low-density CDM spectrum or a CHDM spectrum; see Figure 16.4. The power spectrum of Abell cluster correlations has also been computed; the results are consistent with a rather large value for the correlation length, $r_0 \simeq 21h^{-1}$ Mpc, and indicate that the clustering strength does depend on the cluster richness, as one might expect from the discussion in Section 16.5.

16.9 Polyspectra

Since the power spectrum is the Fourier transform of the two-point correlation function, it would seem likely that similar transforms of the N-point functions for $N > 2$ would also prove to be useful descriptors of galaxy clustering. For example, the Fourier transform of the three-point correlation function is known as the *bispectrum*. The use of higher-order spectra is not yet widespread, but they will probably turn out to be a very effective way of detecting non-Gaussian fluctuation statistics on very large scales and of constraining the gravitational instability picture generally.

To see why, consider the application of the power spectrum to a continuous density contrast field as in Chapters 10–15, i.e. $\delta(\boldsymbol{x})$ defined by

$$\delta(\boldsymbol{x}) = [\rho(\boldsymbol{x}) - \rho_0]/\rho_0, \tag{16.9.1}$$

where ρ_0 is the average density and $\rho(\boldsymbol{x})$ is the local matter density. Because the initial perturbations evolve linearly, it is useful to expand $\delta(\boldsymbol{x})$ as a Fourier superposition of plane waves:

$$\tilde{\delta}(\boldsymbol{k}) = \int \mathrm{d}^x \delta(\boldsymbol{x}) \exp(-\mathrm{i}\boldsymbol{k} \cdot \boldsymbol{x}). \tag{16.9.2}$$

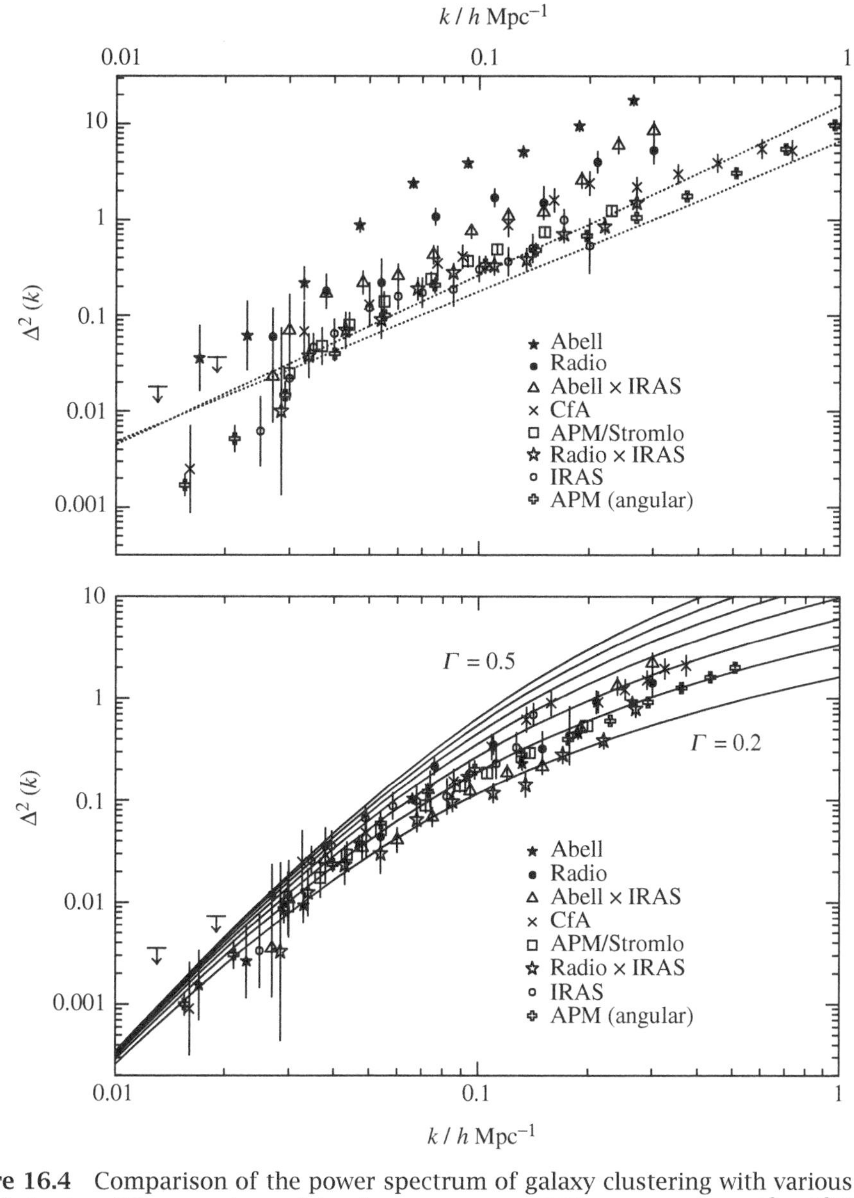

Figure 16.4 Comparison of the power spectrum of galaxy clustering with various CDM models having different values of the shape parameter Γ. The y-axes show $\Delta^2 = k^3 P(k)$ as a function of k; the data points are from a compilation of redshift surveys before (upper panel) and after (lower panel) allowances are made for bias and velocity effects. Picture courtesy of John Peacock.

The Fourier transform $\tilde{\delta}(\boldsymbol{k})$ is complex and therefore possesses both amplitude $|\tilde{\delta}(\boldsymbol{k})|$ and phase $\phi_{\boldsymbol{k}}$, where

$$\tilde{\delta}(\boldsymbol{k}) = |\tilde{\delta}(\boldsymbol{k})| \exp(i\phi_{\boldsymbol{k}}). \tag{16.9.3}$$

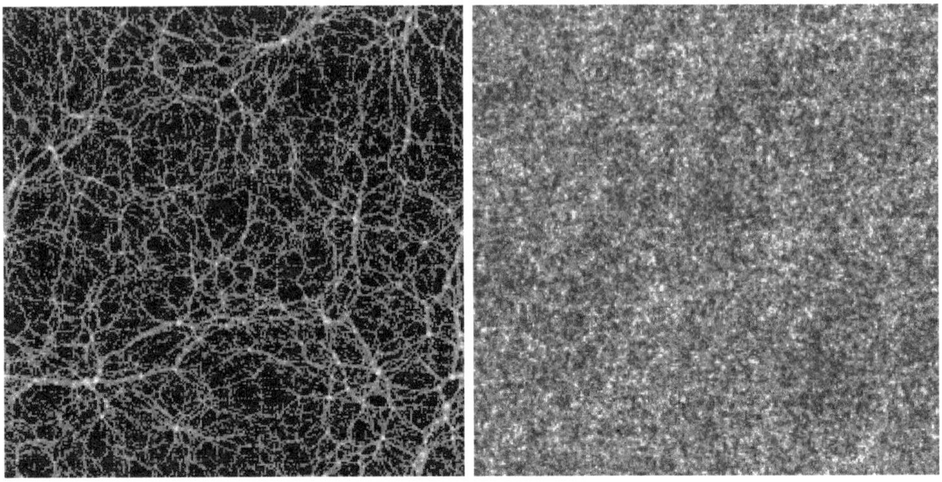

Figure 16.5 Numerical simulation of galaxy clustering (left) together with a version generated randomly reshuffling the phases between Fourier modes of the original picture (right). The reshuffling operation preserves the reality of the original image.

Gaussian random fields possess Fourier modes whose real and imaginary parts are independently distributed. In other words, they have phase angles ϕ_k that are independently distributed and uniformly random on the interval $[0, 2\pi]$. When fluctuations are small, i.e. during the linear regime, the Fourier modes evolve independently and their phases remain random. In the later stages of evolution, however, wave modes begin to couple together. In this regime the phases become non-random and the density field becomes highly non-Gaussian (Coles and Chiang 2000). Phase coupling is therefore a key consequence of nonlinear gravitational processes if the initial conditions are Gaussian. Such phenomena consequently display a potentially powerful signature to exploit in statistical tests of this class of models.

A graphic demonstration of the importance of phases in patterns generally is given in Figure 16.5. The power spectrum $P(k)$ is formally defined by an expression of the form

$$\langle\delta(\boldsymbol{k}_1)\delta(\boldsymbol{k}_2) = (2\pi)^3 P(k)\delta^{\mathrm{D}}(\boldsymbol{k}_1 + \boldsymbol{k}_2); \tag{16.9.4}$$

to take account of the fact that the density field is real we have that $\delta_{\boldsymbol{k}} = \delta^*_{-\boldsymbol{k}}$. Since the amplitude of each Fourier mode is unchanged in the phase-reshuffling operation shown in Figure 16.5, the two pictures have exactly the same power spectrum, $P(k)$. In fact, they have more than that: they have exactly the same amplitudes for all $\boldsymbol{k}$. They also have totally different morphology. The shortcomings of $P(k)$ as a descriptor of pattern can be partly ameliorated by defining higher-order quantities such as the bispectrum (Peebles 1980; Matarrese *et al.* 1997; Scoccimarro *et al.* 1999). The bispectrum is simply a three-point correlation function in redshift space. By analogy with (16.9.5) we have

$$\langle\delta(\boldsymbol{k})\delta(\boldsymbol{k}') = (2\pi)^3 B(\boldsymbol{k}_1, \boldsymbol{k}_2, \boldsymbol{k}_3)\delta^{\mathrm{D}}(\boldsymbol{k}_1 + \boldsymbol{k}_2 + \boldsymbol{k}_3). \tag{16.9.5}$$

The bispectrum is zero unless the three vectors $\boldsymbol{k}_{\mathrm{i}}$ form a triangle. The function $B(\boldsymbol{k}_1, \boldsymbol{k}_2, \boldsymbol{k}_3)$ is particularly useful in redshift space, a fact we shall revisit in more detail in Chapter 18.

This idea can be generalised to arbitrary order correlations in Fourier space – the polyspectra. Alternatively, one can study correlations of quantities like $\tilde{\delta}(\boldsymbol{k})^2$ (Stirling and Peacock 1996). This is a special case of a four-point correlation function in Fourier space.

16.10 Percolation Analysis

Useful though the correlation functions and related quantities undoubtedly are, their interpretation is problematic, except perhaps in the framework of a model such as the hierarchical model. In particular, it is difficult to give a geometrical interpretation to the correlation functions. For this reason, it is useful to develop a different kind of statistical description of galaxy clustering which is more directly related to geometry. We would be interested particularly in a descriptor which revealed whether the distribution has a significant tendency to cluster in sheets, filaments or isolated clumps.

One possible such description is furnished by *percolation analysis*, which we now describe (Shandarin 1983; Dekel and West 1985). Imagine we have a cubic sample of the Universe of side L, containing $N \gg 1$ points (galaxies, clusters, etc.). Let us trace a sphere around each point of diameter $d = b\bar{l}$, where $\bar{l} = L/N^{1/3}$ is the mean interparticle distance. If the spheres around two points overlap with each other, then we connect the two points: they become 'friends'. If one of the spheres connects with another point, then those two points become 'friends' also. Applying the principle 'the friend of my friend is also my friend', all three points now become connected. At a given value of b, therefore, the distribution will consist of some isolated points and some connected 'clusters' (sets of 'friends of friends'). For very small b all points will be isolated (nobody has any friends), while, for large b, all points will be connected (everybody is friends with everybody else). As b increases the number of clusters therefore decreases from N to 1, while the typical number of points per cluster increases from 1 to N. For a particular value, say b_{c} (at least) one cluster forms which can connect two opposite faces of the cube. At this point the system is said to have percolated, and b_{c} is the *percolation parameter*. (Sometimes in the literature the quantity $B_{\mathrm{c}} = 4\pi b_{\mathrm{c}}^3/3$ is called the percolation parameter.) The value of b_{c} depends on the geometry of the spatial distribution of the points, on N and on L. Let us illustrate this with some simple examples.

For a uniform distribution of points on a cubic lattice it is clear that $b_{\mathrm{c}} = 1$. For a uniform distribution of particles in parallel planes of thickness $h \ll L$, separated from each other by a distance λ, percolation will be completed in each plane at a value of the percolation parameter

$$b_{\mathrm{c}} = \left(\frac{h}{\lambda}\right)^{1/3} < 1. \tag{16.10.1}$$

For a regular distribution on bars of square cross-section with side $h \ll L$, separated by a distance λ, percolation again occurs simultaneously along each bar at a value of b_c given by

$$b_c = \left(\frac{h}{\lambda}\right)^{2/3} \ll 1. \qquad (16.10.2)$$

Compared with a uniform distribution within a cube of side L, percolation occurs more easily, i.e. at a smaller value of b_c, for a distribution on parallel planes and even more easily for a distribution on parallel bars.

For a uniform distribution in small cubes of side $h \ll L$, separated by a distance λ, clearly the critical distance $d_c = b_c\bar{l}$ is given by $\lambda - h$, so that

$$b_c = \frac{\lambda - h}{\bar{l}} = \frac{\lambda}{\bar{l}}\left(1 - \frac{h}{\lambda}\right) \simeq \frac{\lambda}{\bar{l}} > 1: \qquad (16.10.3)$$

in this case percolation is more difficult than in the uniform case, or in the case of planes or bars.

It has been shown that, if the points are distributed randomly, the values of b_c from sample to sample are distributed according to a Gaussian distribution with a mean value and dispersion which decrease as N increases; in particular we have $b_{c,N\to\infty} \simeq 0.87$.

A percolation analysis of the Local Supercluster has given an estimate $b_c \simeq 0.67$, less than that expected for a random distribution. This is some empirical confirmation of the existence of some kind of geometrical structure, though it is difficult to say whether it means filaments or sheets. Indeed, according to N-body experiments, it seems that the values of b are not particularly sensitive to different choices of power spectrum, even for extremes such as HDM and CDM. This does not, however, mean that percolation analysis is not useful. There are many other diagnostics of the transition into the percolated regime in addition to b_c. For example, it has been suggested that a useful method might be to look at the increase in the number of members of the second largest cluster as a function of b; the largest cluster essentially determines b_c, but there will be many smaller clusters whose behaviour might be more sensitive to details of the spectrum than b_c. One might also look at the distribution function of the sizes of percolated regions. Despite its simple geometrical interpretation and apparent effectiveness, percolation theory is relatively neglected in cosmological studies, although it is used extensively, for example, in condensed matter physics; see Stauffer and Aharony (1992). An example of the effective use of percolation methods is given in Sahni *et al.* (1997).

Incidentally, a variant of percolation analysis is used in N-body simulations and in the making of catalogues of galaxy groups to identify overdense regions. In this context, particles are connected together by a friends-of-friends algorithm in the same way as was discussed above, but for these studies a value of b in the range 0.2–0.4 is usually used to define clusters and b is called the *linking parameter* in such applications.

We should also mention that many other statistics have been suggested for detecting and quantifying sheets and filaments in the galaxy distribution using techniques from many diverse branches of mathematics, including graph theory and combinatorics; see, for example, Sahni *et al.* (1998). Although these have yet to yield dramatically interesting results, their likely sensitivity to high-order correlations makes it probable that they will come into their own when the next generation of very large-scale redshift surveys are available for analysis.

16.11 Topology

Interesting though the geometry of the galaxy distribution may be, such studies do not tell us about the *topology* of clustering or, in other words, its connectivity. One is typically interested in the question of how the individual filaments, sheets and voids join up and intersect to form the global pattern. Is the pattern cellular, having isolated voids surrounded by high-density sheets, or is it more like a sponge in which under- and over-dense regions interlock?

Looking at 'slice' surveys gives the strong visual impression that we are dealing with bubbles; pencil beams (deep galaxy redshift surveys with a narrow field of view, in which the volume sampled therefore resembles a very narrow cone or 'pencil') reinforce this impression by suggesting that a line of sight intersects at more-or-less regular intervals with walls of a cellular pattern. One must be careful of such impressions, however, because of elementary topology. Any closed curve in two dimensions must have an inside and an outside, so that a slice through a sponge-like distribution will appear to exhibit isolated voids just like a slice through a cellular pattern. It is important therefore that we quantify this kind of property using well-defined topological descriptors.

In an influential series of papers, Gott and collaborators have developed a method for doing just this (Gott *et al.* 1986; Hamilton *et al.* 1986; Gott *et al.* 1989, 1990; Melott 1990). Briefly, the method makes use of a topological invariant known as the *genus*, related to the *Euler–Poincaré characteristic*, of the isodensity surfaces of the distribution. To extract this from a sample, one must first smooth the galaxy distribution with a filter (usually a Gaussian is used; see Section 14.3) to remove the discrete nature of the distribution and produce a continuous density field. By defining a threshold level on the continuous field, one can construct excursion sets (sets where the field exceeds the threshold level) for various density levels. An excursion set will typically consist of a number of regions, some of which will be simply connected, e.g. a deformed sphere, and others which will be multiply connected, e.g. a deformed torus is doubly connected. If the density threshold is labelled by ν, the number of standard deviations of the density away from the mean, then one can construct a graph of the genus of the excursion sets at ν as a function of ν: we call this function $G(\nu)$. The genus can be formally expressed as an integral over the intrinsic curvature K of the excursion set surfaces, S_ν, by means of the Gauss–Bonnet theorem.

The general form of this theorem applies to any two-dimensional manifold $\mathcal{M}$ with any (one-dimensional) boundary $\partial\mathcal{M}$ which is piecewise smooth. This latter condition implies that there are a finite number n vertices in the boundary at which points it is not differentiable. The Gauss–Bonnet theorem states that

$$\sum_{i=1}^{n}(\pi - \alpha_i) + \int_{\partial\mathcal{M}} k_{\rm g}\,{\rm d}s + \int_{\mathcal{M}} k\,{\rm d}A = 2\pi\chi_{\rm E}(\mathcal{M}), \tag{16.11.1}$$

where the α_i are the angle deficits at the vertices (the n interior angles at points where the boundary is not differentiable), $k_{\rm g}$ is the geodesic curvature of the boundary in between the vertices and k is the Gaussian curvature of the manifold itself. Clearly ${\rm d}s$ is an element of length taken along the boundary and ${\rm d}A$ is an area element within the manifold $\mathcal{M}$. The right-hand side of Equation (16.11.1) is the Euler–Poincaré characteristic, $\chi_{\rm E}$, of the manifold.

This probably seems very abstract but the definition (16.11.1) allows us to construct useful quantities for both two- and three-dimensional examples. If we have an excursion set as described above in three dimensions, then its surface can be taken to define such a manifold. The boundary is just where the excursion set intersects the limits of the survey and it will be taken to be smooth. Ignoring this, we see that the Euler–Poincaré characteristic is just the integral of the Gaussian curvature over all the compact bits of the surface of the excursion set. Hence, in this case,

$$2\pi\chi_{\rm E} = \int_{S_\nu} K\,{\rm d}S = 4\pi[1 - G(\nu)]. \tag{16.11.2}$$

Roughly speaking, the quantity G is the genus, which for a single surface is the number of 'handles' the surface possesses; a sphere has no handles and has zero genus, a torus has one and therefore has a genus of one. For technical reasons to do with the effect of boundaries, it has become conventional not to use G but $G_S = G - 1$. In terms of this definition, multiply connected surfaces have $G_S \geqslant 0$ and simply connected surfaces have $G_S < 0$. One usually divides the total genus G_S by the volume of the sample to produce g_S, the genus per unit volume.

One of the great advantages of using the genus measure to study large-scale structure, aside from its robustness to errors in the sample, is that all Gaussian density fields have the same form of $g_S(\nu)$:

$$g_S(\nu) = A(1 - \nu^2)\exp(-\tfrac{1}{2}\nu^2), \tag{16.11.3}$$

where A is a spectrum-dependent normalisation constant. This means that, if one smooths the field enough to remove the effect of nonlinear displacements of galaxy positions, the genus curve should look Gaussian for any model evolved from Gaussian initial conditions, regardless of the form of the initial power spectrum, which only enters through the normalisation factor A. This makes it a potentially powerful test of non-Gaussian initial fluctuations, or of models which invoke non-gravitational physics to form large-scale structure. The observations support the interpretation that the initial conditions were Gaussian, although the distribution looks non-Gaussian on smaller scales. The nomenclature for the non-Gaussian distortion one sees is a 'meatball shift': nonlinear clustering tends to produce an excess of high-density simply connected regions, compared with the

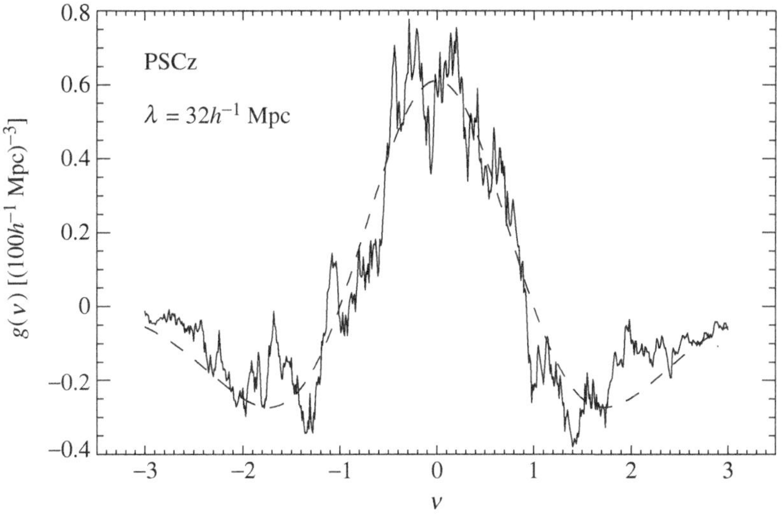

Figure 16.6 Genus curve for galaxies in the IRAS PSCz survey. The noisy curve is the smoothed galaxy distribution, while the solid line is the best-fitting curve for a Gaussian field, Equation (16.11.3). Picture courtesy of the PSCz team.

Gaussian curve. The opposite tendency, usually called 'Swiss cheese', is to have an excess of low-density simply connected regions in a high-density background, which is what one might expect to see if cosmic explosions or bubbles formed the large-scale structure. What one would expect to see in the standard picture of gravitational instability from Gaussian initial conditions is a 'meatball' topology when the smoothing scale is small, changing to a sponge as the smoothing scale is increased. This is indeed what seems to be seen in the observations so there is no evidence of bubbles; an example is shown in Figure 16.6.

The smoothing required also poses a problem, however, because present redshift surveys sample space only rather sparsely and one needs to smooth rather heavily to construct a continuous field. A smoothing on scales much larger than the scale at which correlations are significant will tend to produce a Gaussian distribution by virtue of the central limit theorem. The power of this method is therefore limited by the smoothing required, which, in turn, depends on the space-density of galaxies. An example is shown in Figure 16.6, which shows the genus curve for the PSCz survey of IRAS galaxies.

Topological information can also be obtained from two-dimensional data sets, whether these are simply projected galaxy positions on the sky (such as the Lick map, or the APM survey) or 'slices' (such as the various Center for Astrophysics (CfA) compilations). Here the excursion sets one deals with are just regions of the plane where the (surface) density exceeds some threshold. In this case we imagine the manifold referred to in the statement of the Gauss–Bonnet theorem to be not the surface of the excursion set but the surface upon which the set is defined (i.e. the sky). For reasonably small angles this can be taken to be a flat plane so that the Gaussian curvature of $\mathcal{M}$ is everywhere zero. (The generalisation to large angles is trivial; it just adds a constant-curvature term.) The Euler characteristic is then simply an integral of the line curvature around the boundaries of the excursion set:

$$2\pi\chi_{\mathrm{E}} = \int k_{\mathrm{g}}\, \mathrm{d}s. \tag{16.11.4}$$

In this case the Euler–Poincaré characteristic is simply the number of isolated regions in the excursion set minus the number of holes in such regions.

This is analogous to the genus, but has the interesting property that it is an odd function of ν for a two-dimensional Gaussian random field, unlike $G(\nu)$ which is even. In fact the mean value of χ per unit area on the sky takes the form

$$\chi(\nu) = B\nu \exp(-\tfrac{1}{2}\nu^2), \tag{16.11.5}$$

where B is a constant which depends only on the (two-dimensional) power spectrum of the random field. Notice that $\chi < 0$ for $\nu < 0$ and $\chi > 0$ for $\nu > 0$. A curve shifted to the left with respect to this would be a meatball topology, and to the right would be a Swiss cheese.

There are some subtleties with this. Firstly, as discussed above, two-dimensional topology does not really distinguish between 'sponge' and 'Swiss cheese' alternatives. Indeed, there is no two-dimensional equivalent of a sponge topology: a slice through a sponge is topologically equivalent to a slice through Swiss cheese. Nevertheless, it is possible to assess whether, for example, the mean density level ($\nu = 0$) is dominated by underdense or overdense regions so that one can distinguish Swiss cheese and meatball alternatives to some extent. The most obviously useful application of this method is to look at projected catalogues, the main problem being that, if the catalogue is very deep, each line of sight contains a superposition of many three-dimensional structures. This projection acts to suppress departures from Gaussian statistics by virtue of the central limit theorem. Nevertheless, useful information is obtainable from projected data simply because of the size of the data sets available; as is the case with three-dimensional studies, the analysis reveals a clear meatball shift, which is what one expects in the gravitational instability picture. The methods used for the study of two-dimensional galaxy clustering can also be used to analyse the pattern of fluctuations on the sky seen in the cosmic microwave background.

More recently, this approach has been generalised to include not just the Euler–Poincaré distribution but all possible topological invariants, i.e. all characteristic quantities that satisfy the requirements that they be additive, continuous, translation invariant and rotation invariant. For an excursion set defined in d dimensions there are $d + 1$ such quantities that can be regarded as independent. Any characteristic satisfying these invariance properties can be expressed in terms of linear combinations of these four independent quantities. These are usually called *Minkowski functionals*. Their use in the analysis of galaxy-clustering studies was advocated by Mecke *et al.* (1994) and has become widespread since then.

In three dimensions there are four Minkowski functionals. One of these is the integrated Gaussian curvature (equivalent to the genus we discussed above).

Another is the mean curvature, H, defined by

$$H = \frac{1}{2}\int \left(\frac{1}{R_1} + \frac{1}{R_2}\right) \mathrm{d}A. \tag{16.11.6}$$

In this expression R_1 and R_2 are the principal radii of curvature at any point in the surface; the Gaussian curvature is $1/(R_1 R_2)$ in terms of these variables. The other two Minkowski functionals are more straightforward. They are the surface area of the set and its volume. These four quantities give a 'complete' topological description of the excursion sets.

16.12 Comments

In this chapter we have attempted to give a reasonably complete, though by no means exhaustive, overview of the statistical analysis of galaxy clustering. In addition to those we have described here, many other statistical descriptors have been employed in this field, particularly with respect to the problem of detecting filaments, sheets and voids in the large-scale distribution. More are sure to be developed in the future, and the next generation of galaxy redshift surveys will surely furnish more accurate estimators of those statistics we have had space to describe here. By way of a summary, it is useful to delineate some common strands revealed by the various statistical approaches described in this chapter.

To begin with, a variety of methods give relatively direct constraints on the power spectrum of the matter fluctuations; the two-point correlation function, the galaxy power spectrum and the variance of the counts-in-cells distribution are all related in a relatively simple way to this. Two problems arise here, however. One is the ubiquitous problem of bias we discussed in Chapter 15. In the simplest conceivable case of a linear bias, the various statistics extracted from galaxy clustering, $\xi(r)$, $\Delta^2(k)$ and σ^2, are all a factor b^2 higher than the corresponding quantities for the mass fluctuations. In a more complicated biasing model, the relationship between galaxy and mass statistics may be considerably more obscure than this. The second problem is that we have dealt almost exclusively with the distribution of galaxies in redshift space. The existence of peculiar motions makes the relationship between real space and redshift space rather complicated. This problem is, however, potentially useful in some cases, because the distortion of various statistics in redshift space relative to real space can, at least in principle, give information indirectly about the peculiar velocities and hence about the distribution of mass fluctuations through the continuity equation; we return to this matter in Chapter 18. Within the uncertainties introduced by these factors, a consensus has emerged from these studies that the power spectrum of galaxy clustering is consistent with the shape described by Equation (16.8.4), i.e. with a different shape to the standard CDM scenario, but approximately fitted by a low-density CDM transfer function.

Measures of the topology and geometry of galaxy clustering are less effective at constraining the power spectrum, but relate to different ingredients of models of structure formation. Percolation analysis, and other pattern descriptors not mentioned here, give qualitative confirmation of the existence of Zel'dovich pancakes and filaments as expected in gravitational instability theories. The behaviour of higher-order moments lends further credence to the this picture. Large-scale topology has failed to show up any significant departures from Gaussian behaviour. It seems reasonable therefore to describe all this evidence as being consistent with the basic scenario of structure formation by gravitational instability which we have sought to describe in this book. We shall see that further support for this general picture is furnished by fluctuations in the CMB temperature (Chapter 17) and studies of galaxy-peculiar motions (Chapter 18).

Bibliographic Notes on Chapter 16

The classic reference work for statistical cosmology is Peebles (1980). A more modern survey of statistical methods for cosmology applications is given by Martínez and Saar (2002). Further useful sources are Saslaw (1985) and Peacock (1992). Fall (1979) is also full of interesting ideas.

Problems

1. Suppose the Universe consists of a spherically symmetric distribution of galaxies with density profile $n = n_0 r^{-\alpha}$. Using an appropriate definition of the two-point correlation function, $\xi(r)$, show that

$$\xi(r) \propto r^{3-2\alpha}.$$

2. Assume the galaxy distribution consists of a collection of spherical clusters containing different numbers of galaxies n. Let the number of clusters per unit volume as a function of n be proportional to $n^{-\beta}$ and assume all clusters containing exactly n galaxies have a radius $r_{\mathrm{n}} \propto n^{\alpha}$. Show that, for $\xi(r) \gg 1$ and r small,

$$\xi(r) \propto r^{-3+(3/\alpha)-\beta/\alpha}.$$

3. Enumerate the twelve distinct snake graphs and the four distinct star graphs for $N = 4$, as shown in Figure 16.3.

4. Show that, for a hierarchical distribution, the skewness of the cell-count fluctuations, γ, is related to the variance, σ^2, via $\gamma = 3Q\sigma^4$.

5. Identify the three Minkowski functionals needed to characterise an excursion set in two dimensions.

17

The Cosmic Microwave Background

17.1 Introduction

The detection of fluctuations in the sky temperature of the cosmic microwave background (CMB) in 1992 by the COBE team led by George Smoot was an important milestone in the development of cosmology (Bennett *et al.* 1992; Smoot *et al.* 1992; Wright *et al.* 1992). Aside from the discovery of the CMB itself, it was probably the most important event in this field since Hubble's discovery of the expansion of the Universe in the 1920s. The importance of the COBE detection lies in the way these fluctuations are supposed to have been generated. As we shall explain in Section 17.4, the variations in temperature are thought to be associated with density perturbations existing at $t_{\rm rec}$. If this is the correct interpretation, then we can actually look back directly at the power spectrum of density fluctuations at early times, before it was modified by nonlinear evolution and without having to worry about the possible bias of galaxy power spectra.

The search for anisotropies in the CMB has been going on for around 35 years. As the experiments got better and better, and the upper limits placed on the possible anisotropy got lower and lower, theorists concentrated upon constructing models which predicted the smallest possible temperature fluctuations. The baryon-only models were discarded primarily because they could not be modified to produce low enough CMB fluctuations. The introduction of dark matter allowed such a reduction and the culmination of this process was the introduction of bias, which reduces the expected temperature fluctuation still further. It was an interesting experience to those who had been working in this field for many years to see this trend change sign abruptly in 1992. The $\Delta T/T$ fluctuations seen by COBE were actually larger than predicted by the standard version of the CDM model. This must have been the first time a theory had been rejected because it did not produce high enough temperature fluctuations!

Searches for CMB anisotropy would be (and have been), on their own, enough subject matter for a whole book. In one chapter we must therefore limit our scope quite considerably. Moreover, COBE marked the start, rather than the finish, of this aspect of cosmology and it would have been pointless to produce a definitive review of all the ongoing experiments and implications of the various upper limits and half-detections for specific theories, when it is possible that the whole picture will change within a year or two. Therefore, we shall mainly concentrate on trying to explain the physics responsible for various forms of temperature anisotropy. We shall not discuss any specific models in detail, except as illustrative examples, and our treatment of the experimental side of this subject will be brief and non-technical. Finally, we shall be extremely conservative when it comes to drawing conclusions. As we shall explain, the situation with respect to CMB anisotropy as a function of angular scale is still very confused and we feel the wisest course is to wait until observations are firmly established before drawing definite conclusions.

17.2 The Angular Power Spectrum

Let us first describe how one provides a statistical characterisation of fluctuations in the temperature of the CMB radiation from point to point on the celestial sphere.

The usual procedure is to expand the distribution of T on the sky as a sum over spherical harmonics

$$\frac{\Delta T(\theta,\phi)}{T} = \sum_{l=0}^{\infty}\sum_{m=-l}^{m=+l} a_{lm}Y_{lm}(\theta,\phi), \tag{17.2.1}$$

where θ and ϕ are the usual spherical angles; $\Delta T/T$ is defined by Equation (4.8.1). The $l = 0$ term is a monopole correction which essentially just alters the mean temperature on a particular observer's sky with respect to the global mean over an ensemble of all possible such skies. We shall ignore this term from now on because it is not measurable. The $l = 1$ term is a dipole term which, as we shall see in Section 17.3, is attributable to our motion through space. Since this anisotropy is presumably generated locally by matter fluctuations, one tends to remove the $l = 1$ mode and treat it separately. The remaining modes, from the quadrupole ($l = 2$) upwards, are usually attributed to intrinsic anisotropy produced by effects either at $t_{\rm rec}$ or between $t_{\rm rec}$ and t_0. For these effects the sum in Equation (17.2.1) is generally taken over $l \geqslant 2$. Higher l modes correspond to fluctuations on smaller angular scales ϑ according to the approximate relation

$$\vartheta \simeq 60^\circ/l. \tag{17.2.2}$$

The expansion of $\Delta T/T$ in spherical harmonics is entirely analogous to the plane-wave Fourier expansion of the density perturbations δ; the Y_{lm} are a complete orthonormal set of functions on the surface of a sphere, just as the plane-wave modes are a complete orthonormal set in a flat three-dimensional space. The a_{lm} are generally complex, and satisfy the conditions

$$\langle a^*_{l'm'} a_{lm} \rangle = C_l \delta_{ll'} \delta_{mm'}, \tag{17.2.3}$$

where δ_{ij} is the Kronecker symbol and the average is taken over an ensemble of realisations. The quantity C_l is the *angular power spectrum*,

$$C_l \equiv \langle |a_{lm}|^2 \rangle, \tag{17.2.4}$$

which is analogous to the power spectrum $P(k)$ defined by Equation (14.2.5). It is also useful to define an *autocovariance function* for the temperature fluctuations,

$$C(\vartheta) = \left\langle \frac{\Delta T}{T}(\hat{\boldsymbol{n}}_1) \frac{\Delta T}{T}(\hat{\boldsymbol{n}}_2) \right\rangle, \tag{17.2.5}$$

where

$$\cos\vartheta = \hat{\boldsymbol{n}}_1 \cdot \hat{\boldsymbol{n}}_2 \tag{17.2.6}$$

and the $\hat{\boldsymbol{n}}_i$ are unit vectors pointing to arbitrary directions on the sky. The expectation values in (17.2.3) and (17.2.5) are taken over an ensemble of all possible skies. One can try to estimate C_l or $C(\vartheta)$ from an individual sky using an *ergodic hypothesis*: an average over the probability ensemble is the same as an average over all spatial positions within a given realisation. This only works on small angular scales when it is possible to average over many different pairs of directions with the same ϑ, or many different modes with the same l. On larger scales, however, it is extremely difficult to estimate the true $C(\vartheta)$ because there are so few independent directions at large ϑ or, equivalently, so few independent l modes at small l. Large-angle statistics are therefore dominated by the effect of *cosmic variance*: we inhabit one realisation and there is no reason why this should possess exactly the ensemble average values of the relevant statistics.

As was the case with the spatial power spectrum and covariance functions, there is a simple relationship between the angular power spectrum and covariance function:

$$C(\vartheta) = \frac{1}{4\pi} \sum_{l=2}^{\infty} (2l+1) C_l P_l(\cos\vartheta), \tag{17.2.7}$$

where $P_l(x)$ is a Legendre polynomial. We have written the sum explicitly to omit the monopole and dipole contributions from (17.2.1).

It is quite straightforward to calculate the cosmic variance corresponding to an estimate obtained from observations of a single sky, $\hat{C}(\vartheta)$, of the 'true' autocovariance function, $C(\vartheta)$:

$$\hat{C}(\vartheta) = \frac{1}{4\pi} \sum_{l=2}^{\infty} \sum_{m=-l}^{l} |\hat{a}_{lm}|^2 P_l(\cos\vartheta), \tag{17.2.8}$$

where the $\hat{a}_{lm}$ are obtained from a single realisation on the sky. The statistical procedure for estimating these quantities is by no means trivial, but we shall not describe the various possible approaches here: we refer the reader to the bibliography for more details. In fact the variance of the estimated $\hat{a}_{lm}$ across an ensemble of skies will be $|a_{lm}|^2$ so that the $\hat{C}(\theta)$ will have variance

$$\langle|\hat{C}(\vartheta) - C(\vartheta)|^2\rangle = 2\left(\frac{1}{4\pi}\right)^2 \sum_{l=2}^{\infty}(2l+1)C_l^2 P_l^2(\cos\vartheta). \tag{17.2.9}$$

We have again explicitly omitted the monopole and dipole terms from the sums in (17.2.8) and (17.2.9).

In Sections 17.4–17.6 we shall discuss the various physical processes that produce anisotropy with a given form of C_l (we mentioned these briefly in Section 4.8); the dipole is discussed in Section 17.3. Generally the form of C_l must be computed numerically, at least on small and intermediate scales, by solving the transport equations for the matter–radiation fluid through decoupling in the manner discussed in Chapter 13. We shall make some remarks on how this is done later in this chapter. As we shall see, the comparison of a theoretical C_l against an observed $\hat{C}_l$ or $\hat{C}(\vartheta)$ in principle provides a powerful test of theories of galaxy formation. Before discussing the physics, however, it is worth making a few remarks about observations of the CMB anisotropy.

The fluctuations one is looking for generally have an amplitude of order 10^{-5}. One is therefore looking for a signal of amplitude around 30 μK in a background temperature of around 3 K. One's observational apparatus, even with the aid of sophisticated cooling equipment, will generally have a temperature much higher than 3 K and one must therefore look for a tiny variation in temperature on the sky against a much higher thermal noise background in the instrument. From the ground, one also has the problem that the sky is a source of thermal emission at microwave frequencies. Noise of these two kinds is usually dealt with by integrating for a very long time (thermal noise decreases as $\sqrt{t}$, where t is the integration time) and using some kind of beam-switching design in which one measures not ΔT at individual places but temperature differences at a fixed angular separation (double beam switching) or alternate differences between a central point and two adjacent points (triple beam switching). Recovering the ΔT at any individual point (i.e. to produce a map of the sky) from these types of observations is therefore not trivial. Moreover, any radio telescope capable of observing the microwave sky will have a finite beamwidth and will therefore not observe the temperature point by point, but would instead produce a picture of the sky convolved with some smoothing function, perhaps a Gaussian:

$$F(\vartheta) = \frac{1}{2\pi\vartheta_{\rm f}^2}\exp\left(-\frac{\vartheta^2}{2\vartheta_{\rm f}^2}\right). \tag{17.2.10}$$

It is generally more convenient to work in terms of l than in terms of ϑ so we shall express the response of the instrument as F_l; the relationship between F_l

and $F(\vartheta)$ is the same as between C_l and $C(\vartheta)$ given by Equation (17.2.7). In the case of (17.2.10), for example, we get

$$F_l = \exp[-(l + \tfrac{1}{2})^2 \tfrac{1}{2} \vartheta_f^2]. \tag{17.2.11}$$

The observed (smoothed) temperature autocovariance function can then be written

$$C(\vartheta; \vartheta_f) = \frac{1}{4\pi} \sum_{l=2}^{\infty} (2l + 1) F_l C_l P_l(\cos\vartheta). \tag{17.2.12}$$

One must also allow for the effect of beam switching upon the measured temperature fluctuations. Here we shall just illustrate the effect on the mean square temperature fluctuation. For a single beam experiment this is just

$$\left\langle \left(\frac{\Delta T}{T}\right)^2 \right\rangle = \frac{1}{4\pi} \sum (2l + 1) C_l F_l^2 = C(0; \vartheta_f), \tag{17.2.13}$$

while for a double-beam experiment, where each beam has a width ϑ_f and the *beam throw*, i.e. the angular separation of the two beams, is α, we have

$$\left\langle \left(\frac{\Delta T}{T}\right)^2 \right\rangle = \left\langle \frac{(T_1 - T_2)^2}{T^2} \right\rangle = 2[C(0; \vartheta_f) - C(\alpha; \vartheta_f)]. \tag{17.2.14}$$

The case of a triple-beam experiment is rather more complicated; here

$$\left\langle \left(\frac{\Delta T}{T}\right)^2 \right\rangle = \left\langle \frac{[T_1 - (T_2 + T_3)/2]^2}{T^2} \right\rangle = \tfrac{3}{2} C(0; \vartheta_f) - 2C(\alpha; \vartheta_f) + \tfrac{1}{2} C(2\alpha; \vartheta_f), \tag{17.2.15}$$

where T_1 is the central beam. One can extend the relations (17.2.13)–(17.2.15) to calculate the full-sky autocovariance function measured by the experiment, and hence the effective F_l taking into account smoothing and switching.

The function F_l provides the best way of describing the response of any particular experiment. Of course, different experiments are designed to respond to different angular scales or different ranges of l. For example, the COBE DMR experiment we shall describe in Section 17.4 (the first experiment to detect significant fluctuations other than the dipole) has a beam-switching configuration with a beam width of a few degrees and a beam throw of around 60°; this experiment is sensitive to relatively small l. Single-dish ground-based experiments operate at the other end of the spectrum and can be sensitive to l modes of order several thousand.

17.3 The CMB Dipole

It has been known since the 1970s that the cosmic microwave background is not exactly isotropic, but has a *dipole anisotropy* on the sky, i.e. a variation with angle θ proportional to $\cos\theta$. This is usually interpreted as being due to the motion of our Galaxy with respect to a cosmologically comoving frame in which the CMB is isotropic. The angle θ is the angle between the observation and the direction of motion of the observer. The effect is not a simple Doppler effect. The actual level of anisotropy is of order $\beta = v/c \simeq 10^{-3}$, so for the derivation of the result we shall ignore relativistic corrections. The point is that the Doppler effect will increase the energy of photons seen in the direction of motion relative to that of a static observer in an isotropic background. However, the interval of frequencies $\mathrm{d}\nu$ is also increased by the same factor of $(1 + \beta\cos\theta)$. Since the temperature is defined in terms of energy per unit frequency, the net Doppler effect on the temperature is zero. There are, however, two other effects. The first is that the moving observer actually sweeps up more photons. In a direction θ the observer collects $(c\,\mathrm{d}t + v\cos\theta\,\mathrm{d}t)/c\,\mathrm{d}t$ more photons than an observer at rest, which gives rise to an increase in the temperature by a factor of $(1 + \beta\cos\theta)$. The second effect is aberration: the solid angle for a moving observer gets smaller by a factor $(1 + \beta\cos\theta)^{-2}$, so the flux goes up by the reciprocal of this factor. Hence the spectral intensity seen by a moving observer is

$$I'(\nu') = (1 + \beta\cos\theta)^3 I(\nu). \tag{17.3.1}$$

Inserting all the factors in (9.5.1) gives the Planck spectrum with $T(\theta) = T_0(1 + \beta\cos\theta)$. Including all the relativistic effects, to leading order in β, gives

$$T(\theta) = T_0(1 - \beta^2)^{1/2}(1 + \beta\cos\theta); \tag{17.3.2}$$

cf. Equations (4.8.2) and (11.7.3). The reason why this is accepted to be due to our motion is that the quadrupole moment (variation on 90° scale; $l = 2$) is much less: if it were generated by intrinsic anisotropy, one should expect these two scales to contribute roughly the same order of magnitude to $\Delta T/T$. By making a map of $T(\theta, \phi)$ on the sky, one can determine the velocity vector that explains the dipole. The measured velocity is $390 \pm 30\ \mathrm{km\ s^{-1}}$. After subtracting the Earth's motion around the Sun, the Sun's motion around the Galactic centre and the velocity of our Galaxy with respect to the centroid of the Local Group, this dipole anisotropy tells us the speed and direction of the Local Group through the cosmic reference frame. The result is a velocity of about $600\ \mathrm{km\ s^{-1}}$ in the direction of Hydra-Centaurus ($l = 268°$, $b = 27°$) (Rowan-Robinson *et al.* 1990).

In the gravitational instability picture this velocity can be explained as being due to the net gravitational pull on the Local Group generated by the inhomogeneous distribution of matter around it. In fact the net gravitational acceleration is just

$$\boldsymbol{g} = G\int \frac{\rho(\boldsymbol{r})\boldsymbol{r}}{r^3}\,\mathrm{d}V, \tag{17.3.3}$$

where the integral should formally be taken to infinity. As we shall see in Section 18.1, the linear theory of gravitational instability predicts that this gravitational acceleration is just proportional to, and in the same direction as, the net velocity. Moreover, the constant of proportionality depends on $f \simeq \Omega^{0.6}$. If one can measure ρ from a sufficiently large sample of galaxies, then one can in principle determine Ω. Of course, the ubiquitous bias factor intrudes again, so that one can only determine f/b, and that only as long as b is constant.

The technique is simple. Suppose we have a sample of galaxies with some well-defined selection criterion so that the selection function, the probability that a galaxy at distance r from the observer is included in the catalogue, proportional to the function ψ in Section 16.3, has some known form $\phi(r)$. Then the acceleration vector $\boldsymbol{g}$ at the origin of the coordinates can be approximated by

$$\boldsymbol{g} = \tfrac{4}{3}\pi G \boldsymbol{D} = GM_* \sum_i \frac{1}{\phi(r_i)} \frac{\boldsymbol{r}_i}{r_i^3}, \tag{17.3.4}$$

where the $\boldsymbol{r}_i$ are the galaxy positions, M_* is a normalisation factor with the dimension of mass to take into account the masses of the galaxies at $\boldsymbol{r}_i$, and the factor $1/\phi(r_i)$ allows for the galaxies not included in the survey. The sum in Equation (17.3.4) is taken over all the galaxies in the sample. The dipole vector $\boldsymbol{D}$ can be computed from the catalogue and, as long as it is aligned with the observed CMB dipole anisotropy, one can estimate $\Omega_0^{0.6}$. It must be emphasised that this method measures only the inhomogeneous component of the gravitational field: it will not detect a mass component that is uniform over the scale probed by the sample. This technique has been very popular over the last few years, mainly because the various IRAS galaxy catalogues are very suitable for this type of analysis. There are, however, a number of difficulties which need to be resolved before the method can be said to yield an accurate determination of Ω.

First, and probably most importantly, is the problem of convergence. Suppose one has a catalogue that samples a small sphere around the Local Group, but that this sphere is itself moving in roughly the same direction. For this to happen, the Universe must be significantly inhomogeneous on scales larger than the catalogue can probe. In this circumstance, the actual velocity explained by the dipole of the catalogue is not the whole CMB dipole velocity but only a part of it. It follows then that one would overestimate the $\Omega^{0.6}$ factor by attributing all of the observed velocity to the observed local dipole $\boldsymbol{D}$ when, in reality, this dipole is only responsible for part of this velocity. One must be sure, therefore, that the sample is deep enough to sample all contributions to the Local Group motion if one is to determine Ω with any accuracy. Analyses of the dipole properties of the IRAS catalogues seem to indicate a rather high value of f/b, consistent with $\Omega = 1$. On the other hand, catalogues of rich clusters, which have a selection function $\phi(r)$ that falls less steeply on large scales than that of IRAS galaxies, seem to indicate $\Omega \simeq 0.3$ (Plionis *et al.* 1993).

Another problem is that, because of the weighting in Equation (17.3.4), one must ensure that the selection function is known very accurately, especially at large r. This essentially means knowing the luminosity function extremely well, particularly for the brightest objects (the ones that will be seen at great distances). There is also the problem that galaxy properties may be evolving with time so the luminosity function for distant galaxies may be different from that of nearby ones. There is also the problem of bias. We have assumed a linear bias throughout the above discussion. The ramifications of nonlinear and/or non-local biases have yet to be worked out in any detail.

Finally, we should mention the effect of redshift-space distortions, cf. Section 18.5. On the scales needed to probe large-scale structure, it is not practicable to obtain distances for all the objects, so one uses redshifts to estimate distances. One might expect this to be a good approximation at large r, but working in redshift space rather than real space introduces alarming distortions into the analysis. One can illustrate some of the problems with the following toy example. Suppose an observer sits in a rocket and flies through a uniform distribution of galaxies. If he looks at the distribution in redshift space, even if the galaxies have no peculiar motions, he will actually see a dipole anisotropy caused by his motion. He may, if he is unwise, thus determine Ω from his own velocity and this observed dipole: the answer would, of course, be entirely spurious and would have nothing whatsoever to do with the mean density of the Universe.

The combination of redshift-space effects, bias and lack of convergence is difficult to unravel. We therefore suggest that determinations of Ω by this method be treated with caution. For the latest developments in dipole analysis, see Rowan-Robinson *et al.* (2000).

17.4 Large Angular Scales

17.4.1 The Sachs–Wolfe effect

Having dealt with the dipole, we should now look at sources of intrinsic CMB temperature anisotropy. On large scales the dominant contribution to $\Delta T/T$ is expected to be the *Sachs-Wolfe effect* (Sachs and Wolfe 1967). This is a relativistic effect due to the fact that photons travelling to an observer from the last scattering surface encounter metric perturbations which cause them to change frequency. One can understand this effect in a Newtonian context by noting that metric perturbations correspond to perturbations in the gravitational potential, $\delta\varphi$, in Newtonian theory and these, in turn, are generated by density fluctuations, $\delta\rho$. Photons climbing out of such potential wells suffer a gravitational redshift but also a time dilation effect so that one effectively sees them at a different time, and thus at a different value of a, to unperturbed photons. The first effect gives

$$\frac{\Delta T}{T} = \frac{\delta\varphi}{c^2}, \tag{17.4.1}$$

while the second contributes

$$\frac{\Delta T}{T} = -\frac{\delta a}{a} = -\frac{2}{3}\frac{\delta t}{t} = -\frac{2}{3}\frac{\delta\varphi}{c^2}; \tag{17.4.2}$$

the net effect is therefore

$$\frac{\Delta T}{T} = \frac{1}{3}\frac{\delta\varphi}{c^2} \simeq \frac{1}{3}\frac{\delta\rho}{\rho}\left(\frac{\lambda}{ct}\right)^2, \tag{17.4.3}$$

where λ is the scale of the perturbation. This argument is not rigorous, as the split into potential and time-delay components is not gauge invariant but does explain why (17.4.1) is not the whole effect.

So far we have considered only adiabatic fluctuations. Since the Sachs-Wolfe effect is generated by fluctuations in the metric, then one might expect that isocurvature fluctuations (perturbations in the entropy which leave the energy density unchanged and therefore, one might expect, produce negligible fluctuations in the metric) should produce a very small Sachs-Wolfe anisotropy. This is not the case, for two reasons. Firstly, initially isocurvature fluctuations do generate significant fluctuations in the matter component and hence in the gravitational potential, when they enter the horizon; this is due to the influence of pressure gradients. In addition, isocurvature fluctuations generate significant fluctuations in the radiation density after z_{eq}, because the initial entropy perturbation is then transferred into the perturbation of the radiation. The total anisotropy seen is therefore the sum of the Sachs-Wolfe contribution and the intrinsic anisotropy carried by the radiation. The upshot of all this is that the net anisotropy is six times larger for isocurvature fluctuations than for adiabatic ones. This is sufficient on its own to rule out most isocurvature models since the level of anisotropy detected is roughly that expected for adiabatic perturbations.

According to Equation (17.4.3), the temperature anisotropy is produced by gravitational potential fluctuations sitting on the last scattering surface. In fact this is not quite correct, and there are actually two other contributions arising from the Sachs-Wolfe effect. The first of these is a term

$$\frac{\Delta T}{T} \simeq 2 \int \frac{\delta\dot{\varphi}}{c^2}\,\mathrm{d}t, \tag{17.4.4}$$

where the integral is taken along the path of a photon from the last scattering surface to the observer. This effect, usually called the *Rees–Sciama effect*, is due to the change in depth of a potential well as a photon crosses it. If the well does not deepen, a photon does not suffer a net shift in energy from falling in and then climbing up. If the potential changes while the photon moves through it, however, there will be a net change in the frequency. In a flat universe, $\delta\varphi$ is actually constant in linear theory (see Section 18.1 for a proof) so one needs to have nonlinear evolution in order to produce a nonlinear Sachs-Wolfe effect. Since the potential fluctuations are of order $\delta\varphi \simeq \delta(\lambda/ct)^2$ one requires nonlinear evolution of δ on very large scales to obtain a reasonably large contribution. To calculate the effect in detail for a background of perturbations is quite difficult because of the inherent nonlinearity involved. On the other hand, it is possible to calculate the effect using simplified models of structure. For example, a large void region can be modelled as an isolated homogeneous underdensity (the inverse of the spherical top hat discussed in Section 14.1) which can be evolved analytically into the nonlinear regime. It turns out that, for a spherical void of the same diameter as the large void seen in Bootes, one expects to see a cold spot corresponding to $\Delta T/T \simeq 10^{-7}$ on an angular scale around 15°. Large clusters or superclusters can be modelled using top-hat models, the Zel'dovich approximation or perturbative techniques. The Shapley concentration of clusters, for example, is expected to produce a hotspot with $\Delta T/T \simeq 10^{-5}$ on a scale around 20°. In general these effects are smaller than the intrinsic CMB anisotropies we have described, but may be detectable in large, sensitive sky maps: the position on the sky of these features should correspond to known features of the galaxy distribution.

The second additional contribution comes from tensor metric perturbations, i.e. *gravitational waves*. These do not correspond to density fluctuations and have no Newtonian analogue but they do produce redshifting as a result of the perturbations in the metric. As we shall see at the end of this section, gravitational waves capable of generating large-scale anisotropy of this kind are predicted in many inflationary models, so this is potentially an important effect.

For the moment, we shall assume that we are dealing with temperature fluctuations produced by potential fluctuations of the form (17.4.3). What is the form of C_l predicted for fluctuations generated by this effect? This can be calculated quite straightforwardly by writing $\delta\varphi$ as a Fourier expansion and using the fact that the power spectrum of $\delta\varphi$ is proportional to $k^{-4}P(k)$, where $P(k)$ is the power spectrum of the density fluctuations. Expanding the net $\Delta T/T$ in spherical harmonics and averaging over all possible observer positions yields, after some work,

$$C_l = \langle |a_{lm}|^2 \rangle = \frac{1}{2\pi}\left(\frac{H_0}{c}\right)^4 \int_0^\infty P(k) j_l^2(kx)\frac{\mathrm{d}k}{k^2}, \tag{17.4.5}$$

where j_l is a spherical Bessel function and $x = 2c/H_0$. One can also show quite straightforwardly that, for an initial power spectrum of the form $P(k) \propto k$, the quantity $l(l+1)C_l$ is independent of the mode order l for the Sachs–Wolfe perturbations. In any case the shape of C_l for small l is determined purely by the shape of $P(k)$, the shape of the primordial fluctuation spectrum before it is modified by the transfer function. The reason for this is easy to see: the scale of the horizon at z_{rec} is of order

$$\vartheta_{\mathrm{H}}(z_{\mathrm{rec}}) \simeq \left(\frac{\Omega}{z_{\mathrm{rec}}}\right)^{1/2} \mathrm{rad}, \tag{17.4.6}$$

so that $\vartheta_{\mathrm{H}} \simeq 2°$ for $z_{\mathrm{rec}} \simeq 1000$, which is the usual situation. Fluctuations on angular scales larger than this will retain their primordial character since they will not have been modified by any causal processes inside the horizon before z_{rec}. One must therefore be seeing the primordial (unprocessed) spectrum. This is particularly important because observations of C_l at small l can then be used to normalise $P(k)$ in a manner independent of the shape of the power spectrum, and therefore independent of the nature of the dark matter.

One simple way to do this is to use the quadrupole perturbation modes which have $l = 2$. There are five spherical harmonics with $l = 2$, so the quadrupole has five components a_{2m} $(m = -2, -1, 0, 1, 2)$ that can be determined from a map of the sky even if it is noisy. From (17.4.5), we can show that, if $P(k) \propto k$, then

$$\langle |a_{2m}|^2 \rangle = C_2 \simeq \frac{\pi}{3}\left(\frac{H_0 R}{c}\right)^4 \left(\frac{\delta M}{M}\right)_R^2. \tag{17.4.7}$$

This connects the observed temperature pattern on the sky with the mass fluctuations $\delta M/M = \sigma_M$ observed at the present epoch on a scale R.

17.4.2 The COBE DMR experiment

Such is the importance of the COBE discovery that it is worth describing the experiment in a little detail. The COBE satellite actually carried several experiments on it when it was launched in 1989. One of these (FIRAS) measured the spectrum displayed in Figure 9.1. The anisotropy experiment, called the DMR, yielded a positive detection of anisotropy after one year of observations. The advantage of going into space was to cut down on atmospheric thermal emission and also to allow coverage of as much of the sky as possible (ground-based observations are severely limited in this respect). The orbit and inclination of the satellite is controlled so as to avoid contamination by reflected radiation from the Earth and Moon. Needless to say, the instrument never points at the Sun. The DMR detector consists of two horns at an angle of 60°; a radiometer measures the difference in temperature between these two horns. The radiometer has two channels (called A and B) at each of three frequencies: 31.5, 53 and 90 GHz, respectively. These frequencies were chosen carefully: a true CMB signal should be thermal and therefore have the same temperature at each frequency; various sources of galactic emission, such as dust and synchrotron radiation, have an effective antenna temperature which is frequency dependent. Combining the three frequencies therefore allows one to subtract a reasonable model of the contribution to the observed signal which is due to galactic sources. The purpose of the two channels is to allow a subtraction of the thermal noise in the DMR receiver. Assuming the sky signal and DMR instrument noise are statistically independent, the net temperature variance observed is

$$\sigma_{\rm obs}^2 = \sigma_{\rm sky}^2 + \sigma_{\rm DMR}^2. \tag{17.4.8}$$

Adding together the input from the two channels and dividing by two gives an estimate of $\sigma_{\rm obs}^2$; subtracting them and dividing by two yields an estimate of $\sigma_{\rm DMR}^2$, assuming that the two channels are independent. Taking these two together, one can therefore obtain an estimate of the RMS sky fluctuation. The first COBE announcement in 1992 gave $\sigma_{\rm sky} = 30 \pm 5\ \mu$K, after the data had been smoothed on a scale of 10°.

In principle the set of 60° temperature differences from COBE can be solved as a large set of simultaneous equations to produce a map of the sky signal. The COBE team actually produced such a map using the first year of data from the DMR experiment. It is important to stress, however, that, because the sky variance is of the same order as the DMR variance, it is not correct to claim that any features seen in the map necessarily correspond to structures on the sky. Only when the signal-to-noise ratio is much larger than unity can one pick out true sky features with any confidence. The first-year results should therefore be treated only as a statistical detection.

The value of $\langle a_{lm}^2\rangle^{1/2}$ obtained by COBE is of order 5×10^{-6}. This can also be expressed in terms of the quantity $Q_{\rm rms}$, which is defined by

$$Q_{\rm rms}^2 = \frac{T_0^2}{4\pi}\sum_{\rm m}\langle |a_{2m}|^2\rangle = \frac{5T_0^2}{4\pi}\langle |a_{2m}|^2\rangle \simeq (17\ \mu{\rm K})^2. \tag{17.4.9}$$

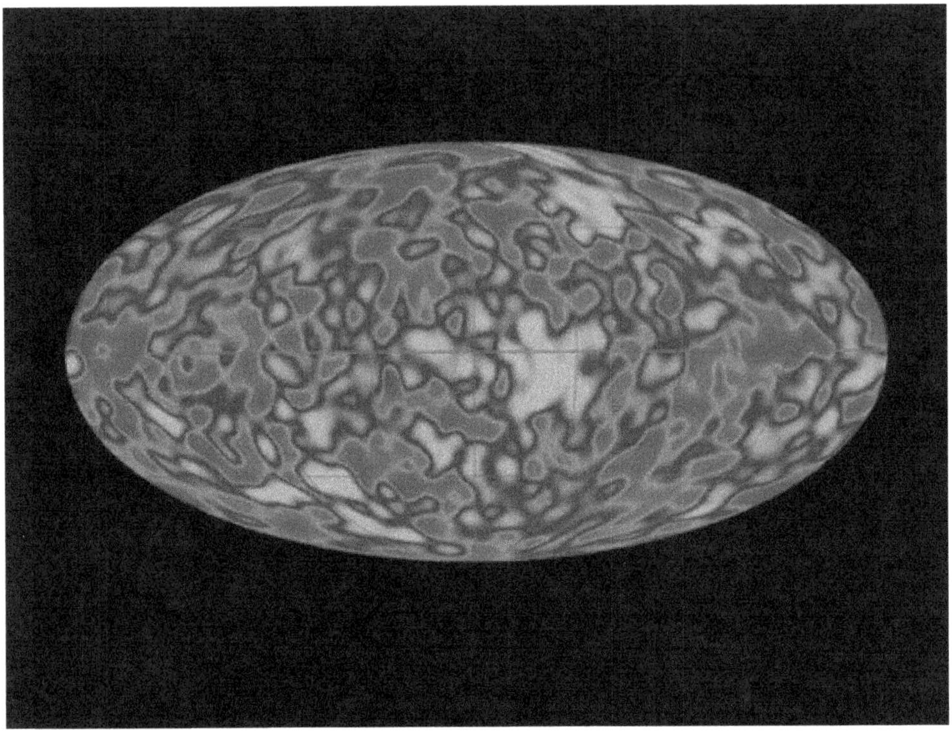

Figure 17.1 Black and white representation of the COBE DMR four-year data map. The typical angular scale of fluctuations is around 10° and the typical amplitude is around 30μ K. Picture courtesy of George Smoot and NASA.

Translated into a value of σ_8(mass) using (17.4.7) with $n = 1$ and a standard CDM transfer function, this suggests a value of $b \simeq 1$, which does not seem to allow the option of a linear bias for removing discrepancies between clustering and peculiar motions, such as those we shall discuss in Chapter 18. We should say that normalising everything to the quadrupole in this way is not a very good way of using the COBE data, which actually constitute a map of nearly the whole sky with a resolution of about 10°. The RMS temperature anisotropy obtained from the whole map is of order 1.1×10^{-5}. (Both this value and the quadrupole value are changed as more data from this experiment were analysed.) The quadrupole mode is actually not as well determined as the C_l for higher l, so a better procedure is to fit to all the available data with a convolution of the expected C_l for some amplitude with the experimental beam response and then determine the best fitting amplitude for the data. The results of more sophisticated data analysis like this are, however, in rough agreement with the simpler method mentioned above. Notice also that one can in principle determine the primordial spectral index n from the data by calculating $C(\vartheta)$ and comparing this with the expected form using Equation (17.4.5) for a given $P(k) \propto k^n$. The results obtained from this type of analysis are rather noisy, and do differ significantly depending on the type of analysis technique used, but they do seem consistent with $n = 1$.

Four years' worth of data from the DMR experiment have now been published; the experiment was turned off in 1994. An independent detection of fluctuations on a slightly smaller scale than COBE was later announced by a team working at Tenerife using a ground-based beam-switching experiment (Hancock *et al.* 1993). The level and form of fluctuations detected in this experiment are consistent with those found by COBE.

17.4.3 Interpretation of the COBE results

At this stage, let us return to a point we raised above: the possible contribution of tensor perturbation modes to the large-scale CMB anisotropy. Gravitational waves do involve metric fluctuations and therefore do generate a Sachs–Wolfe effect on scales larger than the horizon. Once inside the horizon, however, they redshift away (just like relativistic particles) and play no role at all in structure formation. Gravitational waves produce an effect similar to scalar perturbations on large angular scales but have negligible influence upon $\Delta T/T$ on scales inside the horizon at $z_{\rm rec}$. Clearly, normalising the power spectrum $P(k)$ to the observed C_l using (17.4.5) is incorrect if the tensor signal is significant.

One can define a power spectrum of gravitational wave perturbations in an analogous fashion to that of the density perturbations. It turns out that inflationary models also generically predict a tensor spectrum of power-law form, but with a spectral index

$$n_{\rm T} = 1 - 2\epsilon_*, \tag{17.4.10}$$

instead of equation (13.6.10). Since ϵ_* is a small parameter the tensor spectrum will be close to scale invariant. It is also possible to calculate the ratio, $\mathcal{R}$, between the tensor and scalar contributions to C_l:

$$\mathcal{R} = \frac{C_l^{\rm T}}{C_l^{\rm S}} \simeq 12\epsilon_*. \tag{17.4.11}$$

To get a significant value of the gravitational wave contribution to C_l one therefore generally requires a significant value of ϵ_* and therefore both scalar and tensor spectra will usually be expected to be tilted away from $n = 1$. If $\mathcal{R} = 1$, then one can reconcile the COBE detection with a CDM model having a significantly high value of b. Because one cannot use Sachs–Wolfe anisotropies alone to determine the value of $\mathcal{R}$, there clearly remains some element of ambiguity in the normalisation of $P(k)$.

The Equations (17.4.10) and (17.4.11) are true for inflationary models with a single scalar field. More contrived models with several scalar fields can allow the two spectral indices and the ratio to be given essentially independently of each other. The shape of the COBE autocovariance function suggests that n cannot be much less than unity, so the prospects for having a single-field inflationary model producing a large tensor contribution seem small. On the other hand, we have no *a priori* information about the value of $\mathcal{R}$ so it would be nice to be able to constrain it using observations. It turns out that to perform such a test requires, at the very least, observations on a different (i.e. smaller) angular scale. From Figure 17.1 one can see that the scalar contribution increases around degree scales, while the tensor contribution dies away completely. We shall discuss the reasons for this shortly. In principle, one can therefore estimate $\mathcal{R}$ by comparing observations of C_l at different values of l although, as we shall see, the result is rather model dependent.

We should also mention that, if the CMB fluctuations are generated by primordial density perturbations which are Gaussian (Section 13.8), then the fluctuations $\Delta T/T$ should be Gaussian also. The nonlinear Sachs–Wolfe effect generally produces a non-Gaussian temperature pattern, as do various extrinsic anisotropy sources we shall discuss in Section 17.6. To be precise, the prediction is that individual a_{lm} should have Gaussian distributions so that the actual sky pattern will only be Gaussian if one adds a significant number of modes for the central limit theorem to come into play. In principle it is possible to use statistical properties of sky maps to test the hypothesis that the fluctuations were Gaussian, though this task will have to wait for better data than are available at present. Notice that instrumental noise is almost always Gaussian, so if there is a lot of noise superimposed on the sky signal one can have problems detecting any non-Gaussian features which may be generated by extrinsic effects, or non-Gaussian perturbations such as cosmic strings. At the moment, all we can say is that the COBE and Tenerife results are at least consistent with Gaussian primordial fluctuations.

17.5 Intermediate Scales

As we have already explained, the large-scale features of the microwave sky are expected to be primordial in origin. Smaller scales are closer to the size of the Hubble horizon at $z_{\rm rec}$ so the density fluctuations present there may have been modified by various damping and dissipation processes. Moreover, there are physical mechanisms other than the Sachs–Wolfe effect which are capable of generating anisotropy in the CMB on these smaller scales. We shall concentrate upon intrinsic sources of anisotropy in this section, i.e. those connected with processes occurring around $t_{\rm rec}$; we mention some extrinsic (line-of-sight) sources of anisotropy in Section 17.6.

Let us begin with some naive estimates. For a start, if the density perturbations are adiabatic, then one should expect fluctuations in the photon temperature of the same order. Using $\rho_{\rm r} \propto T^4$ and the adiabatic condition, $4\delta_{\rm m} = 3\delta_{\rm r}$, we find that

$$\frac{\Delta T}{T} \simeq \frac{1}{3}\frac{\delta\rho}{\rho}, \tag{17.5.1}$$

which is also stated implicitly in Section 12.2. Another mechanism, first discussed by Zel'dovich and Sunyaev, is simply a Doppler effect. Density perturbations at $t_{\rm rec}$ will, by the continuity equation, induce streaming motions in the plasma. This generates a temperature anisotropy because some electrons are moving towards the observer when they last scatter the radiation and some are moving away. It turns out that the magnitude of this effect for perturbations on a scale λ at time t is

$$\frac{\Delta T}{T} \simeq \frac{v}{c} \simeq \frac{\delta\rho}{\rho}\left(\frac{\lambda}{ct}\right), \tag{17.5.2}$$

where ct is of order the horizon scale at t.

The actual behaviour of the background radiation spectrum is, however, much more complicated than these simple arguments might suggest. The detailed computation of fluctuations originating on these scales is consequently much less straightforward than was the case for the Sachs–Wolfe effect. In general one therefore resorts to a full numerical solution of the Boltzmann equation for the photons through recombination, taking into account the effect of Thomson scattering, as described briefly in Section 11.10. The usual approach is to expand the distribution function of the radiation in spherical harmonics thereby generating a coupled set of equations for different l-modes of the distribution function; in Section 12.10 we used the *brightness function*, $\delta^{(\mathrm{r})}$, to represent the perturbation to the radiation and wrote down a set of equations (11.9.7) for the l-modes, σ_l, defined by

$$\delta_k^{(\mathrm{r})}(\mu, t) = \sum_l (2l+1) P_l(\mu)\sigma_l(k, t); \tag{17.5.3}$$

$\mu = \cos\vartheta$ is the cosine of the angle between the photon momentum and the wave vector $\boldsymbol{k}$. The solution of (11.9.7) is a fairly demanding numerical task. Given a set of σ_l, however, it is straightforward to show that the autocovariance function $C(\vartheta)$ of the sky at the present time is just

$$C(\vartheta) = \frac{1}{2\pi^2}\int_0^\infty \sum_l (2l+1)(\tfrac{1}{4}\sigma_l(k, t_0))^2 P_l(\cos\vartheta) k^2 \, dk, \tag{17.5.4}$$

where the integral takes the distribution from Fourier space back to real space and the factor of 4 is due to the fact that $\delta_{\mathrm{r}} = 4\Delta T/T$. Fortunately, it is now possible to perform computations of both the transfer functions we described in Chapter 15 and the predicted temperature fluctuations rapidly and accurately using an approach that bypasses the complex hierarchy we described above. The code that does this, CMBFAST (Seljak and Zaldarriaga 1996), is available freely on the web so that anyone interested in computing the predicted pattern of fluctuations for their favourite model may download it.

As mentioned above, one can also allow for the effect of different beam profiles and experimental configurations. For example, a double-beam experiment of the form (17.2.14) would have

$$\begin{aligned}\left(\frac{\Delta T}{T_0}\right)^2_{\alpha;\sigma} = \frac{1}{64\pi^2}\int_0^\infty k^2 \int_{-1}^{1} |\delta_k^{(\mathrm{r})}(\mu, t_0)|^2 \\ \times \{1 + \tfrac{1}{3}J_0[2\alpha k r_0(1-\mu^2)^{1/2}] - \tfrac{4}{3}J_0[\alpha k r_0(1-\mu^2)^{1/2}]\} \\ \times \exp[-k^2\sigma^2 r_0^2(1-\mu^2)^{1/2}]\, dk\, d\mu,\end{aligned} \tag{17.5.5}$$

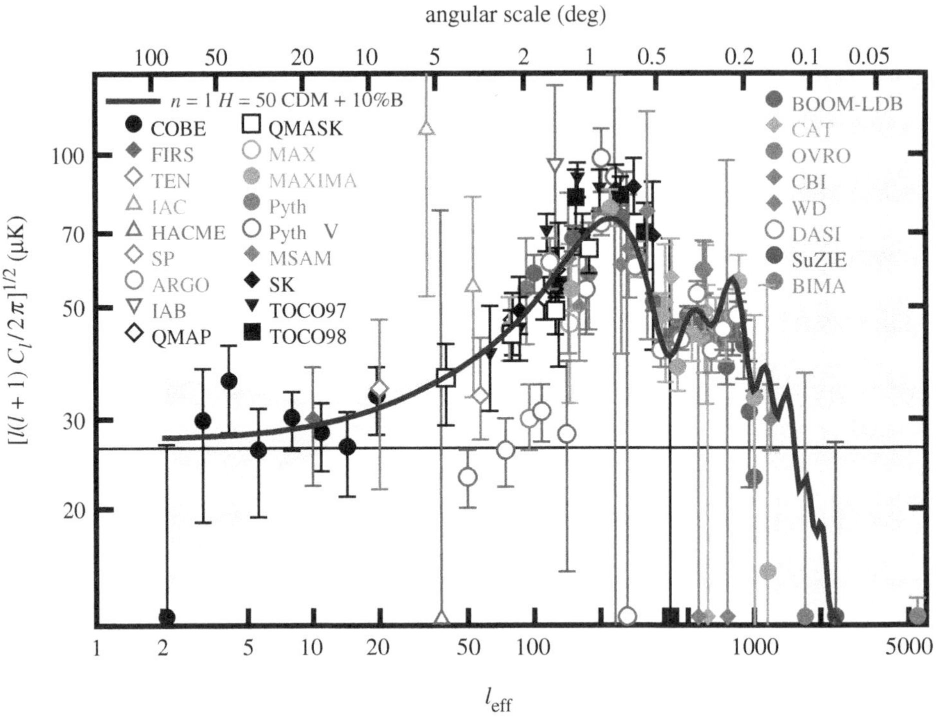

Figure 17.2 A compilation of experimental measurements of C_l along with a theoretical curve for standard CDM. Picture courtesy of Ned Wright.

for a Gaussian beam of width σ and a beam throw of α. In the previous equation J_0 is a Bessel function and $r_0 \simeq 2c/\Omega H_0$.

An example of a numerical computation of the C_l for a CDM model over the range of interest here is given in Figure 17.2 (solid line) along with a morass of points that represents various experimental results. Note the flat behaviour at small l owing to the Sachs–Wolfe effect. After this one notices a steep increase in the angular power spectrum for $l \sim 100$–200. This angular scale corresponds to the horizon scale at z_{rec}. The shape of the spectrum beyond this peak is complicated and depends on the relative contribution of baryons and dark matter. For example, the small 'bumps' at large l change position if Ω_b is changed.

Although these theoretical results are computed numerically, it is important to understand the physical origin of the features of the resulting C_l at least qualitatively. The large peak around the horizon scale is usually interpreted as being due to velocity perturbations on the last scattering surface, as suggested by Equation (17.5.2), and is consequently sometimes called the *Doppler peak*. The features at higher l are connected with a phenomenon called *Sakharov oscillations*. Basically what happens is that perturbations inside the horizon on these angular scales oscillate as acoustic standing waves with a particular phase relation between density and velocity perturbations. These oscillations can be seen in Figures 11.1 and 11.2 and in the transfer function in Figure 15.1. After recombination, when pressure forces become negligible, these waves are left with phases which depend on their wavelength. Both the photon temperature fluctuations (17.5.1) and the velocity perturbations (17.5.2) are therefore functions of wavelength (both contribute to $\Delta T/T$ in this regime) and this manifests itself as an almost periodic behaviour of C_l. The use of the term 'Doppler peak' to describe only the first maximum of these oscillations is misleading because it is actually just the first (and largest-amplitude) Sakharov oscillation. Although velocities are undoubtedly important in the generation of this feature, it is wrong to suggest that the physical origin of the first peak in the angular power spectrum is qualitatively different from the others.

The power spectrum of the matter fluctuations is also expected to display oscillations relating to this phase effect but with a much lower contrast. The reason for this is that most of the matter in standard models is neither baryonic nor collisional. Consequently it neither interacts by scattering with radiation nor produces restoring forces to support induced oscillations. Essentially the CMB anisotropy is influenced by the baryonic component only so the oscillations are dominant, while the power spectrum of the dark matter is smooth with only small baryonic oscillations superimposed upon it.

The physical origin of these oscillations is interesting enough, but their importance in present and future cosmological investigations is paramount. The reason for this is that the position and relative amplitudes of the Doppler peak and its 'harmonics' are a sensitive diagnostic not just of the precise mix of dark matter and baryons, but also the values of the principal cosmological parameters. For instance, the position of the first peak is a direct route to the density parameter Ω_0 or, rather, the global curvature k. The physical length scale at which this peak occurs corresponds to the size of the sound horizon ($c_s t_{rec}$, where c_s is the sound speed) at the surface of last scattering roughly defined by t_{rec}. This does not vary much with cosmological parameters. However, this length scale subtends an angle that depends on the geometry of the Universe. Consequently the spherical harmonic l that corresponds to the Doppler peak changes if the background curvature changes. In a flat universe the peak occurs around $l \simeq 200$. If the universe has positive curvature, geodesics converge towards the observer so the angle subtended by a 'rod' of fixed size is larger than in a flat universe. The peak therefore moves to smaller l in this case. If spatial sections are negatively curved, then the peak moves to higher l; see Figure 2.3 to see why the angle looks smaller in an open universe. This shows how important the first peak is, but the detailed shape of the power spectrum has a strong dependence on the other parameters too. An accurate measurement of these features promises to nail many of the uncertainties facing cosmology in one fell swoop. For further discussion of open universes see Kamionkowski and Spergel (1994).

There are complications, of course. One is the relatively slow rate of recombination. One effect of this is that the optical depth to the last scattering surface can be quite large, and small-scale features can be smoothed out. For example, as we discussed in Section 9.4 in the context of the standard theory of recombination, the last scattering surface can have an effective 'width' up to $\Delta z \simeq 400$, which corresponds to a proper distance now of $\Delta L \simeq 40h^{-1}$ Mpc, and to an angular scale $\simeq$ 20 arcmin. The finite thickness of the last scattering surface can mask anisotropies on scales less than ΔL in the same way that a thick piece of glass prevents one from seeing small-scale features through it. This causes a damping of the contribution at high l and thus a considerable reduction in the $\Delta T/T$ relative to the photon temperature fluctuations (17.5.1).

High angular frequency fluctuations are also quite sensitive to the possibility that the Universe might have been reionised at some epoch. As we shall see in Chapter 20, we know that the intergalactic medium is now almost completely ionised. If this happened early enough, it could smear out the fluctuations on scales less than a few degrees, rather than the few arcminutes for standard recombination, the case shown in Figure 17.2. Some non-standard cosmologies involve such a late recombination so that Δz might be much larger. The minimum allowable redshift is, however, $z \simeq 30$ because an optical depth $\tau \simeq 1$ requires enough electrons (and therefore baryons) to do the scattering; a value $z < 30$ would be incompatible with $\Omega_b < 0.1$; we discussed this in Chapter 9. In any case, if some physical process caused the Universe to be reheated after t_{rec}, then it might smooth out anisotropy on scales less than the horizon scale at the time when the reionisation occurred. Recall from Equation (17.4.6) that the angular scale corresponding to the particle horizon at z is of order $(\Omega/z)^{1/2}$, so late reionisation at $z \simeq 30$ could smooth out structure on scales of 10° or less, but not scales larger than this. We shall see in Section 17.6 that, if this indeed occurred, one might expect to see a significant anisotropy on a smaller angular scale, generated by secondary effects.

The message one should take from these comments is that the fluctuations on these scales are much more model dependent than those on larger scales. In principle, however, they enable one to probe quite detailed aspects of the physics going on at t_{rec} and are quite sensitive to parameters which are otherwise hard to estimate. Moreover, tensor modes do not produce any Doppler motions and their contribution to C_l should therefore be small for high l. Although these oscillatory features are potentially a very sensitive diagnostic of the perturbations generating the CMB anisotropy, it is difficult to resolve them.

The problem with these experiments, which are all either balloon borne or ground based, is twofold. Firstly, they usually probe a relatively small part of the sky and the signal they see may not be representative of the whole sky, i.e. they are dominated by '*sample variance*'. The second problem is that, until recently, they generally did not have the ability to remove point sources (because of the smaller beam) or non-thermal emission (because of the smaller number of frequency channels) as effectively as COBE. Observational programmes aimed at improving the situation have been pursued with great vigour over the last few years, as indicated by the forest of error bars in Figure 17.2.

Over the last few years the situation has changed dramatically with two long-duration balloon flights bearing sensitive bolometers finally giving convincing measurements of the Doppler peak and its first one or two overtones (Hanany

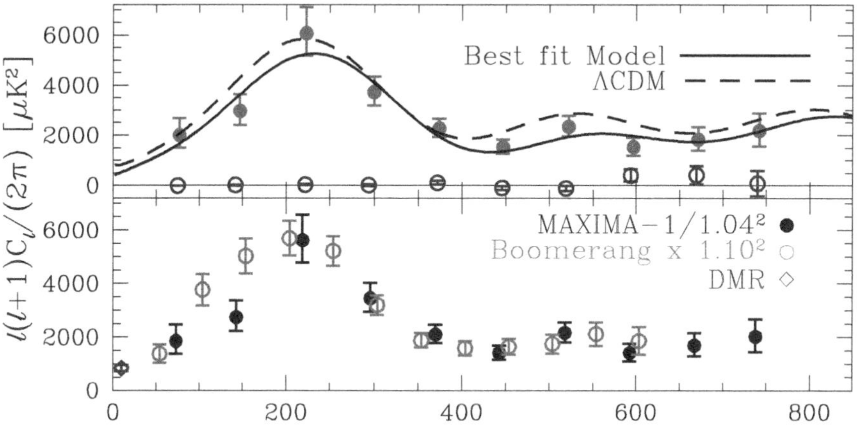

Figure 17.3 The angular power spectrum of the CMB estimated by MAXIMA-1 and Boomerang. Picture courtesy of Andrew Jaffe.

et al. 2000; Jaffe *et al.* 2001); see Figure 17.3. The crucial point about this result is the position of the first peak. This tightly constrains the curvature to be very small. Taken together with the supernova results and the relatively low apparent matter density discussed in Chapter 4, this strongly suggests the existence of a cosmological constant in the Einstein field equations.

These measurements still come from relatively small patches of the sky but show how strong the constraints on cosmological models are likely to become in the near future when all-sky satellites are launched. As we write, in 2002, a US-led mission called MAP (Microwave Anisotropy Probe) is already in space collecting data from which high-resolution whole-sky maps will be constructed. In 2007 the European Space Agency's Planck Surveyor will do a similar job at even higher resolution.

As a final remark, we should stress that intrinsic CMB temperature anisotropy is expected to be Gaussian on these scales, since it is generated by linear processes from density perturbations which are themselves Gaussian. As with the Sachs-Wolfe effect, one can in principle use the properties of $\Delta T/T$ to test the Gaussian hypothesis on these scales also. For example, in the cosmic-string scenario the dominant contribution to the CMB anisotropy is generated by cosmic strings lying between the observer and the last scattering surface which distort the photon trajectories. The detailed statistical properties of the pattern of temperature maps on intermediate and large scales in this scenario will be very different from those in Gaussian scenarios.

17.6 Smaller Scales: Extrinsic Effects

As explained in the introduction to this chapter, one of the main motivations for studying the temperature anisotropy of the cosmic microwave background is that one can, in principle, look directly at the effects of primordial density fluctuations and therefore probe the initial conditions from which structure is usually

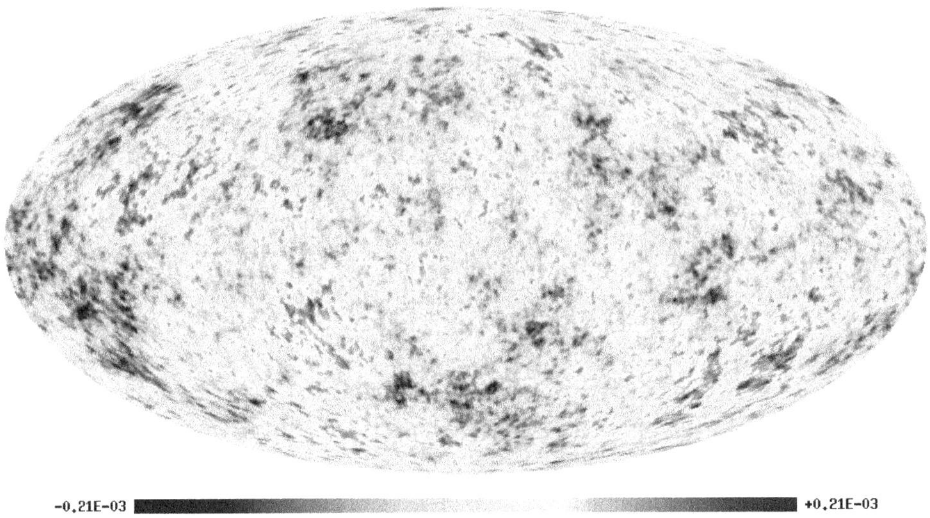

Figure 17.4 A simulation of the CMB sky as it might be seen by MAP or Planck.

supposed to have grown. In the previous two sections we have elucidated the physical mechanisms responsible for generating intrinsic anisotropy and shown that these do indeed involve the primordial density perturbations. The problem is that the length scales probed by these anisotropies are much larger than those of direct relevance to galaxy and cluster formation. In fact, there is a simple rule relating a given (comoving) length scale to the angle that scale subtends on the last scattering surface:

$$1h^{-1}\text{ Mpc} \simeq \tfrac{1}{2}\Omega \text{ arcmin.} \tag{17.6.1}$$

As we explained in Section 17.5, temperature anisotropies due to fluctuations on length scales up to $40h^{-1}$ Mpc will probably be smoothed out by the finite thickness of the last scattering surface. One cannot therefore probe scales of direct relevance to cluster and galaxy formation using measurements of intrinsic CMB anisotropy. COBE and related experiments can only constrain theories of structure formation if there is a continuous spectrum of density fluctuations with a well-defined shape so that a measurement of the amplitude on the scale of a thousand Mpc or so, corresponding to COBE, can be extrapolated down to smaller scales. Because these experiments do not in themselves supply a test of the shape of the power spectrum on smaller scales, theories must be constrained by combining CMB anisotropy measurements with galaxy-clustering data or peculiar velocity data; the latter will be discussed in the next chapter.

There are various ways, however, in which small-scale anisotropy measurements can yield important information on short-wavelength fluctuations due to extrinsic effects, rather than the intrinsic effects we have discussed so far. We shall discuss some possible mechanisms of this type in this section. Because these are highly model dependent and, in some cases, involve complicated physical processes, we shall restrict ourselves to a qualitative discussion without many technicalities. The interested reader is referred to the bibliography for further details.

One important consideration on scales of arcminutes and less is the contribution of various kinds of extragalactic sources to the CMB anisotropy. Point sources generally have a non-thermal spectrum so they can, in principle, be accounted for using multi-frequency observations, but this is by no means straightforward in practice. The brightest point sources can be removed quite easily as they may be resolved by the experimental beam. An integrated background due to large numbers of relatively faint sources is, however, very difficult to deal with. Many of the intermediate scale measurements mentioned in Section 17.5 also suffer from the difficulty of point-source subtraction. Although CMB measurements may in principle place constraints on the evolution of various kinds of radio source, in practice these are usually treated as a nuisance which is to be removed. Nevertheless, it is useful to calculate the approximate contribution to $\Delta T/T$ from point sources distributed in different ways. Firstly, suppose the objects were actually present before z_{rec}, which seems rather unlikely. The radiation from them would have to be thermalised by some agent, such as grains of dust, otherwise it would lead to a spectral distortion of order q, the fraction of the CMB energy density which they generate. If the sources are randomly distributed in space, then the effective anisotropy is just due to Poisson statistics for $\vartheta > \vartheta_{\mathrm{H}}(z_{\mathrm{rec}}) = \vartheta_*$ given by Equation (17.4.6):

$$\left(\frac{\Delta T}{T}\right)_\vartheta \simeq \frac{q}{N_\vartheta^{1/2}} \propto \frac{q}{\vartheta}, \tag{17.6.2}$$

where N_ϑ is the mean number of sources in a beam of width ϑ. On angles less than ϑ_* the radiation would be smoothed out. For example, if we have a population of sources with (comoving) mean spacing l_{s} at a redshift z_{s}, it is quite easy to show that

$$\left(\frac{\Delta T}{T}\right)_\vartheta \simeq \frac{q}{2}\left(\frac{l_{\mathrm{s}}}{ct_0}\right)^{3/2}(1+z_{\mathrm{s}})^{1/4}\frac{\vartheta}{\vartheta_*^2+\vartheta^2}. \tag{17.6.3}$$

This corresponds to two-dimensional white noise filtered on a scale ϑ_*.

Now consider the case of sources at $1 \ll z < z_{\mathrm{rec}}$. In this case there is no filtering and there will be a spectral distortion because this radiation cannot be thermalised. The resulting $\Delta T/T$ is just like (17.6.3) with $\vartheta_* = 0$. As we remarked above, limits on the departure of the spectrum from a black-body form can therefore constrain the contribution from such sources.

The expression (17.6.3) must be modified considerably if one is dealing with local sources, by which we mean those with $z_{\mathrm{s}} \leqslant 1$ or thereabouts. Local sources are usually referred to as 'contamination', which gives some idea of how astronomers regard them. The contribution from such objects is dominated by the brightest ones found in a solid angle ϑ^2 and is therefore closely connected with the $\log N$-$\log S$ relationship (the radio astronomers equivalent of the number-magnitude relation). One generally has

$$N_\vartheta[> S(\nu)] \propto S(\nu)^{-\beta}, \tag{17.6.4}$$

with $\beta \leqslant 2$, where $N_\vartheta[> S(\nu)]$ is the number of sources per unit solid angle with a measured flux at ν greater than $S(\nu)$; see Chapter 19 for some more details. If their spectrum is proportional to ν^α, then

$$\left(\frac{\Delta T}{T}\right)_\vartheta \propto \vartheta^{2/\beta-2}\nu^{\alpha-2}. \tag{17.6.5}$$

The amplitude due to these sources would depend strongly on wavelength. The wavelength dependence can therefore, in principle, be used to identify the contribution from them, but one needs to know the luminosity function of the sources well to be able to subtract them, especially at higher frequencies. Another problem is that the telescopes used for CMB studies often have considerable 'sidelobes', which may pick up bright objects quite a long way away from the main beam of the telescope; these are also difficult to subtract.

A cosmological background of dust may also affect the microwave background, particularly if it is heated by some energetic source at early times. We shall discuss the effect of this type of process upon the spectrum of the CMB radiation in Chapter 19; here it suffices to note that dust generally emits infrared radiation and this may leak into the wavelength range covered by CMB experiments. Dust is generally a signature of structure formation (it is mainly produced in regions forming massive stars). Inhomogeneities in the dust density can lead to a temperature anisotropy of the CMB. If the dust is clustered like galaxies and the distribution evolves as in a CDM model, then it can be shown that one expects anisotropy up to $\Delta T/T \simeq 10^{-5}$ at 400 μm, rising to 10^{-4} at the peak of the CMB spectrum. Given the lack of observed spectral distortions, however, it seems unlikely that dust will generate a significant CMB anisotropy.

Another way in which secondary anisotropy can be generated is connected with possible reionisation of the intergalactic gas after $z_{\rm rec}$. We have already explained in Section 17.5 how this can smooth out intrinsic anisotropy. Generally, however, reionisation will lead to significant secondary anisotropy on a smaller angular scale than we considered in that section.

Reionisation or reheating may have been generated by many different mechanisms. Theories involving a dark-matter particle which undergoes a radiative decay can lead to wholesale reionisation. Early star formation, active galactic nuclei or quasars could also, in principle, have caused reionisation of the intergalactic medium. Cosmological explosions may heat up the intergalactic medium in a very inhomogeneous way leading to considerable anisotropy. As we shall explain in Chapter 21, we know that something reionised the Universe some time before $z \simeq 4$ so these apparently exotic scenarios are not completely implausible.

Whatever caused the gas to become ionised, there is expected to be an accompanying generation of anisotropy. Suppose the plasma is heated enough to ionise it, but not enough for the electrons to become highly relativistic. If the plasma is inhomogeneous, then it will generally have a velocity field associated with it and a photon travelling through the ionised medium will suffer Thomson scattering off electrons with velocities oriented in different directions. The rate of energy loss due to Thomson scattering is just

$$\frac{\mathrm{d}E}{\mathrm{d}t} = -n_\mathrm{e}\sigma_\mathrm{T}c\left[1 + \hat{\boldsymbol{n}}\cdot\frac{\boldsymbol{v}}{c} + \left(\frac{v}{c}\right)^2\right]E, \tag{17.6.6}$$

where n_e and $\boldsymbol{v}$ are the electron number density and velocity, respectively, and σ_T is the Thomson scattering cross-section; $\hat{\boldsymbol{n}}$ is a unit vector in the direction of photon travel. Since Thomson scattering conserves photons we can write

$$\frac{\Delta T}{T} = -\sigma_\mathrm{T}c\int n_\mathrm{e}\left[\delta + \left(\frac{v}{c}\right)^2 + \hat{\boldsymbol{n}}\cdot\frac{\boldsymbol{v}}{c} + \left(\hat{\boldsymbol{n}}\cdot\frac{\boldsymbol{v}}{c}\right)\delta\right]\mathrm{d}t, \tag{17.6.7}$$

where the integral is taken over a line of sight from the observer to t_rec and δ is the dimensionless density perturbation in the medium.

The net anisotropy produced by the linear terms in (17.6.7) is extremely small. The second-order term which corresponds to the interaction between the perturbation δ and the velocity can be significant, however, particularly if the inhomogeneities are evolving in the nonlinear regime. This nonlinear term is usually called the *Ostriker-Vishniac effect* (Ostriker and Vishniac 1986), although it was actually first discussed by Sunyaev and Zel'dovich (1969). For a spherically symmetric homogeneous cluster moving through the CMB rest frame the effect is particularly simple:

$$\frac{\Delta T}{T} = -2\sigma_\mathrm{T}n_\mathrm{e}R\left(\hat{\boldsymbol{n}}\cdot\frac{\boldsymbol{v}}{c}\right) \tag{17.6.8}$$

for a cluster of radius R moving at a velocity $\boldsymbol{v}$.

There is one other important source of extrinsic anisotropy, called the *Sunyaev-Zel'dovich effect.* We shall, however, devote the whole of Section 17.7 to this because it is important in a wider cosmological context than structure-formation theory.

17.7 The Sunyaev–Zel'dovich Effect

The physics behind the *Sunyaev-Zel'dovich* (SZ) *effect* is that, if CMB photons enter a hot (relativistic) plasma, they will be Thomson-scattered up to higher energies, say X-ray energies. If one looks at such a cloud in the Rayleigh–Jeans (long-wavelength) part of the CMB spectrum, one therefore sees fewer microwave photons and the cloud consequently looks cooler. For a cloud with electron pressure p_e the temperature 'dip' is

$$\frac{\Delta T}{T} = -2\int\frac{p_\mathrm{e}\sigma_\mathrm{T}}{m_\mathrm{e}c^2}\,\mathrm{d}l = -2\int\frac{n_\mathrm{e}k_\mathrm{B}T_\mathrm{e}\sigma_\mathrm{T}}{m_\mathrm{e}c^2}\,\mathrm{d}l, \tag{17.7.1}$$

where $\mathrm{d}l = c\,\mathrm{d}t$ is the distance along a photon path through the cloud. This effect has been detected using radio observations of rich Abell clusters of galaxies. Such clusters contain ionised gas at a temperature of up to 10^8 K (the virial temperature) and are about 1 Mpc across. The effect has been detected at a level

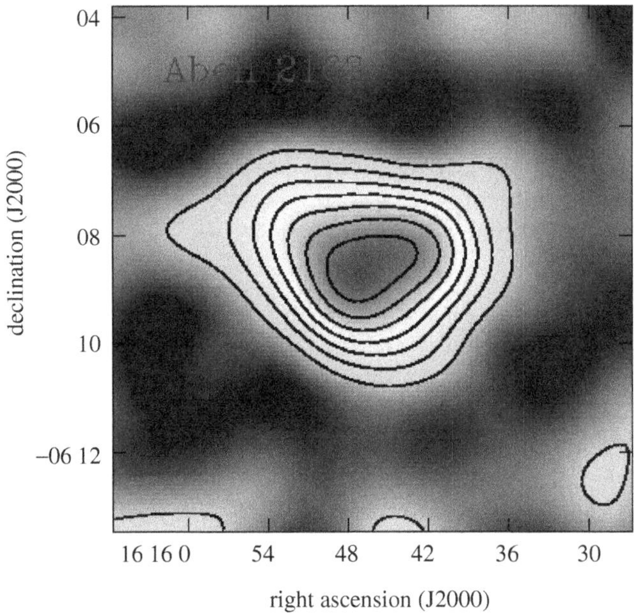

Figure 17.5 A Sunyaev–Zel'dovich (SZ) map of the cluster Abell 2163. Picture courtesy of John Carlstrom.

of order 10^{-4} in several clusters, but a new instrument called the Ryle Telescope, recently built in Cambridge, has improved the technique and substantially reduced the observational difficulties. This instrument is very different from most devices used to search for intrinsic CMB anisotropy because it is supposed to map only a small part of the sky around an individual cluster. (The need to cover a large part of the sky is one of the most demanding requirements on CMB anisotropy searches.) It is possible with this instrument to create detailed maps of clusters in the SZ distortion they produce; an example is shown in Figure 17.5.

A particularly interesting aspect of this technique is that, if one has X-ray observations of a cluster, its redshift and an SZ dip, one can, in principle, get the distance to the cluster in a manner independent of the redshift. This is done by combining X-ray bremsstrahlung measurements, which are proportional to $\int n_e^2 T_e^{1/2}\, dl$, the observed X-ray spectrum, which gives T_e, and the Sunyaev–Zel'dovich dip. These three sets of observations allow one to determine T_e and the integrals of $n_e T_e$ and $n_e^2 T_e^{1/2}$ through the cluster. One then assumes that the physical size of the cluster along the line of sight is the same as its size in the plane of the sky. Extracting an estimate of l, the total path length through the cluster, then yields an estimate of R_c, the physical radius. Knowing its angular size, one can thus estimate a value for the proper distance. Comparing this with the cluster redshift yields a direct estimate of the Hubble constant which is independent of the usual distance ladder methods described in Section 4.3. For example, if we model the cluster as a homogeneous isothermal sphere of radius R_c, then, from

Equation (17.7.1), the dip in the centre of the cluster will be

$$\frac{\Delta T}{T} = -\frac{4R_c n_e k_B T_e \sigma_T}{m_e c^2}. \tag{17.7.2}$$

Obviously, more sophisticated modelling than this is necessary to obtain accurate results, but the example (17.7.2) illustrates the principles of the method.

This method, when applied to individual clusters, has so far yielded estimates of the Hubble constant towards the lower end of its accepted range. One should say, however, that many clusters are significantly aspherical, so one should really apply this technique to a sample of clusters with random orientations with respect to the line of sight. An appropriate averaging can then be used to obtain an estimate of H_0 for the sample which is less uncertain than that for an individual cluster.

As well as being detectable for individual clusters, there should be an integrated SZ effect caused by all the clusters in a line of sight from the observer to the last scattering surface. This is another complicated small-scale effect which is rather difficult to model. In principle, however, constraints on the temperature fluctuations produced by this effect place strong limits on the evolutionary properties of clusters of galaxies. We shall discuss this and other constraints on cosmological evolution in Chapter 21.

17.8 Current Status

The last 10 years have seen a tremendous revolution in CMB physics. Starting with the COBE discovery, and its confirmation at Tenerife, increasing sensitivity and resolution have driven observers forward so that all-sky maps of the temperature pattern with arcminute resolution will shortly be available. At the moment the balloon-based results from MAXIMA and Boomerang represent the state of the art. These data strongly suggest we live in a flat universe. Combined with supernova results and other measurements these results have dramatically altered our view of what the standard model of cosmology could be; ΛCDM has emerged from the pack described in Chapter 15 and now replaces SCDM as the front runner for a complete model of structure formation.

When the issue of the intermediate-scale anisotropy is finally resolved by all-sky maps, a number of other questions can be addressed, connected with extrinsic (nonlinear) anisotropies, the detailed statistical properties of high-resolution sky maps and after-effects of reionisation. Another question which will probably become important in a few years' time is connected with the *polarisation* of the CMB radiation. Thomson scattering is important during the processes of decoupling and recombination and it induces a partial linear polarisation in the scattered radiation (Rybicki and Lightman 1979). It has been calculated that the level of polarisation expected in the CMB is about 10% of the anisotropy, i.e. a fractional level of around 10^{-6}. This figure is particularly sensitive to the ionisation history and it may yield further information about possible reheating of the Universe.

Measurement of CMB polarisation is, however, not practicable with the current generation of telescopes and receivers.

Bibliographic Notes on Chapter 17

The field described in this chapter is developing extremely rapidly. To see how rapidly material has become dated, it is useful to read Hogan *et al.* (1982), Vittorio and Silk (1984), Kaiser and Silk (1987), Partridge (1988) and even White *et al.* (1994). Peacock (1999) is a good up-to-date reference for this material. CMB anisotropy studies have come of age during an era dominated by the internet. Two particularly useful resources are the CMBFAST page

`http://www.physics.nyu.edu/matiasz/CMBFAST/cmbfast.html`

(see Seljak and Zaldarriaga 1996) and Wayne Hu's superb compilation of CMB theory and experiment at

`http://background.uchicago.edu/~whu/`

Problems

1. Verify the approximate relations (17.2.2) and (17.6.1).
2. Derive the results (17.2.13), (17.2.14) and (17.2.15).
3. Derive Equation (17.4.5).
4. Use the results of Chapter 11 to computer the evolution of the sound horizon as a function of redshift through matter-radiation equivalence until the point of recombination.
5. Derive the result (17.6.3).
6. A beam of unpolarised radiation is incident upon an electron. Show that the degree of polarisation in the light scattered at an angle θ to the incident beam is Π, where

$$\Pi = \frac{1 - \cos^2\theta}{1 + \cos^2\theta}.$$

18

Peculiar Motions of Galaxies

18.1 Velocity Perturbations

In our treatment of gravitational instability in Chapters 10 and 11 we focused upon the properties of the density field ρ or, equivalently, the density perturbations δ. The equations of motion do, however, contain another two variables, namely the velocity field $\boldsymbol{v}$ and the gravitational potential φ. These two quantities are actually quite simple to derive once the behaviour of the density has been obtained. To show this, let us write the continuity, Euler and Poisson equations again:

$$\frac{\partial \rho}{\partial t} + \nabla \cdot \rho \boldsymbol{v} = 0, \tag{18.1.1 a}$$

$$\frac{\partial \boldsymbol{v}}{\partial t} + (\boldsymbol{v} \cdot \nabla)\boldsymbol{v} + \frac{1}{\rho}\nabla p + \nabla \varphi = 0, \tag{18.1.1 b}$$

$$\nabla^2 \varphi - 4\pi G \rho = 0; \tag{18.1.1 c}$$

cf. Equations (10.2.1). As we suggested in Section 11.2, it now proves convenient to transform to comoving coordinates; here, however, we adopt a slightly different approach. Since we are looking for perturbations about the uniformly expanding solution with $\boldsymbol{v} = H\boldsymbol{r}$, we introduce a peculiar velocity term $\boldsymbol{V} = \boldsymbol{v} - H\boldsymbol{r}$, where $\boldsymbol{v} = \mathrm{d}\boldsymbol{r}/\mathrm{d}t$, and t is the cosmological time. Let us now change the time coordinate to conformal time τ, so that $\mathrm{d}\tau = \mathrm{d}t/a(t)$, where a is the cosmic scale factor. This makes the handling of the comoving equations of motion rather simpler. We also use a comoving distance coordinate $\boldsymbol{x} = \boldsymbol{r}/a$. The equations of motion (18.1.1) are expressed in terms of proper distances $\boldsymbol{r}$ and proper time t; the comoving equations, expressed in conformal time τ and with derivatives now with respect to comoving coordinates, are

$$\frac{\partial \delta}{\partial \tau} + \nabla \cdot [(1+\delta)\boldsymbol{V}] = 0, \tag{18.1.2 a}$$

$$\frac{\partial \boldsymbol{V}}{\partial \tau} + (\boldsymbol{V} \cdot \nabla)\boldsymbol{V} + \frac{\dot{a}}{a}\boldsymbol{V} + \frac{\nabla p}{\rho} + \nabla \varphi = 0, \tag{18.1.2 b}$$

$$\nabla^2 \varphi - 4\pi G \rho a^2 \delta = 0, \tag{18.1.2 c}$$

where δ, $\boldsymbol{V}$ and φ are the density, velocity and gravitational potential perturbations (in the latter case, within this comoving description, the mean value of φ vanishes so φ coincides with $\delta\varphi$). The most important difference between the two sets of Equations (18.1.1) and (18.1.2) is that, in the Euler Equation (18.1.2 *b*), there is a term in $\dot{a}/a$ (remember that $\dot{a} = \mathrm{d}a/\mathrm{d}\tau$) which is due to the fact that our new system of coordinates is following the expansion and is therefore non-inertial. This term, called the 'Hubble drag', causes velocities to decay in comoving coordinates. There is, however, nothing strange about this: it is merely a consequence of the choice of coordinate system.

We have shown how to solve the equations of motion to obtain the behaviour of δ for various types of perturbations in Chapter 10. We shall now concentrate upon longitudinal adiabatic fluctuations (remember that transverse, or vortical, modes are generally decaying with time), and shall ignore the pressure gradient terms in the Euler Equation (18.1.2 *b*) because we assume $k \ll k_{\mathrm{J}}$. We showed in Section 10.8 that the linear solution to the density perturbation in such a situation behaves as a complicated function of the time and the value of Ω. We shall ignore the decaying mode, so that $\delta(\boldsymbol{x}) = D(\tau)\delta_+(\boldsymbol{x})$, and D is the linear growth law for the growing mode which, for $\Omega = 1$ and matter domination, is given by $D \propto a \propto \tau^2$. For $\Omega \neq 1$ the expression for D is complicated but we do not actually need it. In fact, we only need the expression for

$$f(\tau) = \frac{\mathrm{d}\log D}{\mathrm{d}\log a} = \frac{a\dot{D}}{\dot{a}D}, \tag{18.1.3}$$

which has a behaviour as a function of Ω given quite accurately by the approximate form $f \simeq \Omega^{0.6}$. Notice that $f = 1$ for $\Omega = 1$ is exact.

Now, given a solution for the density perturbation δ, one can easily derive the velocity and gravitational potential fields in these coordinates. Because the linear velocity field is irrotational, $\boldsymbol{V}$ can be expressed as the gradient of some velocity potential, Φ_V, i.e.

$$\boldsymbol{V} = -\frac{\nabla \Phi_V}{a}. \tag{18.1.4}$$

It is helpful now to introduce the peculiar gravitational acceleration, $\boldsymbol{g}$, which is simply

$$\boldsymbol{g} = -\frac{\nabla \varphi}{a}. \tag{18.1.5}$$

From the Poisson equation we have

$$\nabla^2 \varphi = \tfrac{3}{2}\Omega H^2 a^2 \delta, \qquad (18.1.6)$$

and, from the linearised equations of motion, it is then quite straightforward to show that

$$\nabla^2 \Phi_V = H f a^2 \delta. \qquad (18.1.7)$$

It therefore follows that $\varphi \propto \Phi_V$,

$$\varphi = \frac{3\Omega H}{2f}\Phi_V, \qquad (18.1.8)$$

so that $\boldsymbol{V} \propto \boldsymbol{g}$:

$$\boldsymbol{V} = \frac{2f}{3\Omega H}\boldsymbol{g}. \qquad (18.1.9)$$

Notice that, for an Einstein–de Sitter universe, this last relation simply becomes $\boldsymbol{V} = \boldsymbol{g}t$. It is also the case that, in this model, φ is constant for the growing mode of linear theory. Regardless of Ω the velocity and gravitational acceleration fields are always in the same direction in linear theory.

It is also helpful to write explicitly the relationship between $\boldsymbol{g}$ (or $\boldsymbol{V}$) and the density perturbation field $\delta(\boldsymbol{x})$ by inverting the relevant version of Poisson's equation:

$$\boldsymbol{V}(\boldsymbol{x}) = aH\frac{f(\Omega)}{4\pi}\int \frac{\delta(\boldsymbol{x}')(\boldsymbol{x}-\boldsymbol{x}')}{|\boldsymbol{x}-\boldsymbol{x}'|^3}\,\mathrm{d}^3\boldsymbol{x}', \qquad (18.1.10)$$

which we anticipated in Section 17.3. The expression for $\boldsymbol{g}$ can be found from (18.1.10) with the aid of (18.1.9).

Suppose now that the density field $\delta(\boldsymbol{x})$ has a known (or assumed) power spectrum $P(k)$. From Equation (18.1.6) it follows immediately that the power spectrum of the field φ can be written

$$P_\varphi(k) = (\tfrac{3}{2}\Omega H^2 a^2)^2 P(k) k^{-4}, \qquad (18.1.11)$$

which we anticipated in Section 13.4. In linear theory the velocity field may be obtained as either the derivative of Φ_V from (18.1.7) or by noting that, from the continuity equation,

$$\delta(\boldsymbol{x}) = -\frac{\nabla\cdot\boldsymbol{V}}{aHf}; \qquad (18.1.12)$$

either way, one finds the velocity power spectrum

$$P_V(k) = (aHf)^2 P(k) k^{-2}. \qquad (18.1.13)$$

Of course, $\boldsymbol{V}$ is a vector field, whereas both δ and φ are scalar fields. The velocity power spectrum (18.1.13) must therefore be interpreted as the power spectrum of the three components of $\boldsymbol{V}$, each of which is a scalar function of position.

We should stress here that knowledge of $P(k)$ is sufficient to specify all the statistical properties of δ, $\boldsymbol{V}$ and φ only if δ is a Gaussian random field, which is the case we shall assume here.

18.2 Velocity Correlations

In the previous section we showed how the gravitational potential and, more importantly, velocity fields are expected to behave in the gravitational instability picture. As we did in Chapter 14 with the density field, it is now necessary to explain how one might try to characterise the properties of $\boldsymbol{V}$ in a statistical manner. We shall concentrate upon generalising the covariance functions of δ we described in Section 14.9 to the case of a vector field $\boldsymbol{V}$ (Gorski 1988; Gorski *et al.* 1989).

The simplest possible statistical characterisation of $\boldsymbol{V}$ is the *scalar velocity covariance function*, defined by

$$\xi_V(r) = \langle \boldsymbol{V}(\boldsymbol{x}_1) \cdot \boldsymbol{V}(\boldsymbol{x}_2) \rangle, \tag{18.2.1}$$

where $r = |\boldsymbol{x}_1 - \boldsymbol{x}_2|$. One can show (we omit the details here) that this function can be expressed as

$$\xi_V(r) = \frac{(H_0 f)^2}{2\pi^2} \int_0^\infty P(k) j_0(kr)\,\mathrm{d}k, \tag{18.2.2}$$

where $j_0(x) = (\sin x)/x$ is the spherical Bessel function of order zero.

This is probably the simplest statistical characterisation of the velocity field but it does not contain information about directional correlations of the different components of $\boldsymbol{V}$. Since velocity information is generally available only in one direction (the radial direction), the scalar correlation function (18.2.1) is of limited usefulness.

To furnish a full statistical description of the field we must define a *velocity covariance tensor*

$$\Psi^{ij}(\boldsymbol{x}_1, \boldsymbol{x}_2) \equiv \langle V^i(\boldsymbol{x}_1) V^j(\boldsymbol{x}_2) \rangle. \tag{18.2.3}$$

Using the assumption of statistical homogeneity and isotropy, we can decompose the tensor Ψ into transverse and longitudinal parts in terms of scalar functions $\Psi_\perp$ and $\Psi_\parallel$,

$$\Psi^{ij}(\boldsymbol{x}_1, \boldsymbol{x}_2) = \Psi_\parallel(r) n^i n^j + \Psi_\perp(r)(\delta^{ij} - n^i n^j), \tag{18.2.4}$$

which are functions only of r;

$$\boldsymbol{n} = (\boldsymbol{x}_1 - \boldsymbol{x}_2)/r. \tag{18.2.5}$$

If $\boldsymbol{u}$ is any unit vector satisfying $\boldsymbol{u} \cdot \boldsymbol{n} = 0$, then one can show that

$$\Psi_\parallel(r) = \langle (\boldsymbol{n} \cdot \boldsymbol{V}_1)(\boldsymbol{n} \cdot \boldsymbol{V}_2) \rangle \tag{18.2.6}$$

and

$$\Psi_\perp(r) = \langle (\boldsymbol{u} \cdot \boldsymbol{V}_1)(\boldsymbol{u} \cdot \boldsymbol{V}_2) \rangle. \tag{18.2.7}$$

In the linear regime $\nabla \times \boldsymbol{V} = \boldsymbol{0}$ and there is a consequent relationship between the longitudinal and transverse functions:

$$\Psi_{\parallel}(r) = \frac{\mathrm{d}}{\mathrm{d}r}[r\Psi_{\perp}(r)]. \tag{18.2.8}$$

One can express the two functions $\Psi_{\parallel,\perp}$ defined in Equations (18.2.6) and (18.2.7) in terms of the power spectrum $P(k)$:

$$\Psi_{\parallel,\perp}(r) = \frac{H^2 f^2}{2\pi^2} \int_0^\infty P(k) K_{\parallel,\perp}(kr)\, \mathrm{d}k, \tag{18.2.9}$$

where

$$K_{\parallel}(x) = j_0(x) - 2\frac{j_1(x)}{x}, \qquad K_{\perp}(x) = \frac{j_1(x)}{x}; \tag{18.2.10}$$

$j_1(x)$ is the spherical Bessel function of order unity,

$$j_1(x) = \frac{\sin x}{x^2} - \frac{\cos x}{x}. \tag{18.2.11}$$

The *total velocity covariance function*, ξ_V, defined by (18.2.2) is

$$\xi_V(r) = \Psi_{\parallel}(r) + 2\Psi_{\perp}(r). \tag{18.2.12}$$

One can also extend this description to quantities involving the shear of the velocity field, but we shall not discuss these here.

In principle one can test a number of assumptions about the velocity field $\boldsymbol{V}$ by estimating the radial and transverse functions from a sample of peculiar velocities. For example, one can compute the expected form of the radial and transverse functions and then compare the results with estimates obtained from the data. There are, however, a number of problems with doing this kind of thing in practice. First, one needs a rather large sample of galaxy-peculiar motions. As we mentioned in Section 4.6, such a sample is difficult to obtain because it requires the independent determination of both redshifts and distances for a large number of galaxies. Moreover, such a sample would in any case only contain information about the radial component of the galaxy-peculiar motion. One can get around this in principle (see Section 18.5), but it does make it difficult to extract information about the $\Psi(r)$ directly from the data. Results from this type of analysis are presently inconclusive, though they may become more useful when the quantity and quality of the data improve.

There is also a deeper problem. Generally one has estimates of the peculiar velocities of galaxies at a set of discrete points (galaxy positions) in space. When dealing with the density field, the assumption that 'galaxies trace the mass' allows one to construct a discrete set of correlation functions which are simply related to the covariance functions of the underlying density field. For the velocity field the situation is not so simple. If one has a continuous velocity field which is sampled at random positions, $\boldsymbol{x}_i$ in Equation (18.2.3), then the two points may be at any position in space (overdense or underdense). Galaxies, however, represent regions of high matter density, so a galaxy sample does not probe all the available density distribution. Any correlations between density and velocity will therefore result in a biased estimate of the velocity field. One can, in principle, construct a continuous velocity field by smoothing over discrete data, but the results depend on exactly how this smoothing is done in a rather subtle way. One therefore has to take care to compare like with like when relating theoretical models of $\boldsymbol{V}$ to quantities extracted from a sample.

18.3 Bulk Flows

A somewhat simpler way to use the peculiar velocity field is to measure *bulk flows* (sometimes called *streaming motions*), which represent the net motion of a large region, usually a sphere centred on the observer, in some direction relative to the pure Hubble expansion. For example, Bertschinger *et al.* (1990) found that a sphere of radius $40h^{-1}$ Mpc is executing a bulk flow of some 388 ± 67 km s^{-1} relative to the cosmological rest frame; a larger sphere of radius $60h^{-1}$ Mpc is moving at 327 ± 84 km s^{-1}. How can one relate this type of measurement to theory?

Recall from Chapter 13 that one can smooth the density perturbation field to define a mass variance in the manner of Equation (13.3.8) or (13.3.12). If the density field is Gaussian, then so will be each component of $\boldsymbol{V}$. The magnitude of the averaged velocity,

$$V = (V_x^2 + V_y^2 + V_z^2)^{1/2}, \tag{18.3.1}$$

will therefore possess a Maxwellian distribution:

$$P(V)\,\mathrm{d}V = \sqrt{\frac{54}{\pi}}\left(\frac{V}{\sigma_V}\right)^2 \exp\left[-\frac{3}{2}\left(\frac{V}{\sigma_V}\right)^2\right]\frac{\mathrm{d}V}{\sigma_V}. \tag{18.3.2}$$

In these equations $\boldsymbol{V}$ represents the filtered velocity field, i.e.

$$\boldsymbol{V} = \boldsymbol{V}(\boldsymbol{x}; R) = \frac{1}{(2\pi)^3}\int \tilde{\boldsymbol{V}}(\boldsymbol{k}) W_V(\boldsymbol{k}; R)\exp(-\mathrm{i}\boldsymbol{k}\cdot\boldsymbol{x})\,\mathrm{d}\boldsymbol{k}, \tag{18.3.3}$$

where $W_V(\boldsymbol{k}; R)$ is a suitable window function with a characteristic scale R; $\tilde{\boldsymbol{V}}(\boldsymbol{k})$ is the Fourier transform on the unsmoothed velocity field $\boldsymbol{V}(\boldsymbol{x}; 0)$. From Equation (18.1.13) we find that

$$\sigma_V^2(R) = \frac{(H_0 f)^2}{2\pi^2}\int_0^\infty P(k) W_V^2(kR)\,\mathrm{d}k, \tag{18.3.4}$$

by analogy with equation (13.3.12). In Equation (18.3.4), σ_V is the RMS value of $V(\boldsymbol{x}; R)$, where the mean is taken over all spatial positions $\boldsymbol{x}$. Clearly the global mean value of $\boldsymbol{V}(\boldsymbol{x}, R)$ must be zero in a homogeneous and isotropic universe. It is a consequence of Equation (18.3.2) that there is a 90% probability of finding a measured velocity satisfying the constraint:

$$\tfrac{1}{3}\sigma_V \leqslant V \leqslant 1.6\sigma_V. \tag{18.3.5}$$

The window function W_V must be chosen to model the way the sample is constructed. This is not completely straightforward because the observational selection criteria are not always well controlled and the results are quite sensitive to the shape of the window function. Top hat (13.3.14) and Gaussian (13.3.15) are the usual choices in this case, as for the density field.

Because the integral in Equation (18.3.4) is weighted towards lower k than the definition of σ_M^2 given by Equation (13.3.8), which has an extra factor of k^2, bulk flows are potentially useful for probing the linear regime of $P(k)$ beyond what can be reached using properties of the spatial clustering of galaxies. The problem is that one typically has one measurement of the bulk flow on a scale R and this does not provide a strong constraint on σ_V or $P(k)$, as is obvious from Equation (18.3.5): if a theory predicts an RMS bulk flow of 300 km s^{-1} on some scale, then a randomly selected sphere on that scale can have a velocity between 100 and 480 km s^{-1} with 90% probability, an allowed error range of a factor of almost five. Until much more data become available, therefore, such measurements can only be used as a consistency check on models and do not strongly discriminate between them. Velocities can, however, place constraints on the possible existence of bias since σ_V is simply proportional to b (in the linear bias model). For example, the standard CDM model predicts a bulk flow on the scale of $40h^{-1}$ Mpc of around 180 km s^{-1} if $b = 1$. This reduces to 72 km s^{-1} if $b = 2.5$, which was, at one time, the favoured value. The observation of a velocity of 388 km s^{-1} on this scale is clearly incompatible with SCDM with this level of bias; it is, however, compatible with a $b = 1$ CDM model.

It is also pertinent to mention that the factor f in Equation (18.3.4) means that high values of V tend to favour higher values of f and therefore higher values of Ω, remembering that $f \simeq \Omega^{0.6}$. We return to this in Section 18.6.

There is an interesting way to combine large-scale bulk flow information with small-scale velocity data. Let us consider the unsmoothed velocity field $\boldsymbol{V}(\boldsymbol{x};0)$. In fact, some smoothing of the velocity field is always necessary because of the sparseness of the velocity field data, but we can assume that this scale, R_S, is so much less than R that its value is effectively zero. Consider the quantity

$$\Sigma_V^2(\boldsymbol{x}_0;R) \equiv \langle|\boldsymbol{V}(\boldsymbol{x};0) - \boldsymbol{V}(\boldsymbol{x}_0;R)|\rangle^2, \tag{18.3.6}$$

where the average is taken over a single smoothing window centred at $\boldsymbol{x}_0$. Clearly this represents the variance of the unsmoothed velocity field calculated with respect to the mean value of the velocity in the window, $\boldsymbol{V}(\boldsymbol{x}_0;R)$. The ratio

$$\mathcal{M}^2(\boldsymbol{x}_0;R) = \frac{|\boldsymbol{V}(\boldsymbol{x}_0;R)|^2}{\Sigma_V^2(\boldsymbol{x}_0;R)} \tag{18.3.7}$$

measures, in some sense, the 'temperature' of the velocity field on a scale R. If $\mathcal{M}^2 > 1$, then the systematic bulk flow in the smoothing volume exceeds the random motions. If, on the other hand, $\mathcal{M}^2 < 1$, these small-scale random 'thermal' motions are larger than the systematic flow. It is appropriate therefore to regard the spatial average of the quantity $\mathcal{M}^2$,

$$\mathcal{M}^2(R) = \langle\mathcal{M}^2(\boldsymbol{x}_0;R)\rangle_{\boldsymbol{x}_0}, \tag{18.3.8}$$

as defining a kind of *cosmic Mach number* as a function of scale, $\mathcal{M}(R)$ (Ostriker and Suto 1990). In fact, the usual definition of the cosmic Mach number is slightly different from that given in Equation (18.3.8) and is more straightforward to calculate:

$$\mathcal{M}^2(R) = \frac{\sigma_V^2(R)}{\Sigma_V^2(R)}, \tag{18.3.9}$$

where $\Sigma_V^2(R)$ is the spatial average of $\Sigma_V^2(\boldsymbol{x}_0; R)$ taken over all positions $\boldsymbol{x}_0$, by analogy with Equation (18.3.8).

The cosmic Mach number has the advantage that it probes the shape of the primordial power spectrum in a much more sensitive manner than the bulk flow statistics. Its main disadvantage is that $\mathcal{M}^2$ is defined in terms of the ratio of two quantities which are both subject to substantial observational uncertainties. Until the available peculiar velocity data improve, this statistic is therefore unlikely to provide a powerful test of structure-formation theories.

18.4 Velocity–Density Reconstruction

A more sophisticated approach to the use of velocity information is provided by a relatively new and extremely ingenious approach developed primarily by Bertschinger *et al.* (1990) which is now known as POTENT; see also Dekel *et al.* (1993). This makes use of the fact that in the linear theory of gravitational instability the velocity field is curl-free and can therefore be expressed as the gradient of a potential. We saw in Section 18.1, Equation (18.1.8), that this velocity potential turns out to be simply proportional to the linear theory value of the gravitational potential. Because the velocity field is the gradient of a potential Φ_V, one can use the purely radial motions, V_r, revealed by redshift and distance information to map Φ_V in three dimensions:

$$\Delta\Phi_V(r, \theta, \phi) = -\int_0^r V_r(r', \theta, \phi)\, dr'. \tag{18.4.1}$$

It is not required that paths of integration be radial, but they are in practice easier to deal with.

Once the potential has been mapped, one can solve for the density field using the Poisson equation in the form (18.1.7). This means therefore that one can compare the density field as reconstructed from the velocities with the density field measured directly from the counts of galaxies. This, in principle, enables one to determine directly the level of bias present in the data. The only other parameter involved in the relation between V and δ is then f, which, in turn, is a simple function of Ω. POTENT holds out the prospect, therefore, of supplying a measurement of Ω which is independent of b, unlike that discussed in Section 17.3 for example. We return to the estimation of Ω from velocity data in Section 18.6.

At this point, however, it is worth mentioning some of the possible problems with the POTENT analysis. As always, one is of course limited by the quality and

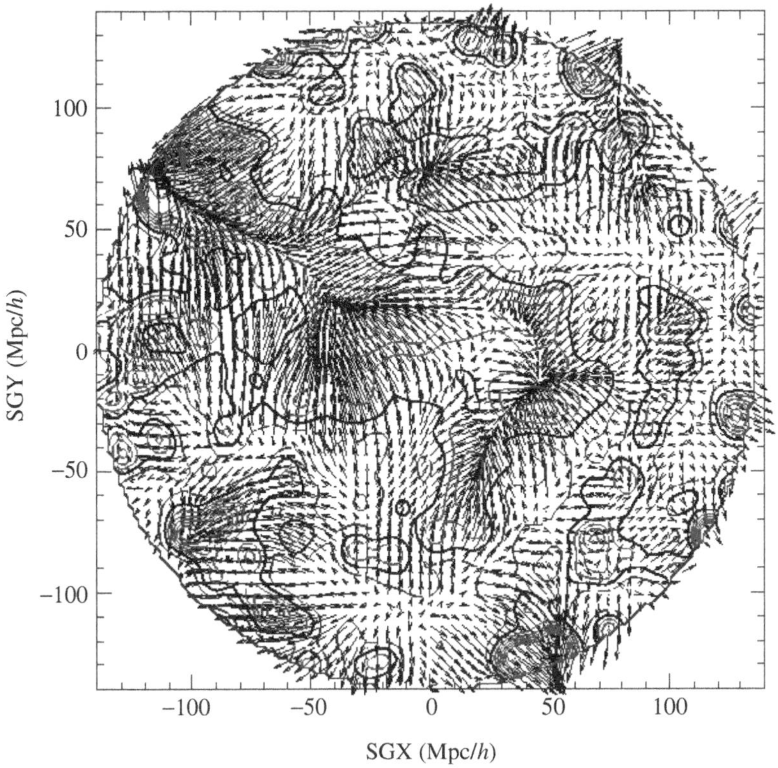

Figure 18.1 Example of a velocity-density reconstruction using the PSCz catalogue, showing the fluctuations of velocity and density in the Supergalactic plane. The vectors are projections of the three-dimensional velocity field and contours show lines of equal δ. Picture courtesy of Enzo Branchini.

quantity of the velocity data available. The distance errors, together with the relative sparseness of the data sets available, combine to produce a velocity field $\boldsymbol{V}$ which is quite noisy. This necessitates a considerable amount of smoothing, which is also needed to suppress small-scale nonlinear contributions to the velocity field. The smoothed field is then interpolated to produce a continuous field defined on a grid. The favoured smoothing is of the form

$$V_{\mathrm{r}}(\boldsymbol{r}) = \sum_i W_i(\boldsymbol{r}) V_{\mathrm{r},i}, \tag{18.4.2}$$

where i labels the individual objects whose radial velocities, $V_{\mathrm{r},i}$, have been estimated and the weighting function $W_i(\boldsymbol{r})$ is taken to be

$$W_i(\boldsymbol{r}) \propto n_i^{-1} \sigma_i^{-2} \exp\left(-\frac{|\boldsymbol{r} - \boldsymbol{r}_i|^2}{2R_{\mathrm{S}}^2}\right); \tag{18.4.3}$$

n_i is the local number density of objects, σ_i is the estimated standard error of the distance to the ith object, and R_{S} is a Gaussian smoothing radius, typically of order $12h^{-1}$ Mpc. If one uses clusters instead of individual galaxies, then σ_i can be reduced by a factor equal to the square root of the number of objects in the cluster, assuming the errors are random. One effect of the heavy smoothing is that the volume probed by these studies consequently contains only a few independent smoothing volumes and the statistical significance of any reconstruction is bound to be poor.

Notice that the potential field one recovers then has to be differentiated to produce the density field which will again exaggerate the level of noise. (It is possible to improve on the linear solution to the Poisson equation by using the Zel'dovich approximation (Section 14.2) to calculate the density perturbation δ from the velocity potential.) The scale of the noise problem can be gauged from the fact that a 20% distance error is of the same order as the typical peculiar velocity for distances beyond $30h^{-1}$ Mpc.

Apart from the problem of noise, there are also other sources of uncertainty in the applicability of this method. In any redshift survey one has to be careful to control selection biases, such as the Malmquist bias (Section 4.2), which can enter in a complicated and inhomogeneous way into this analysis. One also needs to believe that the distance indicators used are accurate. Most workers in this field claim that their distance indicators are accurate to, say, 10–20%. However, if the errors are not completely random, i.e. there is a systematic component which actually depends on the local density, then the results of this type of analysis can be seriously affected. In this case the systematic error in $\boldsymbol{V}$ correlates with density in a similar way to that expected if the velocities were generated dynamically from density fluctuations. There are some suggestions that there is indeed such a systematic error in the commonly used D_n–σ indicator for elliptical galaxies (Guzman and Lucey 1993). What may happen is that old stellar populations produce a different response in the distance indicator compared with young ones. Since older galaxies formed earlier and in higher-density environments, the upshot is exactly the sort of systematic effect that is so dangerous to methods like POTENT. Applying a corrected distance indicator to a sample of elliptical galaxies essentially eliminates all the observed peculiar motions, which means that the motions derived using the uncorrected indicator were completely spurious. Whether this type of error is sufficiently widespread to affect all peculiar motion studies is unclear but it suggests one should regard these results with some scepticism.

18.5 Redshift-Space Distortions

The methods we have discussed in Sections 18.2–18.4 of course require one to know peculiar motions for a sample of galaxies. There is an alternative approach, which does not need such information, and which may consequently be more reliable. This relies on the fact that peculiar motions affect radial distances and not tangential ones. The distribution of galaxies in 'redshift space' is therefore a distorted representation of their distribution in real space. For example, dense clusters appear elongated along the line of sight because of the large radial-velocity component of the peculiar velocities, an effect known as the 'fingers of God'. Similarly, the correlation functions and power spectra of galaxies should be expected to show a characteristic distortion when they are viewed in redshift space rather than in real space. This is the case even if the real-space distribution of matter is statistically homogeneous and isotropic.

Let us first consider the effect of these distortions upon the two-point correlation function of galaxies. The conventional way to describe this phenomenon is to define coordinates as follows. Consider a pair of galaxies with measured redshifts corresponding to velocities $\boldsymbol{v}_1$ and $\boldsymbol{v}_2$. The separation in redshift space is then just

$$\boldsymbol{s} = \boldsymbol{v}_1 - \boldsymbol{v}_2; \tag{18.5.1}$$

an observer's line of sight is defined by

$$\boldsymbol{l} = \tfrac{1}{2}(\boldsymbol{v}_1 + \boldsymbol{v}_2), \tag{18.5.2}$$

and the separations parallel and perpendicular to this direction are then just

$$\pi = \frac{\boldsymbol{s} \cdot \boldsymbol{l}}{|\boldsymbol{l}|} \tag{18.5.3 a}$$

and

$$r_{\mathrm{p}} = \sqrt{\boldsymbol{s} \cdot \boldsymbol{s} - \pi^2}, \tag{18.5.3 b}$$

respectively. Generalising the estimator for $\xi(r)$ given in Equation (16.4.7 *b*) allows one to estimate the function $\xi(r_{\mathrm{p}}, \pi)$:

$$\xi(r_{\mathrm{p}}, \pi) = \frac{n_{\mathrm{DD}}(r_{\mathrm{p}}, \pi) n_{\mathrm{RR}}(r_{\mathrm{p}}, \pi)}{n_{\mathrm{DR}}^2(r_{\mathrm{p}}, \pi)} - 1. \tag{18.5.4}$$

When the correlation function is plotted in the π–r_{p} plane, redshift distortions produce two effects: a stretching of the contours of ξ along the π-axis on small scales (less than a few Mpc) due to nonlinear pairwise velocities, and compression along the π-axis on larger scale due to bulk (linear) motions.

Linear theory cannot be used to calculate the first of these contributions, so one has to use explicitly nonlinear methods. The usual approach is to use the equation

$$\frac{\partial \xi}{\partial t} = \frac{1}{ax^2} \frac{\partial}{\partial x} [x^2 (1 + \xi) v_{12}], \tag{18.5.5}$$

which expresses the conservation of particle pairs; x is a comoving coordinate and $v_{12} = |\boldsymbol{s}|$. The Equation (18.5.5) is actually the first of an infinite set of equations known as the BBGKY hierarchy (Davis and Peebles 1977). To close the hierarchy one needs to make an assumption about higher moments. Assuming that the three-point correlation function has the hierarchical form (16.5.1) and that the real-space two-point correlation function is of the power-law form (16.4.5) leads to the so-called *cosmic virial theorem*:

$$\langle v_{12}^2(r) \rangle \simeq C_\gamma H_0^2 Q \Omega r_{0\mathrm{g}}^{\gamma} r^{2-\gamma}, \tag{18.5.6}$$

where $C_\gamma \simeq 23.8$ if $\gamma = 1.8$. Assuming that the radial anisotropy in $\xi(r_p, \pi)$ is due to the velocities v_{12}, then one can, in principle, determine an estimate of Ω_0 from the small-scale anisotropy. Notice, however, that there is an implicit assumption that the galaxy correlation function and the mass covariance function are identical, so this estimate will depend upon b in a non-trivial way.

On larger scales, the effect of redshift-space distortions is in the opposite sense. One can understand this easily by realising that a large-scale overdensity will tend to be collapsing in real space. Matter will therefore be moving towards a cluster, thus flattening structures in the redshift direction. This both enhances the appearance of walls and filaments and changes their orientation, producing a series of ring-like structures around the observer called the 'bull's-eye effect' (Melott *et al.* 1998).

The effect of these distortions upon the correlation function is actually quite complicated and depends upon the direction cosine μ between the line of sight $\boldsymbol{l}$ and the separation $\boldsymbol{s}$. One can show, however, that the angle-averaged redshift-space correlation function is given by the simple form

$$\bar{\xi}(s) = (1 + \tfrac{2}{3}f + \tfrac{1}{5}f^2)\xi_r(s), \tag{18.5.7}$$

where ξ_r is the real-space correlation function (Kaiser 1987; Hamilton 1992). More instructively one can decompose $\xi(r_p, \pi)$ into spherical harmonics using

$$\xi_l(r) = \frac{2l+1}{2}\int_{-1}^{+1} \xi(r\sin\theta, r\cos\theta) P_l(\cos\theta)\,\mathrm{d}\cos\theta. \tag{18.5.8}$$

A robust diagnostic of the presence of redshift distortions is via the quadrupole-to-monopole ratio:

$$\frac{\xi_2}{\xi_0} = \frac{3+n}{n}\,\frac{\frac{4}{3}f + \frac{4}{7}f^2}{1 + \frac{2}{3}f + \frac{1}{5}f^2}. \tag{18.5.9}$$

In principle, these ideas permit one to estimate Ω (through the f dependence), but this again requires that ξ_r for the matter should be known accurately. Fortunately, with the arrival of redshift surveys like the 2dF GRS such measurements can now be made with confidence (Peacock *et al.* 2001).

Another way to use redshift-space distortions in the linear regime is to study their effect on the power spectrum, where the directional dependence is easier to calculate. In fact, one can show quite easily that

$$P_s(\boldsymbol{k}) = P_r(\boldsymbol{k})[1 + f\mu^2], \tag{18.5.10}$$

where P_s and P_r are the redshift space and real space power spectra, respectively (Kaiser 1987). If one can estimate the power spectrum in various directions of $\boldsymbol{k}$, then one can fit the expected μ dependence to obtain an estimate of f and hence Ω. If galaxy formation is biased, then f in Equations (18.5.9) and (18.5.10) is replaced by $\beta = f/b$. Given the paucity of available peculiar velocity data, it seems that this type of analysis is the most promising approach to the use of cosmological velocity information to estimate Ω.

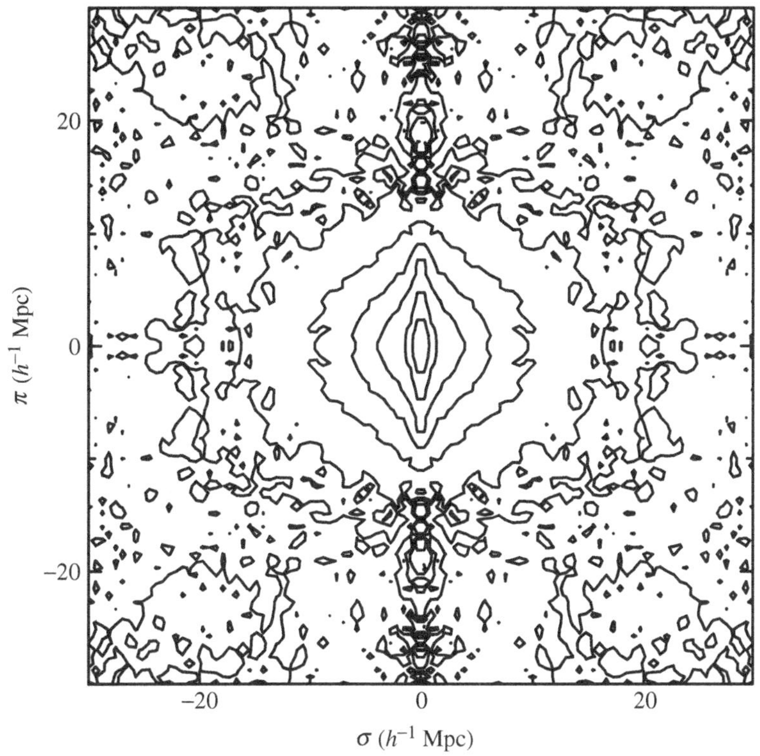

Figure 18.2 The correlation function of galaxies in the 2dF GRS along the line of sight and perpendicular to it. The contours are stretched on small scales along the eye line, but flattened into box shapes on large scales. Picture courtesy of John Peacock.

Other than their possible use in the estimation of the density parameter, the methods we have discussed here are needed to ensure that estimates of $\xi(r)$ or $P(k)$ are not biased by redshift-space distortions. The methods we have discussed here can be used to allow for the velocity-smearing effects and thus yield less biased estimates of these quantities (e.g. Peacock and Dodds 1994).

18.6 Implications for Ω_0

We have already mentioned several times the main problem with relying on a statistical analysis of the spatial distribution of cosmic objects to test theories: the bias. In an extreme case of bias one might imagine galaxies to be just 'painted on' to the background distribution in some arbitrary way having no regard to the distribution of mass. Ideally, one would wish to have some way of studying all the mass, not just that part of it which happens to light up. Since velocities are generated by gravitational instability of all the gravitating material, they provide one way of studying, albeit indirectly, the total distribution of matter. If one uses velocities merely as tracers of the underlying velocity field, it does not matter so much whether they are biased, except if the velocities of galaxies are systematically different from those of randomly selected points.

There are various ways to use the properties of peculiar motions in the estimation of Ω_0. As we have seen, the small-scale anisotropy introduced into statistical measures like the correlation function and power spectrum can be used to estimate the magnitude of the radial component of the typical galaxy-peculiar velocity. The velocities obtained by such methods are around 300 km s^{-1}. One can also use this information to infer the total amount of mass using the statistical mechanics of self-gravitating systems in the form of the cosmic virial theorem (18.5.6). These methods, when applied on small to intermediate scales, consistently yield estimates of Ω_0 in the range 0.1–0.3. These estimates also agree with virial estimates of the masses of rich clusters of galaxies, in which the analysis is considerably simplified if one assumes the clusters are fully relaxed and gravitationally bound systems, as discussed in Chapter 4; as we mentioned there, this value is about an order of magnitude larger than naive estimates of Ω_0 based on the mass-to-light ratios inferred for galaxy interiors. This discrepancy was one of the initial motivations for the introduction of a bias b into the models of galaxy clustering. Typically one compares some statistical measure of the clustering of galaxies with the observed velocity, so what emerges is a constraint on the combination $\beta = f/b \simeq \Omega^{0.6}/b$ if there is a linear bias.

As we have seen in Chapter 17, the COBE detection of microwave background fluctuations casts doubt upon the existence of a bias sufficient to explain the observed peculiar motions if $\Omega = 1$, at least in the context of the CDM model. There is still an escape route for adherents of the critical density. Since direct determinations of Ω from dynamics have been restricted to relatively small volumes which may not be representative of the Universe at large, one can claim that we just live in an underdense part of the Universe. It is probably true that, if one simulates an $\Omega = 1$ CDM model, one will find some places where the local distribution of mass is such as to produce, by the above analyses, a local value of $\Omega \simeq 0.2$ by chance. This does not, however, constitute an argument against the alternative that Ω is actually less than unity.

Recent advances in the accumulation of galaxy redshifts have made it possible to attempt analyses of redshift-space distortions on large scales, which we also discussed in Section 18.5. The recent analysis of the 2dF GRS by Peacock *et al.* (2001) shows that $\beta \simeq 0.4$. If the APM galaxies upon which this survey is based are unbiased, then this means the matter density must be low; redshift distortions are insensitive to the presence of Λ. As we have explained, these measurements probably supply more robust methods for estimating Ω_0 than the relatively local peculiar-motion studies that have always seemed to suggest a high value of $\Omega^{0.6}/b$, consistent with an Einstein–de Sitter universe. In particular, because one can compare the reconstructed density field with the observed galaxy distribution, it is possible, at least in principle, to break the degeneracy between models with a low value of Ω and models having a higher density but a significant bias. This is a relatively new technique for measuring the density parameter, however, and it would be wise to suspend judgement upon it, at least until all possible systematic biases have been investigated. These methods are nevertheless extremely promising and we anticipate that, in the near future, relatively unambiguous determinations of Ω will be forthcoming.

Bibliographic Notes on Chapter 18

Historically interesting reviews of peculiar motions can be found in Rubin and Coyne (1988), Burstein (1990), Bertschinger (1992), Dekel (1994) and Strauss and Willick (1995). A wonderful recent review of linear redshift distortions is given by Hamilton (1998). Other useful references are Vittorio *et al.* (1986), Vittorio and Turner (1987) and Bertschinger and Juszkiewicz (1988).

Problems

1. Derive the cosmic virial theorem (18.5.6).
2. Derive Equations (18.5.7) and (18.5.8).
3. Derive the Kaiser formula (18.5.9).
4. Show that the Zel'dovich displacements in redshift space are a factor $(1 + f)$ larger in the line of sight than at right angles to it. Deduce that caustics form earlier in redshift space than in real space.

19

Gravitational Lensing

In this chapter we shall discuss the cosmological applications of one of the predictions of general relativity. Although it is only recently that the idea of gravitational lensing has found applications in cosmology, the idea that massive bodies could deflect light rays actually furnished the first experimental test of Einstein's theory in 1919. The story of this test has some interesting lessons for modern cosmology so, before going onto the technical applications of gravitational lensing, we begin with a small amount of history.

19.1 Historical Prelude

The idea that gravity might bend light did not originate with Einstein. It had been suggested before, by Isaac Newton for example. In a rhetorical question posed in his Opticks, Newton wrote:

> Do not Bodies act upon Light at a distance, and by their action bend its Rays; and is not this action... strongest at the least distance?

In other words, he was arguing that light rays themselves should feel the force of gravity according to the inverse-square law. As far as we know, however, he never attempted to apply this idea to anything that might be observed. Newton's query was addressed in 1801 by Johann Georg von Soldner. His work was motivated by the desire to know whether the bending of light rays might require certain astronomical observations to be adjusted. He tackled the problem using Newton's corpuscular theory of light, in which light rays consist of a stream of tiny particles. It is clear that if light does behave in this way, then the mass of each particle must be very small. Soldner was able to use Newton's theory of gravity to solve an example of a ballistic scattering problem.

A small particle moving past a large gravitating object feels a force from the object that is directed towards the centre of the large object. If the particle is moving fast, so that the encounter does not last very long, and the mass of the particle is much less than the mass of the scattering body, what happens is that the particle merely receives a sideways kick which slightly alters the direction of its motion. The size of the kick, and the consequent scattering angle, is quite easy to calculate because the situation allows one to ignore the motion of the scatterer. Although the two bodies exert equal and opposite forces on each other, according to Newton's third law, the fact that the scatterer has a much larger mass than the 'scatteree' means that the former's acceleration is very much lower. This kind of scattering effect is exploited by interplanetary probes, which can change course without firing booster rockets by using the gravitational 'slingshot' supplied by the Sun or larger planets. When the deflection is small, the angle of deflection predicted by Newtonian arguments, θ_N, turns out to be

$$\theta_N = \frac{2GM}{rc^2}, \tag{19.1.1}$$

where r is the distance of closest approach between scattering object and scattered body.

Unfortunately, this calculation has a number of problems associated with it. Chief amongst them is the small matter that light does not actually possess mass at all. Although Newton had hit the target with the idea that light consists of a stream of particles, these photons, as they are now called, are known to be massless. Newton's theory simply cannot be applied to massless particles: they feel no gravitational force (because the force depends on their mass) and they have no inertia. What photons do in a Newtonian world is really anyone's guess. Nevertheless, the Soldner result is usually called the Newtonian prediction, for want of a better name.

Unaware of Soldner's calculation, in 1907 Einstein began to think about the possible bending of light. By this stage, he had already formulated the equivalence principle, but it was to be another eight years before the general theory of relativity was completed. He realised that the equivalence principle in itself required light to be bent by gravitating bodies. But he assumed that the effect was too small ever to be observed in practice, so he shelved the calculation. In 1911, still before the general theory was ready, he returned to the problem. What he did in this calculation was essentially to repeat the argument based on Newtonian theory, but incorporating the equation $E = mc^2$. Although photons do not have mass, they certainly have energy, and Einstein's theory says that even pure energy has to behave in some ways like mass. Using this argument, and spurred on by the realisation that the light deflection he was thinking about might after all be measurable, he calculated the bending of light from background stars by the Sun.

For light just grazing the Sun's surface—i.e. with r equal to the radius of the Sun, $R_\odot$, and where M is the mass of the Sun $M_\odot$—Equation (19.1.1) yields a deflection of 0.87 seconds of arc; for reference, the angle in the sky occupied by the Sun is around half a degree. This answer is precisely the same as the Newtonian value obtained more than a century earlier by Soldner. The predicted deflection is tiny, but according to the astronomers Einstein consulted, it could just about be measured. Stars appearing close to the Sun would appear to be in slightly different positions in the sky than they would be when the Sun was in another part of the sky. It was hoped that this kind of observation could be used to test Einstein's theory. The only problem was that the Sun would have to be edited out of the picture, otherwise stars would not be visible close to it at all. In order to get around this problem, the measurement would have to be made at a very special time and place: during a total eclipse of the Sun.

In 1915, with the full general theory of relativity in hand, Einstein returned to the light-bending problem. And he soon realised that in 1911 he had made a mistake. The correct answer was not the same as the Newtonian result, but twice as large. Einstein had neglected to include all effects of curved space in the earlier calculation. The origin of the factor two is quite straightforward when one looks at how a Newtonian gravitational potential distorts the metric of space-time. In flat space (which holds for special relativity), the infinitesimal four-dimensional space-time interval $\mathrm{d}s$ is related to time intervals $\mathrm{d}t$ and distance intervals $\mathrm{d}l$ via

$$\mathrm{d}s^2 = c^2\,\mathrm{d}t^2 - \mathrm{d}l^2; \tag{19.1.2}$$

light rays follow paths in space-time defined by $\mathrm{d}s^2 = 0$, which are straight lines in this case. Of course, the point about the general theory is that light rays are no longer straight. In fact, around a spherical distribution of mass M the metric changes so that, in the weak field limit, it becomes

$$\mathrm{d}s^2 = \left(1 + \frac{2GM}{rc^2}\right)c^2\,\mathrm{d}t^2 - \left(1 - \frac{2GM}{rc^2}\right)\mathrm{d}l^2. \tag{19.1.3}$$

Since the corrections of order GM/rc^2 are small, one can solve the equation $\mathrm{d}s^2 = 0$ by expanding each bracket in a power series.

Einstein's original calculation had included only the first term, which corresponds to the R_{00} part of the field equations. The second doubles the net deflection. Not only does energy gravitate, so does momentum and this appears in the second term in the metric. The angular deflection predicted by Einstein's equations in the Newtonian limit is therefore

$$\theta_{\mathrm{E}} = \frac{4GM}{rc^2}, \tag{19.1.4}$$

which yields 1.74 arcsec for $M = M_{\odot}$ and $r = R_{\odot}$, compared with the 0.87 arcsec obtained using Newtonian theory. Not only is this easier to measure, being larger, but it also offers the possibility of a definitive test of the theory, since it differs from the Newtonian value.

In 1912, an Argentinian expedition had been sent to Brazil to observe a total eclipse. Light-bending measurements were on the agenda, but bad weather prevented them making any observations. In 1914, a German expedition, organised by Erwin Freundlich and funded by Krupp, the arms manufacturer, was sent to the Crimea to observe the eclipse due on 21 August. But when World War I broke out, the party was warned off. Most returned home, but others were detained in Russia. No results were obtained. The war made further European expeditions impossible. One wonders how Einstein would have been treated by history if either of the 1912 or 1914 expeditions had been successful. Until 1915, his reputation was riding on the incorrect value of 0.87 arcsec. As it turned out, the 1919 British expeditions to Sobral and Principe were to prove his later calculation to be right. And the rest, as they say, is history (Dyson *et al.* 1920).

19.2 Basic Gravitational Optics

In general it is a difficult problem to determine the trajectories of light rays in curved space-times. However, in the cosmological setting, we can simplify the task by applying some assumptions. For a start we assume that the global background geometry is well described by the Robertson–Walker metric we introduced in Chapter 1. Next we make use of a Newtonian approximation for the light trajectories, similar to the discussion of the previous section. We assume that a light ray travels unperturbed from a background source until it is very close to the lens, whereupon it is deflected by some angle we shall assume to be small. It then follows an unperturbed trajectory from the lens to the observer. In doing this we are obliged to require that the effective gravitational potential of the lens Φ is such that $|\Phi^2| \ll c^2$ and that the lens is moving with respect to a cosmological frame with a velocity v which is much less than that of light. If these conditions apply, then the deflection produced by the lens is going to be small.

The deflection of a light ray, $\hat{\boldsymbol{\alpha}}$, will in general be given by

$$\hat{\boldsymbol{\alpha}} = \frac{2}{c^2} \int \nabla_{\perp} \Phi \, \mathrm{d}l, \tag{19.2.1}$$

where the gradient of the Newtonian potential is taken perpendicular to the light path and the integral is taken along photon trajectory. With the simplification mentioned above, the gradient can be taken to be perpendicular to the original (unperturbed) light ray rather than the actual (perturbed) one. In this case we only need to consider the impact parameter b of the light ray as it crosses the lens plane. The relevant potential for a point lens can be written

$$\Phi(b, z) = -\frac{GM}{\sqrt{b^2 + z^2}}, \tag{19.2.2}$$

where z is the distance along the ray. For the case (19.2.2) we therefore find

$$\nabla_{\perp} \Phi(b, z) = \frac{GMb}{(b^2 + z^2)^{3/2}} \tag{19.2.3}$$

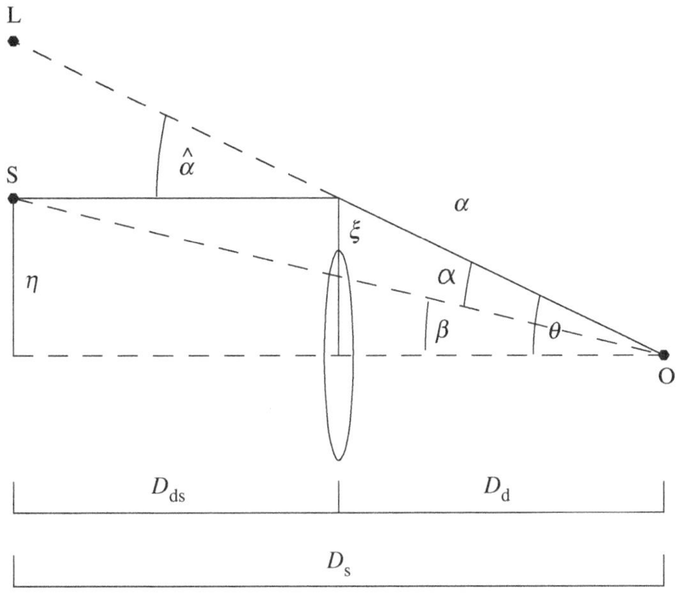

Figure 19.1 Gravitational lensing. A light ray travels from the source S to the observer O passing the lens at an impact parameter ξ. The transverse distance from the optic axis is η. The light ray is deflected through an angle $\hat{\alpha}$; the angular separations of source and image from the optic axis are denoted β and θ, respectively. The angular-diameter distances between observer and source, observer and lens and lens and source are D_s, D_d and D_{ds}, respectively. Picture courtesy of Mathias Bartelmann.

in a direction at right angles to the unperturbed ray. The deflection angle is then

$$\hat{\alpha} = \frac{2}{c^2}\int \nabla_{\perp}\Phi \, dz = \frac{4GM}{c^2 b}. \tag{19.2.4}$$

This is exactly the result we described in Section 19.1.

If we now assume that the deflection occurs as a kind of 'impulse' delivered by the lens within a distance $\pm b$ along the original light ray, then we can simplify matters even further. This approximation corresponds to the assumption that the lens is infinitely thin compared with the distances from source to lens and from observer to lens. One then considers the lens to be a mass sheet lying in a plane usually called the *lens plane*. The relevant property of the sheet is its surface mass density, Σ, where

$$\Sigma(\boldsymbol{\xi}) = \int \rho(\boldsymbol{\xi}, z) \, dz, \tag{19.2.5}$$

in which the integral is taken over the photon path as before. It is then straightforward to show that the net deflection (now written as a vector to show its direction in the lens plane) is given by

$$\hat{\boldsymbol{\alpha}}(\boldsymbol{\xi}) = \frac{4G}{c^2}\int \frac{(\boldsymbol{\xi} - \boldsymbol{\xi}')\Sigma(\boldsymbol{\xi}')}{|\boldsymbol{\xi} - \boldsymbol{\xi}'|^2} \, d^2\boldsymbol{\xi}'. \tag{19.2.6}$$

If the distribution of mass in the lens plane is circularly symmetric, then the deflection angle points towards the centre of symmetry and has modulus

$$\hat{\alpha}(\xi) = \frac{4GM(\xi)}{c^2\xi}, \tag{19.2.7}$$

where ξ is the distance from the centre of the lens and $M(\xi)$ is obviously the mass enclosed within a radius ξ so defined:

$$M(\xi) = 2\pi \int_0^{\xi} \Sigma(\xi')\xi' \, \mathrm{d}\xi'. \tag{19.2.8}$$

We can now put this altogether to look at the geometry of a general lensing system as shown in Figure 19.1. The figure introduces the reduced deflection angle α, which is related to $\hat{\alpha}$ via

$$\boldsymbol{\alpha} = \frac{D_{\mathrm{ds}}}{D_{\mathrm{s}}}\hat{\boldsymbol{\alpha}}. \tag{19.2.9}$$

From the diagram, assuming small angles everywhere, we get

$$\theta D_{\mathrm{s}} = \beta D_{\mathrm{s}} - \hat{\alpha} D_{\mathrm{ds}}, \tag{19.2.10}$$

so that

$$\boldsymbol{\beta} = \boldsymbol{\theta} - \boldsymbol{\alpha}(\boldsymbol{\theta}). \tag{19.2.11}$$

This is called the lens equation; it relates the angular position of images and sources. Note that angular-diameter distances must be used in this and the following.

As an example let us look at a case with constant surface mass density Σ in the lens plane. From Equation (19.2.7) we obtain

$$\alpha(\theta) = \frac{D_{\mathrm{ds}}}{D_{\mathrm{s}}} \times \frac{4G}{c^2\xi} \times \Sigma\pi\xi^2 = \frac{4\pi G\Sigma}{c^2}\frac{D_{\mathrm{d}}}{D_{\mathrm{ds}}}D_{\mathrm{s}}\theta, \tag{19.2.12}$$

where $\xi = D_{\mathrm{d}}\theta$. In this case we can define a critical surface mass density

$$\Sigma_* = \frac{c^2}{4\pi G}\frac{D_{\mathrm{s}}}{D_{\mathrm{d}}D_{\mathrm{ds}}}, \tag{19.2.13}$$

where D is defined by

$$D = \frac{D_{\mathrm{d}}D_{\mathrm{s}}}{D_{\mathrm{ds}}}. \tag{19.2.14}$$

The interpretation of the critical density Σ_* is that the deflection angle $\alpha(\theta) = \theta$ so that $\beta = 0$ for any θ. This is a perfect lens which brings all light rays to focus at a well-defined focal length. Real gravitational lenses are not perfect, but nevertheless display interesting optical properties. Lenses which have $\Sigma > \Sigma_*$ typically produce multiple images of a background source.

Now let us generalise to the case of a circular lens with an arbitrary mass profile. The lens Equation (19.2.10) then becomes

$$\beta = \theta - \frac{D_{\mathrm{ds}}}{D_{\mathrm{d}}D_{\mathrm{s}}}\frac{4GM(\theta)}{c^2\theta}. \tag{19.2.15}$$

If the mass density is sufficient, then a source with $\beta = 0$, i.e. one that lies on the optic axis, is lensed into a ring with radius θ_{E}, where

$$\theta_{\mathrm{E}}^2 = \frac{4GM(\theta_{\mathrm{E}})}{Dc^2}. \tag{19.2.16}$$

This is called the Einstein radius.

For a point mass we obtain

$$\theta_{\mathrm{E}} = \left(\frac{4GM}{Dc^2}\right)^{1/2}. \tag{19.2.17}$$

We can use this to rewrite the lens equation in this case as

$$\beta = \theta - \frac{\theta_{\mathrm{E}}^2}{\theta}, \tag{19.2.18}$$

which has two solutions:

$$\theta_{\pm} = \tfrac{1}{2}(\beta \pm \sqrt{\beta^2 + 4\theta_{\mathrm{E}}^2}). \tag{19.2.19}$$

The two solutions correspond to two images, one lying on either side of the source. One image is always inside the Einstein ring and the other outside it. If the source is moved further from the optic axis (i.e. if β increases), then one image gets closer to the lens and the other gets nearer the source. Gravitational lensing changes the apparent solid angle of the source and therefore results in a magnification by a factor equal to the ratio of the image area to the source area. For a circular lens the magnification factor μ is easily seen to be

$$\mu = \frac{\theta}{\beta}\frac{\mathrm{d}\theta}{\mathrm{d}\beta}. \tag{19.2.20}$$

19.3 More Complicated Systems

The preceding section dealt with simple lens systems. In the following we shall look at some examples of how to deal with the more general case without any special symmetry. To simplify the notation let us start by defining a scaled potential $\psi(\boldsymbol{\theta})$ by

$$\psi(\boldsymbol{\theta}) = \frac{1}{D}\frac{2}{c^2}\int \Phi(D_{\mathrm{d}}\boldsymbol{\theta}, z)\,\mathrm{d}z. \tag{19.3.1}$$

This is useful because the gradient of ψ with respect to θ is just the deflection angle $\boldsymbol{\alpha}$ because

$$\nabla_\theta \psi = D_\mathrm{d} \nabla_\xi \psi = \frac{2}{c^2} \frac{D_\mathrm{ds}}{D_\mathrm{s}} \int \nabla_\perp \Phi \, \mathrm{d}z = \boldsymbol{\alpha}. \tag{19.3.2}$$

Moreover, the Laplacian of ψ with respect to θ is proportional to the surface mass density in the lens plane:

$$\nabla_\theta^2 \psi = \frac{2}{c^2} \frac{D_\mathrm{d} D_\mathrm{ds}}{D_\mathrm{s}} \int \nabla_\xi^2 \Phi \, \mathrm{d}z = \frac{2}{c^2} \frac{D_\mathrm{d} D_\mathrm{ds}}{D_\mathrm{s}} \times 4\pi G \Sigma = 2 \frac{\Sigma}{\Sigma_*}. \tag{19.3.3}$$

It is then convenient to define the convergence κ via

$$\kappa(\boldsymbol{\theta}) \equiv \frac{\Sigma(\boldsymbol{\theta})}{\Sigma_*}, \tag{19.3.4}$$

so that the Laplacian is just twice the convergence in a two-dimensional version of Poisson's equation:

$$\nabla_\theta^2 \psi = 2\kappa. \tag{19.3.5}$$

This means that we can write the potential as a function of κ using

$$\psi(\boldsymbol{\theta}) = \frac{1}{\pi} \int \kappa(\boldsymbol{\theta}) \log |\boldsymbol{\theta} - \boldsymbol{\theta}'| \, \mathrm{d}^2\theta'. \tag{19.3.6}$$

Because the deflection angle is just the gradient of the potential ψ from (19.3.2), we can write

$$\boldsymbol{\alpha}(\boldsymbol{\theta}) = \frac{1}{\pi} \int \kappa(\boldsymbol{\theta}) \frac{\boldsymbol{\theta} - \boldsymbol{\theta}'}{|\boldsymbol{\theta} - \boldsymbol{\theta}'|^2} \, \mathrm{d}^2\theta', \tag{19.3.7}$$

which is equivalent to the Equation (19.2.10) we obtained earlier.

In general the lens produces a mapping of the source plane onto the image plane. The local properties of this mapping are best specified by the Jacobian matrix

$$\boldsymbol{A}_{ij} = \frac{\partial \beta_i}{\partial \theta_j} = \left(\delta_{ij} - \frac{\partial \alpha_i(\boldsymbol{\theta})}{\partial \theta_j} \right) = \left(\delta_{ij} - \frac{\partial^2 \psi}{\partial \theta_i \partial \theta_j} \right). \tag{19.3.8}$$

The Jacobian $\boldsymbol{A}_{ij}$ may be thought of as the inverse of a magnification tensor M_{ij}. The local distortion of an image due to the lens given by the determinant of $\boldsymbol{A}$. If a solid angle $\delta\beta^2$ of the source becomes $\delta\theta^2$ in the image, then

$$\frac{\partial \theta^2}{\partial \beta^2} = \det \mathsf{M} = \frac{1}{\det \boldsymbol{A}}. \tag{19.3.9}$$

This is a general form of Equation (19.2.18).

The general properties of the mapping from source to image can be described somewhat more simply than the general form (19.2.18). First define a notation such that

$$\psi_{ij} \equiv \frac{\partial^2 \psi}{\partial \theta_i \partial \theta_j}. \tag{19.3.10}$$

Using (19.3.5) we find that

$$\kappa = \tfrac{1}{2}(\psi_{11} + \psi_{22}). \qquad (19.3.11)$$

We can also use the elements of ψ_{ij} to construct components of a shear tensor. First define

$$\gamma_1 = \tfrac{1}{2}(\psi_{11} - \psi_{22}) \equiv \gamma \cos(2\phi) \qquad (19.3.12\,a)$$

and

$$\gamma_2 = \psi_{12} = \psi_{21} \equiv \gamma \sin(2\phi). \qquad (19.3.12\,b)$$

Using these definitions we can write

$$\boldsymbol{A} = \begin{pmatrix} 1 - \kappa - \gamma_1 & -\gamma_2 \\ -\gamma_2 & 1 - \kappa + \gamma_1 \end{pmatrix}, \qquad (19.3.13)$$

which can also be written

$$\boldsymbol{A} = (1 - \kappa)\begin{pmatrix} 1 & 0 \\ 0 & 1 \end{pmatrix} - \gamma \begin{pmatrix} \cos 2\phi & \sin 2\phi \\ \sin 2\phi & -\cos 2\phi \end{pmatrix}. \qquad (19.3.14)$$

This notation is useful because it allows a simple visual interpretation of the effects of lensing. A pure convergence κ corresponds to an isotropic magnification of the source in such a way that a circular source becomes a larger but still circular image. The components γ_1 and γ_2 represent *shear* in such a way that

$$\gamma = \sqrt{\gamma_1^2 + \gamma_2^2} \qquad (19.3.15)$$

represents the magnitude of the shear and ϕ its orientation. A non-zero shear transforms a circular source into an elliptical image.

In some places the mapping between source and image plane becomes singular. These singularities are normally called *caustics* and they lead to interesting optical effects owing to the non-uniqueness of the mapping between image and source planes to which they correspond. Basically a given (extended) lens will generate a set of caustics in the source plane. When a source crosses such a caustic a new pair of images is produced in the image. An extended lens can produce many images, depending on the mass distribution in the lens plane, while a point-mass lens only produces two. Near the caustics the shape of the images can be complicated, producing near-circular giant arcs. These can be very bright, owing to the magnification effect which is formally infinite at a caustic.

The consequences of these can be spectacular but complicated and, generally, considerable modelling is needed to understand the complex images obtained.

19.4 Applications

19.4.1 Microlensing

Even if a gravitational lens is not strong enough to form two distinct images of a background source it may still amplify its brightness to an observable extent (e.g. Paczynski 1986a,b). This phenomenon is called microlensing. If a star or other object approaches to within an angle θ_E of a lens, then it will be magnified and will consequently brighten. Inside the galactic halo stars will move across the line of sight to a distant source, such as a star in the Large Magellanic Cloud (LMC). As it traverses the lensing region it will brighten and diminish in a symmetrical fashion. Moreover, because gravitational lensing is achromatic, the variation in brightness can be distinguished from intrinsic stellar variability, which is usually different at different wavelengths. The timescale for a microlensing event in our Galaxy is

$$t_* = \frac{D_d \theta_E}{v}, \tag{19.4.1}$$

where v is the transverse velocity of the lens with respect to the source. For solar mass lenses at a distance D_d of order 10 kpc and v of order 200 km s^{-1} this timescale is of order a few months. Continuous monitoring of stars over this timescale is necessary to detect microlensing. Because the probability of a lens crossing the Einstein radius is small, many millions of stars need to be monitored.

The idea that galactic-halo dark matter might lens the light from distant stars has recently born fruit with convincing evidence for microlensing of stars in the LMC by sub-stellar mass objects in the halo of the Milky Way (Alcock *et al.* 1993; Aubourg *et al.* 1993). Although these do not strongly constrain the total amount of dark matter in our Galaxy, the relatively small number of microlenses detected does constrain the contribution to the mass of the halo in brown dwarfs; see Carr (1994).

A more exotic claim by Hawkins (1993) to have observed microlensing on a cosmological scale by looking at quasar variability is much less convincing. To infer microlensing from quasar light curves requires one to exclude the possibility that the variability seen in the light curves be intrinsic to the quasar. One might naively expect the timescale of intrinsic variability to increase to increase with QSO redshift as a consequence of cosmological time dilation. This increase is not seen in the data, suggesting the variation is not intrinsic, but time dilation is only one of many effects that could influence the timescale of intrinsic variability in either direction. For example, the density of cosmological material surrounding a QSO increases by a factor of eight between $z = 1$ and $z = 3$. Alexander (1995) gives arguments that suggest that observational selection effects may remove the expected correlation and replace it with the inverse effect that is actually observed. It is not inconceivable therefore that a change in fuelling efficiency could change the timescale of variability in the opposite direction to the time dilation effect. In any case, the classic signature of microlensing is that the variability

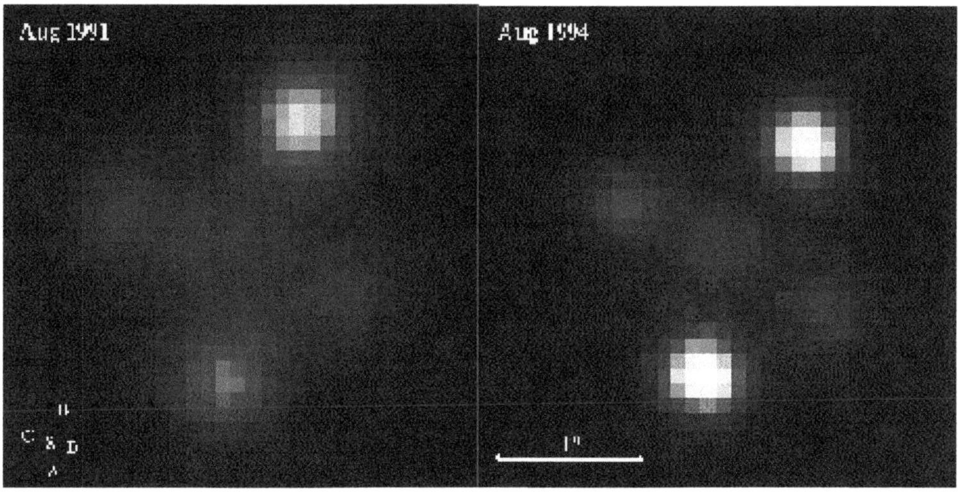

Figure 19.2 William Herschel Telescope images (taken by Geraint Lewis and Michael Irwin) of the 'Einstein cross', a multiply imaged quasar. The two images were taken three years apart and the variation in brightness may be due to microlensing within our Galaxy.

be achromatic: even this is not known about the variability seen by Hawkins. QSOs, and active galaxies in general, exhibit variability on a wide range of timescales in all wavelength regions from the infrared to X-rays. If the lensing interpretation is correct, then one should be able to identify the same timescale of variability at all possible observational wavelengths. An independent analysis of QSO variability by Dalcanton *et al.* (1994) has also placed Hawkins' claim in doubt, so we take the evidence that extragalactic microlensing has been detected to be rather tenuous.

19.4.2 Multiple images

The earliest known instance of gravitational lensing by anything other than the Sun was the famous double quasar 0957 + 561, which upon close examination was found to be a single object which had been lensed by an intervening galaxy (Walsh *et al.* 1979). As time has gone by, searches for similar such lensed systems have yielded more candidates, but the total number of candidate lens systems known is still small.

It has been known for some time that the predicted frequency of quasar lensing depends strongly on the volume out to a given redshift (Turner *et al.* 1984; Turner 1990; Fukugita and Turner 1991) and that the number of lensed quasars observed can consequently yield important constraints on cosmological models. Compared with the Einstein–de Sitter model, both flat cosmologies with a cosmological constant and open low-density ($\Omega_0 < 1$) models predict many more lensed systems. The effect is particularly strong for the flat Λ models: roughly ten times as many lenses are expected in such models than in the $\Omega_0 = 1$ case. Of course, the number of lensed systems also depends on the number and mass of intervening objects in the volume out to the quasar, so any constraints to emerge are necessarily dependent upon assumptions about the evolution of the mass function of galaxies, or at least their massive haloes. Nevertheless, claims of robust constraints have been published (Kochanek 1993; Maoz and Rix 1993; Mao and Kochanek 1994), which constrain the contribution of a Λ term to the total density of a flat universe to $\Omega_\Lambda < 0.5$ at 90% confidence, which seems to contradict the results from high-redshift supernovae we discussed in Chapter 4. Constraints on open, low-density models are much weaker: $\Omega_0 > 0.2$. Unless some significant error is present in the modelling procedure adopted in these studies, the QSO lensing statistics appear to rule out precisely those flat Λ-dominated models which have been held to solve the age problem and also allow flat spatial sections, although at a relatively low confidence level and at the expense of some model dependence. If our understanding of galaxy evolution improves dramatically it will be possible to refine these limits. New large-scale QSO surveys will also help improve the statistics of the lensed objects.

19.4.3 Arcs, arclets and cluster masses

There exists a possible independent test of the dynamical and X-ray masses of rich clusters which does not depend on the assumption of virial or hydrostatic equilibrium. Gravitational lensing of the light from background objects depends on the total mass of the cluster whatever its form and physical state, leading to multiple and/or distorted images of the background object.

The possible lensing phenomena fall into two categories: *strong* lensing in rich clusters can probe the mass distribution in the central parts of these objects; and *weak* lensing distortions of background galaxies can trace the mass distribution much further out from the cluster core. The discovery of giant arcs in images of rich clusters of galaxies as a manifestation of strong gravitational lensing (Tyson *et al.* 1990; Fort and Mellier 1994) has led to a considerable industry in using models of the cluster lens to determine the mass profile. Smaller arcs - usually called arclets - can be used to provide more detailed modelling of the lensing mass distribution. For recent applications of this idea, see Kneib *et al.* (1993) and Smail *et al.* (1995); the latter authors, for example, infer a velocity dispersion of $\sigma^2 \simeq 1400\,\mathrm{km\,s^{-1}}$ for the cluster AC114.

Important though these strong lensing studies undoubtedly are, they generally only probe the central parts of the cluster and say relatively little about the distribution of matter in the outskirts. They do, for example, seem to indicate that the total distribution of matter is more centrally concentrated than the gas distribution inferred from X-ray observations. On the other hand, estimates of the total masses obtained using strong lensing arguments are not in contradiction with virial analysis methods described above.

Weak lensing phenomena - the slight distortions of background galaxies produced by lines of sight further out from the cluster core - can yield constraints on the haloes of rich clusters (Kaiser and Squires 1993; Broadhurst *et al.* 1995).

Figure 19.3 HST image of the rich cluster Abell 2218 showing numerous giant arcs and arclets. Picture courtesy of the Space Telescope Science Institute.

It is also possible to use fluctuations in the $N(z)$ relation of the galaxies behind the cluster to model the mass distribution.

The technology of these methods has developed rapidly and has now been applied to several clusters. Preliminary results are generally indicative of a larger total mass than is inferred by virial arguments, suggesting that there exists even more dark matter than dynamics would suggest. However, this technique is relatively young and it is possible that not all the systematic errors have yet been ironed out, so we take these results as indicating that this is a good – indeed important – method for use in future studies, rather than one which is providing definitive results at the present time.

19.4.4 Weak lensing by large-scale structure

The idea that clusters of galaxies produce observable distortions in the weak lensing limit suggests it may be possible to observe lensing along any line of sight through the background distribution of clusters. What one will see looking through an arbitrary distribution that lacks the special symmetry of a rich cluster will be correlated distortions of the shapes of galaxies. One needs a wide field in order to see sufficient galaxies to obtain a signal, because the shear of any one galaxy is small compared with the distribution of shapes that exists in

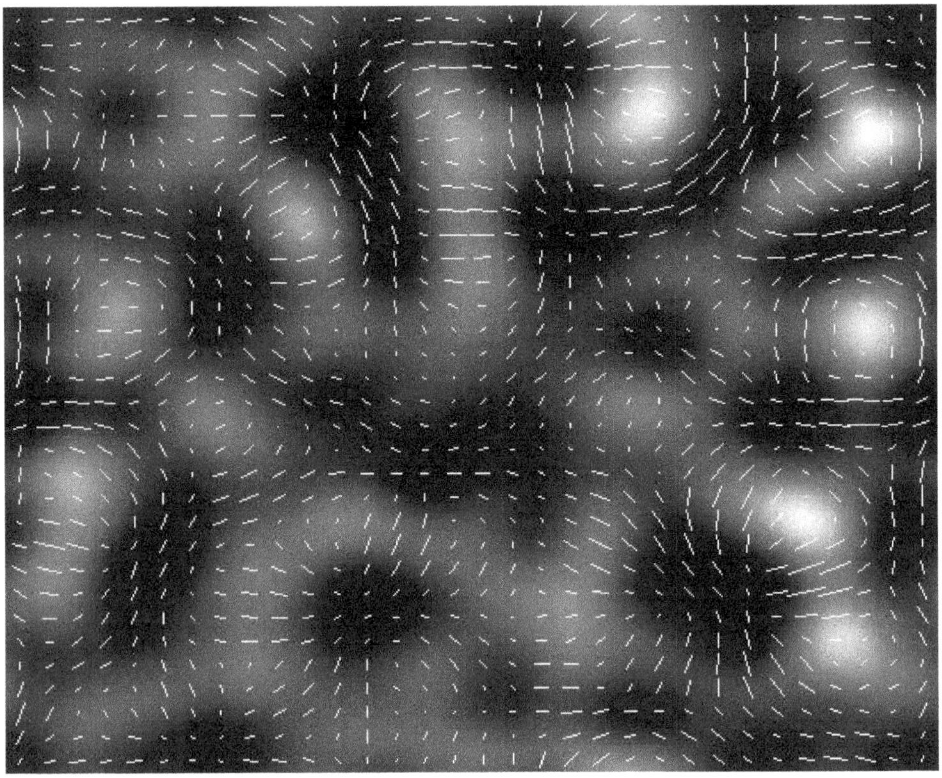

Figure 19.4 Simulation of the weak lensing distortion induced by large-scale structure. The pattern of density perturbations is shown as a greyscale picture upon which lines are superimposed representing the size and angle of the distortions. Picture courtesy of Alex Refregier.

the unlensed galaxy population. However difficult this may be, the payoff is large because one can in principle obtain, from maps of sheared galaxy images, maps of the projected dark-matter distribution. This is a new field, but feasibility studies already show that the signal is measurable (Bacon *et al.* 2000; van Waerbeke *et al.* 2000; Wilson *et al.* 2001; Wittman *et al.* 2000). When larger CCD arrays go online, we will have maps of the evolved dark-matter distribution that can complement maps of the galaxy distribution obtained from redshift surveys and maps of the primordial fluctuations obtained from the cosmic microwave background.

19.4.5 The Hubble constant

One of the consequences of gravitational lensing is that the paths traversed by photons coming from the same source but forming different images may have different lengths. If the source happens to be variable, then one can hope to recognise a pattern in its output in more than one image at different times. If one understands the structure of the lens, then one can estimate the distances involved. Knowing the redshift allows one to obtain an estimate of H_0. This idea, of course, rests on the correct identification of the time delay. An example is the quasar 0957 + 561 which has a measured time lag of 415 days between features seen in the two images it presents to the observer. The lens seems to be dominated by a single galaxy sitting inside a cluster and the modelling is consequently fairly straightforward. Preliminary estimates by Grogin and Narayan (1996) yield a rather high value of the Hubble constant around 80 km s^{-1} Mpc^{-1} but with considerable theoretical uncertainty in the model parameters needed to reproduce the known images. In principle, such studies can yield accurate estimates of the Hubble constant but the technique is relatively young and clearly needs more work to develop it.

19.5 Comments

It is rather ironic that the oldest known observational consequence of general relativity should produce one of the newest and most dynamic areas in cosmology. Now that observational technology is so advanced and wide-field cameras are becoming increasingly available, it seems likely that weak lensing will have a particularly strong impact on cosmology in the relatively near future. In particular we should be able to understand the extent to which the large-scale structure seen in the galaxy distribution represents genuine fluctuations in the mass density and how much may be attributable to bias. Even the cosmic microwave background offers the possibility for lensing studies.

Bibliographic Notes on Chapter 19

Schneider *et al.* (1992) is now the standard reference book on gravitational lensing. Other useful review articles are Blandford and Narayan (1992) and Fort and Mellier (1994). The paper by Refsdal (1964) is a classic which inspired much work in this area, long before the observational discovery of extragalactic lensed systems. Much of the material for this chapter was gleaned from the lecture notes of Narayan and Bartelmann, which are available on the internet at

`http://www.mpa-garching.mpg.de/Lenses/Preprints/JeruLect.html`

Problems

1. If D_d, D_s and D_{ds} are angular-diameter distances, show that, in general, $D_{ds} \neq D_s - D_d$.
2. Derive Equation (19.2.14).
3. Obtain estimates of the Einstein radius for (i) a point lens of mass $M_\odot$ and $D = 10$ kpc, and (ii) a lens of mass $10^{11} M_\odot$ and $D = 1$ Gpc.

4. Show that, for a point-mass lens, the magnifications of the two images are given by

$$\mu_{\pm} = \frac{u^2 + 2}{u\sqrt{u^2 + 4}} \pm \tfrac{1}{2},$$

where $u = \beta/\theta_{\mathrm{E}}$. Hence show that when $\beta = \theta_{\mathrm{E}}$ the total magnification of flux is 1.34.

5. A singular isothermal sphere is defined by a three-dimensional density profile of the form

$$\rho(r) = \frac{\sigma_{\mathrm{v}}^2}{2\pi G r^2}.$$

Show that the deflection produced by such a lens is $4\pi\sigma_{\mathrm{v}}^2/c^2$ and derive an expression for the Einstein radius. Under what circumstances does this system produce multiple images?

6. Show that a circular source of unit radius is mapped into an ellipse with major and minor axes $(1 - \kappa - \gamma)^{-1}$ and $(1 - \kappa + \gamma)^{-1}$, respectively. Show further that the magnification is $[(1 - \kappa)^2 - \gamma^2]^{-1}$.

20

The High-Redshift Universe

20.1 Introduction

In the previous four chapters we have tried to explain how observations of galaxy clustering, the cosmic microwave background, galaxy-peculiar motions and gravitational lensing can be used to place constraints on theories of structure formation in the Big Bang model. In this chapter we shall discuss a number of independent pieces of evidence about the process of structure formation which can also, in principle, shed light upon how galaxies and clusters of galaxies might have formed. The common theme uniting these considerations is that they all involve phenomena occurring after recombination and before the present epoch.

Since galaxy properties are only observable at relatively small distances, and therefore at relatively small lookback times, galaxy clustering and peculiar motions give us information about the Universe here and now. On the other hand, primary anisotropies of the CMB yield information about the Universe as it was at $t \simeq t_{\rm rec}$. In between these two observable epochs lies a 'dark age', before visible structure appeared but after matter was freed from the restraining influence of radiation pressure and viscosity. As we shall see, there are, in fact, a number of processes that can yield circumstantial evidence of various goings-on in this interval and these can, in turn, give us important insights into the way structure formation can have occurred. It should be said at the outset, however, that many of the issues we shall discuss in this chapter are controversial and clouded by observational uncertainties. We shall therefore concentrate upon the questions raised by this set of phenomena, rather than trying to incorporate them firmly in an overall picture of galaxy formation.

We have already mentioned, in Chapter 17, some ways of probing the post-recombination Universe, by exploiting secondary anisotropies in the CMB radiation such as the Sunyaev–Zel'dovich effect. We shall raise some of these issues again here in the context of other observations and theoretical considerations. For the most part, however, this chapter is concerned with early signatures of galaxy formation, sources of radiation at high redshift and constraints on the properties of the intergalactic medium (IGM) at moderate and high redshifts.

20.2 Quasars

The most obvious way to acquire information about the Universe at early times is to locate objects with high redshifts. To be detectable, such objects must be very luminous at frequencies that get redshifted into the observable range of some earthly detector.

The objects with largest known redshifts are the quasars. The current record holder has $z = 6.28$, but quasars with redshifts as high as this are very difficult to detect and/or identify. As we shall see, even the observation of a single high-redshift quasar can place strong direct constraints on models of structure formation. There are many more quasars at $z \simeq 2$ than at the present epoch. Efstathiou and Rees have estimated that the comoving number density of quasars at this epoch (i.e. scaled to the present epoch), with luminosity greater than $L_{\mathrm{Q}} \simeq 2.5 \times 10^{46}$ erg s^{-1}, is

$$n_{\mathrm{Q}}(> L_{\mathrm{Q}}) \simeq 1.5 \times 10^{-8} (h^{-1}\ \mathrm{Mpc})^{-3}. \tag{20.2.1}$$

At higher redshifts the luminosity function of quasars is very poorly known. It seems unlikely that the number density given in (20.2.1) rises drastically and there is also little evidence that it falls sharply before $z \simeq 3.5$. The existence of the record holder shows that there are at least some quasars with redshifts of order 5.

The usual model for a quasar is that its luminosity originates from matter accreting onto a central black hole embedded within a host galaxy. The central mass required depends on the luminosity, the lifetime of the quasar t_{Q} (which is poorly known) and the efficiency ϵ with which the rest-mass energy is released as radiation. For quasars with the luminosity given above, the required mass is

$$M_{\mathrm{Q}} \simeq 5 \times 10^{7} h^{-2} \epsilon^{-1} \left(\frac{t_{\mathrm{Q}}}{10^{8}\ \mathrm{years}} \right) M_{\odot}. \tag{20.2.2}$$

The number density of quasars given in (20.2.1) is, of course, very much less than the present value for galaxies. In a hierarchical clustering model, however, the number of bound objects on a given mass scale decreases at earlier times. It is an interesting exercise therefore to see if the existence of objects on the mass scale required to house a quasar contradicts theories of galaxy formation. To do this we first need to calculate how big the parent galaxy of a quasar has to be. There are three factors involved: the fraction f_{b} of the matter in baryonic form which is subject to the constraints discussed in Chapter 8; the fraction f_{r} of the baryons retained in a halo and not blown out by supernova explosions when star formation begins; the fraction f_{h} of the baryons which participate in the fuelling of the quasar. All these factors are highly uncertain, so one can define a single quantity $F = f_b f_r f_h$ to include them all. It is unlikely that F can be larger than 0.01.

To model the formation of haloes we can use the Press–Schechter theory discussed in Section 14.5 (Efstathiou and Rees 1988). The z-dependence of the mass function of objects can be inserted into equation (14.5.7) by simply scaling the RMS density fluctuation by the factor $1/(1+z)$ coming from linear theory. Recall that the parameter δ_c in equation (14.5.7) specifies a kind of threshold for collapse and that $\delta_c \simeq 1.68$ is the appropriate value for isolated spherical collapse; numerical experiments suggest this analytic formula works fairly well, but with a smaller $\delta_c \simeq 1.33$. Anyway, the number density of quasars is

$$n_Q(>L, z) \simeq \int_{t_{\min}}^{t(z)} \int_{M_{\min}}^{\infty} \frac{\partial n(M, z)}{\partial t} \, dM \, dt. \tag{20.2.3}$$

The lower limit of integration $M_{\min}$ is the minimum mass capable of housing a quasar, which is estimated to be

$$M_{\min} \simeq 2 \times 10^{11} \left(\frac{t_Q}{10^8 \text{ years}}\right) \left(\frac{\epsilon}{0.1}\right)^{-1} \left(\frac{F}{0.01}\right)^{-1} \left(\frac{L}{L_Q}\right) M_\odot, \tag{20.2.4}$$

and $t_{\min}$ is either 0 or $[t(z) - t_Q]$, whichever is the larger. Using equation (14.5.7) with $\delta_c = 1.33$ and defining

$$\beta = \left(\frac{L}{L_Q}\right) \left(\frac{t_Q}{10^8 \text{ years}}\right) \left(\frac{\epsilon}{0.1}\right)^{-1} \left(\frac{F}{0.01}\right)^{-1}, \tag{20.2.5}$$

Efstathiou and Rees (1988) obtained, for a spectrum with $n \simeq -2.2$ (appropriate to a CDM model on the relevant scales),

$$n_Q(>L, z) \simeq 1 \times 10^{-3} (1+z)^{5/2} \left(\frac{t_Q}{10^8 \text{ years}}\right) \beta^{-0.866} \exp[-0.21\beta^{0.266}(1+z)^2], \tag{20.2.6}$$

in the same units as Equation (20.2.1). Notice above all that this falls precipitously at high z because of the exponential term. This can place strong constraints on models where structure formation happens very late, such as in the biased CDM picture. The result (20.2.6) is not, however, incompatible with (20.2.1) for this model. A similar exercise could be attempted for clusters of galaxies and absorption-line systems in quasar spectra, but we shall not discuss this possibility here.

As we explained in Chapter 4 there are also other types of active galaxy that can be observed at high redshifts, although not as high as quasars. One of these types is particularly interesting in the present context: steep-spectrum radio sources. In recent years, samples of these objects have been studied in the optical wavelength region. Many of them are associated with galaxies having redshifts greater than two, and one, called 4C41.17, has a redshift of 3.8, which is the largest known redshift of a galaxy. These objects may yield important clues about the relationship between activity, such as jets, and star formation in galaxies. One popular idea for the peculiar optical morphology of these objects and the alignment between their radio jets and optical emission is that a radio jet may have triggered star formation in the parent galaxy. The fact that these objects have considerable optical emission allows one to study their stellar populations to figure out possible ages. This is difficult because of the high redshift, which means that interesting features of the optical spectrum are shifted into the infrared K-band, which is notoriously problematic to work in. It has been claimed that these objects have relatively old stellar populations: if true, this would be a significant problem for some theories. At the moment, however, it is best to keep an open mind about these claims; we shall mention these objects again in Section 20.6.

20.3 The Intergalactic Medium (IGM)

We now turn our attention to various constraints, not on objects themselves, but on the medium between them: the IGM.

20.3.1 Quasar spectra

Observations of quasar spectra allow one to probe a line of sight from our Galaxy to the quasar. Absorption or scattering of light during its journey to us can, in principle, be detected by its effect upon the spectrum of the quasar. This, in turn, can be used to constrain the number and properties of absorbers or scatterers, which, whatever they are, must be associated with the baryonic content of the IGM. Before we describe the possibilities, it is therefore useful to write down the mean number density of baryons as a function of redshift:

$$n_{\mathrm{b}} \simeq 1.1 \times 10^{-5} \Omega_{\mathrm{b}} h^2 (1+z)^3 \ \mathrm{cm}^{-3}. \qquad (20.3.1)$$

This is an important reference quantity for the following considerations.

20.3.2 The Gunn–Peterson test

Neutral hydrogen has a resonant scattering feature associated with the Lyman-α atomic transition. This resonance is so strong that it is possible for a relatively low neutral-hydrogen column density (i.e. number-density per unit area of atoms, integrated along the line of sight) to cause a significant apparent absorption at the appropriate wavelength for the transition. Let us suppose that light travels towards us through a uniform background of neutral hydrogen. The optical depth for scattering is

$$\tau(\lambda_0) = \frac{c}{H_0} \int \sigma(\lambda_0 a/a_0) n_{\mathrm{I}}(t) \Omega^{-1/2} \left(\frac{a_0}{a}\right)^{-3/2} \frac{\mathrm{d}a}{a}, \qquad (20.3.2)$$

where $\sigma(\lambda)$ is the cross-section at resonance and $n_{\rm I}$ is the proper density of neutral hydrogen atoms at the redshift corresponding to this resonance. (The usual convention is that HI refers to neutral and HII to ionised hydrogen.) We have assumed in (20.3.2) that the Universe is matter dominated. The integral is taken over the width of the resonance line (which is very narrow and can therefore be approximated by a delta function) and yields a result for τ at some observed wavelength λ_0. It therefore follows that

$$\tau = \frac{3\Lambda\lambda_\alpha^3 n_{\rm I}}{8\pi H_0 \Omega^{1/2}}(1+z)^{-3/2}, \tag{20.3.3}$$

where $\Lambda = 6.25 \times 10^8\ {\rm s}^{-1}$ is the rate of spontaneous decays from the 2p to the 1s level of hydrogen (the Lyman-α emission transition); λ_α is the wavelength corresponding to this transition, i.e. 1216 Å. Equation (20.3.3) can be inverted to yield

$$n_{\rm I} = 2.4 \times 10^{-11}\Omega^{1/2}h(1+z)^{3/2}\tau\ {\rm cm}^{-3}. \tag{20.3.4}$$

This corresponds to the optical depth τ at $z = (\lambda_0/\lambda_\alpha) - 1$, when observed at a wavelength λ_0.

The Gunn–Peterson test (Gunn and Peterson 1965) takes note of the fact that there is no apparent drop between the long-wavelength side of the Lyman-α emission line in quasar spectra and the short-wavelength side, where extinction by scattering might be expected. Observations suggest a (conservative) upper limit on τ of order 0.1, which translates into a very tight bound on $n_{\rm I}$:

$$n_{\rm I} < 2 \times 10^{-12}\Omega^{1/2}h(1+z)^{3/2}\ {\rm cm}^{-3}. \tag{20.3.5}$$

Comparing this with Equation (20.3.1) with $\Omega_{\rm b} = 1$ yields a constraint on the contribution to the critical density due to neutral hydrogen:

$$\Omega(n_{\rm I}) < 2 \times 10^{-7}\Omega^{1/2}h^{-1}(1+z)^{-3/2}. \tag{20.3.6}$$

There is no alternative but to assume that, by the epoch one can probe directly with quasar spectra (which corresponds to $z \simeq 4$), the density of any uniform neutral component of the IGM was very small indeed.

One can translate this result for the neutral hydrogen into a constraint on the plasma density at high temperatures by considering the balance between collisional ionisation reactions,

$$\mathrm{H} + \mathrm{e}^- \to \mathrm{p} + \mathrm{e}^- + \mathrm{e}^-, \tag{20.3.7a}$$

and recombination reactions of the form

$$\mathrm{p} + \mathrm{e}^- \to \mathrm{H} + \gamma. \tag{20.3.7b}$$

The physics of this balance is complicated by the fact that the cross-sections for these reactions are functions of temperature. It turns out that the ratio of neutral hydrogen to ionised hydrogen, $n_{\rm I}/n_{\rm II}$, has a minimum at a temperature around 10^6 K, and at this temperature the equilibrium ratio is

$$\frac{n_{\rm I}}{n_{\rm II}} \simeq 5 \times 10^{-7}. \tag{20.3.8}$$

Since this is the minimum possible value, the upper limit on $n_{\rm I}$ therefore gives an upper limit on the total density in the IGM, which we can assume to be made entirely of hydrogen:

$$\Omega_{\rm IGM} < 0.4\Omega^{1/2}h^{-1}(1+z)^{-3/2}. \tag{20.3.9}$$

If the temperature is much lower than 10^6 K, the dominant mechanism for ionisation could be electromagnetic radiation. In this case one must consider the equilibrium between radiative ionisation and recombination, which is more complex and requires some assumptions about the ionising flux. There are probably enough high-energy photons from quasars at around $z \simeq 3$ to ionise most of the baryons if the value of $\Omega_{\rm b}$ is not near unity, and there is also the possibility that early star formation in protogalaxies could also contribute substantially. Another complication is that the spatial distribution of the IGM might be clumpy, which alters the average rate of recombination reactions but not the mean rate of ionisations. One can show that, for temperatures around 10^4 K, the constraint emerges that

$$\Omega_{\rm IGM} < 0.4 I_{21}\Omega^{1/2}h^{-3/2}(1+z)^{9/4}, \tag{20.3.10}$$

if the medium is not clumpy and the ionising flux, I_{21}, is measured in units of 10^{-21} erg cm^{-2} s^{-1} Hz^{-1} ster^{-1}. The limit (20.3.10) is reduced if there is a significant clumping of the gas.

These results suggest that the total IGM density cannot have been more than $\Omega_{\rm IGM} \simeq 0.03$ at $z \simeq 3$, whatever the temperature of the plasma. This limit is compatible with the nucleosynthesis bounds given in Section 8.6.

20.3.3 Absorption line systems

Although quasar spectra do not exhibit any general absorption consistent with a smoothly distributed hydrogen component, there are many absorption lines in such spectra which are interpreted as being due to clouds intervening between the quasar and the observer and absorbing at the Lyman-α resonance. An example spectrum is shown in Figure 20.1.

The clouds are grouped into three categories depending on their column density, which can be obtained from the strength of the absorption line. The strongest absorbers have column densities $\Sigma \simeq 10^{20}$ atoms cm^{-2} or more, which are comparable with the column densities of interstellar gas in a present-day spiral galaxy. This is enough to produce a very wide absorption trough at the Lyman-α wavelength and these systems are usually called *damped Lyman-α systems*. These are

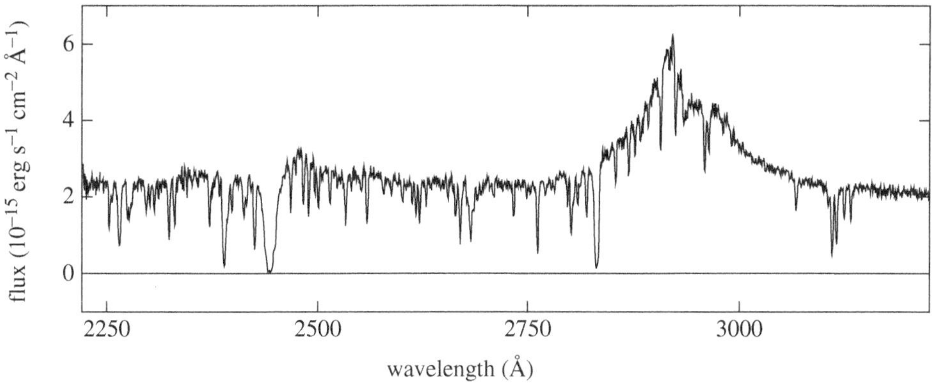

Figure 20.1 An example of a quasar spectrum showing evidence of absorption lines at redshifts lower than the Lyman-α emission of the quasar. Picture courtesy of Sandhya Rao.

relatively rare, and are usually interpreted as being the progenitors of spiral discs. They occur at redshifts up to around 3 (Wolfe *et al.* 1993).

A more abundant type of object is the Lyman limit system. These have $\Sigma \simeq 10^{17}$ atoms cm^{-2} and are dense enough to block radiation at wavelengths near the photoionisation edge of the Lyman series of lines. Smaller features, with $\Sigma \simeq 10^{14}$ atoms cm^{-2} appear as sharp absorption lines at the Lyman-α wavelength. These are very common, and reveal themselves as a 'forest' of lines in the spectra of quasars, hence the term *Lyman-α forest*. The importance of the Lyman limit is that, at this column density, the material at the centre of the cloud will be shielded from ionising radiation by the material at its edge. At lower densities this cannot happen.

As we have already mentioned, the damped Lyman-α systems have surface densities similar to spiral discs. It is natural therefore to interpret them as protogalactic discs. The only problem with this interpretation is that there are about ten times as many such systems at $z \simeq 3$ than one would expect by extrapolating backwards the present number of spiral galaxies. This may mean that, at high redshift, these galaxies are surrounded by gas clouds or very large neutral hydrogen discs which get destroyed as the galaxies evolve. It may also be that many of these objects end up as low-surface-brightness galaxies at the present epoch, which do not form stars very efficiently (e.g. Davies *et al.* 1988): in such a case the present number of bright spirals is an underestimate of the number of damped Lyman-α systems that survive to the present epoch. It is also pertinent to mention that these systems have also been detected in CaII, MgII or CIV lines and that they do seem to have significant abundances of elements heavier than helium. There is some evidence that the fraction of heavy elements decreases at high redshifts.

The Lyman-α forest clouds have a number of interesting properties. For a start they provide evidence that quasars are capable of ionising the IGM. The number densities of systems observed along lines of sight towards different quasars are similar, which strengthens the impression that they are intervening objects and not connected with the quasar. At redshifts near that of the quasar the number density decreases markedly, an effect known as the *proximity effect.* The idea here is that radiation from the quasar substantially reduces the neutral hydrogen fraction in the clouds by ionisation, thus inhibiting absorption at the Lyman-α resonance. Secondly, the total mass in the clouds appears to be close to that in the damped systems or that seen in present-day galaxies. This would be surprising if the forest clouds were part of an evolving clustering hierarchy, but if they almost fill space then one might not see any strong correlations in any case. Thirdly, the comoving number density of such systems is changing strongly with redshift, indicating, perhaps, that the clouds are undergoing dissipation. Finally, and most interestingly from the point of view of structure formation, the absorption systems seem to be only weakly clustered, in contrast to the distribution of galaxies. How these smaller Lyman-α systems fit into a picture of galaxy formation is not absolutely certain, but it appears that they correspond to lines of sight passing through gas confined in the small-scale 'cosmic web' of filaments and voids that corresponds to an earlier stage of the clustering hierarchy than is visible in the local galaxy distribution.

20.3.4 X-ray gas in clusters

We should mention here that there is direct evidence from X-ray observations of hot gas at $T \simeq 10^8$ K in the IGM in rich clusters of galaxies. We mentioned in Chapter 17 that this gas could cause an observable Sunyaev–Zel'dovich distortion of the CMB temperature in the line of sight of the cluster. Direct observations of the gas show that it also has quite high metal abundances and its total mass is of order that contained in the cluster galaxies. Since the cooling time of the gas at these temperatures is comparable with the Hubble time, one expects to see *cooling flows* as the gas dissipates and falls into the potential well of the cluster (a cooling flow occurs whenever the rate of radiative cooling is quicker than the cosmological expansion rate, H). It seems likely, however, that much of the cluster gas is actually stripped from the cluster galaxies, so these observations say nothing about the properties of the primordial IGM.

We discuss the properties of the diffuse extragalactic X-ray background and its implications in Section 20.4.

20.3.5 Spectral distortions of the CMB

The Sunyaev–Zel'dovich effect also allows one to place constraints on the properties of the intergalactic medium. If the hot gas is smoothly distributed, then one would not expect to see any angular variation in the temperature of the CMB radiation as a result of this phenomenon. However, the Sunyaev–Zel'dovich effect is frequency dependent: the dip associated with clusters appears in the Rayleigh–Jeans region of the CMB spectrum. If one measures this spectrum one would expect a smooth gas distribution to produce a distortion of the black-body shape due to scattering as the CMB photons traverse the IGM. The same will happen if gas is distributed in objects at high redshift which are too distant to be resolved. We mentioned this effect in Section 9.5 and defined the relevant parameter, the so-called y-parameter, in Equation (9.5.5). The importance of this effect has been emphasised by the CMB spectrum observed by the FIRAS experiment on COBE, which has imposed the constraint $y < 3 \times 10^{-5}$.

From Equation (9.5.5) the contribution to y from a plasma with mean pressure $n_e k_B T_e$ at a redshift z is

$$y \simeq \sigma_T n_e c t \frac{k_B T_e}{m_e c^2}, \tag{20.3.11}$$

where the suffix e refers to the electrons. Various kinds of object containing hot gas could, in principle, contribute significantly to y. If Lyman-α clouds are in pressure balance at $z \simeq 3$, then they will contribute only a small fraction of the observational limit on y, so these clouds are unlikely to have an effect on the CMB spectrum. Similarly, if galaxies form at high redshifts with circular velocities v, then one can write

$$y \simeq \sigma_T n_e c t \left(\frac{v}{c}\right)^2, \tag{20.3.12}$$

which is of order

$$y \simeq 10^{-8} h \Omega_g \Omega^{-1/2} (1+z)^{3/2} \tag{20.3.13}$$

if $v \simeq 100$ km s^{-1} and Ω_g is the fractional contribution of hot gas to the critical density. The contribution from rich clusters is similarly small, because the gas in these objects only contributes around $\Omega_g \simeq 0.003$. On the other hand, a smooth hot IGM can have a significant effect on y, as we shall see shortly.

20.3.6 The X-ray background

It has been known for some time that there exists a smooth background of X-ray emission. This background actually furnishes an additional argument for the large-scale homogeneity of the Universe because the flux is isotropic on the sky to a level around 10^{-3} in the wavelength region from 2 to 20 keV.

It has been a mystery for some time precisely what is responsible for this background but many classes of object can, in principle, contribute. Clusters of galaxies, quasars and active galaxies at high redshift and even starburst galaxies at relatively low redshift could be significant contributors to it. Disentangling these components is difficult because it may be difficult to locate any counterpart of an X-ray-emitting source in any other waveband. Recently, however, using the sensitive instruments on Chandra, Mushotzky *et al.* (2000) have resolved about three-quarters of the hard X-ray background into sources. The mean X-ray spectrum of these sources is in good agreement with that of the background. The X-ray emission from the majority of the detected sources is unambiguously associated with either the nuclei of otherwise normal bright galaxies or optically faint sources, which could either be active nuclei of dust-enshrouded galaxies or the first quasars at very high redshifts.

The spectrum and anisotropy may well provide strong constraints on models for the origin of quasars and other high-redshift objects. We shall concentrate on the constraints this background imposes on the IGM. A hot plasma produces radiation through thermal bremsstrahlung. The luminosity density at a frequency ν produced by this process for a pure hydrogen plasma is given approximately by

$$J(\nu) = 5.4 \times 10^{-39} n_e^2 T_e^{-1/2} \exp(-h\nu/k_B T_e) \text{ erg cm}^{-3}\text{ s}^{-1}\text{ ster}^{-1}\text{ Hz}^{-1}, \quad (20.3.14)$$

so the integrated background observed now at a frequency ν is

$$I(\nu) = \int cJ(\nu a_0/a, t)(a/a_0)^3 \, \mathrm{d}t, \quad (20.3.15)$$

where the integral is taken over a line of sight through the medium. If the emission takes place predominantly at a redshift z, then

$$I(\nu) = 4 \times 10^{-23} \left(\frac{T_e}{10^4 \text{ K}}\right)^{-1/2} \frac{h^3 \Omega_{\text{IGM}}}{\Omega^{1/2}} (1+z)^{3/2} \text{ erg cm}^{-2}\text{ s}^{-1}\text{ ster}^{-1}\text{ Hz}^{-1} \quad (20.3.16)$$

for $h\nu \ll k_B T_e$. The present surface brightness of the X-ray background is

$$I(\nu) \simeq 3 \times 10^{-26} \text{ erg cm}^{-3}\text{ s}^{-1}\text{ ster}^{-1}\text{ Hz}^{-1} \quad (20.3.17)$$

at energies around 3 keV. Suppose a fraction f of this is produced by a hot IGM with temperature $T \simeq 10^8(1+z)$ K; in this case,

$$\Omega_{\text{IGM}} \simeq 0.3 f \Omega^{1/4} h^{-3/2} \left(\frac{T}{10^8 \text{ K}}\right)^{1/4} (1+z)^{-1/2}, \quad (20.3.18)$$

so that, if the plasma is smooth, the y-parameter is

$$y \simeq 2 \times 10^{-4} (1+z)^2 \left(\frac{f}{h\Omega^{1/2}}\right)^{1/2}. \quad (20.3.19)$$

If the plasma is hot and dense enough to contribute a significant part of the X-ray background, then it would violate the constraints on y.

20.4 The Infrared Background and Dust

We have already discussed the importance of the CMB radiation as a probe of cosmological models. Two other backgrounds of extragalactic radiation are important for the clues they provide about the evolution of gas and structure after recombination.

It has been suggested that various kinds of cosmological sources might also generate a significant background in the infrared (IR) or submillimetre parts of the spectrum, near CMB frequencies. A cosmological IR background is very difficult to detect even in principle because of the many local sources of radiation at these frequencies. Nevertheless, the current upper limits on flux in various wavelength regions can place strong constraints on possible populations of pregalactic objects. For simplicity one can characterise these sources by the contribution their radiation would make towards the critical density:

$$\Omega_R(\lambda) = \frac{4\pi\nu I(\nu)}{c^3\rho_c}, \tag{20.4.1}$$

where $I(\nu)$ is the flux density per unit frequency. The CMB has a peak energy density at $\lambda_{max} = 1400\ \mu m$, corresponding to $\Omega_{CMB} \simeq 1.8 \times 10^{-5}h^{-2}$. The lack of distortions of the CMB spectrum reported by the FIRAS experiment on COBE suggests that an excess background with $500\ \mu m < \lambda < 5000\ \mu m$ can have a density less than 0.03% of the peak CMB value:

$$\Omega_R(\lambda) < 6 \times 10^{-9}h^2\left(\frac{\lambda}{\lambda_{max}}\right)^{-1}. \tag{20.4.2}$$

One obvious potential source of IR background radiation is galaxies. To estimate this contribution is rather difficult and requires complicated modelling. The near-IR background would be generated by redshifted optical emission from normal galaxies. One therefore needs to start with the spectrum of emission as a function of time for a single galaxy, which requires knowledge of the initial mass function of stars, the star-formation rate and the laws of stellar evolution. To get the total background one needs to integrate over a population of different types of galaxies as a function of redshift, taking into account the effect of the density parameter upon the expansion rate. If galaxies are extremely dusty, then radiation from them will appear in the far-IR region. Such radiation can emanate from dusty discs, clouds (perhaps associated with the 'starburst' phenomenon), active galaxies and quasars. The evolution of these phenomena is very complex and poorly understood at present.

More interesting are the possible pregalactic sources of IR radiation. Most of these sources produce an approximate black-body spectrum, because the low density of neutral hydrogen in the IGM is insufficient to absorb photons with wavelengths shorter than the Lyman cut-off. For example, the cooling of gas clouds at a redshift z after they have collapsed and virialised would produce

$$\Omega_R \simeq 2 \times 10^{-7}\left(\frac{\Omega_{clouds}}{0.1}\right)\left(\frac{1+z}{5}\right)^{-1}\left(\frac{v}{300\ \mathrm{km\ s^{-1}}}\right)^2 \tag{20.4.3}$$

at a peak wavelength

$$\lambda_{max} \simeq 0.1\left(\frac{1+z}{5}\right)^{-1}\left(\frac{v}{300\ \mathrm{km\ s^{-1}}}\right)^{-2}\ \mu m, \tag{20.4.4}$$

where v is the RMS velocity of gas in the clouds. In principle, this could therefore place a constraint upon theories of galaxy formation, but the number of objects forming as a function of redshift is difficult to compute in all but the simplest hierarchical clustering scenarios. Pregalactic explosions, often suggested as an alternative to the standard theories of galaxy formation, would produce a much larger background. COBE limits on the spectral distortions (20.4.2) appear to rule out this model quite comfortably. Constraints can also be placed on the numbers of galactic halo black holes, halo brown dwarfs and upon the possibility of a decaying particle ionising the background radiation.

The constraints obtained from this type of study only apply if the radiation from the source propagates freely without absorption or scattering to the observer. Many sources of radiation observed at the present epoch in the IR or submillimetre regions are, however, initially produced in the optical or ultraviolet and redshifted by the cosmological expansion. The radiation may therefore have been reprocessed if there was any dust in the vicinity of the source. Dust grains are generally associated with star formation and may consequently be confined to galaxies or, if there was a cosmological population of pregalactic stars, could be smoothly distributed throughout space. The cross-section for spherical dust grains to absorb photons of wavelength λ is of the form

$$\sigma_{\mathrm{d}} = \frac{\pi r_{\mathrm{d}}^2}{1 + (\lambda/r_{\mathrm{d}})^{\alpha}}, \qquad (20.4.5)$$

where r_{d} is the grain radius and $\alpha \simeq 1$ is a suitable parameter; the cross-section is simply geometrical for small λ but falls as a power law for $\lambda \gg r_{\mathrm{d}}$. If radiation is absorbed by dust (whether galactic or pregalactic), then thermal balance implies that the dust temperature T_{d} obeys the relation

$$T_{\mathrm{d}}(z) = T_{\mathrm{CMB}}(z)\left[1 + \left(\frac{\Omega_{\mathrm{R}}}{\Omega_{\mathrm{CMB}}}\right)\left(\frac{r_{\mathrm{d}}}{0.1\ \mu\mathrm{m}}\right)^{-1}\left(\frac{1+z}{10^4}\right)^{-1}\right]^{1/5}. \qquad (20.4.6)$$

If the radiation density parameter is less than the critical quantity

$$\Omega_* \simeq 2 \times 10^{-7} h^{-2}\left(\frac{r_{\mathrm{d}}}{0.1\ \mu\mathrm{m}}\right)\left(\frac{1+z}{100}\right), \qquad (20.4.7)$$

then the dust temperature will be the same as the CMB temperature at redshift z. On the other hand, if $\Omega_{\mathrm{R}} > \Omega_*$, the dust will be hotter than the CMB and one will expect a far-IR or submillimetre radiation background with a spectrum that peaks at

$$\lambda_{\max} \simeq 700 h^{-2/5}\left(\frac{\Omega_{\mathrm{R}}}{10^{-6}}\right)^{-1/5}\left(\frac{r_{\mathrm{d}}}{0.1\ \mu\mathrm{m}}\right)^{1/5}\left(\frac{1+z}{10}\right)^{1/5}\ \mu\mathrm{m}. \qquad (20.4.8)$$

Notice the very weak dependence on the various parameters, indicating that the peak wavelength is a very robust prediction of these models. This was interesting a few years ago because a rocket experiment by the Nagoya-Berkeley collaboration had claimed a detection of an excess in the CMB spectrum in this wavelength region. Unfortunately, we now know this claim was incorrect and that the experiment had detected hot exhaust fumes from the parent rocket.

Note that if $\Omega_R > \Omega_*$, the total spectrum has three parts: the CMB itself, which peaks at 1400 μm; the dust component, peaking at λ_{max}; and a residual component from the sources. If $\Omega_R < \Omega_*$, the dust and CMB parts peak at the same wavelength, so there are only two components. Nevertheless, the dust component is not a pure black body, so there is some distortion of the CMB spectrum in this case.

We should also mention that a dust background would also be expected to be anisotropic on the sky if it were produced by galaxies or a clumpy distribution of pregalactic dust. One can study the predicted anisotropy in this situation by allowing the dust to cluster like galaxies, for example, and computing the resulting statistical fluctuations. Various experiments have been devised, along the lines of the CMB anisotropy experiments, to detect such fluctuations, with success finally resulting from an analysis of data from the DIRBE measurement on COBE (Wright and Reese 2000). We shall return to this background, and its theoretical importance, shortly.

20.5 Number-counts Revisited

We discussed in Section 1.8 how the number–magnitude and the number–redshift relationships, in the past thought to be good ways to probe the geometry of the Universe, are complicated by the fact that galaxies appear to be evolving on a timescale which is less than or of order the Hubble time; an example is Figure 4.11. While evolution makes it very difficult to obtain the deceleration parameter q_0 from these counts, there is at least the possibility that they can tell us something about how galaxy formation, or at least star formation in galaxies, changes at relatively low redshifts. This, in turn, can yield useful constraints on theories of the origin of structures.

Again, this is an area in which considerable observational advances have been made in recent years. The possibility of obtaining images of extremely faint galaxies using CCD detectors has made it possible to accumulate number-counts of galaxies in a systematic way down to the 28th magnitude in blue light (so-called B-magnitudes). In parallel with this, developments in infrared technology have allowed observers to obtain similar counts of galaxies in other regions of the spectrum, particularly in the K-band. Since these different wavelength regions are sensitive to different types of stellar emission, one can gain important clues from them about how the stellar populations have evolved with redshift. Blue number-counts tend to pick up massive young stars and therefore are sensitive to star formation; longer wavelengths are more sensitive to older stars.

The blue number-counts display a feature at faint magnitudes corresponding to an excess of low-luminosity blue objects compared with what one would expect from straightforward extrapolation of the counts of brighter galaxies. The game is to try to fit these counts using models for the evolution of the stellar content and (comoving) number density of galaxies, as well as the deceleration parameter. The best-fitting model appears to be a low-density-universe model with significant luminosity evolution, i.e. the sources maintain a fixed comoving number density but their luminosities change with time. An independent test of this kind of analysis is afforded by the N–z or M–z relationship for the same galaxies. If pure luminosity evolution explains the excess counts, then one expects a significant number of the faint objects to be at very high redshifts. This actually seems not to be the case: the majority of these sources are at redshifts $z < 0.5$. One ought to admit, however, that the redshift distribution at very faint magnitudes is not well known. This issue is still quite controversial, but it may be that one is seeing a population of dwarf galaxies undergoing some kind of burst of star-formation activity at intermediate redshifts. This is some evidence that galaxies may be forming a significant part of their stars at low redshift, but the sources observed may be localised star formation within a much bigger object. Perhaps the apparent starburst could be induced in a similar way to that usually considered likely for the true 'starburst' galaxies mentioned in Chapter 4; they are somehow induced by mergers.

Number-counts in the infrared K-band appear to be quite different to that of the blue counts shown in Chapter 4. In particular there is an apparent *deficit* of galaxies at faint magnitudes, compared with a straightforward extrapolation of the bright counts. An examination of the colours $(B$–$K)$ of the galaxies suggests that the same population of galaxies is being sampled here as in the blue counts, but that the colours are evolving strongly with redshift.

One possible reconciliation of the blue and infrared counts is that mergers of galaxies have been important in the recent past. Perhaps the faint blue dwarfs merge into massive galaxies by the present epoch. The amount of merging required to achieve this is rather large, but perhaps compatible with that expected in hierarchical models of structure formation. At any rate it seems clear that at least a subset of galaxies have enjoyed a period of star formation, perhaps associated with the formation of a disc. Since the amount of metals produced by the known blue luminosity is comparable with that found in spiral discs, it may be that these objects are somehow related to the damped Lyman-α systems discussed above. Perhaps massive protodiscs, which do not undergo a burst of star formation at such low redshifts and thus appear in the blue population, survive to the present epoch as large galaxies with an extremely low surface brightness. Examples of such systems have been found, but would generally not be included in the normal galaxy surveys. These considerations might reconcile the apparent excess of high-column-density Lyman-α systems at $z \simeq 2$ compared with the number of normal spiral discs at the present epoch.

20.6 Star and Galaxy Formation

The partial and incomplete data we have about galaxies and the IGM at high redshift obviously make it difficult to say for certain at what redshift galaxy formation can have occurred. Obviously, it is unlikely that there is a definite redshift, z_g, at which galaxy formation occurred, particularly in hierarchical theories where structure forms on different scales continuously over a relatively long interval of time.

In fact, there is also considerable confusion about what galaxy formation actually is, and how one should define its epoch. Since galaxies are observed mainly by the starlight they emit, one might define their formation to be when most of the stellar population of the galaxy is formed. Alternatively, since galaxies are assumed to be formed by gravitational instability, one might define formation to have occurred when most of the mass of a galaxy has been organised into a bound object. There is no necessary connection between these two definitions. A galaxy may well have formed as a gas-rich system very early in the Universe, but suffered an intense period of star formation very recently. We shall therefore consider star formation and mass-concentration epochs separately and try to interpret various observations in terms of the epochs at which these can have happened.

Since we know most about the bright central parts of galaxies, say the part within $r_c \simeq 10h^{-1}$ kpc, it makes sense to define the epoch of galaxy formation in the second sense as the redshift by which, say, the mass within this radius reached half of its present value. This will be different for different galaxies, so one picks as a representative epoch the median redshift, z_g, at which this occurs. Spiral galaxies have prominent discs, so one could also usefully define z_d to be the median redshift at which half the mass of a present-day disc had been accumulated. According to most cosmogonical theories, the spiral disc is not the dominant mass within r_c, and the formation of a disc may well take place over an extended period of time. Studies of the dynamics of galaxies suggest that stars contribute a significant fraction of the mass within r_c. Accordingly we define z_* to be the median redshift at which half the stellar content (in long-lived stars of relatively low mass) of a bright galaxy has formed within r_c. We may similarly define z_m to be the redshift at which half the present content of metals, i.e. elements heavier than helium, was formed. In the standard picture, the initial gas content of a protogalaxy would have a chemical composition close to the primordial abundances and therefore a negligible fraction of metals. These would have to be made in stars as the galaxy evolves. Because most stars within r_c are relatively metal rich and most metals are in stars, it seems likely that $z_m > z_*$.

As we have already explained, we cannot give firm model-dependent values for any of the characteristic redshifts z_g, z_d, z_* or z_m. But can we at least place constraints on them, put them in some kind of order or, better still, obtain approximate values? This is what we shall try to do in this section. Although there has been a rapid growth of pertinent observational data, we will find that conclusions are not particularly strong. Notice also that z_g is the epoch which is in principle most closely related to the theoretical models of structure formation by gravitational instability. Unfortunately, it is also probably the furthest removed from observations. Nevertheless, we shall begin with some constraints on z_g.

The most obvious constraint comes from the fact that galaxies, once fully developed, have a relatively well-defined physical size. Galaxies, as we know them, could therefore only have formed after the time at which the volume they now fill occupied all of space. Depending on how one counts them, bright galaxies (which we shall restrict all these considerations to) have a mean separation of $4h^{-1}$ Mpc. The diameter of the bright central parts is $2r_c \simeq 20h^{-1}$ kpc. This suggests an upper limit on z_g of order 200, but this is decreased by the factor C by which a protogalaxy collapses. We therefore have

$$z_g < 200/C. \tag{20.6.1}$$

It is also the case that galaxies could not have existed when the mean cosmological density was greater than the density inside the galaxy. Suppose a galaxy has circular velocity v_c at the present epoch. An estimate of the mean density of its progenitor at maximum expansion is then

$$\rho_m \simeq \frac{r_c v_c^2}{G} \frac{3}{4\pi r_c^3} \frac{1}{C^3}. \tag{20.6.2}$$

According to the spherical collapse model (Section 15.1), this should be given by

$$\rho_m \simeq \tfrac{9}{16}\pi^2 \Omega \rho_c (1 + z_g)^3, \tag{20.6.3}$$

where we have taken z_g to be approximately the turnaround redshift. If $v_c \simeq 250\ \mathrm{km\ s^{-1}}$, then

$$z_g \simeq \frac{30}{\Omega^{1/3} C}, \tag{20.6.4}$$

which is consistent with Equation (20.6.1).

The problem with these estimates is that we do not really know how to estimate the collapse factor C accurately. The simple theory in Section 14.1 suggests $C = 2$, corresponding to dissipationless collapse, but as we already discussed in Section 15.7, this is probably not accurate. If galaxies formed hierarchically, the continuity of clustering properties has led some to argue against a large collapse factor, so that $C < 3$ or so. On the other hand, if our discussion of the origin of angular momentum in Section 15.9 is taken seriously, one seems to require a relatively large collapse factor for spiral galaxies to generate a large enough value of the dimensionless angular momentum parameter λ, while the appropriate factor for ellipticals should be of order unity. In 'top-down' scenarios or in the explosion picture the factor is difficult to constrain.

Now let us turn to z_*. The most obvious constraint on this comes from the fact that the evolutionary timescale for reasonably massive stars is of order 10^7–10^8 years. Since heavy elements need several generations of massive stars, a reasonably conservative bound on t_m, the time when $\rho = \rho_m$, is

$$t_m = \tfrac{2}{3}\Omega^{-1/2}(1 + z_m)^{-3/2} > 10^8\ \text{years}. \tag{20.6.5}$$

In terms of redshift, this gives

$$z_m < 20h^{-2/3}\Omega^{-1/3}, \tag{20.6.6}$$

and, according to the argument given above, we can also conclude that $z_* < z_m$.

At very high redshifts, $z > 10^3$ or so, the temperature is enough to ionise hydrogen and radiation drag ensures that clouds of plasma expand with the radiation background. Although this drag effect decreases after $z \simeq 10^3$ when recombination occurs, any material ionised by stars will still suffer from it; this will prevent any further star formation. After $z \simeq 10^2$ this can no longer occur. This suggests an upper limit of $z \simeq 10^2$ on z_* and probably also on z_g, since star formation is presumably required to halt collapse.

As we mentioned in Section 14.7, the behaviour of a gas cloud is determined by the rate of radiative cooling if Compton scattering off the CMB radiation is negligible, i.e. when $z < 10$. If galaxy formation proceeds hierarchically, the lower mass end of the distribution will cool slowly since the material in such objects will have a relatively low temperature. Higher-mass objects, corresponding to temperatures around 10^4 K and above, will cool rapidly and collisional ionisation will be important; star formation presumably ensues. The mass scale when this becomes important is easily calculated to be around 10^{10}–$10^{12} M_\odot$, in good accord with the typical mass scale of bright galaxies. This agreement would not exist if Compton cooling were important during galaxy formation and this therefore provides a certain amount of motivation for the requirement that z_g and z_* are both less than 10.

These theoretical comments on z_* are disappointingly vague because our understanding of star formation is poor, even for nearby objects. However, once again observations have led the way and a clutch of different programmes have resulted in estimates of star-formation rates from observed colours and synthetic stellar populations; a prominent example of this kind of study is described in Madau *et al.* (1996) using observations of the Hubble Deep Field shown in Figure 4.10. The plot shown as Figure 20.2 is generically known as a 'Madau Plot'. It appears that these observations favour a scenario in which star formation peaks at moderate redshift. The theoretical curve shown in Figure 20.2 also shows that, in a broad-brush sense, this behaviour can be accounted for in hierarchical clustering models (Baugh *et al.* 1998). It is also noteworthy that the integrated star formation that these observations imply is not observed. The infrared background measurements mentioned above perhaps explain why: about half the optical starlight ever produced in the Universe has probably been absorbed by dust and re-radiated in the infrared part of the spectrum.

Although the vaguer arguments we gave above admit the possibility that galaxy formation could occur relatively early, at redshifts up to around 10, in hierarchical models galaxies are expected to form at redshifts much lower than this: $z_g \simeq 1$. The reason for this is that the clustering pattern of galaxies, as measured by the two-point correlation function, evolves very rapidly with time in models based on the Einstein–de Sitter universe. If galaxies formed at redshifts $z_g \simeq 10$, one would expect drastic steepening of the correlation function between $z = 10$ and $z = 0$, which is incompatible with the observed slope. This problem, though rather difficult to quantify, does seem compelling in dark-matter models where light traces mass.

Various other kinds of observations are capable of probing the Universe up to the redshift of quasar formation, so it is interesting to see if these can yield any clues about z_g or z_*.

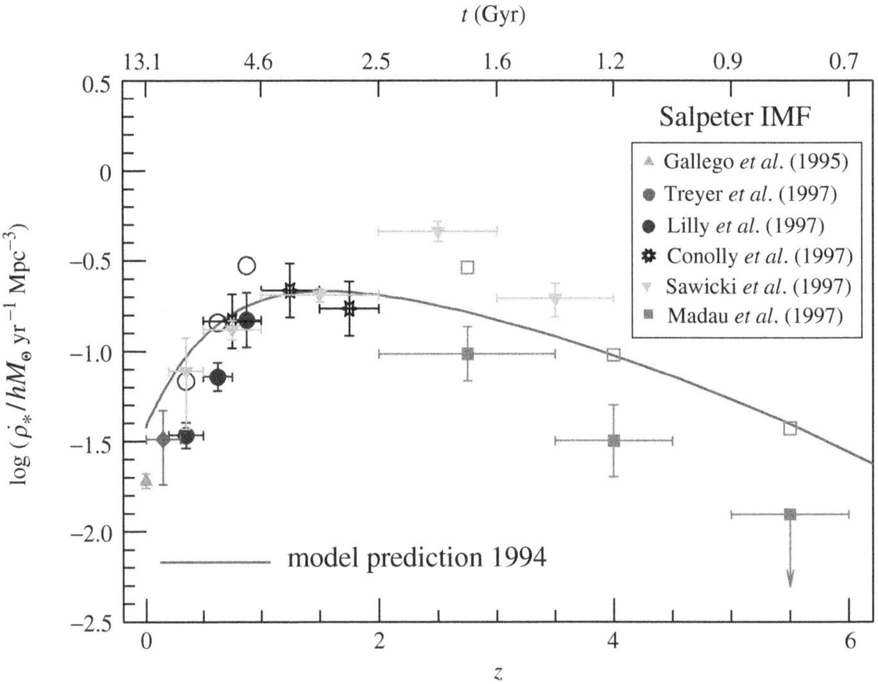

Figure 20.2 The star-formation history of the Universe. Estimates of the star-formation rate as a function of redshift, along with the predictions of a semi-analytic model of galaxy formation described in Chapter 14. Picture courtesy of Carlton Baugh (Baugh *et al.* 1998).

First there is the question of whether some galaxies may have formed at $z < 1$. The number-counts, discussed in Section 20.5, certainly show evidence of strong evolution in galaxy properties at these redshifts. How the faint blue galaxies fit into this picture is still an open question: are they connected with the epoch of galaxy formation, or are they merely a sideshow? There is also the problem posed by the population of starburst galaxies, which, though usually dwarf galaxies, are forming stars at a prodigious rate at the present epoch. It has yet to be established, however, if these sources are really young in the semi-quantitative sense defined above. They could merely have evolved more slowly, as a consequence of their cooling properties. If they are young, however, then they certainly suggest the possibility that larger galaxies may also have formed recently. A more direct argument is based on the relative age of the disc and central spheroid of the Milky Way, as estimated by stellar evolution arguments. If the disc turns out to be half the age of the spheroid, one has $z_d < 1$ regardless of the value of z_g. It appears that there are disc stars as old as 12 Gyr, which might therefore be a reasonable estimate of t_d. In an $\Omega_0 = 1$ Universe we therefore have

$$1 + z_d \simeq \left(\frac{t_0}{t_0 - t_d}\right)^{2/3}, \tag{20.6.7}$$

so that $z_d \simeq 2$ if $t_0 \simeq 15$ Gyr.

At redshifts of order unity and above, galaxies are still reasonably observable and one can attempt therefore to study their stellar populations to see how much evolution there has been between $z \simeq 1$ and the present. There are some notable differences between galaxies then and now: in the past, galaxies were luminous and had younger-looking stellar populations, they were richer in gas and there was also more merging. These differences are, however, not extreme. The giant radio galaxies mentioned in Section 20.2 do seem brighter than would be expected without evolution, but they are only about one magnitude brighter at $z \simeq 1$ than at redshifts much lower than this. This is consistent with relatively slow evolution from a much higher redshift of formation. Many features of 'normal' galaxies at $z \simeq 1$ seem to be characteristic of relatively old stellar populations and there is little evidence for significant evolution in the (comoving) number density of such objects with time. This suggests that both z_g and z_* are rather greater than unity.

The redshift at which galaxies can be observed in large numbers was pushed back further by Steidel *et al.* (1996), who implemented a novel technique for targetting galaxies at high redshift. By choosing appropriate filters they were able to select objects using colours in such a way that preferentially picked out objects in which the ionisation limit of the Lyman series (in the UV part of the spectrum of a galaxy in its rest frame) is redshifted into an optical band. This allows the observer to pick a small subset of galaxies with extreme colours for follow-up spectroscopy. The galaxies thus found tend to have redshifts $z \sim 3$. This has been a remarkably successful approach, but the most interesting thing is that the galaxies found seem to have roughly the same number-density as present-day bright spirals and have similar clustering properties.

Observations at higher redshift are much more difficult and have only become feasible in the last five years or so. We have discussed some of these observations already in Sections 20.2–20.4, so let us now discuss them in the context of structure formation.

First, the damped Lyman-α absorbers discussed in Section 20.2 are usually interpreted as the progenitors of galactic discs. Certainly the mean mass density seems to be of the correct order, but they do seem to be more abundant than one would expect by extrapolating the properties of present-day discs back to redshifts of order 3. If they are not protodiscs, then presumably z_d is relatively low, which again poses problems.

Secondly, as discussed in Section 20.2, there have been a number of indications of relatively old-looking galaxies at high redshifts, $z > 3$. The stellar ages of these objects are difficult to determine because of the redshifting of the optical and UV spectra into the infrared region. None of the objects so far claimed to have been seen has been unambiguously identified as a fully formed galaxy, but if one such object is ever found it will place very important constraints on z_*.

The highest-redshift objects known to observational astronomy are the quasars. Again the evolution of the number density of these objects with time is difficult to quantify, but it seems relatively constant (at least for the brightest ones) from $z \simeq 2$ up to $z \simeq 4$. This suggests that $z_g > 4$, if quasars are housed in galaxies.

Finally, the highest-redshift quasars show that the IGM (Section 20.3) was ionised by $z \simeq 4$. The consequences of this for z_g or z_* are also unclear. One might be led to conclude that $z_g > 4$ on the grounds that galactic stars must have ionised the IGM. On the other hand, a separate population of very massive stars might have formed before galaxies and caused this ionisation.

These arguments are clearly all compatible with $z_* \gg 4$ and $z_g > 4$ but do not rule out more recent epochs. We shall have to wait for further observational breakthroughs before anything more concrete can be said. This is indeed an area where a tremendous observational effort is being directed, and one can expect much to be learned in the next few years.

20.7 Concluding Remarks

In this chapter we have discussed the evolution of the Universe between $t_{\rm rec}$ and the present epoch. Clearly, many questions remain unanswered but we hope we have conveyed to the reader some idea of the intense activity and progress which is taking place in this field. This chapter and the previous three have been aimed at a somewhat more detailed level than the earlier chapters in order to provide a 'bridge' between the fundamentals, covered in Parts 1-3, and some of the areas of particular current research interest. These chapters should make it clear that we still have a long way to go before we can claim to have a complete understanding of the origin and evolution of cosmic structures, but we are making considerable progress both theoretically and observationally. The basic idea that structures form by gravitational instability from small-initial-density perturbations seems to account, at least qualitatively, for most of the observational data we have. Whether this will still be the case when more data are acquired remains to be seen. There is a very good chance that the cosmological parameters H_0 and Ω will be pinned down in the next few years or so. This will also make it easier to construct rigorous tests of these theories. In any event, one thing we can be sure of is that the question of the origin of galaxies and the large-scale structure of the Universe will remain the central problem in cosmology for many years to come.

Bibliographic Notes on Chapter 20

Peebles (1993) contains excellent accounts of the astrophysics of the intergalactic medium. A good review of the properties of Lyman-α absorption systems is given in Wolfe (1993); for some theoretical ideas see Rees (1986). The cosmic X-ray background was reviewed by Boldt (1987). For detailed discussion of the infrared background see Bond *et al.* (1986) and Carr (1994); see also Signore and Dupraz

(1992). The problem of the faint blue galaxies is discussed by Ellis (1993); interpretation of the faint counts in hierarchical models is attempted by Kauffmann *et al.* (1994). Some of this chapter is based on an entertaining discussion described in Frenk *et al.* (1989).

Problems

1. A population of sources in a flat matter-dominated (Einstein–de Sitter) universe has a number-density n_0 at the present epoch and a monochromatic luminosity $P(\nu) \propto \nu^{-\alpha}$ at frequency ν. Show that the flux density $S(\nu_0)$ observed at the present epoch from a source at redshift z satisfies

$$S(\nu_0) = P(\nu_0)(1+z)^{1-\alpha} D_{\mathrm{L}}^{-2},$$

   where D_{L} is the luminosity distance.

2. Following on from Question 1, if sources are neither created nor destroyed as the universe expands, show that the number of sources observed per steradian with redshift $< z$ is

$$N(z) = \tfrac{8}{3} n_0 \left(\frac{c}{H_0}\right)^3 \left[1 - \frac{1}{\sqrt{1+z}}\right]^3.$$

3. Following on from Question 2, show that the integrated background light intensity at frequency ν_0 from this population of sources is

$$I(\nu_0) = \frac{2 c n_0 P(\nu_0)}{H_0(2\alpha + 3)}.$$

21

A Forward Look

21.1 Introduction

From our vantage point at the beginning of the 21st century, we can look back on a hundred years of truly amazing progress in the development of astronomical techniques and technology. Ground-based optical observatories, such as the Keck telescopes and the VLT, offer collecting areas many times larger than their predecessors at Mt Wilson and Mt Palomar, and are equipped with much more sophisticated instrumentation. Perhaps the most important developments, however, have been in the introduction to astronomy of entirely new wavelength regimes. Radio astronomy only came into being after World War II, and X-ray astronomy only in the 1960s with the development of space missions. Some regions of the spectrum, such as the submillimetre region, are still relatively unexplored, but here too progress has been dramatic over the past decade or so.

There are also potentially important phenomena that have not yet emerged as practical possibilities for observation. A prominent example related to cosmology is neutrino astronomy; direct detection of the low-energy neutrino background discussed in Chapter 8 would furnish an important test of the Big Bang. This is as yet a remote possibility. More likely to be feasible in the very near future is the detection of gravitational waves, which we discuss briefly in Section 21.10.

Faced with this continuing revolution driven largely by advances in instrumentation and manufacturing techniques, it seems almost to be inviting ridicule to suggest that the future might be anything like as exciting as the past. But a glance at some of the planned projects and space missions to come over the next two decades suggests that this is indeed very likely to be the case. What is different about the future is that, in contrast to the dawn of the 20th century, we now have a robust theoretical framework within which we can interpret observations and plan future strategies. The bulk of this book has been devoted to this framework.

We are not saying that the emerging consensus model of the Universe is exact in every detail, nor that we are anywhere near a complete understanding of the formation and evolution of cosmic structure. But we do have a much better idea of where the interesting questions lie, and how seemingly disparate pieces of the cosmic jigsaw may be related to each other than was the case even a few years ago.

21.2 General Observations

Before discussing the future of observational cosmology, it is worth taking stock of the present status of this area. Until relatively recently, extragalactic astronomy would have been described in terms of a large number of relatively distinct niches, including, for example,

- cosmography (i.e. surveys);
- distance scale studies (i.e. measurement of H_0);
- the classical cosmological tests (number-counts, angular-diameter and magnitude redshift tests, etc.);
- gravitational lensing (multiple images, arcs and weak lensing);
- studies of galaxy clusters;
- detailed studies of galaxy morphology, stellar populations and kinematics;
- galaxy formation and evolution;
- extragalactic radiation backgrounds (infrared and X-ray);
- active galaxies, AGN, quasars and radio galaxies;
- the intergalactic medium, absorption line studies and the like; and
- element abundances and chemical evolution.

Over the last two decades the overlaps between these areas have become blurred owing to the development of a fairly robust theoretical framework that enables a broad-brush theoretical description of the formation of individual structures such as galaxies and quasars within an overarching cosmological framework.

This framework still has a number of uncertain constituents, but basically involves the hypothesis of a dominant component of collisionless dark matter into which density fluctuations are imprinted in the early Universe. These fluctuations grow until small clumps of dark matter collapse, and begin to merge hierarchically into larger structures. The evolution of the structure thus formed has two particular aspects. One is the formation of *cool matter*, essentially meaning the cooling of baryonic material at high redshift, its incorporation in dark-matter clumps, the fragmentation of gas, the formation of stars and the accompanying generation of dust and complex chemistry. This part of the story can be diagnosed by optical, infrared and submillimetre studies. On the other hand, there is also the *hot* universe involving the formation of very massive black holes and accompanying accretion processes, and the hot intergalactic and intracluster media. The hot universe is typically probed using X-ray studies. Although we have emphasised the coming together of different types of study in recent times, it is still fair to say that the relationship between galaxy formation and nuclear activity and the role of the central black holes in the galaxy-formation process remains poorly understood.

Theoretical developments, including the application of supercomputer simulations, have helped target observational strategies as well as elucidating the possible links between galaxy formation and internal kinematics, and between large-scale structure and galaxy morphology. On the observational side, huge ongoing redshift surveys are mapping the positions of hundreds of thousands of galaxies in representative cosmological volumes. At the other extreme, the development of integral field units, such as SAURON, are displaying unprecedented detail about the internal structure of nearby galaxies. A consensus may also be emerging about the parameters of a cosmological model that describes the evolution of the bulk properties of the Universe, based principally upon the cosmic microwave background and Type Ia supernova searches.

So what are the future directions for observational studies in this area? Some goals are obvious: higher sensitivity, higher angular resolution and higher spectroscopic resolution at existing wavelength ranges will allow more detail to be gleaned and fainter objects to be studied. On the other hand, fields such as weak gravitational lensing lead one to develop wider-field instruments. The desire to probe evolution by moving to extremely high redshift motivates a shift to longer wavelength, as does the desire to avoid excessive extinction of stellar light by dust. On the other hand, the wish to unveil more of the hot universe suggests moving to shorter wavelengths and higher-energy X-rays.

In the following sections we will quickly survey a few of the upcoming developments across the electromagnetic spectrum, starting with the hot universe and X-rays.

21.3 X-rays and the Hot Universe

The current scene in extragalactic X-ray astronomy is dominated by two space missions: Chandra (the telescope formerly known as AXAF) and XMM/Newton. Of the two, Chandra produces the sexiest pictures because it has a high-resolution camera capable of resolving sub-arcsecond detail, while the angular resolution of XMM is only around 5 arcsec. Chandra also has a higher sensitivity. The two missions are nevertheless complementary because XMM/Newton is more suitable for survey work than Chandra. They also have different instrumentation. Both work in the range 0.1 keV to around 10 keV.

The particular difficulties of X-ray astronomy are illustrated nicely by these two satellites. The most important aspect of X-ray-telescope design is that the mirrors work at grazing incidence and one is therefore more or less forced to have a very long focal length in order to obtain any reasonable angular resolution. Chandra has four pairs of mirrors and a focal length of about 9 m; XMM/Newton has three sets of nested mirrors and a focal length of about 7.5 m. These require very large platforms in order to operate in space, with consequent implications for expense.

The difficulties associated with X-ray imaging will not be overcome easily. For the time being, the next major developments in this area will be space missions devoted to higher-throughput spectroscopy. Although these missions will have significant gains in sensitivity, these are somewhat incremental and are obtained at an enormous financial cost.

For example, consider the planned ESA mission *XEUS*. Among the performance goals required of XEUS are the following.

1. Spectral capability at flux levels less than 10^{-17} erg cm^{-2} s^{-1}. This is a factor ~ 100 fainter than the XMM/Newton limit and about a factor 10 fainter than Chandra.
2. Deep surveys to a flux limit of 10^{-16} erg cm^{-2} s-1. Typical redshift limits for extragalactic sources would be in the range $z \sim 10$–15.
3. Angular resolution (at 1 keV) of better than about 5 arcsec is required to avoid source confusion at these levels.
4. Energy resolution of 1–10 eV is required to undertake detailed spectroscopic studies of redshifted line profiles.

XEUS beats the focal-length problem by being made from two spacecraft, called the MSC (which contains the mirrors) and the DSC (which holds the detectors). These are held in station about 50 m apart producing a telescope about five times longer than Chandra. (This idea is taken further by the NASA mission Constellation-X, which is a flotilla of spacecraft rather than two.) XEUS will enable much more detailed spectroscopy of fainter objects than is presently possible. Its imaging capability will, however, still be restricted with a resolution at 1 keV of about 2 arcsec.

It will be a very long time before X-ray imaging can match the standards of optical telescopes, but when it does the results promise to be spectacular. For example, a NASA proposal called MAXIM (MicroArcsecond X-ray Imaging Mission) introduces the concept of interferometry to the X-ray region of the spectrum. With a planned baseline of only 1.4 m it should achieve angular resolution of about 100 μarcsec, about a factor 5000 better than Chandra. (This resolution will be enough to resolve the event horizon of the black hole at the centre of M87.) The major obstacle is that the two vehicles making up MAXIM – it is similar to XEUS in this regard – must be held in station by telemetry to this accuracy although separated by a staggering 500 km. If this can be achieved, it may be possible eventually to obtain resolution measured in hundreds of nanoarcseconds by interferometry.

21.4 The Apotheosis of Astrometry: GAIA

We could not resist the opportunity presented by this invitation to say a few words about GAIA. This mission is a direct descendent of the highly successful ESA astrometry mission Hipparcos, which measured accurate parallaxes and proper motions for stars inside our Galaxy. GAIA's principal aim is to make an accurate three-dimensional map of more than a billion stars in the Milky Way, including detailed photometric studies to characterise luminosities, temperatures and chemical compositions for these stars. GAIA will work by continually scanning the whole sky and repeatedly measuring the positions of all objects it detects down to a limiting V magnitude of 20. Positions will be measured to an astonishing 10 μarcsec (for sources at 15th magnitude) and on-board software will allow variable and bursting sources to be catalogued. In short, GAIA will produce a vast galactic census. Ostensibly this makes GAIA a galactic mission rather than an extragalactic one, but GAIA will in fact make enormous contributions to extragalactic astronomy in a range of environments.

Within the Local Group of galaxies, GAIA will analyse millions of stars within the Large and Small Magellanic Clouds, allowing the internal dynamics and interactions of these galaxies to be studied by stellar kinematics, as well as accurate calibration of the stellar luminosities in these galaxies. This is important in order to compare the information we have about such properties in a large disc galaxy (the Milky Way) to small or medium-sized irregular galaxies. Beyond the SMC and LMC there are eight known dwarf satellite galaxies of the Milky Way. These allow the mass distribution of the galactic halo to be traced, as well as having interesting internal dynamics in their own right. Further afield, stars in M33 and M31 should be amenable to proper motion studies, so that rotation curves of these galaxies can be constructed in a manner independent of line-of-sight velocity data. In effect, GAIA will see the Andromeda Nebula rotate on the sky. As far as the Local Group as a whole is concerned, accurate positions and transverse velocities of all its members will allow detailed studies of the mass distribution and possibly its formation history.

GAIA will also have lessons to teach us about the distribution of galaxies on scales larger than the Local Group. One of the principal science goals relates to the distribution of structures in the local Universe. Very-large-scale structures in the galaxy distribution are already being mapped in great detail by redshift surveys such as the Sloan Digital Sky Survey (SDSS) and the Anglo-Australian 2dF Galaxy Redshift Survey, but GAIA will complement these studies by producing an all-sky magnitude-limited survey including multicolour photometry of around a million galaxies.

Because GAIA will be able to detect any object with an I band magnitude less than about 20, it should be possible to detect supernovae with distance moduli up to about 39 in magnitude. This corresponds to a distance of around 500 Mpc or redshift $z \sim 0.1$. It is therefore anticipated that around 100 000 supernovae will be detected in 4 years of GAIA operation. A particular benefit will be the discovery of supernovae in galaxies of very low surface brightness, which are typically excluded from present surveys.

The limiting V magnitude of 20 will yield a census of around 5 million quasars. Since multicolour information will be available it ought to be possible to identify quasars efficiently by colour selection, and since the objects would be expected to have redshifts in the range $z \sim 0.2$–0.3 it is expected that redshifts to an accuracy of about 0.01 will be obtained. To get the whole idea in perspective, GAIA will provide a quasar catalogue about 50 times larger than that resulting from SDSS.

The quasar catalogue is interesting in itself, but it should also allow a direct link between GAIA's astrometric references and an inertial frame so that Mach's 'fixed stars' will be superseded by GAIA's 'fixed quasars'.

21.5 The Next Generation Space Telescope: NGST

The obvious success of the Hubble Space Telescope obviously lends strong support to the idea of future space telescopes operating around the optical part of the spectrum. The NGST was originally conceived to be an optical/near-IR telescope with a mirror of diameter around 8 m placed in space with a mission lifetime of around 10 years. The idea of an 8 m class telescope in space is undoubtedly appealing. After all, it is not that long since 8 m ground-based telescopes came on the scene.

The addition of better IR capability also results in great advances over HST. However, there are obviously difficulties in getting a mirror as large as 8 m into space, certainly if it is constructed in a manner anything like the mirrors at ground-based facilities. It is generally believed that NGST will have a deployable mirror of some kind, although the final design is not finalised. Moreover, it seems likely that the NGST may be 'de-scoped' to involve a mirror of smaller diameter, perhaps 6 m or thereabouts. Since the chief improvement over the HST is collecting area, these cost-cutting moves do eat into some aspects of the science case for NGST as opposed to, say, the ultra-large ground-based optical telescopes discussed in the next section.

The instrumentation to be carried by the NGST is also uncertain, but it seems likely that it will involve at least a near-IR/visible camera capable of operating from about 0.6 to 5 μm, a multi-object spectrograph functioning in the range 1–5 μm, and possibly a camera/slit spectrograph working at longer wavelengths than 5 μm. The prospect of adding integral field units to the NGST's battery are truly awesome, but whether this will be practically possible remains to be seen.

Some of the principal areas of extragalactic astronomy in which the NGST would be expected to be particularly important are

- weak lensing studies;
- studies of the IGM at high z;
- high-z supernovae searches;
- studies of gamma-ray-burst hosts;
- microlensing in the Virgo cluster;
- very deep imaging and spectroscopic surveys;
- cluster-galaxy evolution;
- the galaxy–AGN connection; and
- obscured star formation at high-z.

Just to take the first of these as an example, the principal benefit of the NGST to weak lensing is the ability to study lensing distortions as a function of redshift to extremely faint flux limits (owing to the large collecting area). It is likely that ground-based survey telescopes (such as VISTA) will have much larger fields than the NGST but will clearly lack the ability to go as deep. Such studies will show how the dark-matter distribution evolves with redshift in a robust fashion that should complement CMB experiments.

21.6 Extremely Large Telescopes

Although the development of the NGST seems to be the obvious step forward in optical extragalactic astronomy, one should not forget the enormous strides that have been taken in traditional optics. As far as ground-based optical telescopes are concerned, the diameter of the 'next' telescope has doubled roughly every 30 years over four centuries since Galileo. The last three notable 'big things' (Mt Wilson, Mt Palomar and the Keck Observatories) fit this rule very well. We now live in the era of 10 m diameter facilities.

The immediate advantage of moving into space, exploited successfully by the HST and anticipated by the NGST, is that one can avoid the blurring effect of the Earth's atmosphere and reach the diffraction limit of a relatively large aperture. On Earth, large telescopes are limited by atmospheric 'seeing' effects long before they reach the famous $1.22\lambda/D$. However, the construction of 10 m class telescopes has been accompanied by impressive developments in adaptive optics (AO) that may allow diffraction-limited performance to be reached even for monstrous mirrors of order 100 m in diameter. It may therefore be that the next generation of incredibly large telescopes will vie with the NGST for scientific predominance. At the very least, 100 m class ground-based telescopes will be complementary to the NGST.

We can illustrate this sort of development with the European Southern Observatory's proposed Overwhelmingly Large Telescope (OWL) (other suggestions are on the table). OWL has a diameter of 100 m, which is about ten times the total collecting area of all telescopes ever built (assuming it has no competitors). This immense collecting area is its real asset compared with the NGST. It should achieve limiting visual magnitudes of 38 and have angular resolution measurable in milliarcseconds in the V band. Such a performance will only be realised with full AO involving around 500 000 active elements moved by actuators to counteract the irritation of seeing. The scale of OWL limits the field of view to about 3 arcmin^2, otherwise the detectors needed would be enormous; each arcsecond pixel occupies about 3 mm in the focal plane.

A 100 m telescope with seeing-limited performance would be little more than an enormous light bucket, useful perhaps for spectroscopy but not for imaging. In any event, even spectroscopy requires one to avoid source confusion. On the other hand, developments in optical interferometry may allow comparable resolution with OWL but without the sensitivity arising from the huge collecting area. It would appear, therefore, that overwhelmingly large filled-aperture telescopes are definitely on the horizon, and for good reasons.

It is clear that the construction of OWL requires the solution of numerous engineering problems, such as the flexure properties of the structure required to house it, the figuring of the mirrors, and the design and implementation of the AO system. It has been claimed that OWL would represent a technological milestone comparable with the invention of the telescope itself, in that it will have to break the scaling law that has since 1600 related cost C to aperture diameter D, in the form of a relation $C \propto D^{2.6}$. This does, however, appear at this stage to be feasible but with a price tag of at least \$1 billion and a timescale of at least 15 years. Assuming it can be done, what is the payoff from OWL for extragalactic astronomy?

For a start, the exquisite imaging potential of OWL in the optical spectrum would enable an unparalleled opportunity to study star formation directly at enormous distances. Individual HII regions could be resolved in galaxies at redshifts $z \sim 2$–3 (i.e. in galaxies seen in the Hubble Deep Field). Today high-redshift stellar populations are probed by measuring integrated quantities produced by unresolved objects (such as emission line fluxes). With OWL these unresolved components could be resolved into their stellar constituents. High-redshift supernovae ($z \sim 10$) also fall within the range of OWL. Studies of star-formation rates as a function of redshift using supernovae of various types are therefore feasible.

There is also an obvious synergy with the NGST, for the reasons I alluded to above. While the NGST can perform all-sky surveys to find interesting objects, telescopes like OWL could perform detailed spectroscopy, much as the Keck telescopes are used today to follow up HST observations.

Yet it is probably with regard to the extragalactic distance scale that OWL will be most revolutionary. For example, Cepheid variables with distance modulus $m - M \simeq 43$ (corresponding to a redshift $z \sim 0.8$) could be measured and calibrated. This allows not only the measurement of H but also its dependence on redshift without the need to use Virgo as a stepping stone. This, of course, assumes that the fields involved are not too crowded.

OWL will also be able to resolve individual solar-type stars in Virgo galaxies, study white dwarfs in M31 and possibly also detect brown dwarfs in external galaxies. There are also a host of galactic topics to which it could be applied, including extra-solar planet searches, but these are beyond the scope of my brief for this review.

21.7 Far-IR and Submillimetre Views of the Early Universe

Such is the attention lavished in cosmological circles upon the Planck Surveyor, to be launched in 2007, that one might forget that another important mission is to be launched at the same time. The Far-Infrared Space Telescope (FIRST), soon to be renamed Herschel, will share the launch but will part company with Planck in order to carry out its own independent scientific programme. Equipped with a 3.5 m diameter passively cooled mirror and operating in the wavelength range between 60 and 670 μm, it pushes sensitivity in the far-IR region to levels comparable with that reached by ground-based facilities like the VLT in the optical. It will be able to study continuum emission from extragalactic dust sources, as well as molecular and atomic line emission.

The most exciting developments at long wavelengths over the next 20 years, however, will come from the Atacama Large Millimetre Array (ALMA), which will operate in the millimetre to submillimetre region of the spectrum. ALMA will take some time to assemble, but is hoped to begin operations in a partial sense within a decade. Although existing submillimetre facilities, especially SCUBA, have demonstrated the interest likely to be found at these wavelengths, but even the possible upgrades of this facility will be strongly limited by the poor angular resolution that makes source identification well nigh impossible.

ALMA is an interferometer, and it beats the resolution problem plaguing single-dish observations at such long wavelengths by combining 64 antennae in a variety of configurations with baselines from 150 m to 10 km. Operating at wavelengths from 10 mm to around 350 μm (providing substantial overlap with FIRST), its sensitivity will be about 10 times greater than FIRST's optimal performance or indeed the peak sensitivity of large optical telescopes like the VLT. Added to this sensitivity is the exquisite angular resolution of 10 milliarcsec, which is about 10 times better than the HST or the nearest directly comparable radio telescope, the Very Large Array (VLA). The spectral performance is likewise impressive. A velocity resolution of about 0.05 km s^{-1} is anticipated, allowing detailed kinematic studies.

Not only will ALMA be able to use its sensitivity at long wavelengths to beat dust extinction, but it will also be able to probe molecular emission at very high redshift. Among the major science goals for ALMA in the extragalactic arena are

- kinematics of obscured nuclei and starbursts;
- detailed mapping of C, N, O and S in galactic discs;
- detection of H_2O and O_2 in galaxies;
- imaging thermal dust at $z \sim 10$;
- kpc-scale resolution of dust in AGN and QSOs;
- Sunyaev–Zel'dovich measurements (complementary to Chandra); and
- studies of radio galaxies.

It is likely that ALMA, perhaps in tandem with X-ray studies, will finally resolve the question of what kinds of sources make up the extragalactic X-ray background.

In the longer term, perhaps one can imagine ALMA forming the core of a millimetre-wave VLBI network, in a similar vein to radio VLBI.

21.8 The Cosmic Microwave Background

We devoted all of Chapter 17 to the cosmic microwave background so we shall comment only briefly upon it here. The much-vaunted satellite missions MAP (NASA) and Planck Surveyor (ESA) are the next developments in this field; MAP is in fact already in space. One of the problems with space missions is that the design tends to be 'frozen-in' many years before launch. One of the consequences of this for CMB studies is that, while waiting for the satellites to be developed and launched, detector technology (particularly bolometers) has surged ahead. Balloon-borne experiments using this new technology have consequently beaten the satellites to the detection of acoustic peaks in the CMB temperature pattern. This is not to say that MAP and Planck are now redundant. Not only will they provide important independent tests of the balloons experiments, they will also allow more detailed studies of foregrounds, Sunyaev–Zel'dovich measurements and, in the case of Planck at least, measurements of the polarisation pattern. In this field the medium-term future is likely to be dominated by these aspects of the CMB sky.

21.9 The Square Kilometre Array

Our gradual move to longer wavelengths has now brought us firmly into the radio region of the spectrum, and to perhaps the most impressive development of all, the Square Kilometre Array (SKA). This facility will operate at frequencies from about 0.15 to 20 GHz, and have at least 100 interferometer beams. It will probably involve about 30 individual radio telescopes of effective diameter about 200 m, adding up to approximately 10^6 m^2 of collecting area. These will be spread over a synthetic aperture about 1000 km in diameter. The central region of the array is close-packed to achieve high sensitivity, while an extended set of outriggers provides higher resolution through aperture synthesis. The resulting performance parameters are astonishing:

- angular resolution less than 0.1 arcsec at 1.4 GHz (comparable with the Hubble Space Telescope);
- a spectral coverage of more than 50% ($\nu/\Delta\nu < 2$);
- a spectral resolution good enough for detailed kinematics ($\nu/\delta\nu > 10^4$);
- a huge field of view ($\sim$ 1 square degree, i.e. larger than the full Moon); and
- a sensitivity more than 100 times better than anything currently available.

In some respects the SKA will be an enormous integral field device, achieving imaging and spectroscopy simultaneously both at great sensitivity. For these reasons alone the SKA could fairly objectively be called the world's premier astronomical imaging instrument.

Many technological, financial and political hurdles will have to be overcome before the SKA is built, but the payoff for science is enormous. Among the extragalactic science tasks it could undertake are

- probing the ionisation history of the Universe using 21 cm radiation;
- large-scale structure via redshift surveys in neutral hydrogen;
- extragalactic star-formation studies;
- redshifted molecular lines, e.g. CO at $z > 4$;
- galaxy rotation curves and Tully–Fisher studies using 21 cm radiation;
- mapping the Lyman-α forest in 21 cm radiation; and
- lensing surveys and dark-matter probes.

Let us expand on some of these items.

The key physics behind many of these tasks relates to 21 cm (HI) radiation produced by hyperfine transitions in hydrogen which, even highly redshifted, can be detected by SKA. Heating of the intergalactic medium (IGM) resulting from the first generation of stars will result in a coupling of the spin temperature of the IGM to the kinetic temperature of the gas, so that it differs from the temperature of the cosmic microwave background. This situation produces a characteristic pattern of 21 cm emission and absorption superimposed on the Cosmic Microwave Background which can be used to map the effects of the 'first light' to form in the Universe. Although high-redshift objects such as quasars have already been detected, and IR measurements may allow some very-high-redshift sources to be detected, it is always going to be difficult to beat the effect on surface brightness due to cosmological expansion with such observations. Studying the distribution of cosmic HI will avoid this difficulty. Among the key questions to be answered by such studies will be the following.

- When did the first stars form?
- What are the first energy sources?
- How large were the primordial density perturbations?
- How did collapsing objects evolve?

In this era of large galaxy redshift surveys it is also worth expanding upon the capabilities that SKA has in that direction too. One of the principal uncertainties in understanding how galaxies and large-scale structure form and evolve is relating the distribution of optical light (through which galaxy surveys are constructed) to that of gravitating mass (which is by and large what theory can predict). Ongoing surveys include on the order of a million galaxy redshifts. In 12 months of observing time, one could expect to detect around 10^7 galaxies in HI in a volume of order 10^8 Mpc^3, which is about a factor of ten increase in both volume and number. Being detected in neutral gas, such a survey would also furnish information about the clustering of matter which complements that provided by optical emission from stellar populations. Accompanied by detailed HI kinematics of the galaxies (e.g. Tully–Fisher studies), the possibilities for constraining galaxy-formation theory are revolutionary.

As a final comment on SKA science, it is worth mentioning the enormous advantage it has for gravitational lensing studies. Not only does it have a much larger field than comparable optical/IR facilities but it also has a very well-defined point-spread function which will enable higher signal-to-noise measurement of individual galaxy ellipticities.

21.10 Gravitational Waves

We have touched briefly on gravitational waves a few times during the course of this book, largely in connection with their possible production during inflation and role in the production of anisotropies in the cosmic microwave background. Most physicists think that gravitational radiation must exist, although they are yet to be detected directly. One of the important results to emerge from Maxwell's theory of electromagnetism was that it was possible to obtain solutions to Maxwell's equations that describe the propagation of an electromagnetic wave through a vacuum. Analogous solutions can be obtained in Einstein's theory, and these represent what are known as gravitational waves or, sometimes, gravitational radiation. The properties of, and searches for, gravitational radiation constitute a rich field all of their own so we cannot give a complete picture here (see, for example, Thorne 1987). What we will do is give a quick summary of their properties and focus on some of the possibilities for gravitational wave cosmology, if and when such radiation is directly detected.

Gravitational waves represent distortions in the metric of space–time in much the same way that fluctuations in the density of matter induce distortions of the metric in perturbation theory. The metric fluctuations induced by density fluctuations are usually called scalar perturbations, whereas those corresponding to gravitational waves are generally described as tensor perturbations. The reason for this different nomenclature is that gravitational waves do not result in a local expansion or contraction of the space–time. Scalar perturbations can do this because they are longitudinal waves: the compression and rarefaction in different parts of the wave correspond to slight changes in the metric such that some bits of space–time become bigger and some smaller. Gravitational waves instead represent a distortion of the geometry that does not change the volume. In technical terms, they are transverse-traceless density fluctuations. (Vector perturbations correspond to vortical motions which are transverse, but not trace free.) Gravitational waves are similar to the shear waves one finds in elastic media: they involve a twisting distortion of space–time rather than the compression seen in longitudinal scalar waves.

Gravitational waves are produced by accelerating masses and in situations of rapidly changing tidal fields. The more violent the accelerations involved the higher the amplitude of the gravitational waves. Because Einstein's theory of general relativity is nonlinear, however, the waves become very complicated when the amplitude gets large: the wave begins to feel the gravitational effect produced by its own energy. These waves travel at the speed of light, just as electromagnetic radiation does. The problem with detecting gravitational waves, however, is that gravity is very weak. Even extremely violent events like a supernova explosion produce only a very slight signal. Gravitational-wave detectors have been built that attempt to look, for example, for changes in the length of large metal blocks when a wave passes through. The expected signal is much smaller than thermal fluctuations or background noise, however, so such experiments are extremely difficult. In fact, the typical fractional change in length associated with gravitational waves is less than 10^{-21}. Despite claims by Weber in the 1960s that he had detected signals that could be identified with gravitational radiation, no such waves have yet been unambiguously observed. The next generation of gravitational wave detectors such as GEO (a UK-German collaboration), Virgo (France/Italy) and LIGO (USA) should reach the desired sensitivity using interferometry rather than solid metal bars. The LIGO experiment, for example, involves an interferometer with arms 4 km in length. Moreover, plans exist to launch satellites into space that should increase the baseline to millions of km and thus increase the sensitivity to a given fractional change in length. One such proposal called LISA is pencilled in for launch by the European Space Agency sometime before 2020.

Although these experiments have not yet detected gravitational radiation, there is very strong circumstantial evidence for its existence. The period of the binary pulsar 1913 + 16 is gradually decreasing at a rate which matches to great precision relativistic calculations of the expected motion of a pair of neutron stars. In these calculations the dominant form of energy loss from the system is via gravitational radiation, so the observation of the 'spin-up' in this system is tantamount to an observation of the gravitational waves themselves (Taylor *et al.* 1979). Hulse and Taylor were awarded the Nobel Prize for studies of this system in 1993.

As we mentioned above, it is also possible that gravitational waves have already been seen directly. The temperature fluctuations seen in the cosmic microwave background radiation are usually attributed to the Sachs–Wolfe effect produced by scalar density perturbations; see primordial density fluctuations. But if these fluctuations were generated in the inflationary Universe phase by quantum fluctuations in a scalar field, they are expected to be accompanied by gravitational waves which in some cases could contribute an observable Sachs–Wolfe effect of their own. It could well be that at least part of the famous ripples seen by the Cosmic Background Explorer (COBE) satellite is caused by gravitational waves with wavelengths of the same order as the cosmological horizon.

It can be speculated that in a theory of quantum gravity the quantum states of the gravitational field would be identified with gravitational waves in much the same way that the quantum states of the electromagnetic field are identified with photons. The hypothetical quanta of gravitation are thus called gravitons. It has been argued that gravitational wave astronomy could push back the frontiers of the observable universe from the epoch of recombination to the Planck epoch, since gravitons are expected to decouple at the latter energy scale.

21.11 Sociology, Politics and Economics

We hope it is apparent that there are many exciting developments on the horizon, and that cosmology can look forward to a vigorous and challenging future. But as well as the forthcoming developments in technology, the years to come will probably also lead to changes on the human side of the subject. Science, after all, is a very human kind of activity and we could not resist the temptation to speculate a little about the likely impact on how astronomy is performed.

The new technology that has driven observational astronomy at the breakneck pace it has enjoyed over the last decades has also led to changes in the way the accompanying human resources are organised. Collaborations are now very much larger than they were even 20 years ago, leading to difficulties in bringing younger scientists through to prominence and assigning credit to individual contributions. This trend is likely to continue, with monolithic survey projects involving dozens if not hundreds of scientists becoming the rule rather than the exception for leading-edge research in astronomy. This is also becoming the case in theory, especially in respect of the large collaborations involved in supercomputer simulations of structure formation.

The organisation and control of access to astronomical facilities may also change dramatically, as more dedicated high-cost facilities take the place of multipurpose facilities whose time is allocated by peer-review processes of various kinds. With more and more observational programs being constructed in response to specific science goals, often strongly informed by theoretical ideas within a specific framework, the role of serendipitous discovery seems set to diminish. Altogether these factors conspire against the creative maverick and in favour of the conformist team player. Whether one thinks this is a good thing or a bad thing depends on one's own personality.

There is also a more subtle change of emphasis, which can be seen even in the structure of this chapter. More for presentational purposes than anything else, we organised the discussion by wavelength region. This is an increasingly outdated way of thinking. Future science programmes are likely to be much more organised by science goal than by wavelength region. Traditional communities, such as radio astronomy and X-ray astronomy, will see their boundaries blurred by the growing number of scientists driven by an interest in particular objects rather than particular kinds of photon.

So much for sociology, how about politics and economics? The main point that comes to mind relates to the cost of these facilities. By any criteria, all the missions and facilities we have discussed are extremely expensive. For this reason, as much as any intrinsic transnationalism between scientists, upcoming developments are likely to be multinational in character. ALMA is a true world astronomy project, involving substantial financial investments from many countries including the ESO member countries and the USA. The SKA has an even broader distribution of likely contributors. These coalitions are brought together by the impossible strain on budgets of individual countries that would be caused if they took on projects of such a scale on their own. But even if global collaborations are possible, there must be some limit to the amount of cash that can be assembled for scientific studies, especially in times of global recession. When will we reach that limit, and what will be able to learn before we do?

21.12 Conclusions

This has been a very superficial and biased review, but we hope it has given some insights into the way extragalactic astronomy might head over the next decade or two. We have refrained from attempting to give accurate dates, because these are so likely to be revised as to make such guesses worthless.

It is astonishing how much things have changed over the last decade and a half. In 1985 the largest redshift survey available comprised a thousand galaxies or so and fluctuations in the cosmic microwave background were not yet detected. In some sense, that was a very good time to be a theorist but it was clear then that, compared with other sciences, cosmology was extremely immature. Now, with a steadily growing empirical foundation and an exciting interplay between theory and observation, it is has come of age as a science. Its future development promises much and, rightly, it is observation that will drive it forward.

Appendix A Physical Constants

Gravitational constant	G	6.7×10^{-11} N m^2 kg^{-2}
Speed of light	c	3.00×10^{8} m s^{-1}
Planck constant	h	6.63×10^{-34} J s
Boltzmann constant	k_B	1.38×10^{-23} J K^{-1}
Gas constant	$\mathcal{R}$	8.32×10^{3} J mol^{-1} K^{-1}
Radiation density constant	σ_r	7.56×10^{-16} J m^{-3} K^{-4}
Stefan–Boltzmann constant	$\sigma = \frac{1}{4}\sigma_r c$	5.6×10^{-8} J m^{-2} K^{-4}
Electron charge	e	1.6×10^{-19} C
Electron mass	m_e	9.11×10^{-31} kg
Mass of hydrogen atom	m_H	1.66×10^{-27} kg
Mass of proton	m_p	1.6726×10^{-27} kg
Mass of neutron	m_n	1.67492×10^{-27} kg
Electronvolt	eV	1.60×10^{-19} J
Thomson scattering cross-section	σ_T	6.65×10^{-29} m^2
Weak coupling constant	g_{wk}	1.4×10^{36} J m^3

The usual symbol for the radiation density constant is a, but this would clash too frequently with our use of a for the cosmic scale factor in this book, so we have chosen to call it σ_r.

The fine-structure constant is

$$\alpha = \frac{e^2}{4\pi\epsilon_0\hbar c} \simeq \frac{1}{137}$$

in SI units.

The cross-section for weak interactions in thermal equilibrium at temperature T is given by

$$\sigma_{\mathrm{wk}} = g_{\mathrm{wk}}^2 \left[\frac{k_{\mathrm{B}} T}{(\hbar c)^2} \right]^2 \mathrm{m}^2,$$

where g_{wk} is the weak coupling constant. These are mediated by the $W^{\pm}$ and Z^0 bosons which have masses 80.6 GeV and 91.18 GeV, respectively.

Appendix B Useful Astronomical Quantities

Properties of the Sun		
Solar mass	$M_\odot$	1.99×10^{30} kg
Solar radius	$R_\odot$	6.98×10^{8} m
Luminosity	$L_\odot$	3.9×10^{26} W

Other astronomical quantities	
Parsec (pc)	3.09×10^{16} m
Kiloparsec (kpc)	3.09×10^{19} m
Megaparsec (Mpc)	3.09×10^{22} m
Day	8.64×10^{4} s
Year	3.16×10^{7} s
Light year	9.46×10^{15} m

Appendix C Particle Properties

The standard model of particle physics has three families of quarks organised in doublets (u,d), (s,c) and (b,t). These have the following properties:

Quark	Charge (in units of e)	Mass (in GeV)
d	$-\frac{1}{3}$	0.310
u	$+\frac{2}{3}$	0.310
s	$-\frac{1}{3}$	0.483
c	$+\frac{2}{3}$	1.5
b	$-\frac{1}{3}$	4.7
t	$+\frac{2}{3}$	177

Quarks are confined in hadrons, which are either mesons ($q\bar{q}$ pairs) or baryons ($q_1q_2q_3$ triplets). Familiar examples of the baryons are the proton (uud) and the neutron (ddu) both with masses around 940 MeV. The π mesons are likewise formed from d, $\bar{d}$, u and $\bar{u}$. Hence $\pi^- = d\bar{u}$, $\pi^+ = u\bar{d}$ and $\pi^0 = (u\bar{u} - d\bar{d})/\sqrt{2}$. The pions have masses around 136 MeV. Since the quarks carry a colour charge (either red, green or blue) these can be constructed to be either colour-anticolour combinations (mesons) or mixtures of three colours (baryons). Either way the resulting states are colourless.

There are also three families of leptons, organised in doublets (e, ν_e), (μ, ν_μ) and (τ, μ_τ). These have the following properties:

Lepton	Charge (in units of e)	Mass (in GeV)
e	−1	0.0005
ν_e	0	?
μ	−1	0.106
ν_μ	0	?
τ	−1	1.784
ν_τ	0	?

References

Abell GO 1958 The distribution of rich clusters of galaxies. *Astrophys. J. Suppl.* 3, 211–288.

Adler RJ 1981 *The geometry of random fields.* Wiley, Chichester.

Albrecht A and Steinhardt PJ 1982 Cosmology for grand unified theories with radiatively induced symmetry breaking. *Phys. Rev. Lett.* **48**, 1220–1223.

Alcock C *et al.* 1993 Possible gravitational microlensing of a star in the large magellanic cloud. *Nature* **365**, 621–623.

Alexander T 1995 The signature of microlensing in QSO variability-redshift correlations. *Mon. Not. R. Astr. Soc.* **274**, 909–918.

Alpher RA and Herman RC 1948 Evolution of the Universe. *Nature* **162**, 774–775.

Alpher RA, Bethe HA and Gamow G 1948 The origin of chemical elements. *Phys. Rev.* **73**, 803–804.

Applegate JH and Hogan CJ 1985 Relics of cosmic quark condensation. *Phys. Rev.* D **31**, 3037–3045.

Arp HC, Burbidge G, Hoyle F, Narlikar JV and Wickramasinghe NC 1990 The extragalactic universe: an alternative view. *Nature* **346**, 807–812.

Aubourg E *et al.* 1993 Evidence for gravitational microlensing by dark objects in the galactic halo. *Nature* **365**, 623–625.

Avelino PP, Shellard EPS, Wu JHP and Allen B 1998 Non-Gaussian features of linear cosmic string models non-Gaussian features of linear cosmic string models. *Astrophys. J.* **507**, L101–L104.

Bacon DJ, Refregier A and Ellis RS 2000 Detection of weak gravitational lensing by large-scale structure. *Mon. Not. R. Astr. Soc.* **318**, 625–640.

Bahcall NA 1988 Large-scale structure in the Universe indicated by galaxy clusters. *A. Rev. Astr. Astrophys.* **26**, 631–686.

Bahcall SR and Tremaine S 1988 Evolutionary corrections to the redshift-volume measurement of the density parameter. *Astrophys. J.* **326**, L1–L4.

Bardeen JM, Bond JR, Kaiser N and Szalay AS 1986 The statistics of peaks of Gaussian random fields. *Astrophys. J.* **304**, 15–61.

Barrow JD 1983 Cosmology and elementary particles. *Fund. Cosmic Phys.* **8**, 83–200.

Barrow JD and Tipler FJ 1986 *The anthropic cosmological principle.* Clarendon Press, Oxford.

Baugh CM, Cole SM, Frenk CS and Lacey CG 1998 The epoch of galaxy formation. *Astrophys. J.* **498**, 504–521.

Bennett C *et al.* 1992 Preliminary separation of the galactic and cosmic microwave background emission for the COBE differential microwave radiometer. *Astrophys. J.* **396**, L7–L12.

Bernstein J 1988 *Kinetic theory in the expanding universe.* Cambridge University Press.

Bernstein J, Brown LS and Feinberg G 1988 Cosmological helium production simplified. *Rev. Mod. Phys.* **61**, 25–39.

Berry MV 1989 *Principles of cosmology and gravitation.* Adam Hilger, Bristol.

Bertschinger E 1992 Large-scale structure and motions: linear theory and statistics. In *New insights into the Universe* (ed. Martinez VJ, Portilla M and Saez D). Springer, Berlin.

Bertschinger E and Gelb JM 1991 Cosmological N-body simulations. *Comput. Phys.* **5**, 164–179.

Bertschinger E and Juszkiewicz R 1988 Searching for the great attractor. *Astrophys. J.* **334**, L59–L62.

Bertschinger E, Dekel A, Faber SM and Burstein D 1990 Potential, velocity and density fields from redshift-distance samples. Application to cosmography within 6000 km sec^{-1}. *Astrophys. J.* **364**, 370–395.

Binney J and Merrifield MR 1998 *Galactic astronomy.* Princeton University Press.

Binney J and Tremaine S 1987 *Galactic dynamics.* Princeton University Press.

Birkhoff G 1923 *Relativity and modern physics.* Harvard University Press, Cambridge, MA.

Blandford R and Narayan R 1992 Cosmological applications of gravitational lensing. *A. Rev. Astr. Astrophys.* **30**, 311–358.

Blumenthal GR, Faber SM, Primack JR and Rees M 1984 Formation of galaxies and large-scale structure with cold dark matter. *Nature* **311**, 517–525.

Boldt E 1987 The cosmic X-ray background. *Phys. Rep.* **146**, 215–257.

Bond JR, Carr BJ and Hogan CJ 1986 The spectrum and anisotropy of the cosmic infra-red background. *Astrophys. J.* **306**, 428–450.

Bond JR, Cole S, Efstathiou G and Kaiser N 1991 Excursion set mass functions for hierarchical Gaussian fluctuations. *Astrophys. J.* **379**, 440–460.

Bondi H 1947 Spherically symmetrical models in general relativity. *Mon. Not. R. Astr. Soc.* **107**, 410–425.

Bondi H and Gold T 1948 The steady state theory of the expanding universe. *Mon. Not. R. Astr. Soc.* **108**, 252–270.

Bonnor WB 1957 Jeans' formula for gravitational instability. *Mon. Not. R. Astr. Soc.* **117**, 104–117.

Bonometto SA and Pantano O 1993 Physics of the cosmological quark–hadron transition. *Phys. Rep.* **228**, 175–252.

Börner G 1988 *The early universe: facts and fiction.* Springer, Berlin.

Bower RG, Coles P, Frenk CS and White SDM 1993 Co-operative galaxy formation and large-scale structure. *Astrophys. J.* **405**, 403–412.

Branch D and Tammann GA 1992 Supernovae as standard candles. *A. Rev. Astr. Astrophys.* **30**, 359–389.

Brandenberger RH 1985 Quantum field theory methods and inflationary universe models. *Rev. Mod. Phys.* **57**, 1–60.

Brandenberger RH 1990 Inflationary universe models and cosmic strings. In *Physics of the Early Universe. Proc. 36th Scottish Universities Summer School in Physics* (ed. Peacock JA, Heavens AF and Davies AT). Edinburgh University Press.

Brans CH and Dicke RH 1961 Mach's principle and a relativistic theory of gravitation. *Phys. Rev.* **124**, 925–935.

Briel UG, Henry IP and Bohringer H 1992 Observation of the Coma cluster of galaxies with rosat during the all-sky survey. *Astr. Astrophys.* **259**, L31–L34.

Broadhurst TJ, Taylor AN and Peacock JA 1995 Mapping cluster mass distributions via gravitational lensing of background galaxies. *Astrophys. J.* **438**, 49–61.

Burles S and Tytler D 1996 The cosmological density and ionization of hot gas: OVI absorption in quasar spectra. *Astrophys. J.* **460**, 584-600.

Burstein D 1990 Large-scale motions in the universe: a review *Rep. Prog. Phys.* **53**, 421-481.

Caditz D and Petrosian V 1989 Cosmological parameters and the evolution of the galaxy luminosity function. *Astrophys. J.* **337**, L65-L68.

Carr BJ 1994 Baryonic dark matter. *A. Rev. Astr. Astrophys.* **32**, 531-590.

Carswell RF *et al.* 1994 Is there deuterium in the $z = 3.32$ complex in the line 0014+813? *Mon. Not. R. Astr. Soc.* **268**, L1-L4.

Carswell RF *et al.* 1996 The high-redshift deuterium abundance: the $z = 3.086$ absorption complex towards Q 0420-388, *Mon. Not. R. Astr. Soc.* **278**, 506-518.

Cen R 1992 A hydrodynamic approach to cosmology - methodology. *Astrophys. J. Suppl.* **78**, 341-364.

Chaichian M and Nelipe NF 1984 *Introduction to gauge field theories.* Springer, Berlin.

Charlton JC and Turner MS 1987 Kinematic tests of exotic flat cosmological models. *Astrophys. J.* **313**, 495-504.

Chibisov GV 1972 Damping of adiabatic perturbations in an expanding universe. *Sov. Astr.* **16**, 56-63.

Coles P 1993 Galaxy formation with a local bias. *Mon. Not. R. Astr. Soc.* **262**, 1065-1075.

Coles P and Chiang L-Y 2000 Characterizing the non-linear growth of large-scale structure in the Universe. *Nature* **406**, 376-378.

Coles P and Ellis GFR 1994 The case for an open universe. *Nature* **370**, 609-615.

Coles P and Ellis GFR 1997 *Is the Universe open or closed?* Cambridge University Press.

Coles P, Melott AL and Shandarin SF 1993a Testing approximations for nonlinear gravitational clustering. *Mon. Not. R. Astr. Soc.* **260**, 765-776.

Coles P, Moscardini, L, Lucchin, F, Matarrese, S and Messina, A 1993b Skewness as a test of non-Gaussian primordial density fluctuations. *Mon. Not. R. Astr. Soc.* **264**, 749-757.

Collins PB, Martin AD and Squires EJ 1989 *Particle physics and cosmology.* Wiley, New York.

Copi CJ, Schramm DN and Turner MS 1995a Big-bang nucleosynthesis and the baryon density of the Universe. *Science* **267**, 192-199.

Copi CJ, Schramm DN and Turner MS 1995b Assessing Big-Bang nucleosynthesis. *Phys. Rev. Lett.* **75**, 3981-3984.

Coulson D, Ferreira P, Graham P and Turok N 1994 Microwave anisotropies from cosmic defects. *Nature* **368**, 27-31.

Cowie LL 1988 Protogalaxies. In *The Post Recombination Universe, Proceedings of NATO Advanced Study Institute 'The Post-Recombination Universe'* (ed. Kaiser N and Lasenby AN), pp. 1-18. Kluwer, Dordrecht.

Cowsik R and McClelland J 1972 An upper limit on the neutrino rest mass. *Phys. Rev. Lett.* **29**, 669-670.

Dalcanton JJ, Canizares CR, Granados A, Steidel CC and Stocke JR 1994 Observational limits on Omega in stars, brown dwarfs, and stellar remnants from gravitational microlensing. *Astrophys. J.* **424**, 550-568.

Davies JI, Phillipps S and Disney MJ 1988 The lowest surface-brightness disk galaxy known. *Mon. Not. R. Astr. Soc.* **231**, 69P-74P

Davis M and Peebles PJE 1977 On the integration of the BBGKY equations for the development of strongly nonlinear clustering in the expanding universe. *Astrophys. J. Suppl.* **34**, 425-450.

Davis M, Efstathiou G, Frenk CS and White SDM 1985 The evolution of large-scale structure in the Universe dominated by cold dark matter. *Astrophys. J.* **292**, 371-394.

Davis M, Summers FJ and Schlegel D 1992 Large-scale structure in a universe with mixed hot and cold dark matter. *Nature* **359**, 393-396.

de Lapparent V, Geller MJ and Huchra JP 1986 A slice of the Universe. *Astrophys. J.* **302**, L1-L4

de Sitter W 1917 On Einstein's theory of gravitation and its astronomical consequences. Third paper. *Mon. Not. R. Astr. Soc.* **78**, 3-28.

De Witt BS 1967 Quantum theory of gravity. I. The canonical theory. *Phys. Rev.* **160**, 1113-1148.

Dekel A 1994 Dynamics of cosmic flows. *A. Rev. Astr. Astrophys.* **32**, 371-478.

Dekel A and Lahav O 1999 Stochastic non-linear galaxy biasing. *Astrophys. J.* **520**, 24-34.

Dekel A and West MJ 1985 On percolation as a cosmological test. *Astrophys. J.* **288**, 411-417.

Dekel A, Bertschinger E, Yahil A, Strauss MA, Davis M and Huchra JP 1993 IRAS galaxies vs POTENT mass-density fields, biasing and Omega. *Astrophys. J.* **412**, 1-21.

Dicke RH 1961 Dirac's cosmology and Mach's principle. *Nature* **192**, 440-441.

Dicke RH, Peebles PJE, Roll PG and Wilkinson DT 1965 Cosmic black-body radiation. *Astrophys. J.* **142**, 414-419.

Dirac PAM 1937 The cosmological constant. *Nature* **139**, 323.

Dirac PAM 1974 Cosmological models and the large number hypothesis. *Proc. R. Soc. Lond.* A **338**, 439-446.

Dominguez-Tenreiro R and Quiros M 1987 *An introduction to cosmology and particle physics.* World Scientific, Singapore.

Doroshkevich AG, Zel'dovich YaB and Novikov ID 1967 Origin of galaxies in an expanding universe. *Sov. Astr.* **11**, 233-239.

Duff H and Isham C (eds) 1982 *Quantum structure of space and time.* Cambridge University Press.

Dyson FW, Eddington AS and Davison C 1920 A determination of the deflection of light by the Sun's gravitational field, from observations made at the total eclipse of May 29, 1919. *Phil. Trans. R. Soc. Lond.* A **220**, 291-333.

Efstathiou G 1990 Cosmological perturbations. In *Physics of the Early Universe. Proc. 36th Scottish Universities Summer School in Physics* (ed. Peacock JA, Heavens AF and Davies AT). Edinburgh University Press.

Efstathiou G and Jones BJT 1979 The rotation of galaxies: numerical investigations of the tidal torque theory. *Mon. Not. R. Astr. Soc.* **186**, 133-144.

Efstathiou G and Rees MJ 1988 High redshift quasars in the CDM cosmogony. *Mon. Not. R. Astr. Soc.* **230**, 5P-11P

Efstathiou G and Silk J 1983 The formation of galaxies. *Fund. Cosmic Phys.* **9**, 1-138.

Efstathiou G, Davis M, Frenk CS and White SDM 1985 Numerical techniques for large cosmological *N*-body simulations. *Astrophys. J. Suppl.* **57**, 241-260.

Efstathiou G, Sutherland WJ and Maddox SJ 1990 The cosmological constant and cold dark matter. *Nature* **348**, 705-706.

Einstein A 1917 Kosmologische Betrachtugen zur allgemeinen Relativitatstheories. *Sitzungsberichte der Preuss. Akad. Wiss.* 142-152. (English translation: Lorentz HA, Einstein A, Minkowski H and Weyl H (eds) 1950 *The principle of relativity*, pp. 177-188. Methuen, London.)

Einstein A 1950 In *The principle of relativity* (eds Lorentz HA, Einstein A, Minkowski H and Weyl H). Methuen, London.

Ellis GFR 1987 Alternatives to the Big Bang. *A. Rev. Astr. Astrophys.* **22**, 157-184.

Ellis GFR and Tivon G 1985 Observational relationships in inflationary universes and other cosmologies. *Observatory* **105**, 189–198.

Ellis RS 1993 Galaxy evolution. *Ann. NY Acad. Sci.* **688**, 207–216.

Evrard AE 1988 Beyond *N*-body: 3D cosmological gas dynamics. *Mon. Not. R. Astr. Soc.* **235**, 911–934.

Faber SM and Gallagher JS 1979 Masses and mass-to-light ratio of galaxies. *A. Rev. Astr. Astrophys* **17**, 135–187.

Fall SM 1979 Galaxy correlations and cosmology. *Rev. Mod. Phys.* **51**, 21–42.

Field GB 1971 Instability and waves driven by radiation in interstellar space and in cosmological models. *Astrophys. J.* **165**, 29–40.

Fort B and Mellier Y 1994 Arc(let)s in clusters of galaxies. *Astr. Astrophys. Rev.* **5**, 239–292.

Frenk CS, White SDM, Davis M and Efstathiou G 1988 The formation of dark halos in a universe dominated by cold dark matter. *Astrophys. J.* **327**, 507–525.

Frenk CS, Ellis RS, Shanks T, Heavens AF and Peacock JA (eds) 1989 *The epoch of galaxy formation.* Kluwer, Dordrecht.

Friedmann A 1922 Über die Krümmung des Raumes. *Z. Phys.* **10**, 377–386.

Fukuda Y *et al.* 1999 Constraints on neutrino oscillation parameters from the measurement of day–night solar neutrino fluxes at Super-Kamiokande. *Phys. Rev. Lett.* **82**, 1810–1814.

Fukugita M and Turner EL 1991 Gravitational lensing frequencies – galaxy cross-sections and selection effects. *Mon. Not. R. Astr. Soc.* **253**, 99–106.

Fukugita M, Hogan CJ and Peebles PJE 1992 The cosmic distance scale and the Hubble constant. *Nature* **366**, 309–312.

Gamow G 1946 Expanding universe and the origin of elements. *Phys. Rev.* **70**, 572–573.

Georgi H and Glashow SL 1974 Unity of all elementary-particle forces. *Phys. Rev. Lett.* **32**, 438–441.

Gorski KM 1988 On the pattern of perturbations of the Hubble flow. *Astrophys. J.* **332**, L7–L11.

Gorski KM, Davis M, Strauss MA, White SDM and Yahil A 1989 Cosmological velocity correlations: observations and model predictions. *Astrophys. J.* **344**, 1–19.

Gott III JR 1980 The growth of structure in the Universe. In *Les Houches, Session XXXII, 1979 – Cosmologie Physique/Physical Cosmology* (ed. Balian R, Audouze J and Schramm DN). North-Holland, Amsterdam.

Gott III JR 1982 Creation of open universe from de Sitter space. *Nature* **295**, 304–307.

Gott III JR, Melott AL and Dickinson M 1986 The sponge-like topology of large-scale structure in the Universe. *Astrophys. J.* **306**, 341–347.

Gott III JR *et al.* 1989 The topology of large-scale structure. III. Analysis of observations. *Astrophys. J.* **340**, 625–646.

Gott III JR, Park C, Juszkiewicz R, Bies WE, Bennet DP and Stebbins A 1990 Topology of microwave background fluctuations: theory. *Astrophys. J.* **352**, 1–14.

Gottlober S, Mucket JP and Starobinsky AA 1994 Confrontation of a double inflationary cosmological model with observations. *Astrophys. J.* **434**, 417–423.

Grogin N and Narayan R 1996 A new model of the gravitational lens 0957+561 and a limit on the Hubble constant. *Astrophys. J.* **464**, 92–113.

Gunn JE and Peterson BA 1965 On the density of neutral hydrogen in intergalactic space. *Astrophys. J.* **142**, 1633–1636.

Gurbatov SN, Saichev AI and Shandarin SF 1989 The large-scale structure of the universe in the frame of the model equation of non-linear diffusion. *Mon. Not. R. Astr. Soc.* **236**, 385–402.

Gurvits LI, Kellermann KI and Frey S 1999 The 'angular size-redshift' relation for compact radio structures in quasars and radio galaxies. *Astron. Astrophys.* **342**, 378-388.

Guth AH 1981 Inflationary universe: a possible solution to the horizon and flatness problem. *Phys. Rev.* D **23**, 347-356.

Guth AH and Pi S-Y 1982 Fluctuations in the new inflationary universe. *Phys. Rev. Lett.* **49**, 1110-1113.

Guzman R and Lucey JR 1993 A new, age-independent distance estimator for elliptical galaxies. *Mon. Not. R. Astr. Soc.* **263**, L47-L50.

Halley E 1720 On the infinity of the sphere of fixed stars. *Phil. Trans. R. Soc. Lond.* **31**, 22-24.

Hamilton AJS 1992 Measuring Omega and the real correlation function from the redshift correlation function. *Astrophys. J.* **385**, L5-L8.

Hamilton AJS 1998 Linear redshift distortions: a review. In *The evolving universe* (ed. Hamilton D), pp. 185-275. Kluwer, Dordrecht.

Hamilton AJS, Gott III JR and Weinberg D 1986 The topology of large-scale structure of the Universe. *Astrophys. J.* **309**, 1-12.

Hamilton AJS, Kumar P, Lu E and Mathews A 1991 Reconstructing the primordial spectrum of fluctuations of the Universe from the observed non-linear clustering of galaxies. *Astrophys. J.* **374**, L1-L4.

Hamuy M, Phillips MM, Maza J, Suntzeff NB, Schommer RA and Avilez R 1995 A Hubble diagram of distant Type Ia supernovae. *Astron. J.* **109**, 1-13.

Hanany S *et al.* 2000 MAXIMA-1: a measurement of the cosmic microwave background anisotropy on angular scales of 10 arcminutes to 5 degrees. *Astrophys. J.* **545**, L5-L8.

Hancock S *et al.* 1993 Direct observation of structure in the cosmic microwave background. *Nature* **367**, 333-337.

Harrison ER 1970 Fluctuations at the threshold of classical cosmology. *Phys. Rev.* D **1**, 2726-2730.

Harrison ER 1973 Standard model of the early Universe. *A. Rev. Astr. Astrophys.* **11**, 155-186.

Harrison ER 1981 *Cosmology.* Cambridge University Press.

Hartle JB 1988 Quantum cosmology. In *Highlights in gravitation and cosmology* (ed. Iyer BR, Kembhavi A, Narlikar JV and Vishveshwara CV), pp. 144-155. Cambridge University Press.

Hartle JB and Hawking SW 1983 Wave function of the Universe. *Phys. Rev.* D **28**, 2960-2975.

Hawking SW 1982 The development of irregularities in a single bubble inflationary universe. *Phys. Lett.* B **115**, 295-297.

Hawking SW and Israel W (eds) 1979 *General relativity, an Einstein centenary survey.* Cambridge University Press.

Hawking SW and Israel W (eds) 1987 *300 years of gravitation.* Cambridge University Press.

Hawkins MRS 1993 Gravitational microlensing, quasar variability and missing matter. *Nature* **366**, 242-245.

Hockney RW and Eastwood JW 1988 *Computer simulations using particles.* Adam Hilger, Bristol.

Hogan CJ, Kaiser N and Rees MJ 1982 Interpretation of the anisotropy of the cosmic background radiation. *Phil. Trans. R. Soc. Lond.* A **307**, 97-110.

Holtzmann JA 1989 Microwave background anisotropies and large-scale structure in universes with cold dark matter, baryons, radiation and massive and massless neutrinos. *Astrophys. J. Suppl.* **71**, 1-24.

Hoyle F 1948 A new model for the expanding universe. *Mon. Not. R. Astr. Soc.* **108**, 372-382.

Hoyle F 1949 The origin of the rotation of galaxies. In *Problems of cosmical aerodynamics* (eds Burgers JM and van de Hulst HC). Central Air Documents Office, Dayton.

Hoyle F 1959 The Relation of Radio Astronomy to Cosmology. In Bracewell RN (ed) *Paris Symposium on Radio Astronomy*, IAU No. 9, p. 529. Stanford University Press, Stanford CA

Hoyle F and Narlikar JV 1963 Mach's principle and the creation of matter. *Proc. R. Soc. Lond.* A **273**, 1-11.

Hoyle F and Narlikar JV 1964 The avoidance of singularities in *C*-field cosmology. *Proc. R. Soc. Lond.* A **278**, 465-478.

Hoyle F and Tayler RJ 1964 The mystery of the cosmic helium abundance. *Nature* **203**, 1108-1110.

Hubble E 1929 A relation between distance and radial velocity among extragalactic nebulae. *Proc. Nat. Acad. Sci.* **15**, 168-173.

Hughes IS 1985 *Elementary particles.* Cambridge University Press.

Izotov YI, Thuan T-X and Lipovetsky VA 1994 The primordial helium abundance from a new sample of metal-deficient blue compact galaxies. *Astrophys. J.* **435**, 647-667.

Jaffe AH *et al.* 2001 Cosmology from MAXIMA-1, Boomerang and COBE/DMR CMB observations. *Phys. Rev. Lett.* **86**, 3475-3479.

Jain B, Mo H-J and White SDM 1995 The evolution of correlation functions and power spectra in gravitational clustering. *Mon. Not. R. Astr. Soc.* **276**, L25-L29.

Janis AI 1986 Angular size in an expanding universe. *Am. J. Phys.* **54**, 1008-1011.

Jeans JH 1902 The stability of spiral nebula. *Phil. Trans. R. Soc. Lond.* A **199**, 1-53.

Jenkins A *et al.* 1998 Evolution of structure in CDM universes. *Astrophys. J.* **499**, 20-40.

Kaiser N 1984 On the spatial correlation of Abell clusters. *Astrophys. J.* **284**, L9-L12.

Kaiser N 1987 Clustering in real space and redshift space. *Mon. Not. R. Astr. Soc.* **227**, 1-21.

Kaiser N and Silk J 1987 Cosmic microwave background anisotropy *Nature* **324**, 529-537.

Kaiser N and Squires G 1993 Mapping the dark matter with weak gravitational lensing. *Astrophys. J.* **404**, 441-450.

Kaluza T 1921 Zum Unitäls problem der Physik. *Sitzungsber Preuss Akad. Wiss.* K **1**, 966-972.

Kamionkowski M and Spergel DN 1994 Large-angle cosmic microwave background anisotropies in an open universe. *Astrophys. J.* **432**, 7-16.

Kauffmann G, Guiderdoni B and White SDM 1994 Faint galaxy counts in a hierarchical universe. *Mon. Not. R. Astr. Soc.* **267**, 981-999.

Kellermann KI 1993 The cosmological deceleration parameter estimated from the angular-size redshift relation for compact radio sources. *Nature* **361**, 134-136.

Kenyon R 1990 *General relativity.* Oxford University Press.

Klapdor-Kleingrothaus HV and Zuber K 1997 *Particle astrophysics.* IOP, Bristol.

Klein O 1926 Quantentheorie und fünfdimensionaler Relativitätstheorie. *Z. Phys.* **37**, 895-906.

Klypin A, Holztman JA, Primack J and Regos E 1993 Structure formation with cold plus hot dark matter. *Astrophys. J.* **416**, 1-16.

Kneib J-P, Mellier Y, Fort B and Mathez G 1993 The distribution of dark matter in distant cluster lenses - modelling A370. *Astr. Astrophys.* **273**, 367-376.

Kochanek CS 1993 The analysis of gravitational lens surveys. 2. Maximum likelihood models and singular potentials. *Astrophys. J.* **419**, 12-29.

Kodama H and Sasaki M 1984 Cosmological perturbation theory. *Prog. Theor. Phys.* **78**, 1-166.

Kolb EW and Turner MS 1990 *The early Universe.* Addison-Wesley, Redwood City, CA.

Lacey C and Cole S 1993 Merger rates in hierarchical models of galaxy formation. *Mon. Not. R. Astr. Soc.* **262**, 627-649.

Landau LD and Lifshitz EM 1975 *Classical theory of fields.* Pergamon Press, Oxford.

Lee BW and Weinberg S 1977 Cosmological lower bound on heavy-neutrino masses. *Phys. Rev. Lett.* **39**, 165-168.

Lemaître G 1927 Un univers homogène de masse constante et de rayon croissant, rendant compte de la vitesse radiale des nébuleuses extra-galactiques. *Ann. Soc. Sci. Brux.* A **47**, 49-59. (English translation: Lemaître G 1931 A homogeneous universe of constant mass and increasing radius accounting for the radial velocity of the extra-galactic nebulae. *Mon. Not. R. Astr. Soc.* **91**, 483-490.)

Liddle AR and Lyth DH 1993 The cold dark matter density perturbation. *Phys. Rep.* **231**, 1-105.

Liddle AR and Lyth DH 2000 *Cosmological inflation and large-scale structure.* Cambridge University Press.

Lidsey JE and Coles P 1992 Inflation, gravitational waves and the cosmic microwave background: reconciling CDM with COBE? *Mon. Not. R. Astr. Soc.* **258**, 57P-62P.

Lifshitz EM 1946 On the gravitational instability of the expanding universe. *Sov. Phys. JETP* **10**, 116-122.

Lima JAS, Zanchin V and Brandenberger R 1997 On the Newtonian cosmology equations with pressure. *Mon. Not. R. Astr. Soc.* **291**, L1-L4.

Limber DN 1953 The analysis of counts of the extragalactic nebulae in terms of a fluctuating density field. I. *Astrophys. J.* **117**, 134-144.

Limber DN 1954 The analysis of counts of the extragalactic nebulae in terms of a fluctuating density field. II. *Astrophys. J.* **119**, 655-681.

Lin CC, Mestel L and Shu F 1965 The gravitational collapse of a uniform spheroid. *Astrophys. J.* **142**, 1431-1446.

Linde AD 1982a A new inflationary universe scenario: a possible solution of the horizon, flatness, homogeneity, isotropy and primordial monopole problems. *Phys. Lett.* B **108**, 389-393.

Linde AD 1982b Scalar field fluctuations in the expanding universe and the new inflationary universe scenario. *Phys. Lett.* B **116**, 335-339.

Linde AD 1983 Chaotic inflation. *Phys. Lett.* B **129**, 177-181.

Linde AD 1984 The inflationary Universe. *Rep. Prog. Phys.* **47**, 925-986.

Linde AD 1990 *Particle physics and inflationary cosmology.* Harwood, London.

Linde AD, Linde D and Mezhlumian A 1994 From the Big Bang theory to the theory of a stationary universe. *Phys. Rev.* D **49**, 1783-1826.

Linsky JL *et al.* 1993 Goddard high-resolution spectrograph observations of the local interstellar medium and the deuterium hydrogen ratio along the line of sight to capella. *Astrophys. J.* **402**, 694-709.

Linsky JL *et al.* 1995 Deuterium and the local interstellar medium - properties for the procyon and capella lines of sight. *Astrophys. J.* **451**, 335-351.

Loh ED and Spillar EJ 1986 A measurement of the mass density of the Universe. *Astrophys. J.* **307**, L1-L4.

Loys de Chéseaux J-P 1744 *Traité de la comète.* Bousequet, Lausanne. (English translation: see Kenyon (1990).)

Lucchin F and Matarrese S 1985 Kinematic properties of generalized inflation. *Phys. Lett.* B **164**, 282–286.

MacCallum MAH 1993 Anisotropic and inhomogeneous cosmologies. In *The renaissance of general relativity and cosmology* (ed. Ellis GFR, Lanza A and Miller JC), pp. 213–233. Cambridge University Press.

Madau P, Ferguson HC, Dickinson M, Giavalisco M, Steidel CC and Fruchter A 1996 High redshift galaxies in the Hubble deep field. color selection and star formation history to $z = 4$. *Mon. Not. R. Astr. Soc.* **283**, 1388–1404.

Maddox S, Efstathiou G, Sutherland W and Loveday J 1990 Galaxy correlations on large scales. *Mon. Not. R. Astr. Soc.* **242**, 43P–47P.

Magueijo J 2000 Covariant and locally lorentz-invariant varying speed of light theories. *Phys. Rev.* D **62**, 103521.

Mao S and Kochanek CS 1994 Limits on galaxy evolution. *Mon. Not. R. Astr. Soc.* **268**, 569–580.

Maoz D and Rix H-W 1993 Early-type galaxies, dark halos and gravitational lensing statistics. *Astrophys. J.* **416**, 425–443.

Martínez VJ and Saar E 2002 *Statistics of the Galaxy distribution.* Chapman & Hall, London.

Matarrese S, Verde L and Heavens AF 1997 Large-scale bias in the universe: bispectrum method. *Mon. Not. R. Astr. Soc.* **290**, 651–662.

Mather JC *et al.* 1994 Measurement of the cosmic background spectrum by the COBE FIRAS instrument. *Astrophys. J.* **420**, 439–444.

Mecke KR, Buchert T and Wagner H, 1994 Robust morphological measures for large-scale structure in the Universe. *Astr. Astrophys.* **288** 697–704.

Melott AL 1990 The topology of large-scale structure in the Universe. *Phys. Rep.* **193**, 1–39.

Melott AL, Coles P, Feldman HA and Wilhite B 1998 The bull's-eye effect as a probe of Ω. *Astrophys. J.* **496**, L85–L88.

Merchant Boesgaard A and Steigman G 1985 Big Bang nucleosynthesis: theories and observations. *A. Rev. Astr. Astrophys.* **23**, 319–378.

Meszaros P 1974 The behaviour of point masses in an expanding cosmological substratum. *Astr. Astrophys.* **37**, 225–228.

Metcalfe N, Shanks T, Campos A, McCracken H and Fong R 2001 Galaxy number counts. V. Ultradeep counts: the Herschel and Hubble deep fields. *Mon. Not. R. Astr. Soc.* **323**, 795–830.

Milne AE 1935 *Relativity, gravitation, and world structure.* Clarendon Press, Oxford.

Misner CW 1968 The isotropy of the Universe. *Astrophys. J.* **151**, 431–457.

Misner CW, Thorne KS and Wheeler JA 1972 *Gravitation.* Freeman, San Francisco.

Moffat J 1993 Quantum gravity, the origin of time and time's arrow. *Found. Phys.* **23**, 411–437.

Molaro P, Primas F and Bonifacio P 1995 Lithium abundance of halo dwarfs revised. *Astr. Astrophys.* **295**, L47–L50.

Mushotzky RF, Cowie LL, Barger AJ and Arnaud KA 2000 Resolving the extragalactic hard X-ray background. *Nature* **404**, 459–464.

Narlikar JV 1993 *Introduction to cosmology.* Cambridge University Press.

Narlikar JV and Padmanabhan T 1991 Inflation for astronomers. *A. Rev. Astr. Astrophys.* **29**, 325–362.

Nugent P, Phillips M, Baron E, Branch D and Hauschildt P 1995 Evidence for a spectroscopic sequence among type 1a supernovae. *Astrophys. J.* **455**, L47–L150.

Olbers HWM 1826 Uber die Durchsichtigheit des Weltraumes. In *Astronomische Jahrbuch fur das Jahr 1826* (ed. Bode JE). Spathen, Berlin. (English translation: Olbers HWM 1826 On the transparency of space. *Edinb. New Phil. J.* **1**, 141.)

Olive KA and Scully S 1995 The deuterium abundance and nucleocosmochronology. *Astrophys. J.* **446**, 272-278.

Olive KA and Schramm DN 1992 Astrophysical Li-7 as a Product of Big-Bang nucleosynthesis and galactic cosmic ray spallation. *Nature* **360**, 439-442.

Olive KA and Steigman G 1995 On the abundance of primordial helium. *Astrophys. J. Suppl.* **97**, 49-58.

Ostriker JP and Cowie LL 1981 Galaxy formation in an inter-galactic medium dominated by explosions. *Astrophys. J.* **243**, L127-L131.

Ostriker JP and Suto Y 1990 The Mach number of the cosmic flow: a critical test for current theories. *Astrophys. J.* **348**, 378-382.

Ostriker JP and Vishniac ET 1986 Reionization and small-scale fluctuations in the microwave background. *Astrophys. J.* **306**, L5-L8.

Paczynski B 1986a Gravitational microlensing at large optical depth. *Astrophys. J.* **301**, 503-516.

Paczynski B 1986b Gravitational microlensing by the galactic halo. *Astrophys. J.* **304**, 1-5.

Padmanabhan T 1993 *Structure formation in the Universe.* Cambridge University Press.

Pagel BEJ *et al.* 1992 The primordial helium abundance from observations of extragalactic H-II regions. *Mon. Not. R. Astr. Soc.* **255**, 325-345.

Partridge RB 1988 The angular distribution of the cosmic microwave background radiation. *Rep. Prog. Phys.* **51**, 647-706.

Peacock JA 1992 Statistics of cosmological density fields. In *New insights into the Universe* (ed. Martinez VJ, Portilla M and Saez D). Springer, Berlin.

Peacock JA 1999 *Cosmological physics.* Cambridge University Press.

Peacock JA and Dodds S 1994 Reconstructing the linear power spectrum of cosmological mass fluctuations. *Mon. Not. R. Astr. Soc.* **267**, 1020-1034.

Peacock JA and Dodds S 1996 Non-linear evolution of cosmological power spectra. *Mon. Not. R. Astr. Soc.* **280**, L19-L26.

Peacock JA *et al.* 2001 A measurement of the cosmological mass density from clustering in the 2dF Galaxy Redshift Survey. *Nature* **410**, 169-173.

Peebles PJE 1971 *Physical cosmology.* Princeton University Press.

Peebles PJE 1980 *The large-scale structure of the Universe.* Princeton University Press.

Peebles PJE 1986 The mean mass density of the Universe. *Nature* **321**, 27-32.

Peebles PJE 1987 Cosmic background temperature anisotropy in a minimal isocurvature model for galaxy formation. *Astrophy. J.* **315**, L73-L76.

Peebles PJE 1993 *Principles of physical cosmology.* Princeton University Press.

Peebles PJE 1999 An isocurvature cold dark matter cosmogony. I. A worked example of evolution through inflation. *Astrophys. J.* **510**, 523-530.

Peebles PJE and Yu JT 1970 Primeval adiabatic perturbations in an expanding universe. *Astrophys. J.* **162**, 815-836.

Peebles PJE, Schramm DN, Turner EL and Kron RG 1991 The case for the hot relativistic Big Bang cosmology. *Nature* **352**, 769-776.

Penzias AA and Wilson RW 1965 Measurement of excess antenna temperature at 4080 Mc/sec. *Astrophys. J.* **142**, 419-421.

Perlmutter S *et al.* 1999 Measurements of Omega and Lambda from 42 high-redshift supernovae. *Astrophys. J.* **517**, 565-586.

Pierpaoli E, Coles P, Bonometto SA and Borgani S 1996 Large-scale structure in mixed dark matter models with a non-thermal volatile component. *Astrophys. J.* **470**, 92–101.

Plionis M, Coles P and Catelan P 1993 The QDOT and cluster dipoles: evidence for a low-Ω_0 universe? *Mon. Not. R. Astr. Soc.* **262**, 465–474.

Press WH and Schechter P 1974 Formation of galaxies and clusters of galaxies by self-similar gravitational condensation. *Astrophys. J.* **187**, 425–438.

Randall L and Sundrum R 1999 An alternative to compactification. *Phys. Rev. Lett.* **83**, 4690–4693.

Raychaudhuri AK 1979 *Theoretical cosmology.* Oxford University Press.

Rees MJ 1986 Lyman absorption lines in quasar spectra: evidence for gravitationally-confined gas in dark minihalos. *Mon. Not. R. Astr. Soc.* **218**, 25P–30P.

Rees MJ and Ostriker JP 1977 Cooling, dynamics and fragmentation of massive gas clouds: clues to the masses and radii of galaxies and clusters. *Mon. Not. R. Astr. Soc.* **179**, 541–559.

Rees MJ and Sciama DW 1968 Large-scale density inhomogeneities in the Universe. *Nature* **217**, 511–516.

Refsdal S 1964 The gravitational lens effect. *Astrophys. J.* **128**, 295–306.

Ribeiro MB 1992 On modeling a relativistic hierarchical (fractal) cosmology by Tolman's spacetime. *Astrophys. J.* **395**, 29–33.

Riess AG, Press WH and Kirshner RP 1995 Using Type Ia supernova light-curve shapes to measure the Hubble constant. *Astrophys. J.* **438**, L17–L20.

Riess AG *et al.* 1998 Observational evidence from supernovae for an accelerating universe and a cosmological constant. *Astr. J.* **116**, 1009–1038.

Rood HJ 1988 Voids. *A. Rev. Astr. Astrophys.* **26**, 245–294.

Roos M 1994 *Introduction to cosmology.* Wiley, Chichester.

Rowan-Robinson M 1985 *The cosmological distance ladder: distance and time in the Universe.* Freeman, New York.

Rowan-Robinson M *et al.* 1990 A sparse-sampled redshift survey of IRAS galaxies. I. The convergence of the IRAS dipole and the origin of our motion with respect to the Local Group. *Mon. Not. R. Astr. Soc.* **247**, 1–18.

Rubin VC and Coyne GV (eds) 1988 *Large-scale motions in the Universe.* Princeton University Press.

Rugers M and Hogan CJ 1996a High deuterium abundance in a new quasar absorber. *Astr. J.* **111**, 2135–2140.

Rugers M and Hogan CJ 1996b Confirmation of high deuterium abundance in quasar absorbers. *Astrophys. J.* **459**, L1–L4.

Rybicki GB and Lightman AP 1979 *Radiative processes in astrophysics.* Wiley, New York.

Sachs RK and Wolfe AM 1967 Perturbations of cosmological model and angular variations of the microwave background. *Astrophys. J.* **147**, 73–90.

Sahni V and Coles P 1995 Approximation methods for non-linear gravitational clustering. *Phys. Rep.* **262**, 1–136.

Sahni V, Sathyaprakash BS and Shandarin SF 1997 Probing large-scale structure using percolation and genus curves. *Astrophys. J.* **477**, L1–L5.

Sahni V, Sathyaprakash BS and Shandarin SF 1998 Shapefinders: a new diagnostic for large-scale structure. *Astrophys. J.* **495**, L5–L8.

Sakharov AD 1966 The initial stage of an expanding Universe and the appearance of a nonuniform distribution of matter. *Sov. Phys. JETP* **22**, 241–249.

Salam A 1968 Weak and electromagnetic interactions. In *Elementary particle physics* (ed. Swartholm N). Almquist and Wiksells, Stockholm.

Sandage A 1961 The ability of the 200-inch telescope to discriminate between selected world models. *Astrophys. J.* **133**, 355–392.

Sandage A 1968 Observational cosmology. *Observatory* **88**, 91–106.

Sandage A 1970 Cosmology: the search for two numbers. *Phys. Today* **23**, 34–43.

Sandage A 1972 Distances to galaxies, the Hubble constant and the edge of the world. *Q. J. R. Astr. Soc.* **13**, 282–296

Sandage A 1988 Observational tests of world models. *A. Rev. Astr. Astrophys.* **26**, 561–630.

Sandvik HB, Barrow JD and Magueijo J 2002 A simple cosmology with a varying fine-structure constant. *Phys. Rev. Lett.* **8803**, 031302.

Saslaw WC 1985 *Gravitational physics of stellar and galactic systems.* Cambridge University Press.

Schneider P, Ehlers J and Falco EE 1992 *Gravitational lenses.* Springer, Berlin.

Schramm DN and Turner MS 1996 Deuteronomy and numbers. *Nature* **381**, 193–194.

Schramm DN and Wagoner RV 1979 Element production in the early Universe. *A. Rev. Nucl. Part. Sci.* **27**, 37–74.

Sciama DW 1993 *Modern cosmology and the dark matter problem.* Cambridge University Press.

Scoccimarro R, Couchman HMP and Frieman JA 1999 The bispectrum as a signature of gravitational instability in redshift space. *Astrophys. J.* **517**, 531–540.

Scully S *et al.* 1996 The local abundance of ^{3}He: a confrontation between theory and observation. *Astrophys. J.* **462**, 960–968.

Seljak U and Zaldarriaga M 1996 A line-of-sight approach to cosmic microwave background anisotropies. *Astrophys. J.* **469**, 437–444.

Shandarin SF 1983 Percolation theory and the cell-lattice structure of the Universe. *Sov. Astr. Lett.* **9**, 100–102.

Shandarin SF and Zel'dovich YaB 1989 The large-scale structure of the Universe: turbulence, intermittency, structures in a self-gravitating medium. *Rev. Mod. Phys.* **61**, 185–220.

Shane CD and Wirtanen A 1967 The distribution of galaxies. *Publ. Lick. Obs.* **22**, 1–60.

Shanks T, Hale-Sutton D, Fong R and Metcalfe N 1989 An extended Galaxy redshift survey. III. Constraints on large-scale structure. *Mon. Not. R. Astr. Soc.* **237**, 589–610.

Shapley H and Ames A 1932 A survey of the external galaxies brighter than the 13th magnitude. *Ann. Harvard College Obs.* **88**, 41–75.

Signore M and Dupraz C (eds) 1992 *The infra-red and submillimetre sky after COBE.* Kluwer, Dordrecht.

Silk J 1967 Fluctuations in the primordial fireball. *Nature* **215**, 1155–1156

Silk J 1968 Cosmic black-body radiation and galaxy formation. *Astrophys. J.* **151**, 459–471.

Skillman ED *et al.* 1993 New results of He measurements: implications for SBBN *Ann. NY Acad. Sci.* **688**, 739–744.

Slipher VM 1914 Spectrographic observations of nebulae. Paper presented at the 17th Meeting of the American Astronomical Society. A short summary can be found in 1915 *Popular Astron.* **23**, 21.

Smail I, Couch WJ, Ellis RS and Sharples RM 1995 Hubble Space Telescope measurements of gravitationally lensed features in the rich cluster AC-114. *Astrophys. J.* **440**, 501–509.

Smith MS, Kawano LH and Malaney RA 1993 Experimental, computational and observational analysis of primordial nucleosynthesis. *Astrophys. J. Suppl.* **85**, 219–247.

Smoot GF *et al.* 1992 Structure in the COBE differential microwave radiometer first year maps. *Astrophys. J.* **396**, L1–L5.

Songaila A, Cowie LL, Hogan CJ and Rugers M 1994 Deuterium abundance and background radiation temperature in high redshift primordial clouds. *Nature* **368**, 599–604.

Spite M, Francois P, Nissen PE and Spite F 1996 Spread of the lithium abundance in halo stars. *Astr. Astrophys.* **307**, 172–183.

Starobinsky AA 1979 Spectrum of relict gravitational radiation and the early state of the Universe. *JETP Lett.* **30**, 682–685.

Starobinsky AA 1982 Dynamics of phase transitions in the new inflationary universe scenario and generation of perturbations. *Phys. Lett.* B **117**, 175–178.

Stauffer D and Aharony A 1992 *Introduction to percolation theory.* Taylor & Francis, London.

Steidel CC, Giavilisco M, Pettini M, Dickinson M and Adelberger KL 1996 Spectroscopic confirmation of a population of normal star-forming galaxies at redshifts $z > 3$. *Astrophys. J.* **462**, L17–L21.

Steigman G 1994 Cosmic deuterium or a hydrogen interloper? *Mon. Not. R. Astr. Soc.* **269**, L53–L54.

Steigman G and Tosi M 1995 Generic evolution of deuterium and He-3. *Astrophys. J.* **453**, 173–177.

Steigman G, Fields B, Olive KA, Schramm DN and Walker TP 1993 Population-II Li-6 as a probe of nucleosynthesis and stellar structure and evolution. *Astrophys. J.* **415**, L35–L38.

Stirling AJ and Peacock JA 1996 Power correlations in cosmology: limits on primordial non-Gaussian density fields. *Mon. Not. R. Astr. Soc.* **283**, L99–L104.

Strauss MA and Willick JA 1995 The density and peculiar velocity fields of nearby galaxies. *Phys. Rep.* **261**, 271–431.

Sunyaev RA and Zel'dovich YaB 1969 The observation of relic radiation as a test of the nature of X-ray radiation from the clusters of galaxies. *Commun. Astrophys. Space Phys.* **4**, 173–178.

Sunyaev RA and Zel'dovich YaB 1970 Small-scale fluctuations of relic radiation. *Astrophys. Space Sci.* **7**, 3–19.

Taylor JH, Fowler LA and Weisberg JM 1979 Measurements of general relativistic effects in the binary pulsar PSR1913+16. *Nature* **277**, 437–440.

Thorne KS 1987 Gravitational radiation. In *300 years of gravitation* (ed. Hawking SW and Israel W), pp. 331–458. Cambridge University Press.

Thorne KS, Price RH and Macdonald DA 1986 *Black holes: the membrane paradigm.* Yale University Press, London and New Haven.

Tolman RC 1934 Effect of inhomogeneity on cosmological models. *Proc. Nat. Acad. Sci.* **20**, 169–176.

Totsuji H and Kihara T 1969 The correlation function for the distribution of galaxies. *Publ. Astr. Soc. Jap.* **21**, 221–229.

Trimble V 1987 Existence and nature of dark matter in the Universe. *A. Rev. Astr. Astrophys.* **25**, 425–472.

Turner EL 1990 Gravitational lensing limits on the cosmological constant in a flat universe. *Astrophys. J.* **365**, L43–L46.

Turner EL, Ostiker JP and Gott III JR 1984 The statistics of gravitational lenses – the distributions of image angular separations and lens redshifts. *Astrophys. J.* **284**, 1–22.

Tyson JA 1988 Deep CCD survey – galaxy luminosity and color evolution. *Astr. J.* **96**, 1–23.

Tyson JA and Seitzer P 1988 A deep CCD survey of 12 high latitude fields. *Astrophys. J.* **335**, 552–583.

Tyson JA, Valdes F and Wenk RA 1990 Detection of systematic gravitational lens galaxy image alignments - mapping dark matter in galaxy clusters. *Astrophys. J.* **349**, L1–L4.

Tytler D, Fan X-M and Burles S 1996 Cosmological baryon density derived from the deuterium abundance at redshift $z = 3.57$. *Nature* **381**, 207–209.

van Waerbeke L *et al.* 2000 Detection of correlated galaxy ellipticities from CFHT data: first evidence for gravitational lensing by large-scale structures. *Astr. Astrophys.* **358**, 30–44.

Vanmarcke EH 1983 *Random fields, analysis and synthesis.* MIT Press, Cambridge, MA.

Viana PTP and Liddle AR 1996 The cluster abundance in flat and open cosmologies. *Mon. Not. R. Astr. Soc.* **281**, 323–332.

Vilenkin A 1984 Quantum gravity of universes. *Phys. Rev.* D **30**, 509–511.

Vilenkin A 1986 Boundary conditions in quantum cosmology. *Phys. Rev.* D **33**, 3560–3569.

Vittorio N and Silk J 1984 Fine-scale anisotropy of the cosmic microwave background in a universe dominated by cold dark matter. *Astrophys. J.* **285**, L39–L43.

Vittorio N and Turner MS 1987 The large-scale peculiar velocity field in flat models of the Universe. *Astrophys. J.* **316**, 475–482.

Vittorio N, Juszkiewicz R and Davis M 1986 Large-scale velocity fields as a test of cosmological models. *Nature* **323**, 132–133.

Wald RM 1984 *General relativity.* The University of Chicago Press.

Walker TP, Steigman G, Schramm DN, Olive KA and Kang KS 1991 Primordial nucleosynthesis redux. *Astrophys. J.* **376**, 51–69.

Walker TP, Steigman G, Schramm DN, Olive KA and Fields B 1993 The boron-to-beryllium ratio in halo stars - a signature of cosmic ray nucleosynthesis in the early galaxy. *Astrophys. J.* **413**, 562–570.

Walsh D, Carswell RF and Weymann RJ 1979 0957 + 561A,B: twin quasistellar objects or gravitational lens? *Nature* **379**, 381–384.

Wampler EJ *et al.* 1996 High resolution observations of the QSO BR 1202-0725, deuterium and ionic abundances at redshifts above $z = 4$. *Astr. Astrophys.* **316**, 33–42.

Weinberg S 1967 A model of Leptons. *Phys. Rev. Lett.* **19**, 1264–1266.

Weinberg S 1971 Entropy generation and the survival of protogalaxies in an expanding universe. *Astrophys. J.* **168**, 175–194.

Weinberg S 1972 *Gravitation and cosmology: principles and applications of the general theory of relativity.* Wiley, New York.

Weinberg S 1988 The cosmological constant problem. *Rev. Mod. Phys.* **61**, 1–23.

Weinberg DH and Cole S 1992 Non-Gaussian fluctuations and the statistics of galaxy clustering. *Mon. Not. R. Astr. Soc.* **259**, 652–694.

White SDM 1979 The hierarchy of correlation functions and its relation to other measures of galaxy clustering. *Mon. Not. R. Astr. Soc.* **186**, 145–154.

White SDM, Frenk CS and Davis M 1983 Clustering in a neutrino dominated universe. *Astrophys. J.* **274**, L1–L5.

White SDM, Briel UG and Henry IP 1993a X-ray archaeology of the Coma cluster. *Mon. Not. R. Astr. Soc.* **261**, L8–L11.

White SDM, Navarro JF, Evrard AE and Frenk CS 1993b The baryon content of galaxy clusters: a challenge to cosmological orthodoxy. *Nature* **366**, 429–433.

White M, Scott D and Silk J 1994 Anisotropies in the cosmic microwave background. *A. Rev. Astr. Astrophys.* **32**, 319–370.

White M, Gelmini G and Silk J 1995 Structure formation with decaying neutrinos. *Phys. Rev.* D **51**, 2669–2676.

Wilson ML 1983 On the anisotropy of the cosmological background matter and radiation distribution. II. The radiation anisotropy in models with negative spatial curvature. *Astrophys. J.* **273**, 2–15.

Wilson ML and Silk J 1981 On the anisotropy of the cosmological background matter and radiation distribution. I. The radiation anisotropy in a spatially flat universe. *Astrophys. J.* **243**, 14–25.

Wilson G, Kaiser N and Luppino GA 2001 Mass and light in the Universe. *Astrophys. J.* **556**, 601–618.

Wittman DM, Tyson JA, Kirkman D, Dell'Antonio I and Berstein G 2000 Detection of weak gravitational lensing distortions of distant galaxies by cosmic dark matter at large scales. *Nature* **405**, 143–148.

Wolfe AM 1993 The progenitors of galaxies and the gas content of the universe at large redshifts. *Ann. NY Acad. Sci.* **688**, 281–296.

Wolfe AM, Turnshek DA, Lanzetta KM and Lu L 1993 Damped Lyman-alpha absorption by disk galaxies with large redshifts. IV. More intermediate-resolution spectroscopy. *Astrophys. J.* **404**, 480–510.

Wright EL and Reese ED 2000 Detection of the cosmic infrared background at 2.2 and 3.5 microns using DIRBE observations. *Astrophys. J.* **545**, 43–55.

Wright EL *et al.* 1992 Interpretation of the CMBR anisotropy detected by the COBE differential microwave radiometer. *Astrophys. J.* **396**, L7–L12

Wyse RG and Jones BJT 1985 The ionisation of the primeval plasma at the time of recombination. *Astr. Astrophys.* **149**, 144–150.

Zel'dovich YaB 1965 Survey of modern cosmology. *Adv. Astr. Astrophys.* **3**, 241–379.

Zel'dovich YaB 1970 Gravitational instability: an approximate theory for large density perturbations. *Astr. Astrophys.* **5**, 84–89.

Zel'dovich YaB 1972 A hypothesis unifying the structure and the entropy of the Universe. *Mon. Not. R. Astr. Soc.* **160**, 1P–3P.

Zel'dovich YaB and Barenblatt G 1958 The asymptotic properties of self-modelling solutions of the non-stationary gas filtration equations. *Sov. Phys. Dokl.* **3**, 44–47.

Zel'dovich YaB and Novikov ID 1983 *The structure and evolution of the Universe.* University of Chicago Press.

Zwicky F 1952 *Morphological astronomy.* Springer, Berlin.

Zwicky F, Herzog E, Wild P, Karpowicz M and Kowal CT 1961–1968 *Catalogue of galaxies and clusters of galaxies*, 6 vols. California Institute of Technology, Pasadena.

Index

Printed in the USA/Agawam, MA
February 13, 2023
805790.006